Library of
Davidson College

Oscillations and Waves
in Linear and Nonlinear Systems

Mathematics and Its Applications (*Soviet Series*)

Managing Editor:

M. HAZEWINKEL
Centre for Mathematics and Computer Science, Amsterdam, The Netherlands

Editorial Board:

A. A. KIRILLOV, *MGU, Moscow, U.S.S.R.*
Yu. I. MANIN, *Steklov Institute of Mathematics, Moscow, U.S.S.R.*
N. N. MOISEEV, *Computing Centre, Academy of Sciences, Moscow, U.S.S.R.*
S. P. NOVIKOV, *Landau Institute of Theoretical Physics, Moscow, U.S.S.R.*
M. C. POLYVANOV, *Steklov Institute of Mathematics, Moscow, U.S.S.R.*
Yu. A. ROZANOV, *Steklov Institute of Mathematics, Moscow, U.S.S.R.*

Volume 50

Oscillations and Waves

in Linear and Nonlinear Systems

by

M. I. Rabinovich
Institute of Applied Physics, Academy of Sciences of the U.S.S.R., Gorky, U.S.S.R.

and

D. I. Trubetskov
Saratov State University, U.S.S.R.

KLUWER ACADEMIC PUBLISHERS
DORDRECHT / BOSTON / LONDON

Library of Congress Cataloging in Publication Data

Rabinovich, M. I.
  [Vvedenie v teoriiu kolebaniĭ i voln. English]
  Oscillations and waves in linear and nonlinear systems / by
M.I. Rabinovich and D.I. Trubetskov.
     p.   cm. -- (Mathematics and its applications. Soviet series ;
50)
  Translation of: Vvedenie v teoriiu kolebaniĭ i voln.
  Includes bibliographical references.
  ISBN 0-7923-0445-4
  1. Wave-motion, Theory of. 2. Oscillations. I. Trubetskov, D.
I., 1938-   . II. Title. III. Series: Mathematics and its
applications (Kluwer Academic Publishers). Soviet series ; 50.
QA927.R3313 1989
530.1'24--dc20   89-19796

ISBN 0-7923-0445-4

Published by Kluwer Academic Publishers,
P.O. Box 17, 3300 AA Dordrecht, The Netherlands.

Kluwer Academic Publishers incorporates
the publishing programmes of
D. Reidel, Martinus Nijhoff, Dr W. Junk and MTP Press.

Sold and distributed in the U.S.A. and Canada
by Kluwer Academic Publishers,
101 Philip Drive, Norwell, MA 02061, U.S.A.

In all other countries, sold and distributed
by Kluwer Academic Publishers Group,
P.O. Box 322, 3300 AH Dordrecht, The Netherlands.

Printed on acid-free paper

This is a revised and enlarged edition of the original work
ВВЕДЕНИЕ В ТЕОРИЮ КОЛЕБАНИЙ И ВОЛН
Published by Nauka, Moscow, © 1984.
Translated from the Russian by R.N. Hainsworth

All Rights Reserved
This English edition © 1989 by Kluwer Academic Publishers
No part of the material protected by this copyright notice may be reproduced or
utilized in any form or by any means, electronic or mechanical
including photocopying, recording, or by any information storage and
retrieval system, without written permission from the copyright owner.

Printed in The Netherlands

SERIES EDITOR'S PREFACE

'Et moi, ..., si j'avait su comment en revenir,
je n'y serais point allé.'
                              Jules Verne

The series is divergent; therefore we may be
able to do something with it.
                              O. Heaviside

One service mathematics has rendered the human race. It has put common sense back where it belongs, on the topmost shelf next to the dusty canister labelled 'discarded nonsense'.
                              Eric T. Bell

Mathematics is a tool for thought. A highly necessary tool in a world where both feedback and nonlinearities abound. Similarly, all kinds of parts of mathematics serve as tools for other parts and for other sciences.

Applying a simple rewriting rule to the quote on the right above one finds such statements as: 'One service topology has rendered mathematical physics ...'; 'One service logic has rendered computer science ...'; 'One service category theory has rendered mathematics ...'. All arguably true. And all statements obtainable this way form part of the raison d'être of this series.

This series, *Mathematics and Its Applications*, started in 1977. Now that over one hundred volumes have appeared it seems opportune to reexamine its scope. At the time I wrote

> "Growing specialization and diversification have brought a host of monographs and textbooks on increasingly specialized topics. However, the 'tree' of knowledge of mathematics and related fields does not grow only by putting forth new branches. It also happens, quite often in fact, that branches which were thought to be completely disparate are suddenly seen to be related. Further, the kind and level of sophistication of mathematics applied in various sciences has changed drastically in recent years: measure theory is used (non-trivially) in regional and theoretical economics; algebraic geometry interacts with physics; the Minkowsky lemma, coding theory and the structure of water meet one another in packing and covering theory; quantum fields, crystal defects and mathematical programming profit from homotopy theory; Lie algebras are relevant to filtering; and prediction and electrical engineering can use Stein spaces. And in addition to this there are such new emerging subdisciplines as 'experimental mathematics', 'CFD', 'completely integrable systems', 'chaos, synergetics and large-scale order', which are almost impossible to fit into the existing classification schemes. They draw upon widely different sections of mathematics."

By and large, all this still applies today. It is still true that at first sight mathematics seems rather fragmented and that to find, see, and exploit the deeper underlying interrelations more effort is needed and so are books that can help mathematicians and scientists do so. Accordingly MIA will continue to try to make such books available.

If anything, the description I gave in 1977 is now an understatement. To the examples of interaction areas one should add string theory where Riemann surfaces, algebraic geometry, modular functions, knots, quantum field theory, Kac-Moody algebras, monstrous moonshine (and more) all come together. And to the examples of things which can be usefully applied let me add the topic 'finite geometry'; a combination of words which sounds like it might not even exist, let alone be applicable. And yet it is being applied: to statistics via designs, to radar/sonar detection arrays (via finite projective planes), and to bus connections of VLSI chips (via difference sets). There seems to be no part of (so-called pure) mathematics that is not in immediate danger of being applied. And, accordingly, the applied mathematician needs to be aware of much more. Besides analysis and numerics, the traditional workhorses, he may need all kinds of combinatorics, algebra, probability, and so on.

In addition, the applied scientist needs to cope increasingly with the nonlinear world and the extra mathematical sophistication that this requires. For that is where the rewards are. Linear models are honest and a bit sad and depressing: proportional efforts and results. It is in the nonlinear world that infinitesimal inputs may result in macroscopic outputs (or vice versa). To appreciate what I am hinting at: if electronics were linear we would have no fun with transistors and computers; we would have no TV; in fact you would not be reading these lines.

There is also no safety in ignoring such outlandish things as nonstandard analysis, superspace and anticommuting integration, p-adic and ultrametric space. All three have applications in both electrical engineering and physics. Once, complex numbers were equally outlandish, but they frequently proved the shortest path between 'real' results. Similarly, the first two topics named have already provided a number of 'wormhole' paths. There is no telling where all this is leading - fortunately.

Thus the original scope of the series, which for various (sound) reasons now comprises five sub-series: white (Japan), yellow (China), red (USSR), blue (Eastern Europe), and green (everything else), still applies. It has been enlarged a bit to include books treating of the tools from one subdiscipline which are used in others. Thus the series still aims at books dealing with:

- a central concept which plays an important role in several different mathematical and/or scientific specialization areas;
- new applications of the results and ideas from one area of scientific endeavour into another;
- influences which the results, problems and concepts of one field of enquiry have, and have had, on the development of another.

The present volume in the series is basically a comprehensive text on the topic of waves and oscillations in both the linear and nonlinear case from as broad a point of view as is consistent with rigour and clarity of view or, to put it another way, with a point of view as broad as the current industrial haze - taken literally or figuratively - will allow. (Thank you, Michael Frayn, for inventing this phrase.) It is also based on some 20 years of teaching such material and experience of the authors at the universities of Gorki and Saratov.

It is a thorough and knowledgeable treatment and additionally interesting in that it includes treatment of such matters as shock waves, solitons, and deterministic chaos. Not that there is, since recently, a particular lack of volumes on these topics, particularly the last two named - the first one could do, I feel, with some more attention in the monographic literature -, but it is much rarer to find these subjects treated within the framework of the more classical topics of waves and oscillations. Here, this is done and that is for me one more reason to welcome this book in this series.

The shortest path between two truths in the real domain passes through the complex domain.
J. Hadamard

La physique ne nous donne pas seulement l'occasion de résoudre des problèmes ... elle nous fait pressentir la solution.
H. Poincaré

Never lend books, for no one ever returns them; the only books I have in my library are books that other folk have lent me.
Anatole France

The function of an expert is not to be more right than other people, but to be wrong for more sophisticated reasons.
David Butler

Bussum, September 1989

Michiel Hazewinkel

CONTENTS

PREFACE   xiii

INTRODUCTION   xv

PART ONE. OSCILLATIONS AND WAVES IN LINEAR SYSTEMS.

CHAPTER 1. LINEAR OSCILLATORS.
1.1. General Notes.   1
1.2. Two Examples. The Phase Plane Diagram of an   3
      Oscillator.
1.3. Resonance. The Effect of an Aperiodic   13
      External Force on an Oscillator.
1.4. Normal Oscillations. Analogy with Quantum   20
      Mechanics. Production and Extinction
      Operators.

CHAPTER 2. OSCILLATIONS IN A SYSTEM WITH TWO LINKED
           OSCILLATORS.
2.1. Initial Equations.   23
2.2. The Fundamental Oscillations of Two Linked   25
      Oscillators.
2.3. Disturbance of Two Linked Oscillators by an   35
      External Force. The Reciprocity Principle.

CHAPTER 3. OSCILLATIONS IN AN ENSEMBLE OF NON-
           INTERACTING OSCILLATORS.
3.1. Classical Theory of Dispersion.   39
3.2. Oscillations in an Ensemble of Dissimilar   44
      Noninteracting Oscillators with a Given
      Distribution Function.

CHAPTER 4. OSCILLATIONS IN ORDERED STRUCTURES.
           LIMIT FOR A CONTINUOUS MEDIUM. WAVES.
           DISPERSION.
4.1. General Remarks.   49
4.2. Oscillations in Ordered Structures (Chains of   50
      Linked Particles and Identical Linked
      Oscillators).
4.3. Limiting Transition from an Ordered Structure to   60
      a One-dimensional Medium. Temporal and Spatial
      Dispersion. Physical Nature of Dispersion.
4.4. Typical Dispersion Characteristics for Medium   66
      Models.
4.5. Formal Method for Obtaining the Dispersion   71
      Equation. Waves in a One-Dimensional Resonator.

	Resonance in Wave Systems.	
4.6.	Quasiparticles.	78

CHAPTER 5. PROPERTIES OF WAVES WITH SMALL AMPLITUDES IN CONTINUOUS MEDIA.

5.1.	General Remarks.	81
5.2.	Equations of Hydrodynamics. Dispersion Equations for Sound Waves.	82
5.3.	A Stratified Fluid. Sound in an Ocean.	85
5.4.	Gravity Waves in an Incompressible Liquid. Internal Waves. Rossby Waves.	90
5.5.	Waves in a Superfluid Liquid.	103
5.6.	Waves in a Plasma. Hydrodynamic Description.	110

CHAPTER 6. STABILITY AND INSTABILITY OF LINEAR SYSTEMS WITH DISCRETE SPECTRA.

6.1.	General Notes and Definitions.	121
6.2.	The Raus-Gurvits Criterion and Three-Dimensional Systems.	124
6.3.	The D-Partition Method.	128
6.4.	Stability of Non-Autonomous Systems.	131
6.5.	Instability Mechanisms.	134

CHAPTER 7. STABILITY OF DISTRIBUTED SYSTEMS WITH CONTINUOUS SPECTRA.

7.1.	General Comments.	141
7.2.	Examples of Instability.	144
7.3.	Absolute and Convective Instability. The Characteristics Method.	153
7.4.	Waves in Flows. Electron Beams. Helmholtz Instability.	157
7.5.	Amplification and Filtering. Separation Criteria.	165

CHAPTER 8. PROPAGATION VELOCITY OF WAVES.

8.1.	Various Introductions to the Concept of Group Velocity.	171
8.2.	Group Velocity of Waves in Some Continuous Media.	178

CHAPTER 9. ENERGY AND MOMENTUM OF WAVES.

9.1.	Equation for the Transport of the Average Energy Density by Wave Packets in Dispersing Media.	185
9.2.	Density of the Energy of an Electromagnetic Wave in a Medium with Dispersion.	189
9.3.	Momentum of a Wave Packet.	193

CHAPTER 10. WAVES WITH NEGATIVE ENERGY. LINKED WAVES.

10.1.	General Notes.	195
10.2.	Waves with Positive and Negative Energies.	196
10.3.	Coupled Waves. Synchronicity. Normal and Anomalous Doppler Effects.	203

CHAPTER 11. PARAMETRIC SYSTEMS AND PARAMETRIC INSTABILITY.

11.1. General Comments. 213
11.2. Parametric Resonance. Floquet's (Blokh's) Theorem. Mathieu's Equation. 214
11.3. Waves in Periodic Structures. The Mathieu Zone and the Brillouin Diagram. 228
11.4. Motion in a Rapidly Oscillating Field. Kapitsa's Pendulum. Free Electron Lasers. 232

CHAPTER 12. ADIABATIC INVARIANTS. PROPAGATION OF WAVES IN INHOMOGENEOUS MEDIA

12.1. The Wentsel-Kramers-Brillouin (VCB) Approximation and Adiabatic Invariants. 239
12.2. Equivalence Between a Rotor and an Oscillator. 244
12.3. Propagation of Waves in Inhomogeneous Media. The Approximation of Geometric Optics. 246
12.4. The Propagation of Waves in a Plane-Layer Medium in the Geometric Optics Approximation. 255
12.5. Linear Wave Interaction in an Inhomogeneous Medium. 261

PART TWO. OSCILLATIONS AND WAVES IN NONLINEAR SYSTEMS.

CHAPTER 13. THE NONLINEAR OSCILLATOR.

13.1. Initial remarks. 273
13.2. Qualitative and Analytical Description. Examples of Nonlinear Systems. 275
13.3. Nonlinear Resonance. 286
13.4. Overlap between Nonlinear Resonances. 290

CHAPTER 14. PERIODIC SELF-EXCITED OSCILLATIONS.

14.1. Definitions. 299
14.2. The Van der Pol Generator. Self-Excited Oscillations as a Function of System Parameters. 302
14.3. Relaxational Self-Excited Oscillations. Fast and Slow Motions. 305

CHAPTER 15. GENERAL PROPERTIES OF NONLINEAR DYNAMIC SYSTEMS IN PHASE SPACE.

15.1. Basic Types of Trajectory. The Fundamentals of Dynamic Systems (Structural Stability). 309
15.2. Basic Bifurcations on a Plane. Poincare Indices. 313
15.3. Point Transformations. 317
15.4. Bifurcation of Periodic Motions. 320
15.5. Homoclinic Structures. 323

CHAPTER 16. SELF-EXCITED OSCILLATIONS IN MULTIFREQUENCY SYSTEMS.
16.1. Forced Synchronization. 329
16.2. Competition. 343
16.3. Mutual Mode Synchronization. 349

CHAPTER 17. RESONANCE INTERACTIONS BETWEEN OSCILLATORS.
17.1. Interaction Between Three Coupled Oscillators in a System with Quadratic Nonlinearity. 353
17.2. Resonance Interactions Between Waves in Weakly Nonlinear Media with Dispersion. 364
17.3. Explosive Instability. 372

CHAPTER 18. SIMPLE WAVES AND THE FORMATION OF DISCONTINUITIES.
18.1. Kinematic Waves. 375
18.2. Travelling Waves in a Nonlinear Medium Without Dispersion. 380
18.3. Determining the Discontinuity Coordinates. 387
18.4. Weak Shock Waves. Boundary Conditions at a Discontinuity. 390

CHAPTER 19. STATIONARY SHOCK WAVES AND SOLITONS.
19.1. Structure of a Discontinuity. 395
19.2. Solitary Waves - Solitons. 403
19.3. Solitons as Particles. 409
19.4. Higher-Dimensional Solitons. 411

CHAPTER 20. MODULATED WAVES IN NONLINEAR MEDIA.
20.1. General Remarks. 417
20.2. Self-Modulation. Reversibility. 421
20.3. Self-Focusing. 431
20.4. Interaction Between Wave Beams and Packets. 434
20.5. Interactions Between Waves Having Randomly Modulated Phases. Wave Kinetics. 437

CHAPTER 21. SELF-EXCITED OSCILLATIONS IN DISTRIBUTED SYSTEMS.
21.1. General Remarks. 445
21.2. Medium Without Dispersion. Discontinuous Waves. 447
21.3. Stationary Waves. 447
21.4. The Existence and Role of Limiting Cycles. 451
21.5. Competition Between Stationary Waves in an Active Medium. 453
21.6. Periodic Self-Excited Oscillations in Hydrodynamic Flows. 455

CHAPTER 22. STOCHASTIC DYNAMICS IN SIMPLE SYSTEMS.

22.1. How Randomness Appears in a Dynamic System.	465
22.2. The Stochastic Dynamics of One-Dimensional Mappings.	473
22.3. Noise Generator. Qualitative Description and Experiment.	478
22.4. Statistical Description of a Simple Noise Generator.	482
22.5. Ways in which Strange Attractors Arise.	487
22.6. Dimensionality of Stochastic Sets.	498

CHAPTER 23. THE ONSET OF TURBULENCE.

23.1. General Remarks.	503
23.2. The Occurrence of Stochastic Self-Excited Oscillations in Experimental Fluid Mechanics.	506
23.3. Stochastic Modulation.	513
23.4. Ideal Flow and Turbulence.	518

CHAPTER 24. SELF-ORGANIZATION.

24.1. Main Phenomena, Models, and Mathematical Forms	523
24.2. Travelling Pulsations	528
24.3. Spiral and Cylindrical Waves. Travelling Centers.	531
24.4. Concerning Self-Organization Mechanisms.	533

REFERENCES	537
INDEX	575

PREFACE

There are quite a number of books on oscillation theory and the various topics in wave theory. Some are devoted to the mathematical apparatus, others to a detailed investigation of a relatively narrow range of topics, while yet others are exclusively concerned with individual aspects, be it mathematical, biological, or some other field of science. Moreover, oscillation and wave theory are often separated into different books. The aim of this book is to present to the reader the current state of the theory of oscillations and waves with as wide a purview as is possible without losing clarity or the formal rigor appropriate in physics.

The book is structured so that it is the fundamental oscillation- wave phenomena and effects are brought together rather than formal methods. We have tried to show unity of oscillation phenomena in Nature, taking examples from widely differing fields.

The book is in two parts, the first considering oscillations and waves in linear systems and media, and the second nonlinear systems and media. We believe that this division promotes an understanding of the modern theory of oscillations and waves. For example, the propagation of plane harmonic waves in a periodically stratified medium is described by practically the same mathematical model as the phenomenon of parametric instability in a distributed system with one degree of freedom, and considering them together is clearly natural. An analysis of, for example, self-excited oscillations in a disturbable medium, viz., an ensemble of self- excited generators, is a direct generalization of the equations for interactions between a small number of generators, etc.

A series of lectures given by the authors over a period of twenty years at the physics faculties of the universities of Gorki and Saratov form the basis of the book. Many years of experience of discussing the subject with students has determined what the relation should be between detailed explanation and where we might include phrases such as "it can easily be seen" and "it can be shown", which becomes necessary to omit cumbersome mathematics that threatens to obscure the idea being presented. We hope that this book will not only be found useful by students and postgraduates, but that experienced scientists whose work bring them into contact with the analysis of oscillations and waves of various natures will find something of interest.

In addition to the topics traditionally included in a

course on oscillations and waves, this book contains new material which has until now only been contained in the specialist literature. For example, the material on the appearance of stochasticity in simple systems, the link between hydrodynamic turbulence and stochastic self-excited oscillations (and their mathematical forms, namely, strange attractors), and the main ideas and phenomena of self-organization, which is now a new field in the nonlinear theory of oscillations and waves.

We now have pleasure in expressing our deep gratitude to Andrei Viktorovich Gapono-Grekhov, whose idea it was to have this book written. Many years of working and debating with him has had a formative influence on the life of one of the authors. We are also grateful to our colleagues and coworkers at the Institute of Applied Physics, USSR Academy of Sciences, at the Science and Research Institute of Mechanics and Physics, Saratov State University, at the faculty of Radio Physics at Gorki State University, and at the physics faulty at Saratov State University.
They have aided us during many useful discussions on topic associated with this book. We are also grateful to G.M. Golotovskii for his meticulous editing of the Russian text.

INTRODUCTION

The idea that oscillatory phenomenon which are outwardly dissimilar and arise in a variety of situations (mechanical, electromagnetic, chemical, biological etc.) form a single family now seems natural both to the experienced researcher and the youth who has just finished school. Indeed, most people might cite both a pendulum and an electric circuit containing a capacitor and coil when asked for an example of an harmonic oscillator. Nevertheless, even today it is often difficult to relate the oscillatory phenomena and effects observed in less trivial situations to the dominant elementary processes. This is particularly true of wave problems. We believe there is now a need for a textbook setting out the theory of oscillations and waves and showing how its phenomena and effects arise in diverse applications, but that even so they can be described and understood in the same way. We would emphasize that although the formal unity of the oscillation and wave phenomena found in such diverse fields is based on the similarity of the mathematical models, this is not all there is to it. The "interdisciplinary" nature of the concepts, models, and approximations is just as important because they enable one to orient oneself when dealing with the manifold variety of the oscillation and wave processes that are encountered in nature, science, and engineering.

Perhaps now a brief formulation of the subject should be attempted. This theory is the area of science which investigates oscillation and wave phenomena in various types of system by looking at the main properties of the oscillation properties while ignoring the detailed behavior of the oscillating system as it relates to its actual nature (i.e., physical, biological, etc.) Using models, the theory of oscillations and waves establishes general properties in real systems, thus enabling an investigator to derive a relation between the system's parameters and the system's possible wave or oscillatory behavior in the presence of other effects.

The application of the theory in each case presumes an idealization of the real system, that is the construction of a model and the compilation of equations for it (e.g., ordinary differential, partial differential, or difference equations). The actual idealization may differ for the same system depending on which phenomenon is being studied, i.e., the model must correspond both to the system and to the phenomenon. For example, when studying the conditions for build up in the oscillation of a swing with a periodically changing length, the model can be quite simple, namely a linear oscillator with a variable frequency. However, when it is necessary to determine the conditions under which the oscillation and its form etc. are stabilized, we need a model that involves (as a minimum) a relation between the frequency of the oscillation and its amplitude. Thus we arrive at a model of a nonlinear oscillator with a periodically changing parameter. Another example concerns ocean waves. These too may be described by a linear model (the wave equation for dispersive waves) if we are interested in moderately sized waves far from the shore. However, to describe rolling waves and the formation of foam, we must of course turn to a nonlinear model.

Let us re-emphasize that by using the established concepts of the theory of oscillations and waves it is possible to relate certain phenomena in a real system with its characteristics without actually solving a problem. For example, suppose we are discussing the transformation of the energy of one sort of vibration into that of another in a weakly nonlinear system or medium (we could be considering water waves, electromagnetic waves in the ionosphere, or the oscillations of a pendulum on a spring), we can say at once that the transformation may only occur if certain resonance conditions are met between the harmonic frequencies of the subsystems.

The first clear thought that oscillations due to seemingly dissimilar phenomena have an underlying unity was expressed by Rayleigh in his "Theory of Sound". He added an extra chapter to the book to cover electric oscillations, emphasizing that both these sorts of small oscillation, viz., sound and electric, are in a certain sense identical. Rayleigh's book was in fact the first textbook on the theory of oscillations and waves in linear systems, i.e., a course on linear oscillations.

However, linear oscillations, i.e., vibrations with small amplitudes for which there are additive responses to additive effects (the superposition principle is obeyed), are an approximation. The equations of linear oscillations are obtained by linearizing the initial model around a certain delineated state or around the motion of the system or medium. A deeper investigation reveals that the majority of the phenomenon in our world are nonlinear. The first field of science to face this was celestial mechanics. It was observed that the period of a planet's orbit depended on its energy (Kepler's third law). Nonlinearity is inherent in properties that change in time. For instance, any transition from one quasi-steady state to another arises due to the appearance of nonlinearity, for example, the appearance and evolution of the Universe, the birth, life, and death of stars, the fusion and disintegration of particles and their birth from a vacuum, and the arbitrary formation of complex structures, leading in the end to organic life. The investigation of nonlinear problems is now not the prerogative of mechanicians and physicists, they are being investigated by biologists, economists, chemists, and so on. The "nonlinear" sciences now include elementary particle theory, nonequilibrium thermodynamics, atmospheric and oceanic dynamics, among many other areas of science.

By the first decade of this century nonlinear problems were being studied in mechanics (the three-body problem, water waves, etc.), acoustics, and the physics of solids (e.g., anharmonic vibrations of atoms in the crystal lattice in the theory of thermal conductivity). Nonlinear problems were posed at the start of radio technology (the generation and detection of oscillations) and they are constantly being encountered in other branches of science and technology. However, the "nonlinear difficulties" encountered in each discipline seemed very specific to the subject and unrelated to the difficulties found in other fields. It has only been since the 1920's and largely thanks to the activity of Leonid Isaakovich Mandel'shtam -- the father of the Soviet school of nonlinear physics -- that scientists in various fields have begun to develop a "nonlinear" way of thinking and the experience of nonlinearity gained in one field has been applied in others. The generality of nonlinear phenomena whatever their origins, the generality of their models, and the images and methods for considering them have become almost obvious. A terminology, with terms such as nonlinear resonance, self-excited oscillation, synchronicity, competition, and parametric interaction, has grown up alongside the modern theory of oscillations and waves.[1]

Wave and oscillation problems can now, we emphasize again, be considered the subject of a single theory. However, historically this was not so, and oscillation theory and wave theory were developed independently for a considerable period of time. The reason was that in the mid-1950s the main interest in wave theory centered on linear problems (with the exception of fluid mechanics) while classical oscillation theory concentrated on nonlinear problems. Moreover, the theory of nonlinear oscillation systems with one degree of freedom was by this time almost complete. This advance was made possible by the concept, advanced by A.A. Andronov and L.S. Pontryagin, of crude dynamic systems, i.e., systems which do not change their qualitative behavior due to small

---

[1] A presentation of oscillatory and wave phenomena from a united view point can be found in: J.H. Pain, *The Physics of Vibrations and Waves*, John Willey and Sons, London (1976).

changes in their parameters.[2] A detailed analysis of crude systems together with investigations of the qualitative rearrangements in the two-dimensional phase portraits that occur when small changes are made in the parameters yield a solution to practically any problem concerning the behavior of nonlinear dynamic systems with two-dimensional phase spaces.

Extensive investigations into nonlinear waves only began in the 1960s, which is when subjects such as the nonlinear physics of plasmas, nonlinear optics, acoustics and electrodynamics, and high energy physics (including the physics of explosions and shock waves) all came into being. Nonlinear thermodynamics also appeared to handle transitions in systems (particularly in chemistry and biology) that are far from thermodynamic equilibrium.

In order to illustrate the present state of the theory, let us look at two lines of research, namely the investigation of coherent states and complex determinate structures and the analysis of the random (stochastic) behavior of determinate systems. The relationship between dynamics and statistics has concerned physicists for a century, the main question being, can a statistical description be obtained from a dynamic one? Until recently the answer has been in the negative. The appearance of randomness in a classical (nonquantum) dynamic system (not exposed to noise) has been exclusively associated with its complexity -- the exceedingly large number of degrees of freedom (for example a gas in a vessel). At this stage it simply makes no sense to have a deterministic description, although one is in principle possible. The transfer to a probabilistic description is then made on the basis of some hypothesis, such as an ergodic one. A strict theory that has now appeared confirms that nonlinear dynamic systems may in a direct sense give rise to statistics, i.e., the statistical approach is not an approximation but the only correct way of reflecting the real behavior of the dynamic system. The amazing thing about recent discoveries in physics and mathematics is that, for instance, random behavior is manifested by even very simple nonlinear systems (small numbers of degrees of freedom), such as billiard balls on tables with concave walls or an electron in a field of two sinusoidal waves. Clearly, you might ask how randomness can arise in a system notwithstanding the fact that it has a unique solution. The short answer is that it is the result of instabilities in individual motions through a restricted phase volume. The instability of all finite motions (those

---

[2] *The Theory of Oscillations* by A.A. Andronov, A.A. Vitt, and S.E. Khaikin (Nauka, Moscow, 1981 (1st edition 1937)) was a milestone in the development of nonlinear oscillations.

occurring in a limited region of space) guarantees the complexity of nearly all the individual motions and their infinite variety, which leads naturally to the concept of ensembles and statistical description based on it.

A theory has now been developed, in the main, and experimentally confirmed for the transition of a wide variety of dynamic systems (e.g., fluid mechanic flows, random signal generators, autocatalytic reactions) from deterministic to stochastic behavior.

It would be natural to suggest that if a system with a few degrees of freedom is complicated, then a system with an infinite number of degrees of freedom must manifest random behavior. However, this is not the case in general. It has been hypothesized that the presence of even weak nonlinearity in systems with many degrees of freedom is enough for energy reserves in certain degrees of freedom to be redistributed among all the modes, and hence for a thermodynamic equilibrium to be established. A series of numerical experiments was conducted at the end of the 1940s to test this hypothesis using models with nonlinear chains containing large numbers of particles. However, no thermalization was observed and the system periodically returned to states with the original energy distribution (the Fermi-Pasta-Ulam paradox). There are in reality two types of nonlinear wave system, namely integrable (or nearly integrable) and nonintegrable systems. Integrable systems manifest only one simple periodic or quasiperiodic behavior, while nonintegrable ones are stochastic when the initial energy is large. By coincidence, the chains with which Fermi, Pasta, and Ulam worked were nearly integrable for the parameters they used.

Partial solutions are possible for both integrable and nonintegrable systems and they correspond to coherent formations or spatial structures. Examples of partial solutions are solitons and shock waves.

Coherent nonlinear formations are now being investigated in solid state physics (domains), plasma physics (Langmuir solitons), geophysics and oceanology (cyclones, anticyclones, and rings), planetary atmospheric physics (Jupiter's red spot), and nonlinear optics (ultrashort impulses). There is now the hope that ideas concerning elementary particles, such as solitons of quantum fields, will be confirmed.

There is now great interest in biophysics in coherent formations in dissipative nonequilibrial media, namely dissipative structures and self-excited waves,[3] combustion waves, perturbation impulses propagating in nerve and muscle fibers, space-time variations in organism populations, and

---

[3] This term was introduced by R.V. Khokhlov by analogy with self-excited oscillations

concentration waves in autocatalytic chemical reactions are all examples of self-excited waves. The main feature of these space-time formations is that they are only weakly dependent on the properties of the source of the disequilibrium, the boundary conditions, and the initial state of the medium. Dissipative structures in nonequilibrial media are now a hot subject for investigation, as they are a very typical and natural form of selforganization.

Finally, we must say that it is not possible, within the covers of a single book that seeks to discuss recent findings and present the classical theory in detail, to cover completely the modern theory of oscillations and waves (particularly the nonlinear aspects). However, we have tried to introduce this fascinating branch of science in the final chapters.

PART ONE

OSCILLATIONS AND WAVES IN LINEAR SYSTEMS

CHAPTER 1

LINEAR OSCILLATORS

1.1 General Notes

We have intimated already that a weight on a spring and an oscillating circuit are the same object so far as oscillation theory is concerned. They are both described by a single well-known differential equation and are characterized by the same phase space, that is by a plane containing the trajectories of a family of ellipses lying one inside the other. This would seem to be a trivial statement. "However, it is not trivial that it is trivial," L.I. Mandel'shtam remarked. "That is, it is not trivial that the analogy between the oscillations of a weight on a spring and the charges or current within a circuit has come so far that it is now an accepted method of argument amongst physicists even though these phenomena belong to two quite different branches of science." [1] The philosophy and content of this chapter concerns this analogy, and we consider the properties of a linear oscillator, this being the basic model in the linear theory of oscillations and waves.

The equation of motion of a linear oscillator which describes its free oscillations is

$$\ddot{x} + 2\gamma\dot{x} + \omega_0^2 x = 0. \tag{1.1}$$

Here x is the displacement from the equilibrium position for a mechanical system (e.g., the length coordinate of the weight on a spring), or the charge in an electrical system (e.g., the charge on the plates of a capacitor in an oscillating circuit), or something else depending on the nature of the oscillator; γ is a parameter describing the loss (e.g., resistance or friction); ω_0 is the oscillator's fundamental frequency; and $\dot{x} = dx/dt$ and $\ddot{x} = d^2x/dt^2$ are the appropriate derivatives with respect to time. A linear oscillator is a special (though very important) case of a linear dynamic system which can be described by the functional linear equations $\hat{L}u = 0$, where $\hat{L}$ is a linear operator (remember that if u_1 and u_2 are solutions of $\hat{L}u = 0$, then the combinations $C_1 u_1 + C_2 u_2$, where C_1 and C_2 are constants, are also solutions). Time is not explicit in (1.1). A corollary of this is that the system which (1.1) describes is not acted upon by a variable force and the parameters are constant over time, i.e., the system is independent. If now we assume that $\gamma = 0$ (or that the system's Q-factor, $Q = \omega_0/2\gamma$, is infinite), then we get the equation for an harmonic oscillator, i.e.,

$$\ddot{x} + \omega_0^2 x = 0. \tag{1.2}$$

This sort of oscillator is called conservative because it energy is conserved over time. This assertion is easy to prove, even for a case more general than (1.2), i.e., a nonlinear oscillator:

$$\ddot{x} = f(x) . \tag{1.3}$$

The total energy of the system given by (1.3) is the sum of the kinetic and potential energies

$$W = W_k + W_p = \frac{\dot{x}^2}{2} - \int_{x_0}^{x} f(\zeta) d\zeta . \tag{1.4}$$

The way the potential energy depends on the coordinates (the form of the potential well) $W_p = -\int_{x_0}^{x} f(\zeta) d\zeta$ is determined by the form of f, and consequently the behavior of (1.3). The reason for the sign in the formula for the potential energy is easy to understand if we calculate W_p for (1.2), say, at $x_0 = 0$ and $W_p = \omega_0^2 x^2/2$. The sign was chosen so that the potential energy of a pendulum increases away from equilibrium. By differentiating (1.4) with respect to time, we get $\dot{W} = \dot{x}\ddot{x} - \dot{x}f(x) = \dot{x}[\ddot{x} - f(x)] = 0$. Thus $W(x, \dot{x})$ does not depend on time. The total energy is conserved and so an oscillatory system described by (1.3) is conservative.

Equations (1.2) and (1.3) are second-order ordinary differential equations with constant coefficients. They describe systems with one degree of freedom (the number of degrees of freedom is the number of normal coordinates, or the number of variables -- equal to the number of fundamental frequencies-- which fully and uniquely define the state of the system at a given moment in time). The number of degrees of freedom of a system is half the order of the differential equation needed to describe it [2]. Thus a system with one degree of freedom corresponds to a two-dimensional phase space, viz., a surface, while a system with one-and-a-half degrees of freedom corresponds to a three-dimensional space, and a system with two degrees of freedom corresponds to a four-dimensional space.

1.2 Two Examples. The Phase Plane Diagram of an Oscillator

Before considering the motion of a linear oscillator (a system with one degree of freedom) on the phase plane, let us first look at two more nontrivial cases that have become classical examples of linear oscillators and which are encountered in chemistry and biology.

The simplest example, in chemistry, of an oscillating reaction which takes place in a homogeneous medium is Lotka's model [3,4]. It has the following kinetic scheme (ignoring the back reactions):

$$A \xrightarrow{k_0} X, \quad X + Y \xrightarrow{k_1} 2Y \quad \text{(autocatalytic process)},$$

$$Y \xrightarrow{k_2} B$$

This notation describes the following hypothetical reaction. A compound A is in some volume and is being consumed by a reaction which converts it, at an almost unnoticeable rate (A is said to be in excess), into compound X. This reaction has a zero-order rate constant k_0. Furthermore, X is converted into Y at a rate that increases with increasing concentration of Y, this reaction being second order. Finally, Y decays via an irreversible first-order reaction to form B. Using the rules for compiling kinetic equations [4] and retaining the notations X, Y, and B to denote the concentrations of the compounds, we can write the mathematical model for Lotka's reactions as

$$\dot{X} = k_0 - k_1 XY, \quad \dot{Y} = k_1 XY - k_2 Y, \quad \dot{B} = k_2 Y \qquad (1.5)$$

If the concentrations X and Y do not change over time, then the reaction may occur such that the formation rate of B is constant.

This corresponds to $\dot{X} = \dot{Y} = 0$, or

$$k_0 - k_1 X_0 Y_0 = 0, \qquad k_1 X_0 Y_0 - k_2 Y_0 = 0, \qquad (1.6)$$

where X and Y are the equilibrium concentrations. It follows from (1.6) that

$$X_0 = k_2/k_1, \qquad Y_0 = k_0/k_2. \qquad (1.7)$$

Let us assume that there are small deviations $x(t)$ and $y(t)$ from the equilibrium concentrations X_0 and Y_0, i.e., we have $x(t) = X_0 + x$ and $Y(t) = Y_0 + y$, with $x \ll X_0$, $y \ll Y_0$. Substituting the expressions for $X(t)$ and $Y(t)$ into the first of the two equations in (1.5), and using (1.7), and neglecting products with second or more order variables in the denominator, we get

$$\dot{x} = -k_2 y - (k_1 k_0/k_2)x, \qquad \dot{y} = (k_1 k_0/k_2)x. \qquad (1.8)$$

This system (1.8) can easily be reduced to the equation for a linear oscillator (1.1) by setting $k_1 k_0/k_2 = 2\gamma$ and $k_1 k_0 = \omega_0^2$. Obviously, being a nonlinear system, (1.5) has more solutions than the linear oscillator equation (1.1), which was derived only by assuming that the concentration perturbations were small. We shall return to the nonlinear Lotka model as a component of more complex periodic chemical systems (e.g., the Belousov-Zhabotinskii reaction).

The second example is one from ecology, namely Volterra's predator--prey model [2-5]. The model involves two animals, one of which feeds on the other. The problem can be formulated by asking whether the weasels can eat all the rabbits (or foxes and hares, or pikes and carp, depending on your imagination).

Suppose two species live in a closed habitat, i.e., a predator and a vegetarian prey. The prey species (of which there are $N_1(t)$) eat vegetation, which is in excess, while the predators (or which there are $N_2(t)$) only eat the prey. If the prey species were to live alone in the habitat and they had sufficient food, their population would increase, i.e.,

$$\dot{N}_1 = \varepsilon_1 N_1 \qquad (1.9)$$

(ε_1 is the population growth and is positive and constant).
Note that (1.9) is analogous to the first-order chemical

LINEAR OSCILLATORS

reaction we considered above. If only the predators lived in the habitat, they would die out, i.e.,

$$\dot{N}_2 = -\varepsilon_2 N_2 \tag{1.10}$$

(ε_2 is the species's mortality rate, positive, and constant). It can be assumed that when both species live in the habitat the predator population will increase faster, the more frequently it encounters prey. This frequency is proportional to $N_1 N_2$. Thus we derive the following system of differential equations to describe the populations of the two species when they live in the same habitat:

$$\dot{N}_1 = N_1(\varepsilon_1 - \gamma_2 N_2), \qquad \dot{N}_2 = -N_2(\varepsilon_2 - \gamma_1 N_1), \tag{1.11}$$

where γ_2 is a positive constant that characterizes the mortality of the prey due to encounters with predators, γ_1 is a positive constant characterizing the population growth rate of the predators.

As above for the case of Lotka's model, we find the equilibrium condition for N_1^0 and N_2^0. When $\dot{N}_1 = \dot{N}_2 = 0$ we find from (1.11) that

$$N_1^0 = \varepsilon_2/\gamma_1, \qquad N_2^0 = \varepsilon_1/\gamma_2. \tag{1.12}$$

For small divergences of the species populations from the equilibrium values ($N_1 = N_1^0 + N_1'(t)$ and $N_2 = N_2^0 + N_2'(t)$) after linearizing (1.11), we obtain

$$\dot{N}_1' = -\gamma_2 N_1^0 N_2' = -(\gamma_2 \varepsilon_2/\gamma_1) N_2', \quad \dot{N}_2' = \gamma_1 N_2^0 N_1' = (\gamma_1 \varepsilon_1/\gamma_2) N_1' \tag{1.13}$$

By differentiating the first of these equations (1.13) with respect to time and using the second equation, we arrive at the equation for an harmonic oscillator

$$\ddot{N}_1' + \omega_0^2 N_1' = 0, \tag{1.14}$$

where $\omega_0^2 = \varepsilon_1 \varepsilon_2$ (we get the same equation for N_2'). If we make the substitution $N_1' = x$ in (1.14), then we get (1.2).

Let us now return to the original model, making the substitution of a variable $y = \dot{x}$ and rewriting (1.2) in the form of a system of two equations, viz.,

$$\dot{x} = y, \quad \dot{y} = -\omega_0^2 x \qquad (1.15)$$

The plane of the variables x and $\dot{x}$ is called the phase plane of equation (1.2). Each point in this plane (called a trace or phase point) corresponds to a completely determined state of the system. The locus of the trace point is called the phase trajectory and one, and only one, trajectory passes through every point on the phase plane. Note that a phase trajectory may consist of a single point, called an equilibrium position. The speed of the trace point is called the diagram velocity, and at an equilibrium point it is zero. The phase trajectory and diagram velocity must not be confused with the trajectory and velocity of motion.

The equation for integral curves in the phase plane has the form

$$dy/dx = -\omega_0^2 x/y . \qquad (1.16)$$

The solutions of (1.16) $y = y(x, C)$ form a family of integral curves. Integral curves on which the direction of motion is defined are called phase trajectories. One and only one trajectory passes through any point on a phase plane.

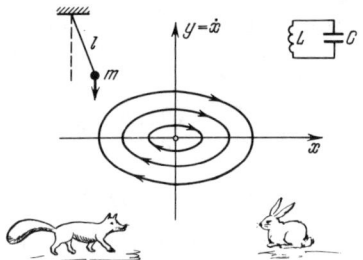

Fig. 1.1. Phase-plane diagram for an harmonic oscillator, as described by (1.2). $\omega_0 = \sqrt{g/l}$ is for a mathematical pendulum, $\omega_0 = 1/\sqrt{LC}$ is for an electric circuit, $\omega_0 = \sqrt{\varepsilon_1 \varepsilon_2}$ is for a linearized predator-prey model. A center-type equilibrium point is at the origin.

By integrating (1.16) we find that the integral curves for an oscillator are a set of ellipses, whose axes coincide with the coordinate axes (Fig. 1.1), i.e.,

$$\frac{y^2}{C} + \frac{x^2}{(C/\omega_0^2)} = 1 . \qquad (1.17)$$

The parameter C is determined by the initial conditions. By adding arrows to the integral curves to determine the direction of motion (in our case, clockwise, $dx/dt > 0$ in the

LINEAR OSCILLATORS 7

upper half-plane), we obtain the full phase-plane diagram of a linear oscillator. One phase trajectory always consists of a single point, which corresponds to the equilibrium condition. The situation when the forces which cause the motion are zero or are absent, i.e., $\dot{x} = \ddot{x} = 0$, arises when there is equilibrium. In our case, the equilibrium is at the origin $(x, y = 0)$. An isolated equilibrium, to which no other trajectory tends, is called a center. The angular velocity of a point along a phase trajectory (the phase velocity) is independent of the trajectory for a harmonic oscillator and the period for a complete turn is always $T = 2\pi/\omega_0$. Let us consider a set of similar oscillators with different initial energies and the same initial phase (on the phase plane these initial conditions would lie on a straight line passing through the origin). After an arbitrary period of time the phases of all the oscillators would still be the same, i.e., the motion of a linear oscillator is isochronous.

What other phase-plane diagrams may linearized systems with one degree of freedom have, you might ask. Let the oscillator equation have the form

$$\ddot{x} - a^2 x = 0 . \tag{1.18}$$

This equation describes, for instance, small deviations from the equilibrium position of the top point of a pendulum; its phase-plane diagram is given in Fig. 1.2 [2].

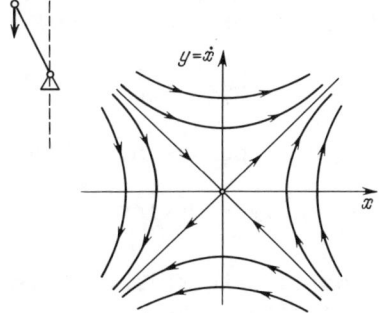

Fig. 1.2. The phase-plane diagram of a linear system with a repulsive force and described by (1.18). The equilibrium is at the saddle, and the asymptotes passing through the origin are separatrices.

It can be seen that as in the previous case, (1.18) can be exchanged for two first-order equations,

$$\dot{x} = y, \quad \dot{y} = a^2 x . \tag{1.19}$$

An equation with separable variables follows from

(1.19), i.e.,

$$dy/dx = a^2x/y ,\qquad(1.20)$$

which can be integrated to yield a family of equilateral hyperbolae whose major axes are

$$y^2 - a^2x^2 = C \qquad(1.21)$$

(C is an integration constant). If $C = 0$ we obtain the straight lines $y = ax$ and $y = -ax$, which are the asymptotes of the hyperbola family (Fig. 1.2) and pass through the equilibrium position at the origin ($x = 0$, $y = 0$). The equilibrium position in this case is called a saddle. The name comes from geography [4] in which a saddle or pass in the mountains is the lowest point between two summits to which water flows from the summits. The flow from the saddle may then go to different valleys. The line passing through the saddle and determining which valley a flow will go to is called a watershed. In our phase-plane diagram, the asymptotes of the hyperbolae pass through the saddle, and are called separatrices. Note that motion in the vicinity of a saddle is clearly unstable[1]. Small deviations have large consequences (a strict definition of stability is given later).

Let us see what happens to the motion of the oscillator and its phase-plane diagram if the losses are significant (due to friction, viscosity, etc.), i.e., when $\gamma \neq 0$ in (1.1). According to (1.1) or its equivalent equation system,

$$\dot{x} = y, \quad \dot{y} = -2\gamma y - \omega_0^2 x . \qquad(1.22)$$

The equation of the integral curves has then form

$$dy/dx = -(2\gamma y + \omega_0^2 x)/y . \qquad(1.23)$$

The equilibrium position in this case is also unique and corresponds to the coordinate origin ($x = 0$, $y = 0$). In order to integrate (1.23), we make the substitution $y = xz$ and rewrite (1.23) in the form

---

[1] It follows from (1.19) that in the upper half-plane the coordinate value x must increase, while in the lower half-plane it must fall off. All the trajectories, except the equilibrium and the two separatrices, correspond to infinite motions of the system (Fig. (1.2).

LINEAR OSCILLATORS

$$\frac{z\,dz}{z^2 + 2\gamma z + \omega_0^2} = -\frac{dx}{x}.$$

After integrating and changing back to the original variables, we can find the link between x and y when the damping is small, i.e., $\gamma^2 < \omega_0^2$, and that is

$$\sqrt{y^2 + 2\gamma xy + \omega_0^2 x^2} = C_1 \exp\left(\frac{\gamma}{\sqrt{\omega_0^2 - \gamma^2}} \arctan \frac{y + \gamma x}{\sqrt{\omega_0^2 - \gamma^2}\, x}\right), \quad (1.24)$$

This can be used to construct the phase-plane diagram of a linear oscillator with damping (C_1 is an arbitrary constant). The angular velocity of a point travelling along a phase trajectory $\sqrt{\dot{x}^2 + \dot{y}^2} = \sqrt{y^2 + (2\gamma y + \omega_0^2 x)^2}$ is not zero anywhere but at the origin. However, we emphasize that when moving along any trajectory the angular velocity tends to zero the closer the point being followed gets to the equilibrium point. In order to make the detail of the phase-plane diagram clearer, we introduce the new variables $u = \sqrt{\omega_0^2 - \gamma^2}\, x$ and $v = y + \gamma x$ and we assume that they are orthogonal coordinates. Whence

$$y^2 + 2\gamma xy + \omega_0^2 x^2 = u^2 + v^2,$$

$$u^2 + v^2 = C_1^2 \exp\left(\frac{2\gamma}{\sqrt{\omega_0^2 - \gamma^2}} \arctan \frac{v}{u}\right),$$

and using polar coordinates ($u = \rho \cos \phi$, $v = \rho \sin \phi$), we finally get in place of (1.24)

$$\rho = C_1 \exp\left[\gamma \phi / \sqrt{\omega_0^2 - \gamma^2}\right]. \quad (1.25)$$

Since the solutions of (1.1) are known, it is easy to show that $\phi = -(\omega t + \alpha)$, $\alpha = \arctan\left[(y_0 + \gamma x_0)/\sqrt{\omega_0^2 - \gamma^2}\, x_0\right]$ (x_0 and y_0 are the values of x and y at $t = 0$) and hence ϕ decays in time while $\rho \to 0$ as $t \to \infty$

Thus the phase trajectories on the u,v-plane are logarithmic spirals rotating around the equilibrium point ($v = 0$, $u = 0$), which is called a stable focus (Fig. 1.3). When $\gamma/\sqrt{\omega_0^2 - \gamma^2}$ is small, the spiral arms are close to the circle $u^2 + v^2 = C_1^2$, which on the x,y-plane is the ellipse $y^2 + 2\gamma xy + \omega_0^2 x^2 = C_1^2$. Consequently, when $\gamma/\sqrt{\omega_0^2 - \gamma^2}$ is small, the arms of the spiral during the first turn are close to the ellipse with the corresponding C_1.

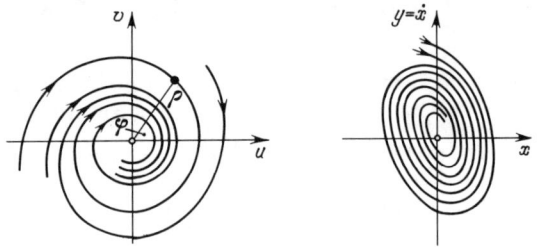

Fig. 1.3. Phase-plane diagram of an oscillator with small damping. The equilibrium is the stable focus.

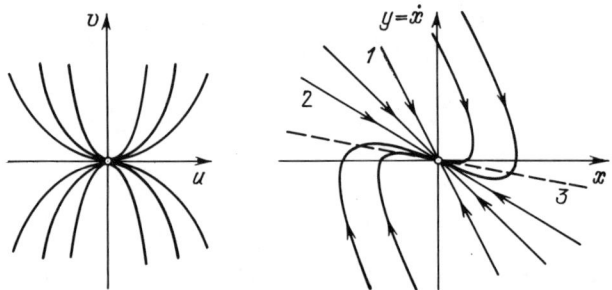

Fig. 1.4. Phase-plane diagram of a linear oscillator with damped aperiodic oscillations. The equilibrium is the stable focus. 1) $y = -q_2 x$; 2) $y = -q_1 x$; 3) $y = -q_1 q_2 x/(q_1 + q_2)$.

Thus the phase diagram on the x,y-plane is, like the diagram on the u,v-plane, a family of logarithmic spirals with a stable focus at the origin. We may thus conclude from the phase diagram that whatever the initial conditions may be (excepting the equilibrium) the motion of our dissipative system is a damped oscillation. All the spirals on the phase-plane diagram asymptotically approach the origin and the radius of a point travelling along a spiral decreases with each turn.

If $\gamma^2 > \omega_0^2$ the oscillation becomes damped aperiodic (Fig. 1.4). The equilibrium becomes the stable focus ($x = 0$, $y = 0$). We leave the reader to prove that in this case the integral curves are given by the equation $v = C_1 u^a$, where

$$v = y + q_1 x, \qquad u = y + q_2 x, \qquad q_1 = \gamma - \sqrt{\gamma^2 - \omega_0^2} > 0, \qquad \text{and}$$

$$q_2 = \gamma + \sqrt{\gamma^2 - \omega_0^2} > 0, \; a = q_2/q_1 > 1 \; [2] \; (\text{Fig. 1.4}).$$

Changing the sign of γ (negative friction, resistance, conductivity etc.) makes the equilibrium unstable (Fig. 1.5). In a system described by (1.19), i.e., one with a repulsive force, the inclusion of friction, whether positive or negative, does not substantially change the phase-plane diagram (Fig. 1.2) because the equilibrium is a saddle.

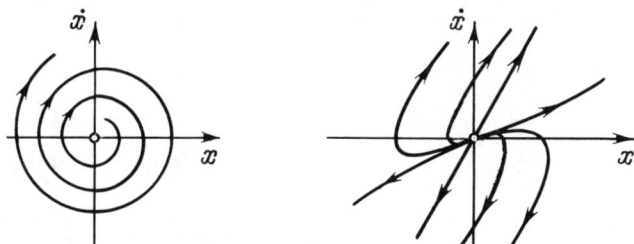

Fig. 1.5. Phase-plane diagrams of a linear oscillator with $\gamma < 0$. The equilibrium is unstable.

The solution of (1.1) has, for $\gamma^2 \neq \omega_0^2$, the form

$$x = A \exp(p_1 t) + B \exp(p_2 t),$$

where p_1 and p_2 are the roots of the characteristic equation $p^2 + 2\gamma p + \omega_0^2 = 0$. The positions of these roots on the complex plane uniquely determines the type of the system's equilibrium, and hence the motion of the oscillator (Fig. 1.6).

When $\gamma = 0$ the roots of the characteristic equation are purely imaginary; as γ ($\gamma > 0$) increases they move toward the left-hand half-plane remaining complex conjugates. When $\gamma^2 = \omega_0^2$ the roots merge on the real axis, and then they split into two real roots as γ increases further. The saddle type equilibrium corresponds to two real roots with different signs. Changing the sign of γ moves the roots into the right-hand half-plane and the equilibria become unstable.

Figure 1.7 shows the 2γ, ω_0^2-plane partitioned into the various types of equilibrium. This diagram contains practically all that must be known about the equilibria on a plane.

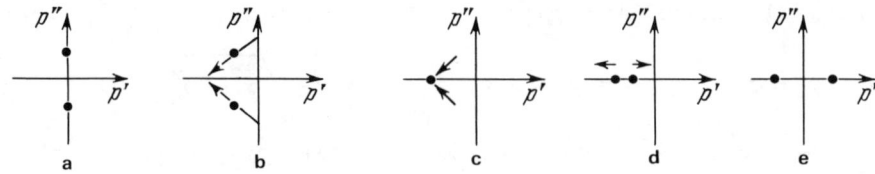

Fig. 1.6. Positions of the roots on the complex plane $p = p' + ip''$ and their relation to the equilibrium type of the system.

a) $\gamma = 0$, $p_{1,2} = \pm i\omega_0$ for a center;

b) $\omega_0 > \gamma > 0$, $p_{1,2} = -\gamma \pm i\sqrt{\omega_0^2 - \gamma^2}$ for a focus;

c) $\gamma = \omega_0$, $p_{1,2} = -\gamma$;

d) $\gamma > \omega_0$, $p_{1,2} = -\gamma \pm \sqrt{\gamma^2 - \omega_0^2}$ for a node;

e) $\gamma = 0$, $p_{1,2} = \pm \omega_0$ for a saddle.

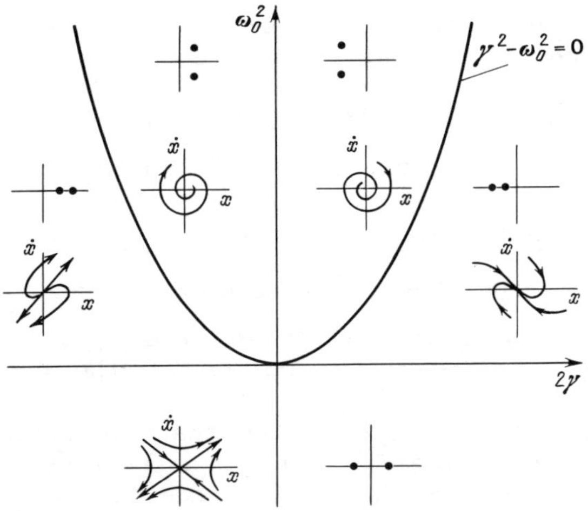

Fig. 1.7. Partitioning of the 2γ, ω_0^2-plane into regions with different types of equilibrium (the positions of the roots on the complex plane and the corresponding phase-plane diagrams are given in each area).

LINEAR OSCILLATORS

1.3 Resonance. The Effect of an Aperiodic External Force on an Oscillator

Up to now we have been discussing an independent oscillator. Suppose now a periodic external force is acting on a linear oscillator. The following equation is the starting point for our analysis

$$\ddot{x} + 2\gamma\dot{x} + \omega_0^2 x = F_0 \cos \omega t , \qquad (1.26)$$

where F_0 is the constant amplitude of the external force, and ω is its frequency.

Resonance is the rapid increase in the amplitude of a steady state oscillation when the frequency ω of the external harmonic force approaches the fundamental frequency ω_0 of the oscillator (or in the more general case to the frequency of one of the harmonics of the system).

Let us again look at an oscillator without damping ($\gamma = 0$). The general solution of (1.26) is in this case

$$x(t) = A \cos \omega_0 t + B \sin \omega_0 t + \rho \cos \omega t ,$$

where $\rho = F_0/(\omega_0^2 - \omega^2)$. Let us choose $x(0) = 0$ and $\dot{x}(0) = 0$ at $t = 0$ for the initial conditions. Whence, $A = -\rho$, $B = 0$ and the motion of such a dependent oscillator will be described by the function

$$x(t) = \frac{F_0}{\omega_0^2 - \omega^2} (\cos \omega t - \cos \omega_0 t) .$$

Now let us trace the growth in the amplitude of the oscillation given resonance, i.e., as $\omega \to \omega_0$. In this case we have $\omega_0^2 - \omega^2 = (\omega_0 - \omega)(\omega_0 + \omega) \approx 2\omega_0(\omega_0 - \omega)$ and

$$x(t) = \frac{F_0}{2\omega_0} \frac{\sin[(\omega_0 - \omega)t/2]}{(\omega_0 - \omega)t/2} t \sin \omega_0 t .$$

When there is complete resonance $x(t) = (F_0/2\omega_0)t \sin \omega_0 t$, i.e., for a periodic perturbation, the amplitude of the oscillation acts as if it were an aperiodic function of time (Fig. 1.8). The factor t leads to the secular increase in the amplitude, the rate of increase depending on F_0.

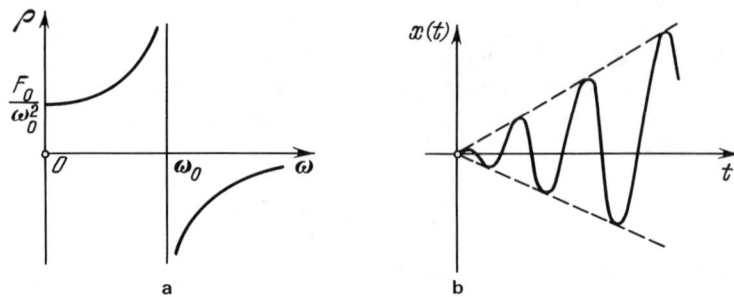

Fig. 1.8. a) Behavior of the solution of (1.26) given resonance. b) Oscillogram for $\gamma = 0$.

A secular growth rate is one of the simplest manifestations of a system's instability to an external disturbance. It is a consequence of the idealization inherent in the original model. Depending on the situation the model must be altered to account for nonlinear effects (the system can then remain conservative) or linear dissipation (viscosity, friction, resistance, etc.). In the first case the nonlinear effects would cause the frequency to shift and hence move the system gradually away from resonance. This can be seen if (1.26) is "corrected" for $\gamma = 0$ thus

$$\ddot{x} + \omega_0^2 x + \mu\omega_n^2(x)x = F \cos \omega t ,$$

where $\omega_n^2(x) = \alpha x + \beta x^2 + \ldots$, and α, β, μ are constants. We leave an analysis of this nonlinear problem for Chapter 13. Let us now consider resonance in an oscillator with a finite Q-factor, using the method of complex amplitudes, i.e., we assume that all the variables are complex and in the solution only consider the real parts of the variables. Thus (1.26) can be written as

$$\ddot{\bar{x}} + 2\gamma\dot{\bar{x}} + \omega_0^2\bar{x} = F_0 \exp(i\omega t) ,$$

where $x = \text{Re}\bar{x}$. If we wait long enough, the oscillator's natural oscillations will be damped and so we need only look at the forced solution $\bar{x} = \rho \exp(i(\omega t - \theta))$, where ρ and θ must be determined. After substituting the solutions back into (1.27) and separating the real and imaginary parts, we get

$$\rho^2 = F_0^2/[(\omega_0^2 - \omega^2)^2 + (2\omega\gamma)^2], \quad \tan\theta = -2\omega\gamma/(\omega_0^2 - \omega^2) .$$

The resonance curve shown in Fig. 1.9 corresponds to a steady state process and defines the relationship between the amplitude of a steady-state oscillation and the frequency of the external force. Note that the maximum amplitude now does not come when the fundamental frequency of the oscillator and the frequency of the forcing oscillation coincide; instead it occurs at some point to the left of this value along the frequency axis, the shift being dependent on γ (Fig. 1.9).

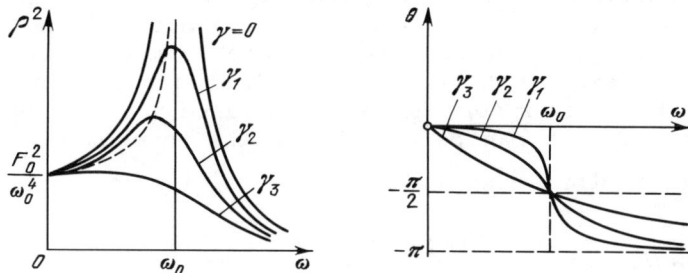

Fig. 1.9. Resonance curves and the phase shift between the external force and the oscillator's shift versus frequency. The dashed line is the trajectory of the shift of the maximum ρ^2 versus γ ($\gamma_3 > \gamma_2 > \gamma_1$).

In fact, if $\omega_0 =$ const we can differentiate the expression for ρ^2 with respect to ω, to find that the maximum in ρ occurs at $\omega = \sqrt{\omega_0^2 - 2\gamma_0^2}$. If $\omega \approx \omega_0$, then clearly the expression for ρ may be rewritten as

$$\rho^2 = F_0^2/4\omega_0^2 [(\omega - \omega_0)^2 + \gamma^2].$$

Whence $\rho^2_{max} = F_0^2/4\omega_0^2\gamma^2$ when $\omega = \omega_0$, the width of the resonance curve at $\rho^2_{max}/2$ being $2(\omega - \omega_0) = 2\gamma$. It goes without saying that $\gamma \ll \omega$ because $\omega \approx \omega_0$. Resonance phenomena occur virtually everywhere, and we give a few examples here.

Resonance phenomena are the basis for ultrahigh frequency electronic devices, which utilize large-Q volume resonators. A typical device of this class is the klystron, the simplest variant of which is the double-cavity amplifying klystron (Fig. 1.10) [6]. The input signal from the external source has a frequency ω which is close to the fundamental frequency of the resonator (ω_0). The signal affects the electron beam inside a high-frequency gap. As a result the electrons at the inlet of the drift tube have different velocities. The drift tube is a space which is free of external high-frequency fields. Because there is a finite transit time within this space, electrons leaving the resonator with fast velocities catch up with those that left earlier with smaller velocities. This causes the electrons to bunch together forming clusters of electrons, and hence the formation of a variable current (see [7], Chap. II). If the frequency of the excitation in the input resonator is close to the fundamental frequency of the output resonator, then the electron clusters will be resonate and hence the signal will be amplified. When the input signal is large, nonlinear effects are added to the beam and harmonics of the current with frequency ω are generated. These harmonics will again effectively excite the oscillation in the output resonator, when the conditions for temporal resonance are met, which for the n-th harmonic can be written in the form $n\omega \approx n\omega_0$ (n is an integer). Hence the device will be a klystron or frequency multiplier.

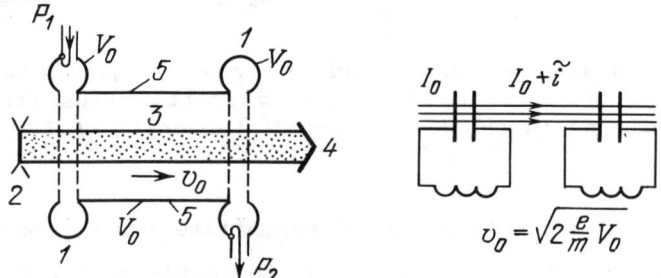

Fig. 1.10. Diagram of a double resonator klystron: 1) input and output volumetric resonators; 2) electron source; 3) electron beam; 4) electron collector; 5) drift tube; V_0 is the drift tube's potential; P_1 and P_2 are the powers of the input and output signals; I_0 is the constant component of the beam current; e and m are the charge and mass of an electron.

The applications of resonance are extremely varied. For instance, it can be used to determine the fundamental oscillations of a molecule of a compound. Some gas molecules, molecules with an electric dipole moment, paramagnetic atoms and ions in an external magnetic field, and so on, have sets of energy levels which correspond to fundamental (resonance) frequencies in the VHF radio band. If such a molecule or atom has VHF electromagnetic oscillations, the frequencies of which satisfy the condition $h\nu = \mathcal{E}_u - \mathcal{E}_l$ (where h is Plank's constant, and $\mathcal{E}_u$ and $\mathcal{E}_l$ are the energies of the upper and lower levels), then resonance absorption may occur.

Radio spectroscopes are used to study VHF absorptions by atoms or molecules (Fig. 1.11) [8]. Radiation from the VHF generator[2] falls on the absorbing cell (a volume resonator or section of a wave guide), which is filled with the sample.

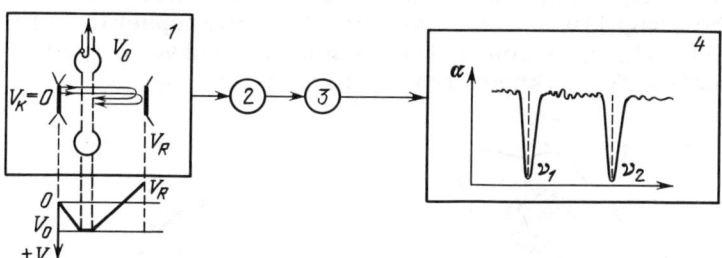

Fig. 1.11. Diagram of a radio spectroscope. 1) Reflex klystron; 2) absorbing cell with the specimen sample; 3) receiver; 4) recording device (α characterizes the absorption, and ν_1 and ν_2 are the resonance frequencies).

When the frequency of the external sources's signal coincides in the resonator or wave guide with the resonance frequency of absorption of the sample material, then the radiation is absorbed and the receiver produces a weaker signal. The resultant peaks in the absorption capacity versus frequency curve are maxima in the absorption spectra. The resonance frequencies, their widths and forms yield information about the structure of the molecule, the structure of the atomic nucleus, and the arrangement of the electron shells of the atoms, and can be used to establish the character of the links between the atoms or the molecules in the material (see [8] for more detail).

Resonance can also be utilized for global measurements. It has been possible, for example, to determine the parameters of the oscillator consisting of the Earth and its atmosphere. The external force in this context is the Moon, which rotates around the Earth and causes a tide in the Earth's atmosphere twice a day with a period of 12 hr 40 min. Clearly, if the atmosphere is moved then because of the restoring gravity force an oscillation will be generated in it relative to the Earth. In order to measure the parameters

---

[2] A reflex klystron can be used as the VHF generator (Fig. 1.11). In this device the electrons are velocity-modulated and energy is transferred from the electron beam to the high-frequency field within a single resonator because the electrons are grouped in a retarding static field in the volume of the reflector resonator (the reflector electrode with potential V_R in the figure) and are then returned to resonator. The oscillation frequency can be smoothly changed by changing the potential on the reflector. The signal from the klystron and incident on the cell is frequency modulated.

γ and ω_0 of this global oscillator, it is sufficient to find ρ and θ (Fig. 1.9) for some value of ω. This has been done, and the magnitude and lag of the atmospheric tide were measured, which allowed the resonance curve $\rho = \rho(\omega)$ to be constructed for a known point (Fig. 1.12).

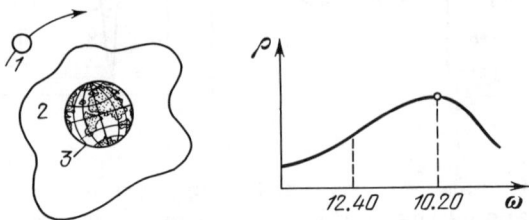

Fig. 1.12. Determining the intrinsic parameters of the Earth-atmosphere oscillator [12]: $T = 2\pi/\omega = 12$hr 40 min; $T_0 = 2\pi/\omega_0 = 10$ hr 20 min. 1) The Moon; 2) the atmosphere; 3) the Earth.

Let us now set a problem: What work is done by the external field on the oscillator? The work done by a force $F = F_0 \cos \omega t$ in a time dt is Fdt, and the power is $P = Fdx/dt$. It follows from (1.26) that $P(t) = F\dot{x} = (\ddot{x} + \omega_0^2 x + 2\gamma\dot{x})\dot{x}$. If we assume that $x = \rho \cos(\omega t - \theta)$ and hence that $\dot{x} = -\rho\omega \sin(\omega t - \theta)$ and $\ddot{x} = -\rho\omega^2 \cos(\omega t - \theta)$, then the average power over a period $T = 2\pi/\omega$ is

$$<P(t)> = T^{-1} \int_0^T [(\ddot{x} + \omega_0^2 x)\dot{x} + 2\gamma \dot{x}^2] dt$$

$$= (2\pi)^{-1} \int_0^{2\pi} \gamma\rho^2\omega^2 [1 - \cos(2\omega t - 2\theta)] d\omega t = \gamma\rho^2\omega^2.$$

Thus we have $<P(t)> = 0$ for $\gamma = 0$. This would seem to be a strange result. However, it should be remembered that this is the power loss in a stationary regime, i.e., when the oscillator conserves all the energy associated with it and the external force is only spent to cover the dissipative losses. This is the explanation for the paradox, that when $\gamma = 0$ the external force does no work on the oscillator.

We now look at how an harmonic oscillator behaves when affected by an arbitrary aperiodic external force $F(t)$, whence the motion is described by the equation

$$\ddot{x} + \omega^2 x = F(t).$$

LINEAR OSCILLATORS

To examine this problem we use the method of undetermined coefficients (Lagrange multipliers), setting

$x(t) = A(t) \cos \omega t + B(t) \sin \omega t$, whence

$\dot{x}(t) = \dot{A}(t) \cos \omega t + \dot{B}(t) \sin \omega t - A(t) \omega \sin \omega t + B(t) \omega \cos \omega t$,

where $\dot{A} = dA/dt$ and $\dot{B} = dB/dt$.
Since we introduced two arbitrary functions and there is only one equation, we may specify an arbitrary relation between them. For arithmetic convenience, we can require that $\dot{A}(t) \cos \omega t + \dot{B}(t) \sin \omega t = 0$, whence

$$\ddot{x}(t) = -\dot{A}(t) \omega \sin \omega t - A(t) \omega^2 \cos \omega t \\ + \dot{B}(t) \omega \cos \omega t - B(t) \omega^2 \sin \omega t .$$

By substituting $\ddot{x}(t)$ and $\omega^2 x(t)$ into the original equation, we finally arrive at

$$-\dot{A}(t) \omega \sin \omega t + \dot{B}(t) \omega \cos \omega t = F(t).$$

We can solve this equation and the link equation as a system of equations for $\dot{A}(t)$ and $\dot{B}(t)$. Elementary transformations yield

$$A(t) = -\frac{1}{\omega} \int_0^t F(\tau) \sin \omega \tau \, d\tau, \quad B(t) = \frac{1}{\omega} \int_0^t F(\tau) \cos \omega \tau \, d\tau.$$

If, for example, $F(t) = F_0 \cos \omega t$, which was the case in previous examples, then $B(t) = \frac{F_0}{\omega} \left\{ \frac{1}{2} t + \frac{1}{4\omega} \sin 2\omega t \right\}$ and as $t \to \infty$ the solution tends to infinity, viz., there is secular growth. Clearly, if $A(t)$ and $B(t)$ remain small as t grows, then there will be no resonance. Hence the condition for the absence of resonance can be rewritten as

$$\lim_{t \to \infty} \frac{1}{T} \int_0^T F(\tau) \begin{pmatrix} \sin \omega \tau \\ \cos \omega \tau \end{pmatrix} d\tau = 0.$$

Mathematically, this relation means that the function $F(t)$ need not contain the eigenfunctions of our problem. If, moreover, $F(t) = \sum_{i=1}^{\infty} F_{oi} \cos \omega_i t$ (the external force can be given as a Fourier series) and one of the ω_i coincides with the fundamental frequency ω of the oscillator, then there will be resonance. All the components of the other frequencies will become insignificant.

1.4 Normal Oscillations. Analogy with Quantum Mechanics. Production and Extinction Operators

Let us look once more at the harmonic oscillator, but this time we start from its Hamiltonian function

$$\mathcal{H} = (p^2 + \omega^2 q^2)/2 , \qquad (1.27)$$

where q and p are the coordinate and momentum, respectively, and ω the fundamental frequency of the oscillator. The equations for the oscillator's motion in Hamiltonian form are

$$\partial q/\partial t = \partial \mathcal{H}/\partial p = p, \qquad (1.28)$$

$$\partial p/\partial t = - \partial \mathcal{H}/\partial q = - \omega^2 q . \qquad (1.29)$$

We multiply (1.28) by $\pm i\omega$ and add the result to (1.29). Whence we get

$$da/dt = - i\omega a, \qquad (1.30)$$

$$da^*/dt = i\omega a^* \qquad (1.31)$$

where the complex conjugates a and a^* are given by

$$a = (\omega q + ip)/\sqrt{2\omega} , \quad a^* = (\omega q - ip)/\sqrt{2\omega} . \qquad (1.32)$$

The solutions of (1.30) and (1.31) may be written as

$$a(t) = a(0)e^{-i\omega t} = \frac{1}{\sqrt{2\omega}} [\omega q(0) + ip(0)]e^{-i\omega t},$$

$$a^*(t) = a^*(0)e^{i\omega t} = \frac{1}{\sqrt{2\omega}} [\omega q(0) - ip(0)]e^{i\omega t}.$$

Equations (1.30) and (1.31) are called the equations of normal oscillations, while $a(t)$ and $a^*(t)$ are often simply called the normal vibrations of an oscillator [9,10].
Note that $a(t)$ and $a^*(t)$ can be represented graphically

LINEAR OSCILLATORS

as vectors of the same length rotating in opposite directions (we shall need this representation later).

It can be seen from (1.32) that

$$q = \left(1/\sqrt{2\omega}\right)(a^* + a), \quad p = i\sqrt{\omega/2}\,(a^* - a),$$

and consequently, from the definition of the oscillator's Hamiltonian (1.27), we have

$$\mathcal{H} = (pp^* + \omega^2 qq^*)/2 = \omega a^* a. \tag{1.33}$$

In quantum mechanics the Hamiltonian of an oscillator is

$$\mathcal{H} = \hbar\omega(a^+ a + 1/2), \tag{1.34}$$

where $\hbar = h/2\pi$ and the quantities $a = \left(1/\sqrt{2\hbar\omega}\right)(\omega q + ip)$ and $a^+ = \left(1/\sqrt{2\hbar\omega}\right)(\omega q - ip)$ are called, respectively, the extinction and production operators [11]. The extra energy $\hbar\omega/2$ over the classical case is called the null energy of the oscillator. At the classical limit, i.e., when $\mathcal{H} \gg \hbar\omega/2$, we often utilize the expression for the number of quanta $N = aa^+$ (it follows from (1.34) that $N = \mathcal{H}/\hbar\omega - 1/2$).

To conclude, let us dwell upon the interpretation of the energy state of an oscillator that Dirac put forward and justified [11]. The Hamiltonian describes a system composed of n identical independent quanta which are in the same state with energy $\hbar\omega$. We have already said that N is the number of particles. Now this becomes understandable; each of the eigenvalues N of the operator gives a certain number of quantum particles, e.g., photons or phonons, about which more will be said in Chapter 3.

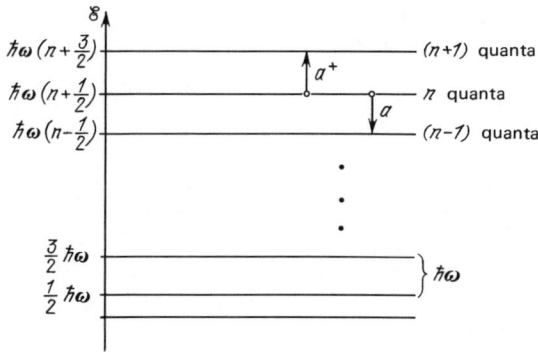

Fig. 1.13. An interpretation of the production (a^+) and extinction (a) operators and the number of particles operator N.

In the quantum case, if an operator existing in a state with n quanta is acted on by the operator a^+, it goes into a state with $(n + 1)$ quanta, whence the name production operator(a^+). If, however, an oscillator in a state with n quanta is acted on by the a operator, it will transfer into a state with $(n - 1)$ quanta, whence the name extinction operator(a). This interpretation of the N, a^+, and a operators is illustrated in Fig. 1.13.

CHAPTER 2

OSCILLATIONS IN A SYSTEM WITH TWO LINKED OSCILLATORS

2.1 Initial Equations

In the last chapter we looked at the phenomenon of resonance in its simplest form, namely external resonance in a linear oscillator. If the system is not that simple, for instance it has several degrees of freedom, another effect is possible. This is the effect of internal resonance, or the resonance between individual subsystems. We shall see how internal resonance between individual (we shall call them partial) subsystems allows them to exchange energy, i.e., it is an interaction between them. Clearly, external resonance can be considered as a special case of internal resonance when the energy of one of the subsystems is infinite, in which case it ceases to be an interaction, and is rather the action of one subsystem upon the other.

In general, systems with two degrees of freedom manifest many of the features of more complicated systems, and hence in this chapter we shall make quite a detailed analysis of the systems of two linked oscillators.

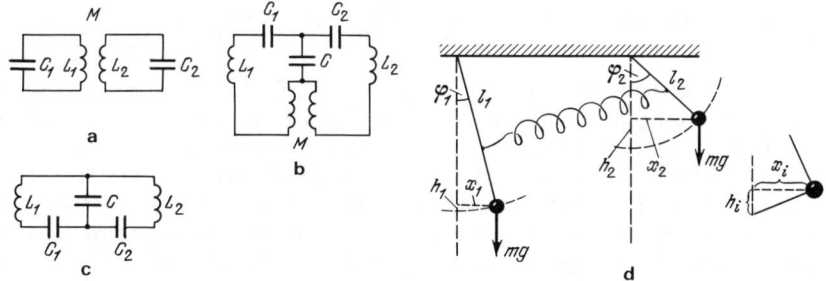

Fig. 2.1. Simple examples of electrical and mechanical systems of two linked oscillators: a) inertial (inductive) connection; b) mixed connection; c) force (capacitive) connection; d) two linked pendulums in a gravity field.

We shall utilize the usually cited simple examples of linked oscillators (Fig. 2.1). One example, in particular, consists of two pendulums of lengths l_1 and l_2 with the same masses $m_1 = m_2 = m$ swinging in a gravity field. They are linked by a weightless spring with a modulus of elasticity k (Fig. 2.1d). The motion of this conservative system with two degrees of freedom is described for the linear approximation by the equations

$$\frac{dv_1}{dt} = -\omega_1^2 x_1 + \frac{k}{m}(x_2 - x_1), \quad \frac{dx_1}{dt} = v_1,$$

$$\frac{dv_2}{dt} = -\omega_2^2 x_2 + \frac{k}{m}(x_1 - x_2), \quad \frac{dx_2}{dt} = v_2,$$

(2.1)

where $\omega_{1,2}^2 = g/l_{1,2}$. These equations may either be derived from the expression for the energy of the system, which for small pendulum displacements has the form

$$\mathcal{H} = \frac{m}{2}\left(v_1^2 + v_2^2\right) + \frac{m}{2}\left(\omega_1^2 x_1^2 + \omega_2^2 x_2^2\right) + \frac{k}{2}\left(x_1^2 - x_2^2\right) \quad (2.2)$$

(the first term in $\mathcal{H}$ is obvious, the second and third terms are the potential energies of the masses in the gravity field and of the spring - the energy of the link), or derived from a physical argument that the acceleration of a pendulum is related to the presence of the restoring force of gravity $(-m\omega_{1,2}^2 x_{1,2})$ and of the spring $(k(x_{2,1} - x_{1,2}))$. Usually, (2.1) is rewritten in the form of the equations of linked oscillators, viz.,

$$\ddot{x}_1 + \omega_1^2 x_1 = (k/m)(x_2 - x_1), \quad \ddot{x}_2 + \omega_2^2 x_2 = (k/m)(x_1 - x_2). \quad (2.3)$$

Before going on to an analysis of (2.3), we wish to present another lesser known example of a system of linked oscillators. It is often encountered in problems associated with vacuum and quantum microwave electronics and involves the induction of resonance in an oscillating system by sources whose properties depend on those of the active medium (the electron beam, gas, paramagnetic crystal, etc.). If the resonator is empty ("cold") and when losses may be neglected, the resonator may behave like a set of unconnected oscillators, i.e., a set of normal modes. Disturbing the complex dielectric permeability of the medium filling the resonator links the modes [1,2]. This can be simply explained in that the modes modulate the medium and hence they act upon each other via the medium. Let us consider the following model. A resonator is filled with a dielectric medium whose dielectric permeability ε is perturbed, $\varepsilon = \varepsilon + \delta\varepsilon$ (a change of this sort corresponds to perturbing current with density $j = i\omega\delta\varepsilon E$, where E is the solenoid part of the electric field). This model was linked in [1] with changes in the active medium in the resonator of a laser being acted on by a pumping field. Then, we obtain the following equation for the expansion coefficients of the solenoid part of the electric field induced in the resonator

$$\frac{d^2 A_s}{dt^2} + \omega_s^2 A_s = -\frac{d}{dt} \sum_r k_{sr} A_r \;, \tag{2.4}$$

where the index s corresponds to the s-th fundamental mode and k_{sr} is the link coefficient which depends on the perturbation in the dielectric permeability [1].

For instance, if we assume that the fundamental oscillations with the indices $s = 1$ and $s = 2$ are in resonance, then we get the following system of two equations from (2.4):

$$\ddot{A}_1 + \omega_1^2 A_1 + k_{12} \dot{A}_2 = 0 \;, \quad \ddot{A}_2 + \omega_2^2 A_2 + k_{21} \dot{A}_1 = 0 \;, \tag{2.5}$$

which is similar to (2.3).

2.2 The Fundamental Oscillations of Two Linked Oscillators

We now present the general theory for small oscillations of two linked oscillators. This is a linear conservative system with two degrees of freedom [3] and hence to describe it we must introduce two generalized coordinates x and y. The equations of motion of the system can be conveniently written in a Lagrangian form [4], i.e.,

$$\frac{d}{dt}\left(\frac{\partial \mathcal{L}}{\partial \dot{x}_i}\right) - \frac{\partial \mathcal{L}}{\partial x_i} = F_i \;, \tag{2.6}$$

where $x_1 = x$, $x_2 = y$, and the F are generalized nonpotential forces (for conservative systems these are external forces acting upon it). The Lagrangian function for this system has the form

$$\mathcal{L} = T - V \;,$$

where

$$T = A\dot{x}^2 + B\dot{y}^2 + 2H\dot{x}\dot{y} \;, \quad V = ax^2 + by^2 + 2hxy \tag{2.7}$$

are respectively the kinetic and potential energies of the system, and H and h are the inertial and force constants of the link.

Equation (2.6) can be rewritten for an independent system ($F_{1,2} = 0$) using (2.7), i.e.,

$$A\ddot{x} + ax + H\ddot{y} + hy = 0, \quad H\ddot{x} + hx + B\ddot{y} + by = 0. \quad (2.8)$$

If we restrict ourselves to the case where T and V are positive definite quadratic forms (this is not the case for the system given Fig. 2.2a, for example), then the following equalities are necessary and sufficient for positive definiteness: $A, B > 0$; $a, b > 0$; $AB - H^2 > 0$; $ab - h^2 > 0$.

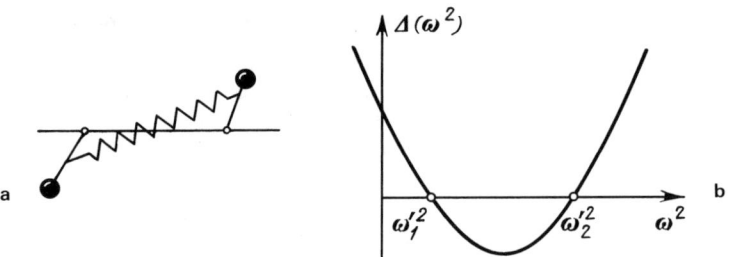

Fig. 2.2. a) A system of two linked pendulums which does not fulfill the conditions for the positive definiteness of T and V. b) The $\Delta(\omega^2)$ function when T and V are positive definite.

If we make the usual assumptions for linear systems, viz., that $(x, y) = (X, Y)e^{i\omega t}$, then substituting these into (2.8) yields

$$(-A\omega^2 + a)X + (-H\omega^2 + h)Y = 0,$$
$$(-H\omega^2 + h)X + (-B\omega^2 + b)Y = 0. \quad (2.9)$$

In order for this system of homogeneous equations (2.9) to have a nontrivial solution, it must have a zero determinant:

$$\Delta(\omega^2) = \begin{vmatrix} -A\omega^2 + a & -H\omega^2 + h \\ -H\omega^2 + h & -B\omega^2 + b \end{vmatrix}. \quad (2.10)$$

The characteristic equation to determine the normal (fundamental) frequencies of the system follows from this condition:

$$\Delta(\omega^2) = \omega^4(AB - H^2) - \omega^2(aB - bA - 2Hh) + ab - h^2 = 0.$$

The condition for the positive definiteness of T and V

graphically means that the parabola $\Delta(\omega^2)$ always intersects the abscissa at two points (Fig. 2.2b). These points correspond to the two normal frequencies of the system ω'_1 and ω'_2. It is easy to show that we can change, using a linear transformation, from the x, y coordinates to new coordinates (called normal coordinates) u_1, u_2, in which (2.8) can be rewritten as the equations of two independent oscillators, viz.,

$$\ddot{u}_1 + (\omega'_1)^2 u_1 = 0, \quad \ddot{u}_2 + (\omega'_2)^2 u_2 = 0 . \quad (2.11)$$

In the same way, any conservative linear system with n degrees of freedom can be represented in the form of a set of n noninteracting oscillators. This means that a linear conservative system with constant parameters can be completely characterized by a spectrum of normal frequencies (obviously, initial conditions must be supplied in order to get a solution).

It is reasonable to compare the normal frequencies characterizing linked oscillators with the partial frequencies. Remember that a partial system corresponding to a given coordinate is what is obtained by fixing all the other coordinates (we assume that the current in the coil and the potential across the capacitor in one of the circuits are zero for the system in Fig. 2.1a, or that one of the pendulums in Fig. 2.1c is fixed). The choice of partial system depends on the choice of coordinate (and vice versa). The equations for the partial frequencies can be obtained from (2.8), for example, by removing the terms for the link between the systems, i.e., by "nullifying" the link constants ($H = h = 0$). Whence $A\ddot{x} + ax = 0$, and $B\ddot{y} + by = 0$ and the partial frequencies will be $n_1 = \sqrt{a/A}$, $n_2 = \sqrt{b/B}$. So what is the relationship between the partial and normal frequencies? It is clear from Fig. 2.2b that if $\Delta(n^2_{1,2}) < 0$, then the n_1 and n_2 frequencies will lie between ω'_1 and ω'_2. For partial frequencies

$$\Delta(n^2_{1,2}) = \begin{vmatrix} 0 & -Hn^2 + h \\ -Hn^2 + h & 0 \end{vmatrix} = -(-Hn^2 + h)^2 < 0 ,$$

i.e., the partial frequencies will always lie between the normal frequencies $\left(\omega'_1 \leq n_{1,2} \leq \omega'_2\right)$.

To sum up, the introduction of a link into a

conservative system may only increase the interval between the fundamental frequencies of the linear system. This is an extremely important result, for instance when evaluating the oscillation constants of a molecule, which is characterized by partial frequencies. The spectrum of the normal frequencies is, however, observed because any compound being studied is a collection of linked systems, and so a correction must be made for the link between the subsystems. The results we have obtained concerning the distance between the fundamental frequencies when there is a link in the system can be used to assess the location of the desired partial frequencies.

By way of example, let us look at the well studied case of two linked sympathetic pendulums, the oscillations of which are described by (2.3) given $\omega_1^2 = \omega_2^2 = \omega_0^2$. By fixing one of the coordinates, we determine the partial frequencies from the relations $n_1^2 = \omega_0^2 + k/m$; $n_2^2 = \omega_0^2 + k/m$. Hence, for this set of subsystems the partial frequencies are equal. In order to see how the pendulums affect each other, let us attempt to guess the normal frequencies. We introduce the new variables $u_1 = x_1 + x_2$ and $u_2 = x_1 - x_2$. In these coordinates (2.3) with $\omega_1^2 = \omega_2^2 = \omega_0^2$ becomes the system for two independent oscillators, i.e.,

$$\ddot{u}_1 + \omega_0^2 u_1 = 0, \quad \ddot{u}_2 + (\omega_0^2 + 2k/m)u_2 = 0. \qquad (2.12)$$

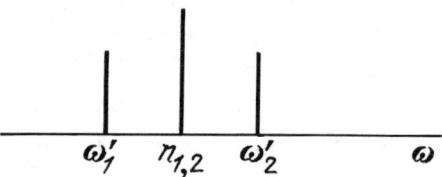

Fig. 2.3. The increase in the interval between the fundamental frequencies of a system consisting of two identical pendulums when they are linked.

Consequently, the normal frequencies are $\omega_1' = \omega_0$ and $\omega_2' = \sqrt{\omega_0^2 + 2k/m}$, i.e., the interval between the fundamental frequencies really does increase when the link is introduced (Fig. 2.3) because $\omega_0 < \sqrt{\omega_0^2 + 2k/m} < \sqrt{\omega_0^2 + 2k/m}$. If $u_2 = 0$, then $x_1 = x_2$ and both pendulums move with the "undisturbed" frequency $\omega_1 = \omega_0$ and the spring in this case does no work (in phase oscillations, Fig. 2.4a). When $u_1 = 0$ we have $x_1 = -x_2$ and the pendulums move out of phase with frequency $\omega_2' = \sqrt{\omega_0^2 + 2k/m}$, which is increased due to the action of the spring (the out-of-phase oscillations in Fig. 2.4b).

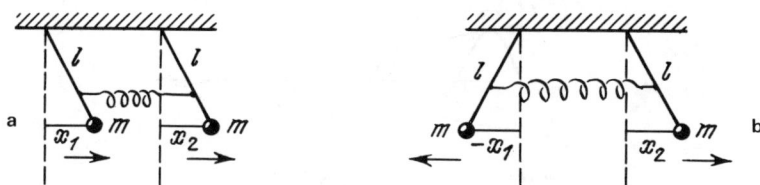

Fig. 2.4. a) In-phase oscillation of two identical pendulums. b) Anti-phase oscillations of the pendulums.

When the link is weak, it can be naturally considered to be a small disturbance and the joint oscillation as the interaction between the oscillators. For a weak link $(k/m\omega_0^2 \ll 1)$ we have $\omega_2' \approx \omega_0 + k/m\omega_0$, and the solutions of (2.12) have the form

$$u_1 = a_1 \cos \omega_0 t + b_1 \sin \omega_0 t,$$
$$u_2 = a_2 \cos (\omega_0 + k/m\omega_0)t + b_2 \sin (\omega_0 + k/m\omega_0)t. \quad (2.13)$$

Let us assume at $t = 0$ that $x_1 = x_2 = u_1 = u_2 = 0$ (the pendulums are at rest), but that one of the pendulums has been given the velocity $\dot{x}_1 = C$, whence $\dot{u}_1 = \dot{u}_2 = C$ because $\dot{x}_2 = 0$. We find from (2.13) and these initial conditions that

$$u_1 = \frac{C}{\omega_0} \sin \omega_0 t, \quad u_2 = \frac{C}{\omega_0 + k/m\omega_0} \sin (\omega_0 + k/m\omega_0)t.$$

We are interested in the behaviors of each pendulum, and so we change to the original coordinates:

$$x_{1,2} = \frac{C}{2\omega_0} \left\{ \sin \omega_0 t \pm \left(1 - k/m\omega_0^2\right) \sin \left[\left(\omega_0 + k/m\omega_0\right)t\right] \right\}.$$

Finally, we find that

$$x_1 \approx \frac{C}{\omega_0} \sin \omega_0 t \cos \frac{k}{2m\omega_0} t ,$$

$$x_2 \approx -\frac{C}{\omega_0} \cos \omega_0 t \sin \frac{k}{2m\omega_0} t .$$

(2.14)

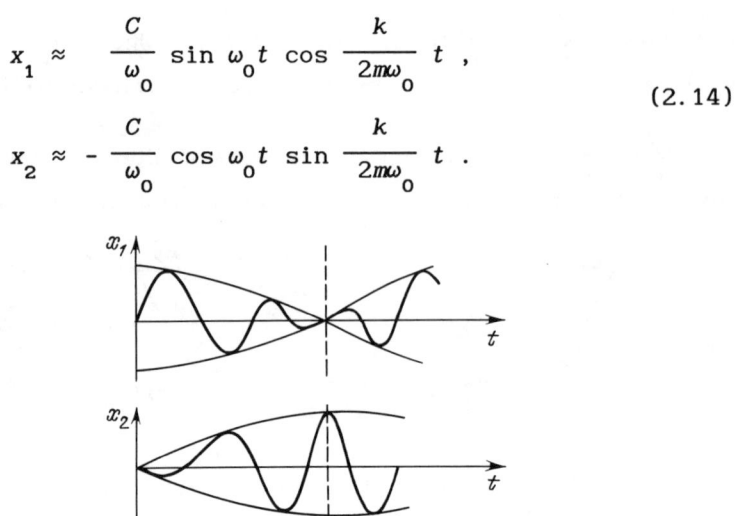

Fig. 2.5. Beats between two identical linked oscillators when the link is weak.

During the derivation of (2.14) we neglected $k/m\omega_0^2$ with respect to unity. Because the quantity $\alpha = k/2m\omega_0$ is small the pendulums oscillate with a frequency ω_0, the amplitude of their swings changing slowly. We thus get beats (Fig. 2.5). It isn't difficult to see that when $\alpha t = \pi/2$, for example, the first pendulum will be stationary ($x_1 = \dot{x}_1 = 0$ ignoring the terms containing $k/m\omega_0^2$) and all the energy will have been transferred to the second pendulum. Hence, the introduction of a link causes the energy to be transferred periodically between the oscillators, the temporal period of the transfer depending on the link's magnitude (periodic transfers of energy take place over periods of time that are multiples of $T \approx \pi/\alpha = 2\pi m\omega_0/k$). When α is small, the interaction energy, i.e., the energy transferred from one oscillator to the other, is small. Nevertheless, there is a complete transfer of energy no matter how weak the link. True, the period of the transfer will grow without limit ($T \sim 1/\alpha$). It would seem that as $\alpha \to 0$ the energy transfer should cease, but instead it simply slows down. Once again we are dealing with resonance, in that the partial frequencies of the pendulums are the same and so no matter how weak the link the interaction leads to the effective transfer of energy. This sort of resonance is called internal resonance (see Chapter 18) when it is considered that the subsystems of a single system are interacting.

When $n_1 \neq n_2$, very weak links do not have any effect. Therefore, in order to determine the level of interaction between several oscillators, we introduce a parameter to express both the strength of the link and the closeness of the partial frequencies. This parameter is called the connectedness and is defined as

$$\rho = \frac{k/m}{\left| n_1^2 - n_2^2 \right|}.$$

In several cases the equations of motion of the system we are analyzing can be more conveniently represented as special forms, called the linked oscillation form and the normal oscillation form [5]. We shall briefly discuss their derivation.

We multiply the second equation in (2.1) by $\mp i\omega_1$ and add the first equation in (2.1). This yields

$$\frac{d}{dt}\left(v_1 \mp i\omega_1 x_1\right) = \mp i\omega_1 \left(v_1 \mp i\omega_1 x_1\right) + \frac{k}{m}\left(x_2 - x_1\right),$$

or

$$\frac{da_1}{dt} = -i\omega_1 a_1 + \frac{k}{2\sqrt{m}}\left(x_2 - x_1\right), \quad \frac{da_1^*}{dt} = i\omega_1 a_1^*$$

$$+ \frac{k}{2\sqrt{m}}\left(x_2 - x_1\right), \qquad (2.15)$$

where

$$a_1 = (v_1 - i\omega_1 x_1)\sqrt{m}/2, \quad a_1^* = (v_1 + i\omega_1 x_1)\sqrt{m}/2. \qquad (2.16)$$

We make all the necessary identity transformations only into the first equation of (2.15) to get

$$\frac{da_1}{dt} = -i\omega_1 a_1 + \frac{k}{2i\omega_2 m}\frac{\sqrt{m}}{2}\left(v_2 + i\omega_2 x_2\right)$$

$$-\frac{k}{2i\omega_2 m}\frac{\sqrt{m}}{2}\left(v_2 - i\omega_2 x_2\right)$$

$$-\frac{k}{2i\omega_1 m}\frac{\sqrt{m}}{2}\left(v_1 + i\omega_1 x_1\right) + \frac{k}{2i\omega_1 m}\frac{\sqrt{m}}{2}\left(v_1 - i\omega_1 x_1\right).$$

By analogy with (2.16), i.e.,

$$a_2 = (v_2 - i\omega_2 x_2)\sqrt{m}/2, \quad a_2^* = (v_2 + i\omega_2 x_2)\sqrt{m}/2, \qquad (2.17)$$

we finally get

$$da_1/dt = c_{11}a_1 + c_{12}a_2 + c_{13}a_1^* + c_{14}a_2^*, \qquad (2.18)$$

where the coefficients have the form

$$c_{11} = -i\omega_1\left(1 + \frac{k}{2m\omega_1^2}\right), \quad c_{12} = i\frac{k}{2m\omega_2},$$

$$c_{13} = i\frac{k}{2m\omega_1}, \quad c_{14} = -i\frac{k}{2m\omega_2}.$$

We leave the reader to transform the other equation in the same way. We shall, however, write out the final result in matrix form, viz.,

$$\frac{dA}{dt} = CA, \quad A = \begin{bmatrix} a_1 \\ a_2 \\ a_1^* \\ a_2^* \end{bmatrix}, \quad C = \begin{bmatrix} c_{11} & c_{12} & c_{13} & c_{14} \\ c_{21} & c_{22} & c_{23} & c_{24} \\ c_{31} & c_{32} & c_{33} & c_{34} \\ c_{41} & c_{42} & c_{43} & c_{44} \end{bmatrix}, \qquad (2.19)$$

moreover, $c_{11} = -c_{33}$, $c_{22} = -c_{44} = -i\omega_2(1 + k/2m\omega_2^2)$, $c_{13} = c_{21} = -c_{23} = -c_{31} = c_{41} = -c_{43}$, $c_{12} = -c_{14} = c_{24} = c_{32} = -c_{34} = -$

$c_{42} = ik/2m\omega_2$. The notation in (2.19) is called the linked oscillation form [5]. This designation emphasizes that the link coefficients $c_{ij} \sim k$ relate the normal oscillations a_1, a_2, a_1^*, a_2^* of the isolated pendulums. If we seek a solution to (2.19) in the form $a_i(t) = a_i(0)\exp(i\omega t)$ and $a_i^*(t) = a_i^*(0)\exp(i\omega t)$ ($i = 1, 2$), we get a system of algebraic equations for $a_i(0)$ and $a_i^*(0)$. In order for them to coincide, we must have

$$\begin{vmatrix} -i\omega + c_{11} & c_{12} & c_{13} & c_{14} \\ c_{21} & -i\omega + c_{22} & c_{23} & c_{24} \\ c_{31} & c_{32} & -i\omega + c_{33} & c_{34} \\ c_{41} & c_{42} & c_{43} & -i\omega + c_{44} \end{vmatrix} = 0. \quad (2.20)$$

The roots of this equation are the frequencies of the normal oscillations of the system of the two linked pendulums. We made no assumptions when deriving (2.19) and (2.20) other than that the amplitudes were small and the coordinates change harmonically over time. In this sense (2.20) and (2.10) are equivalent. We noted in Chapter 1 that two normal oscillations a_i and a_i^* may be graphically represented by two counter-rotating vectors of equal length. It is naturally to suppose, therefore, that the linkage between counter-rotating vectors is weak, i.e., all the terms relating a_i to a_i^* in (2.19) can be neglected. It is intuitively clear that in order for this assumption to be true in reality, the oscillators must be weakly linked, that is, the link energy must be small with respect to the potential energy of each oscillator, i.e., $k/m\omega_{1,2}^2 \ll 1$ (2.2). Moreover, the connectedness of the oscillators must be strong. (A considerable fraction of the energy of one of the oscillators will be transferred to the other.) For this to be so, we must have $\omega_1 \approx \omega_2$, whence we get from (2.19)

$$da_1/dt = c_{11}a_1 + c_{12}a_2, \quad da_2/dt = c_{21}a_1 + c_{22}a_2 \quad (2.21)$$

and two analogous equations for a_1^* and a_2^*, while from (2.20) we get

$$\begin{vmatrix} -i\omega + c_{11} & c_{12} \\ c_{21} & -i\omega + c_{22} \end{vmatrix} = 0 . \tag{2.22}$$

If we solve (2.22), remembering the expressions for the c_{ij}, for a linked system that corresponds to (2.21), we get the following frequencies for the normal oscillations

$$\omega'_{1,2} = \frac{\omega_1 + \omega_2}{2} + \frac{k}{2m\sqrt{\omega_1 \omega_2}} \mp \sqrt{\left(\frac{\omega_1 - \omega_2}{2}\right)^2 + \left(\frac{k}{2m\sqrt{\omega_1 \omega_2}}\right)^2} . \tag{2.23}$$

When $\omega_1 = \omega_2 = \omega_0$ we obtain from (2.23) $\omega'_1 = \omega_0$ and $\omega'_2 = \omega_0 + k/m\omega_0$, which is what we got from the expressions we obtained earlier. The equations for a^*_1 and a^*_2 correspond to the normal oscillations $(-\omega'_1)$ and $(-\omega'_2)$.

The solution (2.23) may be found directly from (2.20) when $k/m\omega^2_{1,2} \ll 1$ and $\omega_1 \approx \omega_2$. Hence, the neglection in (2.19) of the terms linking a_i and a^*_i is indeed equivalent to the assumption that there is weak linkage and strong connectedness.

Another general method is often used to find another form for the equations of a system of linked oscillators, viz., the normal oscillations form. Without dwelling on the detail, we formulate the method as a theorem [5]. For a system of n linked differential equations with constant coefficients:

$$\frac{dX}{dt} = MX , \quad X = \begin{bmatrix} x_1 \\ x_2 \\ \vdots \\ x_n \end{bmatrix} , \quad M = \begin{bmatrix} m_{11} & m_{12} & \cdots & m_{1n} \\ m_{21} & m_{22} & \cdots & m_{2n} \\ \vdots & \vdots & & \vdots \\ m_{n1} & m_{n2} & & m_{nn} \end{bmatrix}$$

there exists a linear transformation of the variables $X(x_1, x_2, \ldots, x_n)$ to another set of variables $Y(y_1, y_2, \ldots, y_n)$ in the form $X = NY$, where N is a constant matrix similar in form to the matrix M and which maps the original equation system to

$$dY/dt = \Omega'Y ,$$

where Ω' is a matrix the elements of whose leading diagonal are the normal frequencies of the oscillations, while the elements of Y are the amplitudes of the normal oscillations.

We note that as was the case for the single oscillator, there is an analogy with quantum mechanics. The linked oscillations form is analogous to a treatment of two linked oscillators by the production and extinction operators [6].

When considering the most mathematically convenient approach for analyzing the vibrations of linked oscillators, then the linked oscillations form seems to be the best.

2.3 Disturbance of Two Linked Oscillators by an External Force. The Reciprocity Principle.

Suppose there is an external periodic force acting on a system of two linked oscillators. Then the equation of motion (2.6) together with (2.7) will be

$$A\ddot{x} + ax + H\ddot{y} + hy = F_1, \quad H\ddot{x} + hx + B\ddot{y} + by = F_2. \quad (2.24)$$

To begin with, let $F_2 = 0$ and $F_1 = F \cos \Omega t$. We are interested in the forced solution of (2.24), i.e., $x, y = (X, Y) \cos \Omega t$. Substituting this into (2.24) yields

$$X(-A\Omega^2 + a) + Y(-H\Omega^2 + h) = F, \quad X(-H\Omega^2 + h) + Y(-B\Omega^2 + b) = 0,$$

whence the equation for the resonance curves $X = X(\Omega)$ and $Y = Y(\Omega)$ are

$$X = F(-B\Omega^2 + b)/\Delta(\Omega^2), \quad Y = F(-H\Omega^2 + h)/\Delta(\Omega^2). \quad (2.25)$$

where Δ is a determinant, see (2.10).

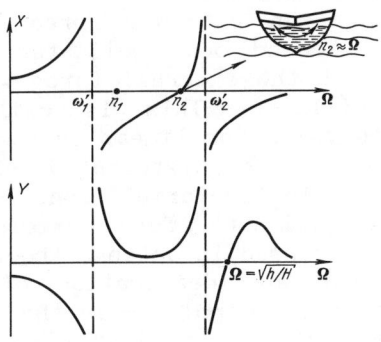

Fig. 2.6. The amplitude of the forced oscillations of X and Y versus the frequency of the external force (the resonance curves).

The resonance curves demonstrate some interesting effects (Fig. 2.6):

1) If the frequency of the external force coincides with one of the fundamental frequencies of the system, resonance sets in and the amplitude of the oscillations of both oscillators grows without limit.

2) If the frequency of the external force acting on the first oscillator coincides with the partial frequency of the second one $\Omega = n_2$, the first oscillator does not oscillate ($X = $)0. This phenomenon is called dynamic damping.

3) When the frequency of the external force is $\Omega_1 = \sqrt{h/H}$, the second oscillator does not oscillate ($Y = 0$), although this phenomenon only occurs if the link is mixed, i.e., it is both force (capacitive) and inertial (inductive). When $\Omega = \Omega_1$, the link is canceled out and the vibrations of one oscillator are not transferred to the other.

Dynamic damping is often used to suppress dangerous vibrations [8]. For example, in order to reduce the rocking of a tanker ship in heavy seas, its tanks are filled with water to a level such that the partial oscillation of the water in the tanks approximates the frequency of the beat of the waves on the bows of the ship (Fig. 2.6). The rocking of the tanker itself is then considerably reduced.

Suppose now the external force acts upon the second oscillator, rather than the first as above, i.e., $F_1 = 0$, $F_2 = F \cos \Omega t$. However, we again look for the forced solution as above, whence

$$X = F(-H\Omega^2 + h)/\Delta(\Omega^2), \quad Y = F(-A\Omega^2 + a)/\Delta(\Omega^2). \quad (2.26)$$

A comparison of (2.25) and (2.26) yields an important fact, namely, that when an external force acts upon the first oscillator, the second one oscillates in the same way as would the first if the external force were being applied to the second oscillator. This is called the reciprocity principle. It is valid for linear systems with any number of degrees of freedom, (e.g., distributed systems) and, given the appropriate change in the formulation, for continuous media.

In electrodynamics, for example, the reciprocity principle is widely used in antenna theory. The principle can be formulated for an idealization of an antenna -- an elementary oscillating dipole -- in the following way [7].

Let a dipole at a point 1 with an electrical moment p_1 excite an electromagnetic field E_1, H_1, and a dipole p_2 at a point 2 excite an electromagnetic field E_2, H_2. The reciprocity principle will then be expressed by the equation

$$p_1 E_2(1) = p_2 E_1(2) , \qquad (2.27)$$

where $E_2(1)$ is the value of the electrostatic field E_2 at point 1, and $E_1(2)$ is the value of the electrostatic field E_1 at point 2. When the absolute values of the dipole moments are equal, dipole 2 acts upon dipole 1 in the same way as dipole 1 acts upon dipole 2.

If p_1 is a ground-based transmitting antenna, for example, and it is necessary to find the field the dipole creates at point 2 high above the ground where there is an aircraft with a receiving antenna on board, then another problem can be solved in which the transmitting antenna is p_2 at point 2 and the receiving antenna is at point 1, and then the reciprocity principle is applied [7].

CHAPTER 3

OSCILLATIONS IN AN ENSEMBLE OF NONINTERACTING OSCILLATORS

3.1 Classical Theory of Dispersion

The two most important cases in application are the almost trivial one in which all the oscillators are identical and the much more interesting one when the parameters of the oscillators are scattered either in frequency or in damping. The consideration of the behavior of an ensemble of identical noninteracting oscillators is the main subject of the classical theory of the dispersion of light. Heterogeneous oscillators must be considered, for example, in the analysis of the scattering of electromagnetic radiation in heated gases, in which the spread of molecular velocities results in Doppler shifts in the radiation's frequencies with respect to the frequency of the field.

The mechanical theory of the dispersion of light was in fact constructed by Maxwell in 1869 in answer to an examination question. In 1871 Zelmayer again obtained the formula linking the refractive index n with the frequency ω in a mechanical theory of ether [1].

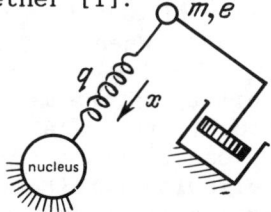

Fig. 3.1. Mechanical model of an atom: m and e are the mass and charge of the electron, x is its coordinate, q is the stiffness of the spring. To the right is a device which generates a force proportional to velocity [2].

Suppose we have a medium consisting of identical oscillators which are not interacting, e.g. the medium can be thought of as an ensemble of independent atoms each of which contains one electron. The mechanical model of an atom is illustrated in Fig. 3.1. The electron in this model is considered to be a damped harmonic oscillator whose oscillations under the influence of the external field are described by

$$m\ddot{x} + g_1\dot{x} + qx = eE \exp(i\omega t), \qquad (3.1)$$

where m and e are the mass and charge of the electron, $-g_1\dot{x}$ is the "friction" force which is introduced to account for the absorption of radiation, $-qx$ is a quasielastic restoring force, and E is the amplitude of the strength of the external electrostatic field acting on the electron.[1] In general, the electrons and ions oscillate in the nonhomogeneous field of the electromagnetic wave $E \sim E_0 \exp(-ikx)$ (k is the wave number).

However, we shall ignore the dependence of the field on the coordinate and consider that the amplitude of the electron's oscillations is smaller even than an optical wavelength. Hence, if we introduce the notations $2\gamma = g_1/m$ (the loss in the atomic oscillator), $q/m = \omega_0^2$ (ω_0 is the fundamental frequency of the oscillator), and $F_0 = (e/m)E$, then we arrive at an equation analogous to (1.26). We are interested in the forced oscillations of the oscillator, hence we find from (3.1) after substituting in our notation

$$x = eE/m \left[\omega_0^2 - \omega^2 + 2i\omega\gamma\right]. \qquad (3.2)$$

We are interested in the dielectric permeability or the polarizability of a medium composed of a collection of oscillators as a function of frequency. The polarizability of an elementary oscillator is $\alpha = p/E$, where $p = ex$ is the dipole moment an atom acquires in an electrostatic field. The vector polarizability of a medium containing N atoms per unit volume is $P = pN = \alpha E N$. Whence by noting that the electrical induction is $D = E + 4\pi P = (1 + 4\pi\alpha N)E$, we obtain the dielectric permeability of the medium $\varepsilon = 1 + 4\pi\alpha N$ (the unity reflects the permeability in a vacuum), the dependence of which on frequency can be found from (3.2).

We get the following for the real and imaginary parts of the dielectric permeability $\varepsilon(\omega)$:

$$\text{Re } \varepsilon(\omega) = 1 + \frac{4\pi e^2 N \left(\omega_0^2 - \omega^2\right)}{m\left[\left(\omega_0^2 - \omega^2\right)^2 + 4\omega^2\gamma^2\right]}, \qquad (3.3)$$

---

[1] It should be noted that the strength E of the electrostatic field acting on the electron differs from the average macroscopic field which figures in Maxwell's equations. However, these fields can be considered equal in certain cases, such as in very dense gases [1].

$$\text{Im }\varepsilon(\omega) = -\frac{4\pi e^2 N \cdot 2\gamma\omega}{m\left[\left(\omega_0^2 - \omega^2\right)^2 + 4\omega^2\gamma^2\right]} \qquad (3.4)$$

If we introduce a complex refractive index $\sqrt{\varepsilon} = n - i\kappa$ (n is the real refractive index, κ is the medium's damping index), then

$$n^2 - \kappa^2 = \text{Re }\varepsilon(\omega), \qquad 2n\kappa = -\text{Im }\varepsilon(\omega).$$

When $2\omega\gamma \ll |\omega_0^2 - \omega^2|$ we may consider that $n^2 = \varepsilon = [\text{Re }\varepsilon(\omega)]_{\gamma=0}$. Whence we arrive at Zelmayer's well known formula, i.e.,

$$n^2 = \varepsilon \quad 1 + 4\pi e^2 N/m(\omega_0^2 - \omega^2).$$

It follows from (3.3) and (3.4) that in a frequency band where there is little absorption (see Fig. 3.2) the refractive index grows with frequency (normal dispersion). In a frequency band in which there is marked absorption, there is anomalous dispersion and $n(\omega)$ decreases with frequency.

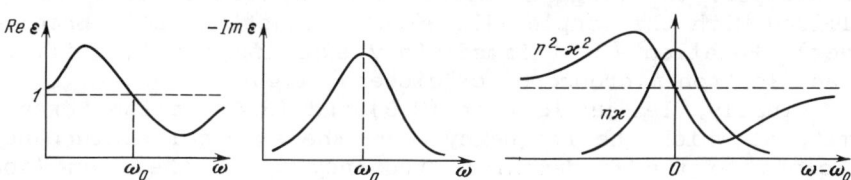

Fig. 3.2. Real and imaginary parts of the dielectric permeability versus frequency.

In the model under study an atom is an oscillating dipole that always emits or scatters electromagnetic radiation with an amplitude A_p and proportional to $\ddot{x} = -\omega^2 x$, i.e.,

$$A_p \sim -e\omega^2 E/m\left[\omega_0^2 - \omega^2\right]$$

when $\gamma = 0$. Since the intensity of a radiation I_p is proportional to the square of its amplitude, we have

$$I_p \sim \text{const} \cdot [(\omega_0/\omega)^2 - 1]^{-2}. \qquad (3.5)$$

We shall investigate (3.5) for several limiting cases, in which the external radiation has a frequency ω which is either much larger or much smaller than ω_0. In the visible spectrum, when $\omega \ll \omega_0$, we have

$$I_p \sim \text{const} \, (\omega^4/\omega_0^4) \, , \tag{3.6}$$

which corresponds to Rayleigh scattering. This formula explains, for instance, why the sky is blue. Because blue light has a greater frequency than red light, the blue is much more strongly scattered even though the blue end of the spectrum of sunlight is weaker than the red end.[2] The molecular scattering of light is widely used to investigate transparent media, for instance to determine the dimensions, form, and average velocity of polymers or large bacteria in solution [6].

At the other limit $\omega \gg \omega_0$, $I_p \sim$ const, which corresponds, for instance, to the collisions of fast electrons or X-rays with matter, we obtain

$$I_p = I_0 \, . \tag{3.7}$$

The intensity of the radiation is independent of the frequency (Thompson scattering). In an electron microscope, for example, the image is obtained after fast electrons have collided with the sample (the electrons acting as de Broglie waves). Equation (3.7) immediately explains the "inability" of an electron microscope to "discern" light.

Finally, let us look at (3.3) and (3.4) at the Lorentz limit, at which the frequency ω of the external disturbance is close to the fundamental frequency ω_0 of the atom (the frequency of the excitation approximates that of the absorption lines in the spectrum). We shall consider that $|\omega - \omega_0| \ll (\omega + \omega_0)$, i.e., $\omega \approx \omega_0$ and $(\omega_0 - \omega)^2 \approx 2\omega_0(\omega_0 - \omega)$. Hence

$$\text{Re} \, \varepsilon(\omega) = 1 + \frac{2\pi e^2 N}{m\omega_0} \left\{ \frac{\omega_0 - \omega}{(\omega_0 - \omega)^2 + \beta^2} \right\} \, , \tag{3.8}$$

---

[2] We note that although (3.6) is correct, light is scattered by inhomogeneities in the medium caused by fluctuations in the densities of the molecules in small volumes and not by the molecules themselves as Rayleigh himself had thought [1].

$$\operatorname{Im} \varepsilon(\omega) = -\frac{2\pi e^2 N}{m\omega_0} \left\{ \frac{\beta}{(\omega_0 - \omega)^2 + \beta^2} \right\}, \qquad (3.9)$$

where $\beta = 1/\tau$ and $\tau = 1/\gamma$ is the characteristic relaxation time. The expressions in the brackets in (3.8) and (3.9) are frequently encountered in spectroscopy and are called the Lorentz contours of the spectral lines. The expression in the curly brackets in (3.9) is called the absorption contour. The width of a Lorentz lines can be used to measure the relaxation time of the sample. We note that this model of a medium consisting of identical noninteracting oscillators is very general. Indeed, if instead of the electrons being on springs, we consider the atoms to be on springs, we can calculate the strength of a chemical bond in a molecule. In order to do so we must obtain the frequency ω_0, which lies in the infrared end of the spectrum [3].

In reality, the atom oscillators are not strictly identical. So does this lead to any new effect? Since the dielectric permeability of a medium is the sum of the permeability of a vacuum and the responses of each of the noninteracting oscillators to the field acting on them, we get the summed response

$$\varepsilon = 1 + \sum_k \frac{4\pi e_k^2 N_k}{m_k \left[\left(\omega_k^2 - \omega^2 \right) + 2i\omega\gamma_k \right]}. \qquad (3.10)$$

In classical models (but not the quantum one) of matter, a medium is assumed to consist of electrons and ions, which are particles with different charges e_k, masses m_k, and concentrations N_k and which comprise an ensemble of noninteracting damped harmonic oscillators with different fundamental frequencies ω_k and damping coefficients γ_k.

The classical theory of dispersion and absorption is clearly a model, and moreover, one in which there are quasielastic forces ($-qx$) and friction forces ($-g\dot{x}$) which do not exist in atoms and molecules as all the forces within an atom or molecule are electrostatic in origin.

What then is the quantum-mechanical theory of dispersion and absorption? Let there be an electromagnetic wave acting on a set of noninteracting neutral atoms which are fixed in space and contain only one electron (the simplest model of a solid). We apply the nonstationary theory of perturbations [4] and obtain the following for the real part of the dielectric permeability

$$\text{Re } \varepsilon(\omega) = 1 + \frac{4\pi N e^2}{m} \sum_k \frac{f_k}{\omega_k^2 - \omega^2}, \qquad (3.11)$$

where f_k is the force required to transfer the oscillator to the k-th state, with

$$f_k = (2m/\hbar^2)\hbar\omega_k |x_{ok}|^2,$$

$ex_{ok} = \int \Psi_0^* ex \Psi_k d\mathbf{r}$ is a matrix element for the component of the electron's dipole moment in the direction of the vector potential of the electrostatic field, $\Psi_0(\mathbf{r})$ is the wave function for the ground state of the atom, $\Psi_k(\mathbf{r})$ is the wave function for the excited state of the atom.

In order to derive (3.11) we neglected the natural radiational damping of the levels, the widening of the spectral lines due to impurities, etc. A phenomenological account of these factors leads back to (3.10) if we replace N_k by Nf_k and assume that $e_k = e$, $m_k = m$.

3.2 Oscillations in an Ensemble of Dissimilar Noninteracting Oscillators with a Given Distribution Function

We now look at the behavior of a medium consisting of noninteracting oscillators which have a given frequency distribution function, e.g., gas molecules with various velocities. If the properties of the oscillators vary continuously, then the medium's permeability may be obtained using an argument similar to the one leading to (3.10) but with an integration step. If we differentiate the oscillators solely by their fundamental frequency ω', the polarizability of a medium containing N atoms per unit volume can be written

$$\chi = \frac{\alpha_0}{\pi} \int_{-\infty}^{\infty} \frac{N(\omega') d\omega'}{[(\omega'^2 - \omega^2) + 2i\omega g(\omega')]}, \qquad (3.12)$$

where $\alpha_0 = \pi e^2/m$, while $N(\omega')$ is the number distribution function of the oscillators with respect to their fundamental frequencies ω', this function also determining the properties of this "gas" of differing elementary oscillators.

In order to calculate χ we must specify the distribution $N(\omega')$. Let $N(\omega')$ be the frequently encountered Lorentz distribution (3.9), viz.,

$$N(\omega') = \beta[(\omega' - \omega_0)^2 + \beta^2]^{-1}, \qquad (3.13)$$

where β is the width of the Lorentz lines (Fig. 3.3). If we include (3.12) and integrate (3.12), we discover that the medium acts like an averaged oscillator, the properties of which are determined by the distribution function.

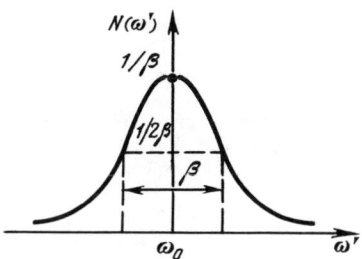

Fig. 3.3. The Lorentz contour of a line: ω_0 is the frequency at which $N(\omega')$ is at a maximum of $1/\beta$; and the width of the Lorentz line ($\Delta\omega' = \beta$) is determined at a level of $1/2\beta$.

To use the residue theorem we chose the integral [5,7]

$$\chi = \frac{\alpha_0}{\pi} \int_{-\infty}^{\infty} \frac{\beta d\omega'}{[(\omega'^2 - \omega^2) + 2i\omega g][(\omega' - \omega_0)^2 + \beta^2]}, \qquad (3.14)$$

integrate along a contour, and assume for simplicity that g is independent of ω'. The integration contour we chose is the one that embraces the upper half-plane. We find the poles of the integrand in (3.14), which coincide with the zeros of the denominator, lying in the upper half-plane. These are 1) $\omega' = \omega_0 - i\beta$ (because the contour only takes in the upper half-plane, the value $\omega' = \omega_0 - i\beta$ does not interest us), and 2) $\omega' = -(\omega^2 - 2i\omega g)^{1/2}$ (the value $\omega' = (\omega^2 - 2i\omega g)^{1/2}$ lies in the lower half-plane).

Now we find the residues for the integral in (3.14). Since

$$p(\omega') = [(\omega'^2 - \omega^2) + 2i\omega g][(\omega' - \omega_0)^2 + \beta^2],$$

and

$$p'(\omega')|_{\omega'=\omega_0+i\beta} = \{2\omega'[(\omega'-\omega_0)^2+\beta^2] + 2(\omega'-\omega_0)[(\omega'^2-\omega^2)$$

$$+ 2i\omega g]\}_{\omega'=\omega_0+i\beta}$$

$$= 2i\beta[(\omega_0^2-\omega^2-\beta^2)+2i\omega(\beta+g)],$$

we get

$$\text{Res}_{\omega'=\omega_0+i\beta} f(\omega') = \frac{\alpha_0}{2\pi i} [(\omega_0^2-\omega^2-\beta^2) + 2i\omega(\beta+g)]^{-1}.$$

Similarly,

$$\text{Res}_{\omega'=-\sqrt{\omega^2-2i\omega g}} f(\omega')$$

$$= -\frac{\beta\alpha_0}{2\pi\sqrt{\omega^2-2i\omega g}} \left[\omega^2+\omega_0^2+\beta^2+2\omega_0\sqrt{\omega^2-2i\omega g}-2i\omega g\right]^{-1}$$

If β is small ($\beta \ll \omega_0$) and $\omega \to \omega_0$, then

$$\text{Res}_{\omega'=\omega_0+i\beta} f(\omega') = \frac{\alpha_0}{2\pi i} \left[(\omega_0^2-\omega^2) + 2i\omega(\beta+g)\right]^{-1},$$

while

$$\text{Res}_{\omega'=-\sqrt{\omega^2-2i\omega g}} f(\omega')$$

$$= -\frac{\beta\alpha_0}{2\pi\sqrt{\omega^2-2i\omega g}} \left[\omega^2+\omega_0^2+2\omega_0\sqrt{\omega^2-2i\omega g}\right.$$

$$\left. - 2i\omega g\right]^{-1}$$

$$\sim \beta \ll \omega_0.$$

For instance, when $\omega = \omega_0$ and $g = 0$

$$\text{Res}_{\omega'=-\omega_0+i\beta} f(\omega') = -\frac{\alpha_0}{2\pi} \frac{1}{2\omega_0 \beta},$$

$$\text{Res}_{\omega'=-\sqrt{\omega^2-2i\omega g}} f(\omega') = -\beta \frac{\alpha_0}{2\pi} \frac{1}{4\omega_0^3}$$

and it can be seen that the second residue can be neglected with respect to the first.

Finally, we have [7]

$$\chi = 2\pi i \sum \text{Res} = \alpha_0 / [(\omega_0^2 - \omega^2) + 2i\omega(g+\beta)] . \qquad (3.15)$$

Two remarkable results follow from these expressions.

1) The polarizability of a medium consisting of oscillators with a Lorentz frequency distribution is the same as the one obtained for an ensemble of identical oscillators with fundamental frequency ω and damping coefficient $g + \beta$.

2) Even if the medium is purely conservative ($g = 0$), the ensemble average oscillator will nevertheless have a damping β.

To explain these effects recall that an oscillator may be represented as a vector rotating in a plane with a frequency ω', like the arm of a clock. If all the oscillators are the same, and we give them the same phase, then the summed response to a field acting upon them will be the product of the number of oscillators and the response of one oscillator. However, if the oscillators differ slightly in frequency, then even if they are started off in a single phase, they will over a long interval of time become evenly distributed around the clock. Hence for any individual response it will be possible to find one with the opposite phase and the overall response to the external disturbance will thus be zero. The characteristic time for the vectors to become dispersed over the interval π and the summed oscillation in the system to be damped is $\tau \sim 1/\beta$.

If the oscillators are identical, but inked together arbitrarily, then when we go over to normal frequencies, we once again get a "gas" of noninteracting oscillators but with differing frequencies. Thus this problem reduces to the previous one.

CHAPTER 4

OSCILLATIONS IN ORDERED STRUCTURES. LIMIT FOR A CONTINUOUS MEDIUM WAVES. DISPERSION

4.1 General Remarks

John Ziman begins the chapter on Lattice Vibrations in his book [1] to the effect that: The simplest solid seems to be solid argon. It consists of regularly spaced neutral atoms with tightly bound electron shells. These atoms are held together by Van der Waals forces, which mainly act on the nearest neighbors in the lattice. The physical processes in such a crystal are associated with the thermal motions of the atoms about their ideal equilibrium positions. The simplest description of this motion is given by Einstein's model, according to which each atom is a simple harmonic oscillator in the potential well formed due to the interaction forces between neighbors.

A lattice is a very simple object which can naturally be called an ordered structure of oscillators. Other simple examples of ordered structures consisting of identical oscillators connected together in a certain way include a linear chain of identical particles spaced at equal distances along a straight line (a one-dimensional lattice of similar particles), a mechanical system consisting of a set of pendulums, a chain of LC-circuits, an infinite series of acoustic resonators, and a chain of magnets.

In the preceding chapter we concluded that a set of independent neutral atoms fixed in space (the simplest model of a solid) would behave, with respect to dispersion and absorption, like an ensemble of classical noninteracting dissimilar oscillators. We now change the model. The crudest model of the solid argon described by Ziman is a system of regularly spaced spheres connected together with springs. If one of the spheres is displaced from its equilibrium position, it will move its neighbor and hence a wave will run along the whole of the ordered structure. Waves in solids or other ordered structures are characterized by a wavelength λ and a frequency ω, which obey a dispersion law

$$\omega = \omega(\mathbf{k}) , \qquad (4.1)$$

$k = 2\pi/\lambda$. It follows from (4.1) that a wave with a given wave vector $\mathbf{k}$ also has a given frequency; this allows us to look at a wave as an oscillator vibrating at a frequency $\omega(\mathbf{k})$. Hence we return once more to the analogy between a solid and a "gas" of oscillators vibrating independently of each other. True, all the atoms of the solid participate in the new elementary oscillation. The wave (oscillator) may be equated with a quasiparticle with energy $\mathcal{E} = \hbar\omega(\mathbf{k})$ and momentum $\mathbf{p} = \hbar\mathbf{k}$. In this way our analysis has brought us to one of the most interesting concepts in modern physics, namely quasiparticles [16].

4.2 Oscillations in Ordered Structures
(Chains of Linked Particles and Identical Linked Oscillators)

We start with the derivation of the equation of motion of an unbounded one-dimensional lattice of similar equally spaced particles (Fig. 4.1). Let us consider the longitudinal oscillations of the chain.

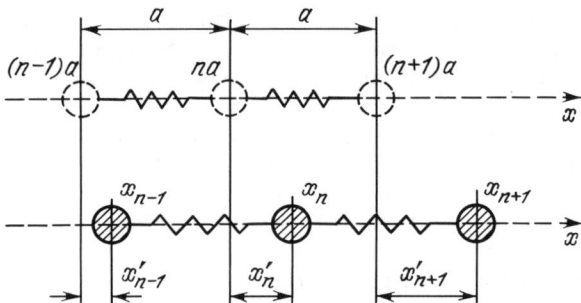

Fig. 4.1. A one-dimensional "lattice" of similar equally spaced particles: above, the lattice before a disturbance; below, after a disturbance (longitudinal oscillation).

It can be seen from Fig. 4.1 that the coordinate of the n-th particle at some moment in time after a disturbance is

$$x_n = na + x'_n , \qquad (4.2)$$

where x'_n is the displacement from the equilibrium position (we shall consider that $x'_n \ll a$). The distance between two longitudinal particles (the n-th and $n + 1$-th) is

$$X_{n, n+1} = x_{n+1} - x_n = 1a + x'_{n+1} - x'_n . \qquad (4.3)$$

OSCILLATIONS IN ORDERED STRUCTURES

If we assume that the potential energy, which we can use to find the interaction force between the two particles, depends only on the distance between the particles $|x_{n+1} - x_n|$ (which we denote as $W(x) = W(|x_{n+1} - x_1|)$), then we can write the overall potential energy of the lattice as

$$W = \sum_n \sum_{l>0} W(|x_{n+1} - x_n|) . \tag{4.4}$$

Let us look at the linear oscillations, i.e., we assume that x'_n is small. By expanding $W(|x_{n+1} - x_n|)$ into a Taylor series and truncating after the second-order terms, we get

$$W(x_{n+1} - x_n) = W(la) + (x'_{n+1} - x'_n)W'(la)$$
$$+ (1/2)(x'_{n+1} - x'_n)^2 W''(la) , \tag{4.5}$$

where

$$W'(la) = dW/dX|_{X=la} , \quad W''(la) = d^2W/dX^2|_{X=la} .$$

After substituting (4.5) into (4.4), we can write an expression for the potential energy of the chain in the form

$$W = \sum_n \sum_{l>0} \{(x'_{n+1} - x'_n)W'(la) + (1/2)(x'_{n+1} - x'_n)^2 W''(la)\} + W_0 ,$$
$$\tag{4.6}$$

where $W_0 = \sum_n \sum_{l>0} W(la)$. If W is known, it is easy to calculate the force acting on the p-th particle because $F_p = -\partial W/\partial x'_p$. The derivative is with respect to the the displacement x'_p of the particle under study, and hence the only terms to contribute to F_p when summing over n are those which depend on x'_p, i.e., those for which $n = p$ and $n + 1 = p$. Therefore it follows from ((4.6) that

$$F_p = -\partial W/\partial x'_p = \sum_{l>0} W''(la)\{x'_{p+1} + x'_{p-1} - 2x'_p\} . \tag{4.7}$$

It can be shown that the quantity $W''(la)$ for the model in Fig. 4.1 is analogous to the elasticity of the springs connecting the spheres. If the particles in the lattice have mass m, then given (4.7) and Newton's second law $md^2x'_p/dt^2 = F_p$ the equation of motion of the p-th particle in the lattice is

$$md^2x'_p/dt^2 = \sum_{l>0} W''_l \left\{ x'_{p+1} + x'_{p-1} - 2x'_p \right\},\qquad(4.8)$$

where $W''(la) = W''_l$. We look for a solution to (4.8) in the form

$$x'_p = C \exp(i\omega t - ikpa).\qquad(4.9)$$

If such a solution exists, we say that a wave is propagated with wave number $k = 2\pi/\lambda$ and constant amplitude C. The quantity ka, moreover, characterizes the change in phase between the p-th and $(p+1)$-th particles, i.e., $x'_{p+1} = x'_p \exp(-ika)$. After substituting (4.9) into (4.8) we can see that a solution in the form of (4.9) exists if ω and k satisfy the transcendental equation

$$\omega^2 = 4 \sum_{l>0} \frac{W''_l}{m} \sin^2 \frac{kla}{2},\qquad(4.10)$$

which is usually called the dispersion equation.

It can be seen from the dispersion equation that the frequency ω is a periodic function of the wave number k with period $2\pi/a$, and so all the oscillations possible can be found by altering k over the interval $-\pi/a \le k \le \pi/a$. Let us now assume that each particle in the lattice only affects its nearest neighbor. Then instead of (4.8) and (4.10), we get

$$m \frac{d^2x'_p}{dt^2} = W'' \left(x'_{p+1} + x'_{p-1} - 2x'_p \right),\qquad(4.11)$$

$$\omega^2 = \frac{4W''}{m} \sin^2 \frac{ka}{2},\qquad(4.12)$$

where $l = 1$ and $W''_1 = W''$. The oscillation frequencies corresponding to (4.12) are given in Fig. 4.2. Note that at small ka ($ka \ll 1$ i.e., $a \ll \lambda$) it follows from (4.12) that

$$\omega = \sqrt{W''/m}\, ka\qquad(4.13)$$

which is the linear dispersion function.

We now return to the more general case of (4.10), in which each particle is affected by the forces of all the other particles closer than a distance equal to the product of the number of such particles and a number a.

OSCILLATIONS IN ORDERED STRUCTURES   53

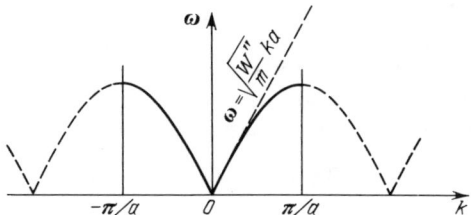

Fig. 4.2. Dispersion function for a one-dimensional chain of identical regularly separated particles. The solid lines are for the fundamental interval for a change with a wave-number k, and the dashed lines are for the periodic continuations.

Note that this situation is typical for carbide chains or the recently discovered spiral polymers. The corresponding dispersion is illustrated in Fig. 4.3 [2]. It follows from the figure that the wave number is a multivalued function of the frequency.

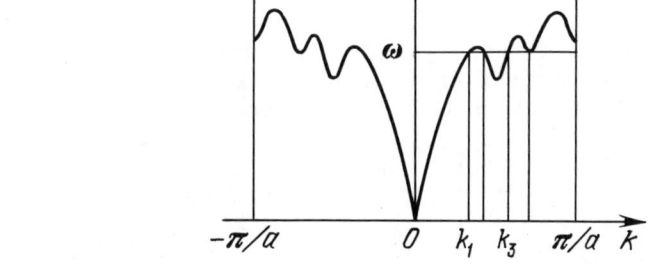

Fig. 4.3. Dispersion diagram of a one-dimensional chain allowing for long-range interactions between the particles.

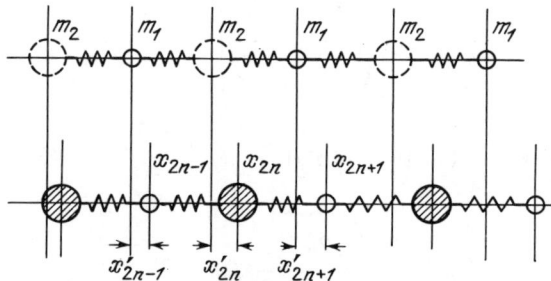

Fig. 4.4. One-dimensional "lattice" of regularly positioned alternating particles with different masses (longitudinal oscillations).

A one-dimensional lattice consisting of two sorts of particle, with masses m_1 and m_2, alternating in a chain is of interest (Fig. 4.4). Suppose that the particles are equally spaced and are placed in the same force field, as in the previous problem. This model corresponds, for example, to the sodium chloride lattice, in which the atoms of sodium and chlorine alternate. If we assume that only neighboring particles affect each other, we can write equations or the oscillations for each type of particle (the even numbers correspond to particles with mass m_2 and the odd numbers to particles with mass m_1) in the form

$$m_2 d^2 x'_{2n}/dt^2 = W''(x'_{2n-1} + x'_{2n+1} - 2x'_{2n}) ,$$

$$m_1 d^2 x'_{2n+1}/dt^2 = W''(x'_{2n} + x'_{2n+2} - 2x'_{2n+1}) . \quad (4.14)$$

If we assume that

$$x'_{2n} = A_2 \exp(i\omega t - i2nka),$$

$$x'_{2n\pm 1} = A_1 \exp[i\omega t - i(2n \pm 1)ka] ,$$

then we find a system of algebraic equations from (4.14), the condition for the consistency of which leads to a dispersion equation that is fourth order in frequency, i.e.,

$$\omega^4 - 2\omega^2 W'' \left[\frac{1}{m_2} + \frac{1}{m_1} \right] + \frac{4(W'')^2}{m_1 m_2} \sin^2 ka = 0 . \quad (4.15)$$

It follows from (4.15) that

$$\omega^2_\pm = W'' \left\{ \left[\frac{1}{m_1} + \frac{1}{m_2} \right] \pm \left[\left(\frac{1}{m_1} + \frac{1}{m_2} \right)^2 - \frac{4\sin^2 ka}{m_1 m_2} \right]^{1/2} \right\} . \quad (4.16)$$

When ka is small, (4.16) yields

$$\bar{\omega}_- \approx \sqrt{\frac{2W''}{(m_1 + m_2)}} \, ka , \quad (4.17)$$

$$\bar{\omega}_- \approx \sqrt{2W'' \left(\frac{1}{m_1} + \frac{1}{m_2} \right)} . \quad (4.18)$$

It can be seen from (4.16)-(4.18) that two types of wave may propagate in such a medium; their dispersion curves are illustrated in Fig. 4.5. The upper line, which corresponds to the high-frequency oscillations of the chain, is called the optical curve (when ka is small, the line is given by (4.18)). The low-frequency, or acoustic, line is quite separate (when ka is small the curve is given by (4.17)). As ka increases in value the two curves approach each other. We suggest that the reader studies the transition from the two-atom chain to a one-atom chain.

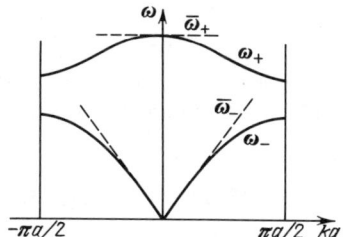

Fig. 4.5. Dispersion curves for a chain of two sorts of particle: the upper curve is the optical one and the lower curve is the acoustic one.

Note that as the number of sorts of different particles increases so too does the number of optical curves.

An electrical analog of a one-dimensional lattice of identical regularly spaced particles is a chain of serially connected LC circuits (Fig. 4.6). This type of chain acts as a low-frequency filter and can be described by an equation for the current, i.e.,

$$d^2 i_n/dt^2 = (LC)^{-1}(i_{n-1} + i_{n+1} - 2i_n) , \qquad (4.19)$$

which becomes the same as (4.11) if we make the substitution $W''/m \leftrightarrow (LC)^{-1}$, $x'_n \leftrightarrow i_n$.

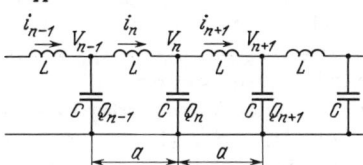

Fig. 4.6. Electrical analog of a one dimensional chain consisting of identical particles: i_n is the current flowing through the inductor between the $(n-1)$-th and n-th capacitor; Q_n and $V_n = Q_n/C$ are the charge on a capacitor and the potential difference across it.

Let us look at another realization of a one-dimensional chain, namely an infinite sequence of identical acoustic resonators with volumes V_r which are connected by tubes with

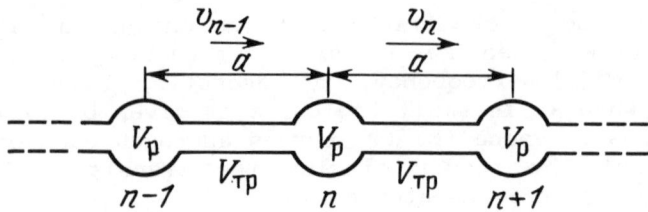

Fig. 4.7. Acoustical analog of a one-dimensional chain consisting of identical particles: v_{n-1} and v_n are the speeds of the masses of gas respectively between the $(n - 1)$ and n-th and between the n and $(n + 1)$-th tubes.

cross sectional areas S and volumes V_t (Fig. 4.7). Suppose a gas with density ρ is flowing through this system. Further, we assume that at any moment in time the gas in the resonators is in equilibrium and that the volume of the resonators is much larger than that of the connecting tubes. Using Newton's second law, we can prove that the following equation is valid:

$$\frac{d^2 p_n}{dt^2} = \frac{S^2}{\chi V_r V_t \rho} \left[p_{n+1} + p_{n-1} 2 p_n \right], \qquad (4.20)$$

where dp_n is the change in pressure in the n-th resonator, and $\chi = -\dfrac{1}{V_r} \dfrac{(v_n - v_{n-1})S}{dp_n/dt}$ is the compressibility of the gas. Equation (4.20) is analogous to (4.11), i.e., the chain is indeed an acoustic analog of a one-dimensional lattice consisting of identical particles, with each particle only affecting its immediate neighbor.[1]

We now turn to the more complicated and more general case when the chain consists of connected identical oscillators, rather than particles. The oscillators may, for example, be pendulums with mass m and having a natural frequency of $\omega_0 = \sqrt{g/l}$. The oscillators are linked by springs with a stiffness γ_1 (Fig. 4.8). The equation for the displacement $\phi_n(t)$ of the n-th pendulum when the oscillation

---

[1] The last two examples correspond to problems 4.1 and 4.44 in [3], which we recommend the reader solves.

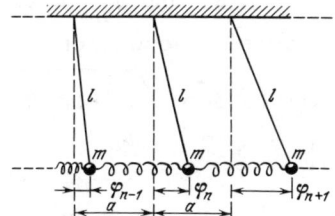

Fig. 4.8. Chain of identical pendulums linked together by springs.

is small and assuming that a disturbance in one oscillator only affects its immediate neighbors may be written in the form

$$\ddot{\phi}_n + \omega_0^2 \phi_n = (\gamma_1/m)(\phi_{n+1} + \phi_{n-1} - 2\phi_n) . \qquad (4.21)$$

The reader can easily derive this equation using the approach outlined for the derivation of (4.8). We look for a solution of this differential-difference equation (4.21), the difference being second order, in the form of single-frequency oscillations (by analogy with the solution of (4.9)), i.e.,

$$\phi_n = A \exp [i\omega t - inka] , \quad \phi_{n-1} = A \exp [i\omega t - i(n-1)ka] ,$$

$$\phi_{n+1} = A \exp [i\omega t - i(n+1)ka] . \qquad (4.22)$$

Substituting (4.22) into (4.21) yields the dispersion law for real k, i.e.,

$$\omega^2 = \omega_0^2 + \frac{2\gamma_1}{m}\left(1 - \cos ka\right) = \omega_0^2 + \frac{4\gamma_1}{m}\sin^2\frac{ka}{2} , \qquad (4.23)$$

and for $k = -i\kappa$ (where κ is a real number) we get

$$\omega^2 = \omega_0^2 + \frac{2\gamma_1}{m}\left(1 - \cosh \kappa a\right) = \omega_0^2 - \frac{4\gamma_1}{m}\sinh^2\frac{\kappa a}{2} . \qquad (4.24)$$

If the frequency ω is set in (4.23) (the external effect on the chain), k may be found. If k is real then a wave with frequency ω will propagate along the chain; whilst if k is imaginary, then the wave will damp exponentially.

In fact because $\phi_n = A \exp (i\omega t - inka)$ then for $k = -i\kappa$

we get $\phi_n = A \exp(i\omega t - n\kappa a)$ and $\phi_n \to 0$ as the number of cells n increases. The dispersion equation (4.23) determines the frequency from $\omega = \omega_0$ to $\omega^* = \sqrt{\omega_0^2 + 4\gamma_1/m}$, which corresponds to values of ka from $ka = 0$ to $ka = \pi$. The frequency region $\omega_0 < \omega < \omega^*$, with the corresponding wave numbers, is the transparency region, in which the waves in the chain propagate without being damped (Fig. 4.9). It follows from (4.23) that the condition $\omega < \omega_0$ may only hold when $\sin^2(ka/2) < 0$, i.e., when k is imaginary. The condition $\omega > \omega^*$ too may only hold for imaginary k. This region corresponds to the equation $\omega^2 = \omega_0^2 + (4\gamma_1/m)\cosh^2(\kappa a/2)$ and the interval $\omega < \omega_0$ corresponds to (4.24). These values of ω and k correspond to the opaque region, within which the amplitudes of waves along the chain due to disturbances at its ends are exponentially damped with increasing n (Fig. 4.10).

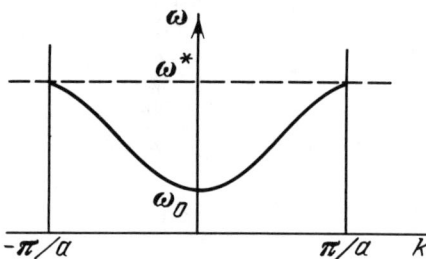

Fig. 4.9. Dispersion curve in the transparency region for the chain illustrated in Fig. 4.8.

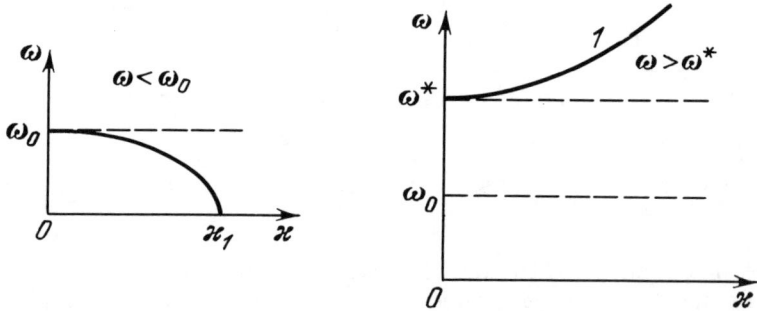

Fig. 4.10. Dispersion curves in the opaque region for the chain illustrated in Fig. 4.8:
1) $\omega^2 = (\omega^*)^2 + (4\gamma_1/m)\sinh^2(\kappa a/2)$.

OSCILLATIONS IN ORDERED STRUCTURES

We shall in conclusion examine another example, viz., a chain of compass needles, which are oscillators with a nonelastic links (Fig. 4.11). The chain is placed in an external magnetic field and each needle is free to rotate in the plane of the paper about its own axis, which is fixed. The notation is given in the figure. We shall assume, as we have done in most cases, that the magnetic interaction only occurs between the poles of immediate neighbors.

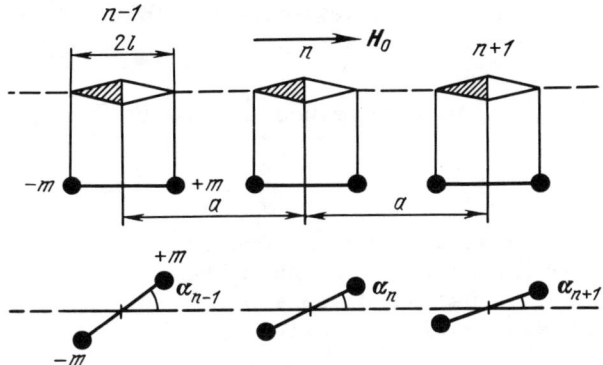

Fig. 4.11. A chain of compass needles: above the chain is in the undisturbed state; below the chain has been disturbed due to a displacement of the $(n - 1)$-th dipole from the equilibrium position through a small angle α_{n-1}.

The propagation of a wave along such a chain was considered by M. Parodi when he was investigating ferromagnetic crystals [2], and it has recently been re-examined in [4] as part of investigations of magnetostatic waves in magnetically ordered media. We leave out the arguments [4] and simply write the equation of motion for the n-th needle, which has the form

$$I\ddot{\alpha}_n = -\frac{2m^2 l^2}{a^3}\left(\alpha_{n+1} + \alpha_{n-1} - 2\alpha_n\right)$$
$$- m^2 l^2\left[\frac{1}{(a - 2l)^3} + \frac{1}{(a + 2l)^3}\right]\left(\alpha_{n+1} + \alpha_{n-1} + 2\alpha_n\right) \quad (4.25)$$
$$- 2m^2 l\left[\frac{1}{(a - 2l)^2} + \frac{1}{(a + 2l)^2}\right]\alpha_n - 2mlH_o\alpha_n ,$$

where I is the moment of inertia of the compass needle with respect to its axis of rotation.

The dispersion equation corresponding to (4.25) when $l \ll a$ can, as is shown in [4], be written as

$$\omega^2 = \omega_m^2 + (4ml/a^3)\gamma \cos ka \, . \qquad (4.26)$$

The quantity $\omega_m^2 = \gamma H_0$ ($\gamma = 2ml/I$) is determined by the parameters of the chain and the external field; it has the dimensions of the frequency squared and so ω_m is the analog of the natural frequency of the magnetization precession. The parameter $(4ml/a^3)\gamma$ characterizes the link between the needle oscillators. If the external field is absent, then for very long waves ($ka \ll 1$) we have $\omega = \omega_M[1 - (ka)^2/4]$ where $\omega_M = \sqrt{8m^2l^2/Ia^3}$, is determined solely by the parameters of the chain.

4.3 Limiting Transition from an Ordered Structure to a One-Dimensional Continuous Medium. Temporal and Spatial Dispersion. Physical Nature of Dispersion

We return to the chain of identical pendulums linked by springs (Fig. 4.8). Let us assume that the natural spatial period of a wave in a discrete chain is much larger than the distance between the pendulums, i.e., much larger than the dimension of the cell. Then the following substitutions are possible:

$$\phi_n(t) \to \phi(x, t) \, ,$$

$$\phi_{n+1}(t) \to \phi\{(x + a), t\} = \phi(x, t) + \frac{\partial \phi(x, t)}{\partial x} a + \frac{1}{2} \frac{\partial^2 \phi(x, t)}{\partial x^2} a^2 + \ldots \, ,$$

$$\phi_{n-1}(t) \to \phi\{(x - a), t\} = \phi(x, t) - \frac{\partial \phi(x, t)}{\partial x} a + \frac{1}{2} \frac{\partial^2 \phi(x, t)}{\partial x^2} a^2 + \ldots$$

By going from discrete coordinates to continuous ones and using these substitutions in (4.21), we obtain a partial differential equation

$$\partial^2 \phi(x, t)/\partial t^2 - v^2 \partial^2 \phi(x, t)/\partial x^2 + \omega_0^2 \phi(x, t) = 0 \, , \qquad (4.27)$$

where $v^2 = \gamma_1 a^2/m$. This is a linearization of the Klein-Gordon equation, which was first derived in field theory.

Let us look in more detail at the meaning behind the assumptions needed to derive (4.27). First, the function $\phi_n(t)$ was defined at discrete points along the x-axis; we have made it continuous. Second, we expanded the $\phi(x,t)$ function into series and discarded the high order terms (this is where (4.27) becomes inaccurate). In addition, when completing these operations, we did not define in comparison with what a is small. When are these assumptions valid? We shall obtain the dispersion equation for (4.27). By substituting $\phi(x, t) \sim \exp(i\omega t - ikx)$ into (4.27) we get

$$\omega^2 = \omega_0^2 + v^2 k^2 \qquad (4.28)$$

or

$$\omega^2 = \omega_0^2 + (\gamma_1/m)(ka)^2 . \qquad (4.29)$$

It is easy to see that (4.29) can be obtained from (4.23) if $\sin^2(ka/2) \approx (ka)^2/4$, i.e., when $ka \ll 1$. Thus, when we say a is small in comparison to the natural spatial period of the wave, we are saying that ka is small, and hence that a is small in comparison to the wavelength. This is because $k = 2\pi/\lambda$ ($ka \ll 1$) or $a \ll \lambda$). For long waves our assumptions are valid, and hence a chain of pendulums can be regarded as a medium described by the Klein-Gordon equation. However, all the approximations are violated when $\lambda \sim a$, i.e., when the wavelength in the structure is commensurable with the structure's period. Thus the transformation of the dispersion equations in Section 4.1 for chains of identical particles when $ka \ll 1$ implies a transition from an ordered structure to a one-dimensional continuous medium.

If ω_0 tends to zero in (4.27), then we obtain the usual wave equation

$$\partial^2\phi(x, t)/\partial t^2 - v^2 \partial^2\phi(x, t)/\partial x^2 = 0 ,$$

the dispersion equation of which is

$$\omega = \pm vk \qquad (4.30)$$

or for the model being analyzed

$$\omega = \pm \sqrt{\gamma_1/m} \; ka . \qquad (4.31)$$

Here $v = a\sqrt{\gamma_1/m}$ is the phase velocity of the wave. The latter equation becomes the same as (4.13) for a chain of identical regularly spaced particles when $ka \ll 1$. Physically this is obvious because as $\omega_0 = \sqrt{g/l} \to 0$ for the pendulums, it is necessary that $l \to \infty$. This means that the length of the pendulum becomes so long that it no longer affects its oscillations and hence we have a chain of balls connected together with springs (but $ka \ll 1$!).

If the relationship between ω an k in the dispersion equation is linear, i.e., (4.30) is valid, then we may say the medium has no dispersion. The phase velocity, defined as ω/k, will be constant and independent of frequency (Fig. 4.12a). For instance, when $ka \ll 1$ a chain of atomic spheres in a one-dimensional lattice behaves like an elastic spring described by the wave equation. In this case, we consider the propagation of elastic waves in a continuous medium with a velocity v, which equals the velocity of sound (and hence the name acoustic for the lower curve in Fig. 4.5). It follows from (4.29) for ω slightly larger than ω_0 that the dispersion curve will be a parabola, i.e.,

$$\omega \approx \omega_0 + Ak^2, \quad \text{if} \quad A = \gamma_1 a^2/2m\omega_0 \ll 1, \qquad (4.32)$$

i.e., dispersion appears close to ω_0. At the same time it is interesting that at large constant ω, there will be no dispersion; $\omega(k)$ is linear. We shall attempt to systematize these results in order to make clear what the appearance of dispersion is associated with in a medium.

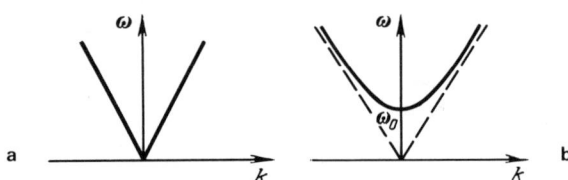

Fig. 4.12. Dispersion curves for a medium: a) with linear dispersion; b) with dispersion given by equation (4.29).

We return to the Klein-Gordon equation, which describes the propagation of a one-dimensional wave in a medium with dispersion, e.g., along a chain of pendulums with natural frequencies ω_0 and spaced a distance $a \ll \lambda$ apart (the dispersion curve is the solid one in Fig. 4.12b). We have mentioned that when $\omega_0 \to 0$ the dispersion disappears, i.e., the pendulum threads are so long that they have no natural oscillatory period and the chain turns in this case into an elastic string. Dispersion disappears when the natural frequency of the scale characterizing the medium disappears. When each pendulum has a natural period of $T = 2\pi/\omega_0$, the "medium" of pendulums will not tolerate a frequency less than the natural one. At this critical frequency all the pendulums will oscillate in phase, that is there will be no wave, only the oscillation. If now we return to (4.21) and (4.23), the relation between the a and λ of which may be of any nature, then it is not difficult to see that the dispersion is conserved in the system, even when $\omega_0 \to 0$. In fact in this case, we get a chain of spheres connected together by springs. The dispersion in this medium is significant whilst a is small in comparison to λ. Thus, in a "lattice" of spheres, the dispersion is determined by the fundamental spatial scale, i.e., the lattice period. The dispersion in a lattice of equally spaced identical masses is associated with the same parameter (see (4.16)). As far as the chain of connected pendulums is concerned, when $\omega_0 \neq 0$ and a is comparable with λ, the dispersion is determined by both the temporal and spatial scales. The dispersion for a chain of compass needles is described in a similar manner, except that here the frequency ω_m figures together with the period a, the frequency being due to the presence of an external magnetic field (see (4.26)). Hence, we may say that the presence of dispersion in a medium is due to the presence in it of natural temporal and spatial scales that are independent of the parameters of the wave.

If there are no characteristic spatial or temporal scales in a medium (for example, when sound waves propagate in water or electromagnetic waves in a vacuum), i.e., there are no characteristic frequencies or periods, then non-sinusoidally propagating waves will not be distorted. There will be no dispersion either.

If, for example, there are bubbles in the water, i.e., there is some form of spatial scale a, i.e., the distance between the bubbles or the dimension of the bubbles, then for waves with $\lambda \gg a$ there will be no distortion, while if $\lambda \sim a$, then waves will be distorted and there will be dispersion in the system. In crystal, say, a low-frequency wave (the wavelength of which is many times greater than the distance between the ions) will propagate without distortion, while for a high-frequency wave the inter-ion separation will become significant and the "medium" will be discrete (Figs. 4.2, 4.3, and 4.5).

Dispersion associated with the presence in a medium of temporal scales is usually called temporal dispersion, but when there are spatial scales it is called spatial dispersion. Note that this classification is only convenient in electrodynamics, in which it is possible to speak separately about the equations of the medium and those of the field. Formally, the dispersion equations are nonlocal relations between the various physical variables in time or space. Hence, in the electrodynamics of continuous media spatial dispersion is associated with the fact that the electrical induction **D** at a given point in space is determined by the strength of the electric field **E** both at that point and in a certain neighborhood, i.e., **D** and **E** are not just locally related in space:

$$D_i(\omega, \mathbf{k}) = \varepsilon_{ij}(\omega, \mathbf{k}) E_j(\omega, \mathbf{k}) ,$$

where $\varepsilon_{ij}(\omega, \mathbf{k})$ is the complex dielectric permeability tensor [5].

It is formally permissible to introduce the following definition: a medium in the electrodynamics of continuous media has spatial dispersion if its dielectric permeability depends on the wave vector; if it depends on the frequency, the dispersion is frequency or temporal.

The latter also arises because of a nonlocal relation between **D** and **E** in time. Moreover, temporal dispersion is often very large because the natural frequencies of the medium fall within the frequency interval being considered [5]. Spatial dispersion must be born in mind, for example, in the physics of isotropic plasmas, where the wavelength is comparable with the Debye radius, and when accounting for collisions in the theory of conducting media when the mean free path is of the order of a wavelength.

Spatial dispersion leads to new effects in crystal optics, such as natural optical activity (gyrotropy) and the optical anisotropy of cubic crystals [5,6]. We can also show that in a plasma, for example, the group velocity of a longitudinal wave becomes nonzero due to spatial dispersion (we shall return to this topic in the next chapter).

We should note that although spatial dispersion is the result of the presence of natural spatial scales in the medium, i.e., the result of the discreteness of the "medium", it can be accounted for using a model of a continuous medium if a phenomenological relation can be found between the parameter which allows for the nonlocal character of their relationship in space. Thus in order to account for spatial dispersion, it is necessary to construct the model of the medium correctly.

Let us consider, by way of example, the propagation of an electromagnetic wave along a long line, as shown in Fig. 4.13 (see problem 4.23 in [3]).

If the connections between the cells are absent, then the telegraph equation is valid, i.e.,

$$\frac{\partial I}{\partial x} = -\frac{\partial Q}{\partial t} = -C\frac{\partial U}{\partial t}, \quad \frac{\partial U}{\partial x} = -\frac{\partial \Phi}{\partial t} = -L\frac{\partial I}{\partial t},$$

which can be easily transformed into the wave equation

$$\frac{\partial^2 I}{\partial t^2} - \frac{1}{LC}\frac{\partial^2 I}{\partial x^2} = 0,$$

whence there is no dispersion in this model of a chain.

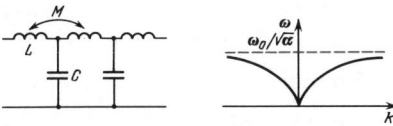

Fig. 4.13. A long line with inductive connections M between cells and the corresponding dispersion curve.

However, if the cells are inductively connected, the magnetic flux Φ and current I will be related by a material equation $\Phi = LI - M\partial^2 I/\partial x^2$, from which it follows that there is a nonlocal relation between these variables (the presence of a spatial derivative of the current). Whence

$$\frac{\partial^2 I}{\partial t^2} - \frac{1}{LC}\frac{\partial^2 I}{\partial x^2} = M\frac{\partial^4 I}{\partial t^2 \partial x^2}.$$

The corresponding dispersion equation has the form

$$\omega^2 = \omega_0^2 k^2/(1 + \alpha k^2), \tag{4.33}$$

where $\omega_0^2 = 1/LC$ and $\alpha = M/L$. (Note that the k in (4.33) is dimensionless because we are not looking at the chain cell-by-cell rather than in units of length, that is, the dimensions of L and C are respectively [henry] and [farad] in the cell and in order to give the quantity dimensions it must be multiplied by a, i.e., the dimension of the cell in the corresponding units of length.) If $\alpha \ll 1$, then we obtain from (4.33) by keeping first-order terms in α

$$\omega^2 = \omega_0^2 k^2 (1 - \alpha k^2). \tag{4.34}$$

Now let us return to equation (4.12) for a one-dimensional lattice of identical particles. We assume that ka is small and expand $\sin^2(ka/2)$ into series, discarding terms of the order $(ka)^4$ and above, i.e.,

$$\omega^2 = (W''/m)(ka)^2[1 - (ka)^2/12] \ . \tag{4.35}$$

By remembering that the k in (4.34) is dimensionless, replacing k with ka, setting $W''/m = \omega_0^2$, and $1/12 = \alpha$, we transform (4.35) into (4.34). Thus, the two approaches -- the discrete and the phenomenological accounts of the nonlocality of the relation between the physical parameters -- leads to a correct description of spatial dispersion (the "bends" in the dispersion curves in Figs. 4.12 and 4.13 are due to spatial dispersion). Spatial dispersion is also manifest close to the frequency ω_0 (see Fig. 4.12b and (4.34)). In (4.33) α may take either sign. Whence if $\omega^2 = \omega_0^2 k^2/(1 - \alpha k^2)$, then for $\alpha \sim k^{-2}$ the phase velocity of the wave $v_{ph} = \omega/k \to \infty$ and the group velocity (the velocity at which energy is transported without loss in a medium) is $v_{gr} = d\omega/dk \to \infty$. (We shall consider the concepts of phase and group velocity in more detail later.) Consequently, information is transferred instantaneously from one point to another. Consider which idealizations of the model are responsible for this error.

4.4 Typical Dispersion Characteristics for Medium Models

Let us look at the more typical dispersion characteristics of various one-dimensional media, using for the sake of clarity equivalent LC-chains. Practically any dispersion relationship can be realized with LC-chains, and so these will be our models in our investigation of the propagation of waves in various media.

We start from the telegraph equations

$$\partial I/\partial x = - \partial Q/\partial t \ , \quad \partial U/\partial x = - \partial\Phi/\partial t \ , \tag{4.36}$$

together with equations linking the charge Q with the potential U and the magnetic flux Φ with the current I, i.e.,

$$Q = \hat{Q}\{U\} \ , \quad \Phi = \hat{\Phi}\{I\} \ . \tag{4.37}$$

In general, $\hat{Q}$ and $\hat{\Phi}$ are either differential or integral operators, and only when there is no dispersion in the medium is the link between the variables instantaneous, viz., $Q = CU$, $\Phi = LI$ (the link relations become algebraic). The charge or flux are not dependent on the potential or the current in neighboring points or during neighboring moments in time. If $\hat{Q}$ and $\hat{\Phi}$ are differential operators containing derivatives with respect to t or x, then the relations between the variables are nonlocal and we say that the medium has temporal or spatial dispersion, respectively.

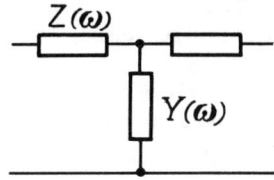

Fig. 4.14. Equivalent circuit for the chain: $Z(\omega)$ is the impedance and $Y(\omega)$ is the shunt admittance.

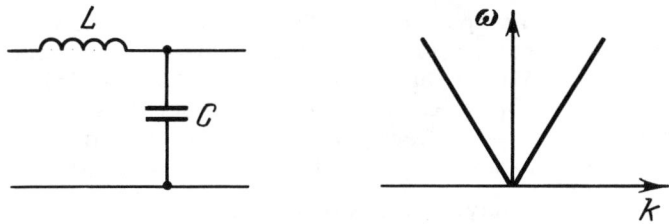

Fig. 4.15. Equivalent circuit for a medium model without dispersion and its dispersion characteristic.

We have already used this approach in the special case of the circuit in Fig. 4.13 at the end of the preceding section. We shall find the dispersion equation corresponding to the very general equivalent circuit illustrated in Fig. 4.14, assuming that $I = \mathcal{I} \exp(i\omega t)$ and $U = \mathcal{U} \exp(i\omega t)$. The equations for the complex amplitude have the form

$$d\mathcal{I}/dx = -Y(\omega)\mathcal{U}, \quad d\mathcal{U}/dx = -Z(\omega)\mathcal{I}.$$

If we assume that $(\mathcal{I}, \mathcal{U}) = (\mathcal{I}_0, \mathcal{U}_0)\exp(-ikx)$, then we can find the dispersion equation from the condition needed to make the system soluble in the amplitudes $\mathcal{I}_0$ and $\mathcal{U}_0$, i.e.,

$$k^2 = -Y(\omega)Z(\omega). \tag{4.38}$$

The actual forms of $Y(\omega)$ and $Z(\omega)$ depend on the link equations.

Let us consider several dispersion characteristics for models using LC chains and using (4.38).

A Medium without Dispersion. We have for the chain in Fig. 4.15 $Y(\omega) = i\omega C$ and $Z(\omega) = i\omega L$, i.e., $\omega^2 = k^2/LC$. We can recover the corresponding differential equation from the dispersion equation. In this case the wave equation is $\partial^2 U/\partial t^2 - (LC)^{-1} \partial^2 U/\partial x^2 = 0$. The phase velocity is $v_{ph} = (LC)^{-1/2}$ = const, and so the model corresponds to a medium without dispersion. This model describes the propagation of an electromagnetic wave through a vacuum, a sound wave through pure water, a low-frequency wave through a solid, and a fundamental direct spatial harmonic through a delay system for electronic amplifiers of travelling waves (e.g., in coils).

A Medium with Dispersion at Low Frequencies (Fig. 4.16). We shall consider a model of a medium whose dispersion is described by the equation

$$\omega^2 = \omega_0^2 + k/LC , \quad \omega_0^2 = 1/L_1 C ,$$

the corresponding partial differential equation being the linear Klein-Gordon equation (4.27). Thus, the chain in Fig. 4.16 is the electrical analog of the model of linked pendulums, when $a \ll \lambda$. This model describes, for instance, the propagation of electromagnetic waves in plasmas, for which $\omega_0 = \omega_p$ (where ω_p is the plasma frequency) and the propagation of a wave in a waveguide.

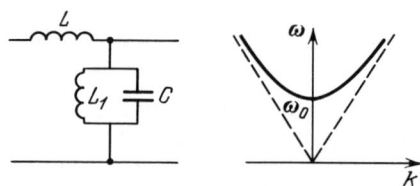

Fig. 4.16. Equivalent circuit for a model medium with low-frequency dispersion and its dispersion characteristic.

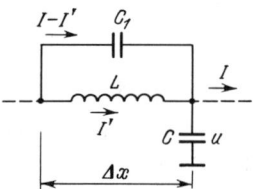

Fig. 4.17. A typical cell in a long line as used in experiments [7]: $L = 182\mu H$, $C_1 = 100pF$, $C = 24pF$, $\Delta x = 1cm$.

A Medium with Dispersion at High Frequencies. The propagation of waves along a long line consisting of the cells shown in Fig. 4.17 is described by the partial differential equation

$$\frac{\partial I}{\partial x} = -C \frac{\partial U}{\partial t} , \quad \frac{\partial U}{\partial x} = -\frac{\partial \Phi}{\partial t} \left(\frac{\partial U}{\partial x} = -L \frac{\partial I'}{\partial t} \right) , \quad (4.39)$$

$$\Phi = LI - L\frac{C_1}{C}\frac{\partial^2 I}{\partial x^2} \left(I' = I - \frac{C_1}{C}\frac{\partial^2 I}{\partial x^2}\right).$$

Assuming that all the variables vary exponentially $\exp(i\omega t - ikx)$, and using the substitution $\omega_0^2 = 1/LC_1$, $k_0^2 = C/C_1$, we obtain from (4.39) the dispersion equation

$$\omega^2 = \omega_0^2 k^2/(k_0^2 + k^2). \qquad (4.40)$$

Clearly, we did not need to write out (4.39). We could have found $Z = i\omega L/(1 - \omega^2 LC_1)$ and $Y = i\omega C$ directly from the equivalent circuit, which together with (4.38) would immediately give (4.40). However, we wanted to demonstrate, at least once, how dispersion arises due to nonlocal links between the variables (note the material equation $\Phi = \hat{\Phi}(I)$ in (4.39)). It is interesting that the dispersion in this model is the same as that for a long line with inductive connections between the cells (Fig. 4.13).

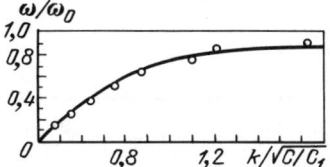
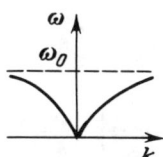

Fig. 4.18. Predicted (solid lines) and experimental (points) dispersion curves for the line shown in Fig. 4.17 ($f_0 = \omega_0/2\pi = 37.3$ MHz) [7].

The dispersion curve in Fig. 4.18 was determined using the usual experiments for this purpose [7], in which one end of the line was loaded by a resistance not equal to the characteristic impedance Z_0 of the line ($Z_0 = \sqrt{L/C}/(1 - \omega^2/\omega_0^2) \sim [L/C]^{1/2} \sim 71 \, \Omega$). Due to reflection in the line, a standing wave is set up. The wavelength can be determined using a probe and a voltmeter by measuring the distance between the waves's minima. The greatest frequency corresponds to a wavelength of $\sim 2\Delta x$. It is shown in [7] that a medium model quantitatively "describes" the propagation of ion acoustic waves (ion sound) in plasmas. This line also models the propagation of sound in a solid (sound waves propagate without dispersion so long as the wave number k is much less than the reciprocal of the lattice vector $q = 2\pi/a$, where a is the inter ion separation, otherwise there is substantial spatial dispersion due to the discreteness of the "medium") and spin waves in a ferromagnetic.

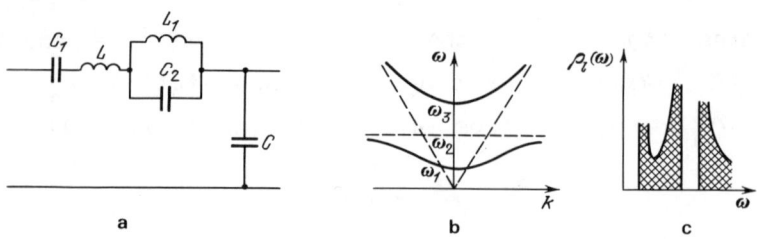

Fig. 4.19. a) Equivalent circuit for a model medium in which there are natural oscillators. b) Its dispersion characteristic (the lower curve is the acoustic one and the upper one is the optical one). c) The number density of oscillators for a medium with two types of oscillation (low frequency and high frequency).

A Medium of Oscillators (Fig. 4.19). The dispersion equation has the form

$$k^2 = \left[(\omega^2 - \omega_2^2)(\omega^2 - \omega_4^2) - \omega^2\omega_5^2\right] / \omega_0^2(\omega^2 - \omega_2^2) ,$$

where $\omega_0^2 = 1/LC$, $\omega_2^2 = 1/L_1 C_2$, $\omega_4^2 = 1/LC_1$, $\omega_5^2 = 1/LC_2$,

$$\omega_1^2 = \frac{1}{2}\left[(\omega_4^2+\omega_2^2+\omega_5^2) - \sqrt{(\omega_4^2 + \omega_2^2 + \omega_5^2) - 4\omega_2^2\omega_5^2}\right]$$

$$\omega_1^2 = \frac{1}{2}\left[(\omega_4^2+\omega_2^2+\omega_5^2) + \sqrt{(\omega_4^2 + \omega_2^2 + \omega_5^2) - 4\omega_2^2\omega_5^2}\right]$$

An example is a medium with elastic dipoles for electromagnetic waves, or an anisotropic plasma for Langmuir or ion-acoustic waves. When $\omega \gg \omega_3$ the wave does not notice the natural oscillations of the dipoles and the medium behaves as if it had no dispersion. When ω is close to ω_1, ω_2, or ω_3, the dispersion becomes considerable.

Model of a Delay System in which a Reverse Spatial Harmonic is Propagating. The dispersion equation has the form $\omega^2 = (k^2 LC)^{-1}$. The gap in the dispersion characteristic in the region $k \approx 0$, ($\lambda \to \infty$) corresponds to a spatially homogeneous region, which clearly is not realized except for the trivial case of $\omega = 0$. Note that this model also describes the propagation of transverse waves in elastic bars (Fig. 4.20).

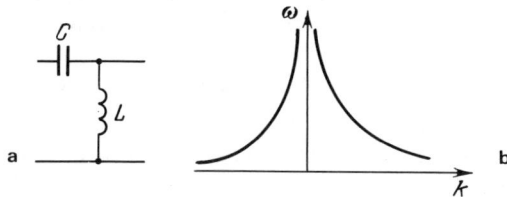

Fig. 4.20. a) Equivalent transmission line corresponding to a propagating reverse wave. b) Its dispersion characteristic. The group velocity $v_{gr} = -\omega^2\sqrt{LC}$ is in the opposite direction to the phase velocity $v_{ph} = \omega^2\sqrt{LC}$.

4.5 Formal Method for Obtaining the Dispersion Equation. Waves in a One-Dimensional Resonator. Resonance in Wave Systems

Suppose the equation describing the propagation of a wave in a medium be written in the form

$$A\partial u/\partial t + B\partial u/\partial x + Cu = 0 , \quad (4.41)$$

where A, B, and C are matrices, and u is a vector. We look for a solution to (4.41) in the form

$$u = \Psi \exp(i\omega t - ikx) , \quad (4.42)$$

where $\Psi\{\Psi_1, \Psi_2, \ldots, \Psi_n\}$ is a complex (polarization) vector, the components of Ψ_i being the distribution coefficients which characterize the ratio between the amplitudes of the various physical variables in a harmonic wave.

Substituting (4.42) into (4.41) yields an algebraic system of equations for Ψ_i. The condition for the existence of nontrivial solutions to this system will be the dispersion equation, viz.,

$$\text{Det}(A\omega - Bk - iC) = D(\omega, k) = 0 . \quad (4.43)$$

Let (4.43) have a solution $\omega = \omega_s(k)$ and $k = k_s(\omega)$, where $s = 1, 2, \ldots, n$. This means that there are n types of wave in the system, i.e.,

$$u(x, t) = \sum_{s=1}^{n} \Psi_s \exp[i\omega_s t - ik_s(\omega)x] + cc ,$$

where cc means the complex conjugate. As was the case for a lumped system (see Chapter 2), it is possible to go over to normal waves

$$a_s(x, t) = \Psi_s \exp[i\omega_s t - ik_s x] .$$

Since there are no links between normal waves, they satisfy the equation

$$\partial a_s/\partial x + ik_s a_s = 0 \quad (s = 1, 2, \ldots, n) . \tag{4.44}$$

This notation is even convenient when there are weak links between the waves. The necessary a_j are simply added to (4.44) with the corresponding link coefficients (we have devoted a separate chapter to linked waves).

For distributed systems, the dispersion equation is one that relates the two complex quantities ω and k. There is a characteristic equation for lumped systems which gives more information about the system, namely its spectrum of complex natural frequencies.

We then should ask whether there is an analog for distributed systems. Up until now we have been considering infinite media. We therefore turn to systems in which there is feedback (we shall call them resonators). Feedback occurs in the simplest case in a ring resonator. Both a travelling wave regime and a regime with superimposed oncoming waves (a special case of which is a standing wave) can be realized in a ring resonator. An example is the Fabry-Perot optical resonator or a transmission line with its ends either shorted or open. The solution is given in the form of a superposition of oncoming waves

$$u(x, t) = \Psi_1 \exp(i\omega t - ikx) + \Psi_2 \exp(i\omega t + ikx) , \tag{4.45}$$

the absolute values of the amplitudes of which must be equal in the simplest case of ideal reflection at the ends of the resonator. For example, in the case of a piece of string clamped at both ends $u(0, t) = 0$ and $u(l, t) = 0$ (l is the length of the string). We find from (4.45) the condition for the amplitudes of the oncoming waves to be $\Psi_1 = -\Psi_2$ and a limitation on the spectrum of the wave numbers $\sin kl = 0$, whence

$$k_n = \pi n/l \quad (n \text{ is an integer}) . \tag{4.46}$$

It is not difficult to verify that in any one-dimensional resonator with limiting reflection at the ends only elementary solutions satisfying (4.46) are possible, i.e., only an integer number of half waves can be fitted into the

OSCILLATIONS IN ORDERED STRUCTURES   73

resonator. In a ring resonator, the periodicity condition for
all the variables serve as boundary conditions. For example,
it is U, $I(x, t) = U$, $I(x + 1, t)$ for a transmission line
closed into a circle, whence follows the condition
$\exp (ikl) = 1$, i.e., we have the spectrum

$$k_n = 2\pi n/l .\qquad (4.47)$$

The physics of this condition is obvious, that is, a ring
resonator may only contain waves that are periodic in space
and which can pack into the resonator an integer number of
times. Once we know the dispersion equation $D(\omega, k) = 0$ of
the medium filling the resonator and the spectrum of the wave
numbers (4.46) or (4.47), we can obtain an equation with
respect to one variable, i.e., $\Delta(\omega) = D(\omega, k_n) = 0$ which
gives the spectrum of the normal frequencies of the
resonator. This equation is, in fact, the analog of the
characteristic equation for lumped systems. For instance, in
the case of a medium without dispersion and for ideal
reflections at the ends $k_n = \pi n/l$ and
$\omega_n = \pi n/l\sqrt{LC} = k_n/\sqrt{LC}$ (Fig. 4.21). We now want to know
what would be the ω spectrum for an equidistant k spectrum if
the medium has dispersion. Its qualitative behavior can be
found, if the dispersion characteristic is known, using an
elementary graphical construction, as is clear from
Figs. 4.22 and 4.23.

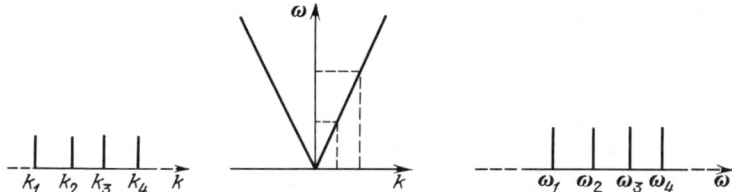

Fig. 4.21. The equidistant spectrum of the natural
frequencies corresponding to the equivalent spectrum of wave
numbers in a medium without dispersion.

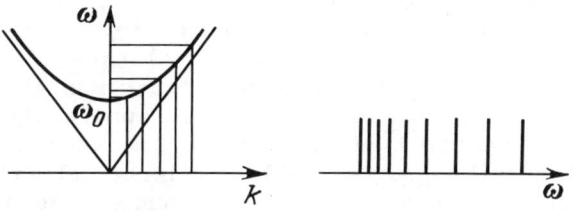

Fig. 4.22. Nonequidistant spectrum of natural frequencies
corresponding to the equivalent spectrum of wave numbers in a
spectrum with low-frequency dispersion.

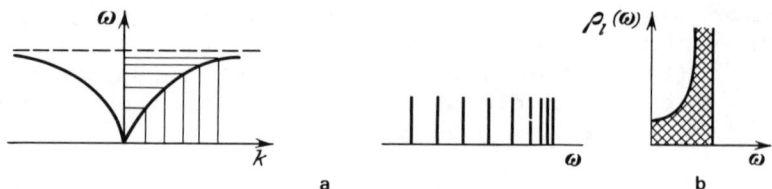

Fig. 4.23. a) Nonequidistant spectrum of natural frequencies corresponding to the equivalent spectrum of wave numbers in a medium with high-frequency dispersion. b) Number density of oscillators for the low-frequency curve.

In a medium with low-frequency dispersion, the spectrum of natural frequencies starts with ω_0 (Fig. 4.22) and becomes dense close to this critical frequency. Far from ω_0 the lines in the spectrum are practically equidistant. As ω tends to ω_0 the spectrum becomes practically continuous. The same situation is seen for a medium with high-frequency dispersion, except that the spectrum gets thinner closer to the null frequency (Fig. 4.23a). If there are two critical frequencies, two dense regions appear in the spectrum.

Note that when we must find the natural frequencies of long lines, which are represented by equivalent circuits with boundary conditions at the ends, then the spectrum of the wave numbers k_n can be found from the well-known characteristic equation $\tan kl = iY(Z_0 + Z_1)/(1 + Z_0 Z_1 Y^2)$, where Y is the characteristic admittance of a long line, and Z_0 and Z_1 are the loads at $x = 0$ and $x = l$, respectively [8,3]. Let us look at another case, namely a line shorted at one end and open at the other, viz., $Z_0 = 0$, $Z_1 = \infty$ (or $Z_0 = \infty$, $Z_1 = 0$), whence $k_n = \pi(2n - 1)/2l$.

Hence if the medium filling the resonator has dispersion, the density $\rho(\omega)$ of the normal modes will be different in different regions even if there is an equidistant k spectrum. This yields a method for measuring the dispersion properties of a one-dimensional medium, and it is especially valuable when investigating chains of linear polymers. Suppose that we can evenly disturb all the degrees of freedom of the chain, then the experimental spectrum of its oscillations will be a simple superposition of the densities of spectral distributions corresponding to the various dispersion curves. The density of the spectral distribution for each curve (the number density of the oscillators) can be reduced to the formula

$$\rho(\omega)d\omega = \text{const} \cdot dk. \quad (4.48)$$

OSCILLATIONS IN ORDERED STRUCTURES 75

It was assumed here that the number of modes in the interval $(k, k + dk)$ is independent of k. We obtain the following from (4.48), with some arbitrary multiplier, for the longitudinal oscillations of a chain consisting of identical particles

$$\rho_l(\omega) = \text{const} \cdot (d\omega/dk)^{-1} = (2C/a)(\omega_{max}^2 - \omega^2)^{-1/2} . \qquad (4.49)$$

This spectrum is illustrated in Fig. 4.23b. It is similarly not difficult to construct the density of the spectral distribution $\rho_l(\omega)$ of a chain consisting of alternating light and heavy molecules [15]. If both longitudinal and transverse waves are excited, then we must add to the $\rho_l(\omega)$ spectrum (4.49) the spectrum of the transverse oscillations, which is described by the dispersion equation $\omega(k) = B \sin^2(ka/2)$. The density of the spectral distribution of the frequencies in a full spectrum is illustrated in Fig. 4.19c [15].

Let us recall a direct space-time analog. We first consider the propagation of a travelling wave $\frac{1}{v_{ph}} \frac{\partial u}{\partial t} + \frac{\partial u}{\partial x} = 0$ in a one-dimensional medium (v_{ph} is the constant phase velocity of the wave in the medium) upon which an external distributed force $G(x, t) = G(x) \exp(i\omega t)$ is acting. Whence, we clearly get

$$\partial u(x)/\partial x + (i\omega/v_{ph})u(x) = G(x) ,$$

if $u(x, t) = u(x) \exp(i\omega t)$.

This equation can be more conveniently rewritten in integral form, with the condition $u(0) = 0$, as

$$u(x) = \exp(- i\omega x/v_{ph}) \int_0^x G(\zeta)\exp(i\omega \zeta/v_{ph})d\zeta , \qquad (4.50)$$

where ζ is the running integration variable.

Assuming that $G(x, t) = G(0) \exp(i\omega t - i\omega x/v_{ex})$, i.e., that the external disturbance is a wave with a constant amplitude and frequency ω travelling with a phase velocity v_{ex} and by integrating (4.50), we find

$$u(x, t) = G(0) \exp(i\omega t - i\omega x/x_{ph}) \frac{\exp(i\omega/v_{ph} - i\omega/v_{ex})x - 1}{i\omega/v_{ph} - i\omega/v_{ex}} .$$

When $\omega/v_{ph} \approx \omega/v_{ex}$ we obtain the secular increase $u(x, t)$ in the x-direction

$$v(x, t) = G(0)x \exp(i\omega t - i\omega x/v_{ph}).$$

There is a space-time analogy, namely that for a rising harmonic wave in a space affected by an external field, it is essential for the spatial periods to coincide, i.e., resonance in the wave numbers. Indeed, there are both frequency and wave-number resonances, which is manifested in the equality of the phase velocity of the natural waves in the medium with the phase velocity of the external wave. This is usually called the condition for the waves to be synchronous. If v_{ph} and V_{ex} are very different, then the system enters a regime of spatial beats (the wavelength of the beats can easily be determined). When many waves may propagate in the medium, i.e.,

$$u(x, t) = \sum_{n=1}^{N} u^{(n)} \exp(i\omega_n t - ik_n x)$$

and the external disturbance may also have many waves, there will be n synchronicity conditions, i.e., there will be n equations in which the phase velocity of the natural wave at a frequency ω_i equals the phase velocity of the external wave at the same frequency. By realizing the synchronicity conditions we have just formulated, microwave frequency electronic devices can be created with long interactions between the electrons and the wave (the best known of these being a travelling wave tube TWT [9]). The duration of the electrons flight through the interaction space in such devices is much longer than the period of microwave frequency field oscillation, in contrast with resonance microwave frequency devices such as the klystron, the short-term interaction devices we described in Chapter 1.

If we consider $u(x, t)$ to be the longitudinal component of the electrical field E in a wave guide and $G(x, t)$ to be a wave of an alternating current I in an electron beam (we omit the dimensioning factor), then the equation for $u(x, t)$ is the disturbance equation for the wave guide by the current [10,11], i.e.,

$$\frac{\partial E}{\partial x} + i\frac{\omega}{v_{ph}} E = -\frac{1}{2}\left(\frac{\omega}{v_{ph}}\right)^2 KI(x),$$

where K has the dimensions of resistance and is called the coupling impedance. If a straight electron beam with a small current density is represented as a stream of independent particles moving at a velocity v_0, then high frequency disturbances will have the form of a current wave $I(x, t) = I(0) \exp(i\omega t - i\omega x/v_0)$ moving with a phase velocity of v_0 ($V_{ex} = v_0$). Thus the simplest synchronicity condition is the equality between the convective velocity of the electrons and the phase velocity of the wave. By the way, it follows from this condition that at the nonrelativistic velocities of the electrons the electromagnetic wave must be decelerated (in most TWTs a spiral delay system is used, see Fig. 4.24). When Coulomb forces are significant in the beam, disturbances propagate in the form of waves of space charge at a velocity different from v_0. In order to get spatial resonance in this case, there must be synchronicity between one of the space-charge waves and the wave in the delay system. It should be noted that we have only considered the effect of the external wave on the natural wave. In most cases this is not the only effect and when there are synchronicity conditions there is also feedback. In a TWT, for example, the field of the wave carrying system modulates the beam with respect to velocity and groups the electrons into clusters. This effect arises in the case of linked waves, which we shall consider in Chapter 10.

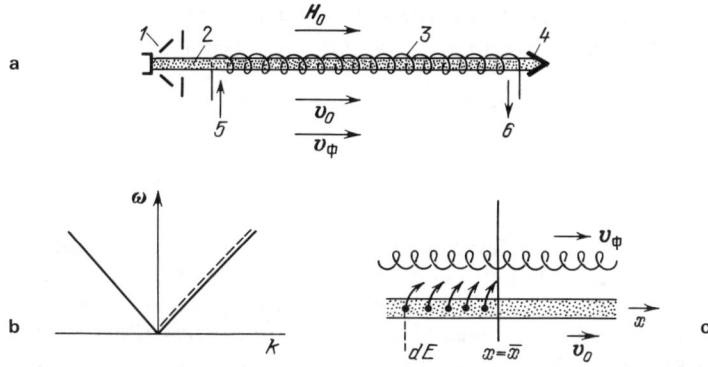

Fig. 4.24. a) Diagram of a TWT: 1) electron gun; 2) electron beam; 3) spiral; 4) collector; 5 and 6) input and output devices; H_0 focusing magnetic field.

b) Dispersion diagram of wave (continuous line) and beam (dashed line) for the model $v_0 \approx v_{ph}$.

c) Illustration of spatial resonance. a field in the section $x = \bar{x}$ is superimposed by the field induced by each element of the exciting beam, which are positioned at $x < \bar{x}$. The field phases add together if $v_0 \approx v_{ph}$.

4.6 Quasiparticles

Starting from the dualism between waves and particles we can introduce a quantum of energy of a field in a macroscopic body, viz., a quasiparticle. We shall make this analogy clearer for those who are familiar with quantum mechanics. Starting from quantum dynamics the Hamiltonian for a chain of identical linked particles (Fig. 4.1) can be written using the notation of Chapter 1 as a sum of terms like (1.34), i.e.,

$$\mathcal{H} = \sum_k \hbar\omega_k (a_k^+ a_k + 1/2) , \qquad (4.51)$$

where $a_k^+ a_k = N_k$ is the operator for the number of bosons in state k, and the sum is taken over all admissible values of the **k** vector. These values of k are usually determined from periodic boundary conditions $q_{i+M} = q_i$ for the coordinate (the condition of closure for the chain; M is the number of particles in the chain). Whence the admissible values of the wave vector are $k = 2\pi n/Ma$, where n is an integer between $-M/2$ and $M/2$. In (4.51)

$$\omega_k = 2\sqrt{W''/m} \, \sin(\pi n/M) , \qquad (4.52)$$

i.e., the frequency is exactly the same as the corresponding formula from the classical theory.

The eigenvalues of the energy are

$$\mathcal{E} = \sum_k \hbar\omega_k (n_k + 1/2) , \qquad (4.53)$$

where k runs over all the values of the positive numbers from M.

Thus, the behavior of a one-dimensional chain of identical particles can be represented as a set of normal oscillations, each of which corresponds to an harmonic oscillator. Each normal oscillation, therefore, has a certain corresponding wave number k and certain natural frequency ω_k, the energy state of the oscillator being given by the quantum number of its normal oscillation. Since the energy of a normal oscillation with a frequency ω_k takes values which can be expressed as an integer multiplied by $\hbar\omega_k$ (if the origin is taken at zero and not the ground energy level $\hbar\omega_k/2$), we come to the idea that the process has a quantum structure. The quanta that are thus obtained are called quasiparticles. If several normal oscillations are simultaneously disturbed, then several quasiparticles are present in the system.

We have just considered the simplest model of a solid (a one-dimensional chain of identical particles) and the particles oscillate elastically in the lattice.

Various types of elastic wave may exist in a crystal and they differ, in the first place, by the signs of their dispersions. The first case is the acoustic wave. The dispersion equation of (4.13) correspond at small ka ($a \ll \lambda$) to acoustic waves, this equation being conveniently rewritten in the form $\omega_k = vk$, where $v = \sqrt{W''a^2/m}$ has the sense of the velocity of sound. From the quantum point of view, as we have seen, energy and momentum are associated with each normal wave and only take discrete values proportional to $\varepsilon = \hbar\omega_k$ and $p = \hbar k$. Each such wave may be considered as a quasiparticle with a quasimomentum p (quasimomentum is not conserved during collisions of quasiparticles in crystals) and energy ε. These quasiparticles are called phonons and are the quanta of sound. It follows from the above that the dispersion equation for long wave phonons is $\varepsilon = vp$. When solving the classical equation of motion of a system in the quantum theory of fields the normal oscillations are first found and then each of these is called by a quantum oscillator with a ground-state energy of $\mathcal{E}_n = \hbar\omega(n + 1/2)$ [12,13].

Quantized normal oscillations (quasiparticles) have various names. We have already spoken of phonons, but the most interesting quasiparticle is the photon, the quantum in the theory of electromagnetic fields. It has an energy $\varepsilon = \hbar c$, where c is the speed of light in a vacuum. Waves in a system of electron spins that are linked by exchange forces are called spinor waves and the corresponding quasiparticle is called the magnon. Collective plasma oscillations of the electron gas in a metal, which are linked by Coulomb forces, are often called plasmons. Neutral quasiparticles called excitons exist in semiconductors. More information about these and many other quasiparticles (e.g., polarons, fluctuons) can be found in [14].

CHAPTER 5

PROPERTIES OF WAVES WITH SMALL AMPLITUDES IN CONTINUOUS MEDIA

5.1 General Remarks

In this chapter we discuss the dispersion characteristics of waves in various real media. We shall restrict ourselves to media in which physical phenomena permit a hydrodynamic description. This clearly includes fluids and gases, but it also includes plasmas and plasma-like media (such as charged particle beams), during the analysis of which kinetic effects can either be ignored or allowed for phenomenologically. We shall examine waves known from general physics courses, such as sound waves, and more specific ones, such as waves in the atmosphere and the oceans associated with the rotation of the Earth, internal waves in a stratified ocean, and ion-sound waves in anisothermal plasmas.

The fairly detailed discussion of the properties of the various waves in this chapter will help later when understanding and describing them from a unified point of view.

In a hydrodynamic description, a fluid is considered to be a continuous medium (see [1,26]), i.e., the motion of some particle of fluid is defined as being the motion of some volume containing many molecules, not just one. Such an element, which is small in comparison to the scale of the phenomenon under study but large in comparison to the intermolecular forces, is considered in hydrodynamics to be a point. It is sufficient for a full description of the behavior of a moving fluid to give the following independent variables at this point: the fluid velocity $\mathbf{v}(x, y, z, t)$, the thermodynamic quantities, such as entropy $S(x, y, z, t)$, per unit mass of fluid [6], and the density of the fluid $\rho(x, y, z, t)$ (where x, y, and z are the coordinates of the elementary volume at time t). The velocity $\mathbf{v}(x, y, z, t)$ in this Eulerian description is not associated with particular particles of fluid, which are moving through space over time; it is instead associated with a certain point in space at the moment in time t. This definition also applies to the values of S and ρ.

The existence of waves in a fluid, which initially is in a state of rest, due to a disturbance in the fluid and a contest between a force that acts to return the fluid to its initial condition and inertial forces, which force the fluid to overshoot it. For example, the restoring forces for a water wave are those due to gravity and surface tension, the restoring force for a rotating fluid is the Coriolis force, while that for a conducting fluid is the force due to the magnetic field.

5.2 Equations of Hydrodynamics.
Dispersion Equations for Sound Waves

We restrict ourselves here to an ideal fluid. This is one whose stress vector when the fluid is moved is perpendicular to every element of the surface irrespective of how the element is oriented in space (i.e., Pascal's law is obeyed). Mathematically, this means that pressure is a scalar and not a tensor [2]. There are therefore no shear forces in the fluid, e.g., the forces due to viscosity. According to Newton's second law, the equation of motion of an element of volume dV of a fluid with density ρ can be written in the form $\rho(d\mathbf{v}/dt)dV = d\mathbf{F}$, where $\mathbf{v}$ is the velocity of the element, and $d\mathbf{F}$ is the force acting on each element of volume dV. A force acts on any volume of fluid V from the surrounding fluid, the force being the integral of the pressure taken over the surface of the volume, i.e., $\oint p d\mathbf{S}$. (It is assumed that the magnitude of the vector $d\mathbf{S}$ is equal to the area of the element of surface and that it is directed along the external normal; whence the minus sign in front of the force). However, from the integral theorem for gradients we have $-\oint p d\mathbf{S} = -\int_V \nabla p dV$. Moreover, an external given force may act with a density $\rho \mathbf{a}_{ex}$ may act on element. Thus $d\mathbf{F} = -\nabla p dV + \rho \mathbf{a}_{ex} dV$, and the equation of motion becomes

$$\rho d\mathbf{v}/dt = -\nabla p + \rho \mathbf{a}_{ex} . \qquad (5.1)$$

Since $d\mathbf{v}/dt = \partial \mathbf{v}/\partial t + (\mathbf{v}\nabla)\mathbf{v}$ in (5.1), we arrive at Euler's equation, the basic one in hydrodynamics, i.e.,

$$\rho\{\partial \mathbf{v}/\partial t + (\mathbf{v}\nabla)\mathbf{v}\} = -\nabla p + \mathbf{a}_{ex}\rho . \qquad (5.2)$$

Clearly, the law of the conservation of mass $\int_V \rho dV$ holds for the volume, i.e., the change in mass in the volume $\dfrac{\partial}{\partial t}\int_V \rho dV$

WAVES WITH SMALL AMPLITUDES

is equal to the negative of the mass flux through the surface bounding the $\oint \rho \mathbf{v} d\mathbf{S}$ volume, i.e.,

$$\frac{\partial}{\partial t} \int \rho dV + \oint \rho \mathbf{v} d\mathbf{S} = 0 , \qquad (5.3)$$

or in differential form

$$\partial \rho / \partial t + \text{div } \rho \mathbf{v} = 0 \qquad (5.4)$$

This is the continuity equation. The vector $\mathbf{j} = \rho \mathbf{v}$ is called the fluid's flux density.

There are five unknowns in (5.2) and (5.4), namely density, the three velocity components, and the pressure, i.e., one more equation is needed. The remaining equation is the thermodynamic equation of state.

We shall consider there is no heat exchange between individual fluid particles (the fluid flows at such a rate that individual elements have no time to exchange heat with each other) and that it does not exchange heat with bodies it comes into contact with. This assumption means that the motion is adiabatic at each element of fluid, i.e., the entropy S per unit mass of fluid is constant during the motion of an element in space. Hence we get

$$dS/dt = \partial S/\partial t + \mathbf{v} \nabla S = 0 . \qquad (5.5)$$

We multiply (5.4) by S and (5.5) by ρ and divide one by the other to get $S \partial \rho / \partial t + \rho \partial S / \partial t + S \text{ div } \rho \mathbf{v} + \rho \mathbf{v} \nabla S = 0$. By using the transformation div $(a\mathbf{f}) = a \text{ div } \mathbf{f} + \mathbf{f} \nabla a$, we get the continuity equation for entropy, i.e.,

$$\partial (\rho S)/\partial t + \text{div}(\rho S \mathbf{v}) = 0 , \qquad (5.6)$$

where $\rho S \mathbf{v}$ is the entropy flux density. If the fluid's entropy distribution at the initial moment in time is spatially homogeneous, then

$$S = \text{const} \qquad (5.7)$$

at any moment in time. An adiabatic process which takes pace with constant entropy is called isentropic. In such a case the equation of state is simply a functional relation between the density and the pressure, viz., $p = p(\rho)$ (or $\rho = \rho(p)$), whence

$$\frac{dp}{dt} = \left(\frac{dp}{d\rho}\right)_S \frac{d\rho}{dt} . \qquad (5.8)$$

By linearizing (5.2) and (5.4) for small disturbances ρ', $\mathbf{v}'$, and p' in the density, velocity, and pressure, respectively against the background of their equilibrium values ρ_0, $\mathbf{v}_0$ and p_0, we get (if we assume that $\mathbf{a}_{ex} = 0$)

$$\frac{\partial \mathbf{v}'}{\partial t} + (\mathbf{v}_0 \nabla)\mathbf{v}' = -\frac{1}{\rho_0} \nabla p' ,$$

$$\frac{\partial p'}{\partial t} + \left(\frac{\partial p}{\partial \rho}\right)_S [\mathrm{div}(\mathbf{v}_0 \rho') + \rho_0 \mathrm{div}\, \mathbf{v}'] = 0 . \qquad (5.9)$$

For a stationary medium $\mathbf{v}_0 = 0$ and with a velocity potential $\mathbf{v} = \nabla \phi$, we obtain the pressure perturbation $p' = -\rho_0 \partial \phi/\partial t$. The following well-known wave equation follows from the second expression in (5.9):

$$\partial^2 \phi/\partial t^2 - c^2 \Delta \phi = 0 , \qquad (5.10)$$

where $c = \sqrt{(\partial p/\partial \rho)_S}$ is the speed of sound. It is clear that in Cartesian coordinates the pressure and each of the velocity components obey the wave equation (to prove this for yourself the grad operator should be applied to the wave equation).

If all the variables in a wave depend solely on one of the Cartesian (flat wave), then (5.10) transforms into the one-dimensional equation $\partial^2 \phi/\partial t^2 - c^2 \partial^2 \phi/\partial x^2 = 0$, which we studied in Chapter 4. It has a general solution in the form of the superposition of two oncoming plane waves:

$$\phi(x, t) = f_1(x - ct) + f_2(x + ct).$$

Since sound waves have no dispersion in this approximation, the dispersion equation is of the form

$$\omega = \pm cki . \qquad (5.11)$$

Travelling sound waves, of whatever form, will be stationary, i.e., their profiles do not change as they propagate. This is easily explained in terms of spectra. Because there is no dispersion, all the spectral components forming the wave move at equal velocities and the phase relations between them are conserved.

In planar acoustic waves only the x-component of the velocity $v_x = \partial \phi/\partial x$ is not zero, i.e., the particles in the wave move either only in the direction of wave propagation or

WAVES WITH SMALL AMPLITUDES

only in the opposite direction. This is why acoustic waves in fluids are only longitudinal.

If the velocity of a medium in which a sound wave is propagating is nonzero, then the dispersion equation (5.11) will be violated. For example, if a plane wave is propagating in a homogeneous stream that is moving in the x-direction at a constant velocity v_0 then a dispersion equation follows from (5.9):

$$\omega = \pm ck + v_0 k . \qquad (5.12)$$

The quantity $v_0 k$ characterizes the extra frequency shift of an acoustic wave in a moving medium with respect to a stationary observer. If the wave is moving in the direction of the stream motion, then its frequency is increased by $v_0 k$, if it is moving against the stream, then the frequency is reduced.

5.3 A Stratified Fluid. Sound in an Ocean

In order to describe waves in an ocean or atmosphere, the fluid mechanics equations must be generalized so as to allow for the rotation of the Earth and for liquid stratification, that is, for the liquid density to be a function of the vertical coordinate. In particular, the density of sea water depends on the pressure, temperature, and relative content of dissolved salt, which all change with depth [5,21,22]. The corresponding generalization results in Euler's equation taking the following form instead of (5.2):

$$\rho\{\partial \mathbf{v}/\partial t + (\mathbf{v}\nabla)\mathbf{v}\} = -\nabla p - 2\rho[\Omega \mathbf{v}] - \rho g \nabla z . \qquad (5.13)$$

Here Ω is the angular velocity of the Earth's rotation, ∇z is the unit vector of the vertical coordinate; $\mathbf{a}_{ex}$ is replaced by $\mathbf{g}$ because the liquid is in the Earth's gravitational field.

Let us assume that the wavelengths of interest to us are much less than the Earth's radius and that (5.13) and (5.4) are solved in a plane tangent to a spherical Earth at a given point. A Cartesian reference frame is established with the z-axis directed vertically up, the x-axis directed from West to East, and the y-axis directed from South to North. We linearize the equations about the state of rest, with the density and pressure being functions of z only. Let $p = p_0(z) + p'(x, y, z, t)$, $\rho = \rho_0(z) + \rho'(x, y, z, t)$, where $p', \rho' \ll p_0 \rho_0$. Note that $\mathbf{v} = \mathbf{v}'(x, y, z, t)$ like both p' and

ρ' is a small quantity of magnitude one. We therefore get from (5.13) and (5.4)

$$\frac{\partial \mathbf{v}'}{\partial t} = -2[\ \mathbf{v}'\ ] - \frac{1}{\rho_0}\nabla p' - g\frac{\rho'}{\rho_0}\nabla z, \qquad (5.14)$$

$$\frac{\partial \rho'}{\partial t} = -\rho_0 \text{div }\mathbf{v}' - \mathbf{v}'\nabla \rho_0. \qquad (5.15)$$

The equation of state (5.8) has the form in the linear approximation of

$$\frac{d}{dt}\left(\rho_0 + \rho'\right) = \frac{1}{c^2}\frac{d}{dt}\left(p_0 + p'\right)$$

or

$$\frac{\partial \rho'}{\partial t} + v_z\frac{\partial \rho_0}{\partial z} = \frac{1}{c^2}\left(\frac{\partial p'}{\partial t} + v_z\frac{\partial p_0}{\partial z}\right),$$

where $c(z) = \sqrt{(\partial p/\partial \rho)_s}$ is the adiabatic speed of sound. Since $\partial p_0/\partial z = -\rho_0 g$, we get finally

$$\frac{\partial \rho'}{\partial t} + v_z\frac{\partial \rho_0}{\partial z} = \frac{1}{c^2}\left(\frac{\partial p'}{\partial t} - \rho_0 g v_z\right). \qquad (5.16)$$

The normal component of the velocity must disappear at the horizontal floor, and hence for $z = -H$ we get

$$v_z = 0, \qquad (5.17)$$

where H is the liquid depth. At the liquid surface the pressure will be $p_0 + p' =$ const and hence $d(p_0 + p')/dt = 0$, which, taken with the right-hand side of (5.16) gives for $z = 0$

$$\partial p'/\partial t - \rho_0 g v_z = 0. \qquad (5.18)$$

We now make use of Bussinesque's approximation in (5.14)-(5.16), i.e., where $\rho_0(z)$ is not a part of a differential, we assume that $\rho_0 =$ const, and let $\rho_0(0) = \rho_{00}$. We look for a solution to (5.14)-(5.16) in the form [3]

WAVES WITH SMALL AMPLITUDES

$$v_x = (P(z)/\rho_{00})V_x(x, y)e^{i\omega t}, \quad v_y = (P(z)/\rho_{00})V_y(x, y)e^{i\omega t},$$

$$v_z = -i\omega V(z)V_z(x, y)e^{i\omega t}, \quad (5.19)$$

$$p = P(z)V_z(x, y)\exp(i\omega t), \quad \rho' = \rho'(x, y, z)\exp(i\omega t),$$

where ω is the frequency of the waves we are interested in.
    By substituting (5.19) into (5.14)-(1.16) and rearranging, we get the following from (5.14):

$$V_x + iqV_y - s\omega\rho_{00}\frac{V(z)}{P(z)}V_z = \frac{i}{\omega}\frac{\partial V_z}{\partial x}, \quad (5.20)$$

$$V_y - iqV_x = \frac{i}{\omega}\frac{\partial V_z}{\partial y}, \quad (5.21)$$

$$\frac{\partial P(z)}{\partial z} + \frac{g}{c^2}P(z) + \rho_{00}(\omega^2 - N^2)V(z) - s\omega P\frac{V_x}{V_z} = 0; \quad (5.22)$$

and from (5.16) we get

$$\frac{1}{c^2} + \frac{\rho_{00}g}{c^2}\frac{V(z)}{P(z)} - \rho_{00}\frac{1}{P(z)}\frac{\partial V(z)}{\partial z} = \frac{i}{\omega}\left(\frac{\partial V_x}{\partial x} + \frac{\partial V_y}{\partial y}\right)\frac{1}{V_z}. \quad (5.23)$$

    To derive (5.20)-(5.23), we used the following expression obtained from (5.15):

$$\rho' = \left[\frac{P(z)}{c^2} - \frac{\rho_{00}}{g}N^2(z)V(z)\right]V_z \quad (5.24)$$

and the definitions of the frequency of free vertical oscillations of the liquid particles (called the Vyaisyalya frequency)

$$N(z) = \left[-\frac{g}{\rho_{00}}\left(\frac{\partial\rho_0}{\partial z} + \frac{\rho_0 g}{c^2(z)}\right)\right]^{1/2}. \quad (5.25)$$

The following dimensionless variables are substituted into

(5.20) and (5.21):

$$q = (2\Omega_z/\omega) = (2\Omega/\omega)\sin\phi,$$
$$s = (2\Omega_y/\omega) = (2\Omega/\omega)\cos\phi,$$
(5.26)

where ϕ is the geographical latitude of the location. Using (5.19), we can rewrite boundary conditions (5.17) and (5.18) thus

$$z = -H, \quad V(z) = 0,$$ (5.27)

$$z = 0, \quad P(z) + \rho_{00}gV(z) = 0.$$ (5.28)

It was shown in [3] that (5.20)-(5.23) admit a separation of variables in two cases: 1) When s and q chosen for given ϕ, which is the geographical latitude, are constant. This approximation is valid for waves along the length of which s and q change weakly, i.e., sound, surface, internal, and inertia waves. 2) The terms containing only Ω_y, e.g., s because $s \sim \Omega_y$, can be neglected.

Suppose, therefore, that

$$\begin{Bmatrix} V_x \\ V_y \\ V_z \end{Bmatrix} = \begin{Bmatrix} V_{ox} \\ V_{oy} \\ V_{oz} \end{Bmatrix} e^{-i(k_x x + k_y y)},$$ (5.29)

and that V_{ox}, V_{oy}, V_{oz} are constant, with $V_{oz} = 1$. This does not hinder the generality of the solution. The solutions to (5.20) and (5.21) then take the form

$$V_{ox} + iqV_{oy} = k_x/\omega + s\omega\rho_{00}V(z)/P(z), \quad V_{oy} - iqV_{ox} = k_y/\omega.$$

We find from this system that

$$V_{ox} = \frac{1}{\omega(1-q^2)}\left\{k_x + s\omega^2\rho_{00}\frac{V(z)}{P(z)} - iqk_y\right\},$$ (5.30)

$$V_{oy} = \frac{1}{\omega(1-q^2)}\left\{k_y + iqk_x + iqs\omega^2\rho_{00}\frac{V(z)}{P(z)}\right\}.$$ (5.31)

Finally, we obtain two equations for $V(z)$ and $P(z)$ from (5.22) and (5.23) using (5.29)-(5.31):

$$\frac{\partial P(z)}{\partial z} + \left[\frac{g}{c^2} - \frac{sk_x - iqsk_y}{1-q^2}\right]P(z) + \rho_{00}\left[\frac{\omega^2 - 4\Omega^2}{1-q^2} - N^2\right]V(z) = 0, \quad (5.32)$$

$$\frac{\partial V(z)}{\partial z} + \left[-\frac{g}{c^2} + \frac{sk_x + iqsk_y}{1-q^2}\right]V(z)$$
$$+ \frac{1}{\rho_{00}}\left[-\frac{1}{c^2} + \frac{\xi^2}{\omega^2(1-q^2)}\right]P(z) = 0, \quad (5.33)$$

where $\xi^2 = k_x^2 + k_y^2$.

Since the frequency ω of sound waves considerably exceeds Ω and N, gravity does not significantly affect these waves in the ocean. Thus it is possible to neglect those terms in (5.32) and (5.33) that contain $(s, q) \sim \Omega$, and N and g. This yields

$$\frac{\partial P(z)}{\partial z} + \rho_{00}\omega^2 V(z) = 0, \quad (5.34)$$

$$\frac{\partial V(z)}{\partial z} - \frac{1}{\rho_{00}}P(z) + \frac{\xi^2}{\rho_{00}\omega^2}P(z) = 0.$$

Excluding $V(z)$ yields

$$\partial^2 P(z)/\partial z^2 + \{\omega^2/c^2(z) - \xi^2\}P(z) = 0, \quad (5.35)$$

which is the main equation for oceanic acoustics.

Although $c(z)$ only changes little with depth, the presence of a minimum in $c(z)$ at some depth leads to the formation of a submarine acoustic waveguide, along which low-frequency sound (damping is small at low frequencies in water) can propagate for several tens of thousands of kilometers from their source [3,23].

5.4 Gravity Waves in an Incompressible Liquid. Internal Waves. Rossby Waves

We note straightaway that an assumption of incompressibility $c^2 = dp/d\rho \to \infty$ can be used to simplify equations (5.32) and (5.33) to reduce them to the form

$$\frac{\partial \mathcal{P}}{\partial z} - \frac{sk_x - iqsk_y}{1 - q^2}\mathcal{P} + \rho_{00}\left[\frac{\omega^2 - 4\Omega^2}{1 - q^2} - N^2\right]V = 0, \quad (5.36)$$

$$\frac{\partial V}{\partial z} + \frac{sk_x + iqsk_y}{1 - q^2}V + \frac{\xi^2}{\omega^2\rho_{00}(1 - q^2)} = 0. \quad (5.37)$$

By excluding $\mathcal{P}$ we arrive at the equation

$$\frac{\partial^2 V}{\partial z^2} + 2i\frac{qsk_y}{1 - q^2}\frac{\partial V}{\partial z} - \left\{\frac{k_x^2 s^2 + q^2 k_y^2 + \xi^2 - (q^2 + s^2)\xi^2}{(1 - q^2)^2}\right.$$

$$\left. - \frac{N^2 \xi^2}{\omega^2(1 - q^2)}\right\}V = 0. \quad (5.38)$$

In order to analyze gravitation waves at the surface of a liquid, as we shall soon see, neither liquid stratification nor the rotation of the Earth are significant, i.e., we can discard terms containing N, q, and s in (5.38), and we arrive at

$$\partial^2 V(z)/\partial z^2 + \xi^2 V(z) = 0 \quad (5.39)$$

with the boundary conditions (5.27) and (5.28), which for the given assumptions can be written in the form

$$V(z)\bigg|_{z=-H} = \left\{gV(z) - \frac{\omega^2}{\xi^2}\frac{\partial V(z)}{\partial z}\right\}_{z=0} = 0. \quad (5.40)$$

It was assumed here that $\partial V(z)/\partial z + (\xi^2/\omega^2\rho_{00})\mathcal{P}(z) = 0$ (see (5.37)).

We can show the validity of these assumptions by using dimensionality arguments. Let us consider surface waves assuming that at equilibrium the surface of a liquid is horizontal. If it leaves this state, then in order for there to be waves at the liquid surface there must be a force

WAVES WITH SMALL AMPLITUDES

restoring the system to equilibrium and an inertia force due to which the liquid overruns the equilibrium. The force acting on a hump appearing on the surface of the liquid to make it disappear so that the surface once again becomes horizontal could, for example, be the force of gravity $F_g \sim g$ or surface tension force $F_\sigma \sim \sigma$ (σ is the coefficient of surface tension). We shall discuss the action of these forces separately.

In moving downwards under the force of gravity the bump moves further than the equilibrium position due to inertia. Next to it another hump will be forced up and so on. A wave, called a gravity wave, begins to propagate in the liquid. A dimensional analysis yields the form of the relation between the phase velocity of the wave v_{ph} and its wave length λ. The value of v_{ph} must depend of $F_g \sim g$ and on the inertia of the oscillating liquid, a measure of which is its density ρ, and may depend on the depth of the liquid H. In this way $v_{ph} = f(\lambda, g, \rho, H)$. It is immediately apparent from dimensionality arguments that the density ρ will not enter the final formula because the dimensions of mass only enter ρ. Physically, this is because both the weight of the hump returning to the equilibrium position and the mass of the hump (its time lag) are proportional to ρ. The dimensionality of ρ and H are identical, the dimensions of time are only contained in g, therefore the velocity at which the wave propagation may be written in two equivalent forms:

$$v_{ph} = \sqrt{gH}\, f_1(\lambda/H), \quad v_{ph} = \sqrt{g\lambda}\, f_2(H/\lambda)\ . \tag{5.41}$$

Let $\lambda \ll H$, which means we are discussing waves in deep water or short waves only moving in the surface layer of the liquid (a thickness of $\sim \lambda$). Then it is clear that the velocity of wave propagation must be independent on the depth of the liquid, i.e., $f_1(\lambda/H) = c_1(\lambda/H)^{1/2}$ and hence

$$v_{ph} = c_1 \sqrt{\lambda g} \tag{5.42}$$

Clearly if we assume that $v_{ph} = f(\lambda, g, \rho)$ and that it does not depend on H, we immediately arrive at (5.42).

If however $\lambda \gg H$ (wave is propagating in shallow water or the wave is long), then the velocity of wave propagation may not depend on λ because the motions of all the particles in a thin layer of liquid are practically identical. In this case $f_2(H/\lambda) = c_2(H/\lambda)^{1/2}$ in (5.36) and

$$v_{ph} = c_2\sqrt{gH} \,. \tag{5.43}$$

Because $k = \omega/v_{ph}$ we obtain the following dispersion laws for gravity waves in both limiting cases from (5.42) and (5.43):

$$kH \gg 1, \quad \omega(k) = c_1\sqrt{2\pi g k} \quad \text{(a deep liquid)}, \tag{5.44}$$

$$kH \ll 1, \quad \omega(k) = c_2 k\sqrt{gH} \quad \text{(a shallow liquid)}. \tag{5.45}$$

This analysis is not strict. We cannot use it to find c_1 or c_2. In order to find them we must use (5.39) and (5.40). If the solution to (5.39) $V(z) = A_1\exp(\xi z) + A_2\exp(-\xi z)$ is substituted into the boundary conditions (5.40), then because the resultant algebraic system of equations in the unknowns A_1 and A_2 are simultaneous we find the dispersion equation for surface waves in a liquid of finite depth:

$$\begin{vmatrix} e^{-\xi H} & e^{-\xi H} \\ g - \omega^2/\xi & g + \omega^2/\xi \end{vmatrix} = 0 \quad \text{or} \quad \omega^2 = \xi g \tanh \xi H \,. \tag{5.46}$$

It is easy to see that

$$\omega = \sqrt{\xi g}, \quad \text{when } \xi H \gg 1, \tag{5.47}$$

$$\omega = \xi\sqrt{gH} \{1 - (\xi H)^2/6 + \ldots\}, \quad \text{when } \xi H \ll 1 \,. \tag{5.48}$$

Thus when, for example $\xi = k_x = k$, we have $c_1 = 1/\sqrt{2\pi}$, $c_2 = 1$ in (5.44) and (5.45). It follows from (5.42)–(5.44) when $\xi = k_x = k$ that when $kH \to 0$ (a shallow liquid) the phase velocity v_ϕ turns to a constant limit $\sqrt{gH}$, i.e., the dispersion is weak. In a deep liquid the dispersion is always ($\omega \sim \sqrt{k}$). This is because the relation between the pressure and depth of the liquid is not local. Gravity waves have a negative dispersion because $v_{ph} = [(g/k)\tanh kH]^{1/2}$ falls as the frequency increases. The group velocity $v_{gr} = d\omega/dk$ also decreases as the frequency increases, and hence waves which travel from a region in which they arise towards the shore of a sea or ocean, are at first long but get shorter. This fact can be used to determine the distance from a storm (the reader might be interested in thinking up a method for determining whether there is a storm and if so the maximum distance away it is, see Chapter 4).

Note that when analyzing gravity waves we started from quite general equations. If we restrict ourselves from the beginning to an analysis of gravity waves at the surface of an ideal incompressible liquid (ρ = const), then it is possible to start from the equation

$$\partial \mathbf{v}/\partial t + (\mathbf{v}\nabla)\mathbf{v} = - \nabla(p/\rho) + \mathbf{g}, \quad \text{div } \mathbf{v} = 0. \qquad (5.49)$$

Assuming further that the motion is potential (curl $\mathbf{v}$ = 0), it is possible to introduce a potential of velocity $\mathbf{v} = \nabla_{ph}$. From vector analysis we have $(\mathbf{v}\nabla)\mathbf{v} = \nabla v^2/2 - [\mathbf{v} \text{ curl } \mathbf{v}]$. Then from an incompressible liquid $\partial \mathbf{v}/\partial t + (\mathbf{v}\nabla)\mathbf{v} = \partial \mathbf{v}/\partial t + \partial v^2/2$ and hence

$$\frac{\partial}{\partial t}\left(\nabla_{ph}\right) + \nabla\left(\frac{v^2}{2}\right) = -\nabla\left(\frac{p}{\rho}\right) + \mathbf{g}.$$

Because $\mathbf{g}$ is a body force acting in gravitational field per unit mass we can say $\mathbf{g} = -\nabla U$, where U is the potential energy per unit mass of liquid in a field of gravity. Whence

$$\nabla\left\{\frac{\partial_{ph}}{\partial t} + \frac{v^2}{2} + \frac{p}{\rho} + U\right\} = 0.$$

from which it is easy to obtain the Cauchy-Lagrange integral [7]:

$$\frac{\partial_{ph}}{\partial t} + \frac{v^2}{2} + \frac{p}{\rho} + U = f(t),$$

where $f(t)$ is a function of time. In a steady state flow of liquid ($\partial_{ph}/\partial t = 0$), when the motion is established and the velocity is independent of time, this integral becomes the Bernoulli equation

$$v^2/2 + p/\rho + U = \text{const}. \qquad (5.50)$$

Moreover the constant in (5.50) is the same for all the liquid for a potential motion. If curl $\mathbf{v} = \varphi \neq 0$ (φ characterizes the vorticity and determines the angular velocity of an element of volume of liquid), then (5.50) is valid along the given line of current (the constant may be different along different lines of current).

It is clear that (5.50) expresses the law of the conservation of energy. This is the meaning of the Bernoulli equation, which relates the velocity and the pressure, because U is known. We shall use (5.50) in Chapter 7 to explain Helmholtz instability without solving the hydrodynamic equations.

We now return to very short waves, when the liquid tries to return to equilibrium under the force of surface tension. These waves are called capillary waves. It is reasonable to assume for these waves that $v_{ph} = f(\lambda, \sigma, \rho)$. Only one combination of these quantities has dimensions of velocity, that is

$$v_{ph} = c_3 \sqrt{\sigma/\rho\lambda} . \qquad (5.51)$$

The dispersion law corresponding to (5.51) has the form

$$\omega = c_3 k^{3/2} \sqrt{\sigma 2\pi \rho} . \qquad (5.52)$$

We now shall study the problem more strictly starting from the Cauchy-Lagrange integral and the equation

$$\nabla^2_{ph} = \Delta_{ph} = 0 , \qquad (5.53)$$

which is obtained from the incompressibility condition div $\mathbf{v} = 0$ and the definition $\mathbf{v} = \nabla_{ph}$. When the surface of an interface between air and liquid, for example, is curved, then the pressure drop across the interface may be determined from the Laplace equation [1, 6]:

$$p_1 - p_2 = \sigma/R .$$

This difference is called the surface pressure; R is the radius of curvature of this surface, where $1/R = \partial^2 \zeta/\partial x^2$ if $\zeta = \zeta(x, t)$ is the equation of the curve corresponding to the interface and the surface is slightly bent. In our case the Laplace formula has the form

$$p - p_0 = - \sigma \partial^2 \zeta/\partial x^2 , \qquad (5.54)$$

where p is the pressure close to the liquid surface, and p_0 = const is the external pressure. In (5.1) the surface is negatively curved, which is taken into account by the sign in (5.54). In the linear approximation the Cauchy-Lagrange integral takes the form

WAVES WITH SMALL AMPLITUDES

$$\partial\varphi_{ph}/\partial t + (p - p_0)/\rho = 0 \qquad (5.55)$$

because the term $v^2/2$ in this approximation may be neglected, the force due to gravity is not allowed for because we are only considering capillary waves, and without loss of generality we may consider $f(t)$ to be zero [1]. By using (5.54) we have the following from (5.55) for $z = 0$:

$$\frac{\partial\varphi_{ph}}{\partial t} - \frac{\sigma}{\rho}\frac{\partial^2\zeta}{\partial x^2} = 0 \qquad (5.56)$$

We shall look a solution to (5.53) in a form $\Phi = \phi(z)\exp[i(\omega t - kx)]$. Then $\partial^2\phi/\partial z^2 - k^2\phi = 0$ and $\phi(z) = B_1\exp(kz) + B_2\exp(-kz)$. However if the liquid is sufficiently deep, then $\phi(z) \approx B_1\exp(kz)$ because $z < 0$ under a surface (the x, y - plane coincides with the undistorted horizontal surface of the liquid). We differentiate (5.56) with respect to t and consider that $\partial\zeta/\partial t = v_z = \partial\varphi_{ph}/\partial z$. We then have

$$\rho\frac{\partial\varphi_{ph}}{\partial t^2} - \sigma\frac{\partial}{\partial t^2}\left(\frac{\partial\varphi_{ph}}{\partial z}\right) = 0. \qquad (5.57)$$

Since $\Phi \approx B_1\exp(kz) \cdot \exp(i\omega t - ikx)$, we obtain the following equations for capillary waves from (5.57):

$$\omega^2 = (\sigma/\rho)k^2 \qquad (5.58)$$

In this case c_3 in (5.52) is equal to $\sqrt{2\pi}$. If we take into account the effect on the liquid of both restoring forces (i.e., the force of gravity and the force of liquid tension), then by assuming that $\Phi = \Phi(x, z, t)$ we obtain the following dispersion equation for a liquid of depth H:

$$\omega^2 = [kg + \sigma k^3/\rho]\tanh kH \qquad (5.59)$$

This equation yields the dispersion law for a gravity and capillary wave (we leave the reader to obtain (5.59) on his own).

For capillary waves $v_{ph} = \sqrt{\sigma k/\rho}$, i.e., the phase velocity grows with increasing ω, which corresponds to a positive dispersion. In Fig. 5.2, ω and v_{ph} are given as functions of k for surface waves; the curves correspond to (5.59).

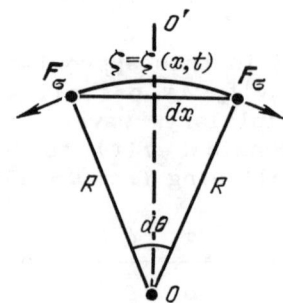

Fig. 5.1. Determining the surface tension force for a surface with a negative curvature.

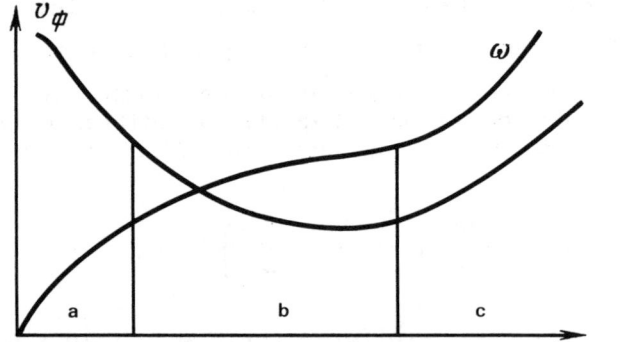

Fig. 5.2. ω and v_{ph} versus k for surface waves: a) long gravity waves ($kH \ll 1$, $kg \gg \sigma k^3/\rho$, $\omega \approx k\sqrt{gH}$, $v_{ph} \approx \sqrt{gH}$); b) short gravity waves ($kH \gg 1$, $kg \gg \sigma k^3/\rho$, $\omega \approx \sqrt{gk}$, $v_{ph} \approx \sqrt{g/k}$); c) capillary waves ($kH \gg 1$, $kgf \ll \sigma k^3/\rho$, $\omega \approx k\sqrt{\sigma k/\rho}$, $v_{ph} = \sqrt{\sigma k/\rho}$).

Some time after the discovery of the fission of uranium, the theory of capillary waves was successfully applied to the stability of atomic nuclei against fission into two approximately equal parts [9]. The theory was based on the fact that the short-range force acting between particles in a nucleus are similar to the surface tension force acting in a liquid (the force is acting between molecules are also short range). This "surface tension" in a nucleus is opposed by the long range Coulomb forces between the protons. A formula similar to (5.59) with $kH \gg 1$ has been derived for the frequency of the vibrations of a spherical nucleus, with the first term on the right-hand side being electrical and not a gravitational in origin and with a minus sign (a Coulomb force is directed along an external normal to the surface). The condition under which the nucleus is unstable to an infinitesimal disturbance to the surface can be found from this relationship.

Construct for yourselves the theory for the fragmentation of charged rain drops assuming that each drop is spherical and a liquid is incompressible (the oscillation should be decomposed into spherical standing waves using Lagrange polynomials) [8].

The simplest examples of internal waves in a stratified liquid are waves propagating along an interface between two homogeneous liquids which differ in density. The waves propagate due to a balance between buoyancy forces and the bulk inertial force of the liquid. A more complicated case involves waves in a liquid with a continuous stratification. In a stratified liquid any height displacement of a part of the liquid disturbs the equilibrium and induces an oscillation. As we have said the density of sea water depends on pressure, temperature, and on the relative content of dissolved salt, which change with depth.

We shall assume initially that $\omega \gg \Omega$ and that the rotation of the Earth can be neglected. In this case (5.38) is considerably simplified:

$$\frac{\partial^2 V(z)}{\partial z^2} - \xi^2 \left(1 - \frac{N^2}{\omega^2}\right) V(z) = 0 . \qquad (5.60)$$

If the medium is unbounded and N = const, then $V(z) = V(0)\exp(\pm ik_z z)$ and $k_z^2 = -\xi^2(1 - N^2/\omega^2)$ and

$$k^2 = k_z^2 + \xi^2 = \xi^2(N^2/\omega^2) \qquad (5.61)$$

or

$$\sin \theta = \mu\omega/N , \qquad (5.62)$$

where θ is the angle of the vector **k** from the vertical, and $\mu = \pm 1$. It follows from (5.62) that waves may only exist when $\omega < N$. If θ is given, then the frequency ω is uniquely determined, even though the wavelength and the phase velocity may be arbitrary.

Note that in an incompressible liquid the condition N = const corresponds to an exponential increase of density with depth.

Let us now consider the propagation of internal waves in a waveguide formed by the surface of the liquid and a horizontal bottom. In this case the solution of (5.60), given the assumption that the Vyaisyalya frequencies are constant, has the form

$$V(z) = c_1 \exp(-ik_z z) + c_2 \exp(ik_z z),$$

$$k_z = \xi[(N/\omega)^2 - 1]^{1/2}.$$

(5.63)

by substituting (5.63) into the boundary conditions (5.40) we attain the following system of equations:

$$c_1 \exp(ik_z H) + c_2 \exp(-ik_z H) = 0,$$

$$(g + ik_z \omega^2/\xi^2)c_1 + (g - ik_z \omega^2/\xi^2)c_2 = 0.$$

(5.64)

We find the following dispersion equation from the condition that system (5.64) is consistent (i.e., the determinant is equal to zero):

$$gk_z \tan k_z H = N^2 - \omega^2.$$

(5.65)

When $k_z H \ll 1$ it is possible to consider that $\tan k_z H \approx k_z H$ and hence when $\omega < N$, one of the solutions of (5.65) can be written as:

$$k_{z0} = [(N^2 - \omega^2)/gH]^{1/2}.$$

(5.66)

Bearing in mind the second relationship in (5.63) we find from (5.66) that $\omega = \xi_0 \sqrt{gH}$, which coincides with (5.49) when $\xi H \to 0$.

It is clear that the wave found in these approximations is a surface wave in shallow water, which propagates at a velocity $\sqrt{gH}$, i.e., the stratification of the liquid does not affect the character of the wave.

We have already mentioned that when $N = $ const the density in an incompressible liquid is a function of depth, i.e., $\rho_0(z) = \rho_{00} \exp(-2\nu z)$. Here $\nu = N^2/2g$. Because $\nu H \ll 1$ (a typical value of $N[c^{-1}]$ for the ocean varies from 0 to 0.001 [5]), the value of $(k_{z0}H)^{-2} = g/H(N^2 - \omega^2)$ when $\omega < N$ will be of the same order of magnitude as $(\nu H)^{-1}$, which is much greater than unity. Rewriting (5.65) in the form $(\cot k_z H)/k_z H = g/H(N^2 - \omega^2) \sim (\nu H)^{-1} \gg 1$, we find that roots of the dispersion equation are fairly close to

$$k_z H \approx n\pi \quad (n = \pm 1, \pm 2, \ldots),$$

(5.67)

or taking into account the second relationship in (5.63) we have

$$\omega = N[1 + (n\pi/\xi_n H)^2]^{-1/2} . \qquad (5.68)$$

This dispersion law for internal waves is typical for a multimode waveguide (Fig. 5.3).

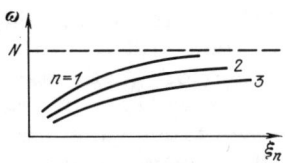

Fig. 5.3. Dispersion law for internal waves in a multimode waveguide.

When N depends on z, more complicated dispersion laws are possible [24]. It is pointed out in [3] that the solutions to (5.60) with boundary conditions (5.40) for $N = N(z)$ describe waves, one of which is close to a surface wave because the maximum $V(z)$ is obtained at $z = 0$, and moreover there is set of internal waves whose maxima are located in the interval $0 < z < H$.

We shall now look briefly at gyroscopic (inertial) waves, whose dispersion law can be obtained from (5.32) and (5.33) for homogeneous ($N = 0$) incompressible ($c \to \infty$) liquids. These waves are typical in the ocean, and they are due to the rotation of the Earth.

After simple but cumbersome transformation we obtain the following for solutions of the form $V(z)$ and $P(z) \sim \exp(\pm ik_z z)$ [3]:

$$\omega^2 = 4\Omega(k/k) \quad \text{or} \quad \omega = 2\mu\Omega \cos\theta \quad (\mu = \pm 1) , \qquad (5.69)$$

where θ is the angle between Ω and k; the value of μ is chosen from the condition $\mu \cos\theta > 0$. It follows from (5.69) that the wavelength may be arbitrary, as was the case for internal waves, because for a given frequency the angle θ is completely defined.

If $N = $ const and $\Omega \neq 0$, then we have gravity gyroscopic waves, the dispersion law for which, as is shown in [3], has the form $\omega^2 = N^2 \sin\kappa + 4\Omega^2 \cos^2(k\Omega)$, where κ is the angle between k and the z-direction.

The dispersion equation for inertial waves propagating in shallow water in a waveguide with the $k_y \Omega_y$ term neglected (this is possible if $k_y \ll k_z$, i.e., if the dimensions of the changes in the z-direction are much smaller than the wavelength in the y-direction) is $\omega^2 = \xi^2 gH + 4\Omega_z^2$. When $\Omega_z \to 0$ we obtain long gravity waves ($\omega = \xi\sqrt{gH}$). Hence the rotation of the Earth leads to the appearance of dispersion in long gravity waves.

Rossby waves may be investigated using the same general equations (5.20)-(5.23), but when $d\Omega/dy = \beta = $ const (the β-plane approximation, see [22], vol. 1, p. 35). Before considering the properties of these waves, we note that they are very important in the study of meteorological oceanic vortices [3,4]. These vortices are similar to cyclones and anticyclones in the atmosphere (hence the term meteorological). Understanding their dynamics and their relationships to processes occurring in the ocean and atmosphere is very important in order to construct a correct mathematical model of circulation in the atmosphere, and hence provide correct, though relatively short range, predictions of the weather.

Linear models of the propagation of Rossby waves turn out to be very useful in describing the median drift of meteorological vortices [4].

The traditional approximation to obtain Rossby waves is to assume that $k_z \gg k_y$. This also allows us to discard terms in equations containing horizontal components of the Ω vector, i.e., terms containing s. The main condition for the existence of these waves is for that to be a change in the vertical component of Ω_z with latitude ϕ, i.e., a change of the horizontal component of the Coriolis force with latitude. In order to account for this we expand $q = (2/\omega)\Omega \sin\phi$ in series in powers of (y/a) about a point $\phi = \phi_0 = \phi_{x=0,y=0}$ and truncate to the first two terms of the expansion. It is clear that

$$q = q(\phi_0) + \frac{dq}{d\phi}\Big|_{\phi=\phi_0} \Delta\phi$$

$$= (2\Omega/\omega)\sin\phi_0 + (2\Omega/\omega)\cos\phi_0 \cdot \Delta\phi$$

$$= y/a,$$

where a is the radius of the Earth. Finally

$$q = (2\Omega/\omega)\sin\phi_0 + \beta y/\omega, \qquad (5.70)$$

$$\beta = (2\Omega/a)\cos\phi_0. \qquad (5.71)$$

WAVES WITH SMALL AMPLITUDES

Allowing for the βy term in (5.70) is said to be allowing for the β-effect. If it is also assumed that $c \to \infty$, then we can rewrite system (5.20)-(5.23) as

$$V_x + iqV_y = \frac{i}{\omega}\frac{\partial V_z}{\partial x}, \quad V_y - iqV_x = \frac{i}{\omega}\frac{\partial V_z}{\partial y},$$

$$\frac{\partial \mathcal{P}(z)}{\partial z} + \rho_{00}(\omega^2 - N^2)V(z) = 0,$$

$$-\rho_{00}\frac{1}{\mathcal{P}(z)}\frac{\partial V(z)}{\partial z} = \frac{i}{\omega}\left(\frac{\partial V_x}{\partial x} + \frac{\partial V_y}{\partial y}\right)\frac{1}{V_z}.$$

We have already pointed out that the variables can be divided in such a system (see section 5.2). The right-hand side of the last equation may only depend on x and the left-hand side may only depend on z. By introducing a division parameter ε and equating both sides of the last equation of the system to this parameter we finally obtain

$$V_x + iqV_y = \frac{i}{\omega}\frac{\partial V_z}{\partial x},$$

$$V_y - iqV_x = \frac{i}{\omega}\frac{\partial V_z}{\partial y}, \quad \frac{\partial V_x}{\partial x} + \frac{\partial V_y}{\partial y} = -i\omega\varepsilon V_z;$$

(5.72)

$$\frac{\partial V(z)}{\partial z} + \frac{\varepsilon}{\rho_{00}}\mathcal{P}(z) = 0,$$

$$\frac{\partial \mathcal{P}(z)}{\partial z} + \rho_{00}(\omega^2 - N^2)V(z) = 0.$$

(5.73)

We differentiate the first equation in (5.72) with respect to y and the second with respect to x. We take the first from the second and set $dq/dy = \beta/\omega$. By using the third equation from (5.72), we find

$$\frac{\partial V_x}{\partial y} - \frac{\partial V_y}{\partial x} + \omega\varepsilon qV_z + i\frac{\beta}{\omega}V_y = 0.$$

(5.74)

By differentiating the third equation from (5.72) with

respect to y and using the second equation we have

$$\partial^2 V_y/\partial y^2 + \omega^2 \varepsilon V_y = -\partial^2 V_x/\partial x\, \partial y + i q \omega^2 \varepsilon V_x \ . \tag{5.75}$$

The expression for $\partial^2 V_x/\partial x\, \partial y$ is easy to find taking the derivative with respect to y from (5.74). By substituting this relation for $\partial^2 V_x/\partial x\, \partial y$ into (5.75) and using the first equation from (5.72), it is possible to exclude V_x and V_z from (5.72).

The equation for V_y has the form

$$\left\{ \frac{\partial^2}{\partial x^2} + \frac{\partial^2}{\partial y^2} - i\frac{\beta}{\omega}\frac{\partial}{\partial x} + \omega^2 \varepsilon (1 - q^2) \right\} V_y = 0 \ . \tag{5.76}$$

Let us assume that $\omega \ll \Omega$ and $q^2 \gg 1$, i.e., $1 - q^2 \approx -q^2$. If this solution to (5.76) has the form of flat waves $V_y = V_{yo}\exp[-i(k_x x + k_y y)]$, then the dispersion equation will be

$$(k_x + \beta/2\omega)^2 + k_y^2 = (\beta/2\omega)^2 - 4\varepsilon\Omega^2 \sin^2\phi_0 \ . \tag{5.77}$$

The division parameter ε is found as the eigenvalue to (5.73) with the boundary conditions (5.27) and (5.28). These equations and the conditions can easily be rewritten as

$$\partial^2 V(z)/\partial z^2 - \varepsilon\omega^2(1 - N^2/\omega^2)V(z) = 0 \ , \tag{5.78}$$

$$V(z)|_{z=-H} = \{gV(z) - \varepsilon^{-1}\partial V(z)/\partial z\}_{z=0} = 0 \ . \tag{5.79}$$

If we set $\varepsilon = \xi^2/\omega^2$, then (5.78) coincides with (5.60) for internal waves, while (5.79) coincides with the boundary conditions (5.40). We obtain the following relation for the mode $n = 0$, which is called the barotropic mode, from the corresponding relations for waveguide waves (viz., the second equation in (5.63) and (5.66) and (5.67)):

$$\varepsilon_0 = (gH)^{-1} \ ; \tag{5.80}$$

For modes with higher orders of n, which are called baroclinic modes, we obtain

$$\varepsilon_n = [(n\pi)^2/H^2](N^2 - \omega^2)^{-1} \quad (n = \pm 1, \pm 2, \ldots) \ . \tag{5.81}$$

We shall now analyze in more detail the dispersion equation

of (5.77). In order for k_x and k_y to be real the right-hand side of (5.77) must be positive, i.e., the condition $\beta/2\omega > 2\sqrt{\varepsilon}\ \Omega \sin \phi_0$ must hold. This condition can be rewritten using definition (5.71) as

$$\tan \phi_0 < [2a\omega\varepsilon^{1/2}]^{-1} . \qquad (5.82)$$

If the latitude of the position ϕ_0 is given, then Rossby waves exist for frequencies $\omega < \omega_{cr}$, where the critical frequency is determined by the formula $\omega_{cr} = [2a\varepsilon^{1/2}\tan \phi_0]^{-1}$ (for instance, for barotropic modes it follows from (5.80) that $\omega_{cr} = (gH)^{1/2}[2a \tan \phi_0]^{-1}$). When ω and ϕ_0 are far from the critical values, it is possible to neglect the last terms in the dispersion equation of (5.77). It is assumed that $\omega > 0$. In this case the dispersion law for Rossby waves will have the form

$$\omega = - k_x\beta/(k_x^2 + k_y^2) . \qquad (5.83)$$

It can be seen from this equation, that it is only satisfied when $k_x < 0$ (as is the case for equation (5.77)). This means that Rossby waves only propagate from East to West. This has been observed by meteorologists where the average oceanic flow is weak [4].

5.5. Waves in a Superfluid Liquid

Our topic in this section is the fluid mechanics of Helium II without the effect of dissipation.

Helium, which becomes liquid at 4.2 K, does not freeze at atmospheric pressure even if the temperature is taken down to absolute zero. However, at the temperature $T \approx 2,19$ K

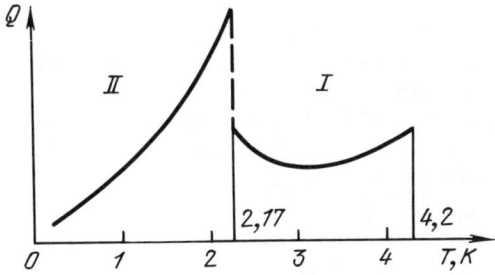

Fig. 5.4. Heat capacity of He as a function of temperature at atmospheric pressure.

there is phase transition and the helium then becomes a new liquid, which possesses properties that are quite different from those it had earlier. For instance, close to this temperature (the λ-point) the heat capacity behaves anomalously as a function of temperature. As can be seen in Fig. 5.4., the helium is in one phase from 4.2 K to 2.17 K, in which it behaves as a normal liquid (He I). At temperatures below the λ-point, the helium exists in another phase called He II and it has a series of remarkable properties.

In 1938 L. Kapitsa discovered the phenomenon of He II superfluidity, which means that He II flows along narrow capillaries (the diameter of which is $\sim 10^{-4}$ cm) as if there were no viscosity η, i.e., below the λ-point $\eta_{He II} < 10^{-12}$ Pa·s, whereas $\eta_{He I} \sim 10^{-6}$ Pa·s. The results of Kapitsa's experiment seem to indicate that He II would behaved as an ideal classical liquid, the behavior of which is described by Euler's equations. However, it was established in a series of experiments that this is not the case. For example, measuring the viscosity by studying the torsional oscillations of a disc immersed in the liquid yielded values for $\eta_{He II}$ that are not very different from $\eta_{He I}$. Superficially, there seemed to be a paradox. In one experiment He II behaved like a superfluid liquid without viscosity while in another experiment it behaved like a normal liquid with a finite viscosity, even though usually

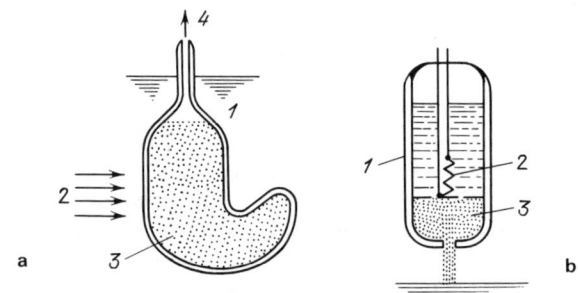

Fig. 5.5. Situations demonstrating the unusual dynamic properties of liquid helium:
a) The fountain effect. A tube filled with emery powder (3) and immersed in a helium bath (1) is irradiated with light (2) ejects a fountain (4) of liquid helium from its upper end.
b) The mechanical caloric effect. When a fast stream of liquid helium flows from a vessel (1), the temperature within the vessel is increased (in the reverse process, the temperature inside the vessel is decreased); (2) is a thermocouple and (3) is the compressed powder.

both tests yield the same result. Moreover, several of the dynamic properties of He II could not be described by Euler's equation even though it was certain that internal friction could be neglected (the fountain effect and the mechanical caloric effect, see Fig. 5.5).

In 1941, Kapitsa performed an experiment (Fig. 5.6) in which a small vessel with a heater and thermocouple were immersed in helium II. The vessel was partially filled with He II and connected to a large volume via a narrow capillary. When the heater was turned on there was a flow from inside the vessel to the outside and a small fountain of helium emerged from the capillary. The fountain was detected by the deviation of the wand from a torsion balance. However the level of liquid in the small vessel did not change. It could only be assumed that there was a countervailing current into the small vessel which did not affect the wand of the torsional balance.

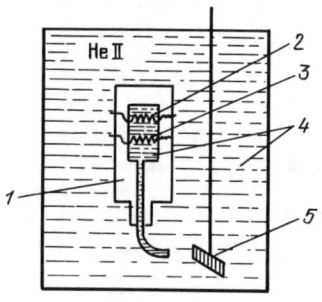
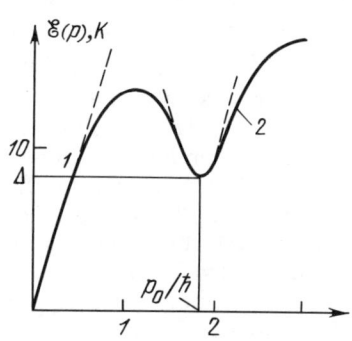

Fig. 5.6. Kapitsa's experiment: 1) vacuum; 2) heater; 3) thermocouple; 4) liquid helium (He II); 5) wand of the torsional balance.

Fig. 5.7. Dispersion of elementary perturbations in liquid helium: 1) the section corresponding to the perturbation of phonons; 2) the section corresponding to the perturbation of rotons.

Kapitsa's experiment, when combined with the available experimental results, led to the two-liquid model for He II. In this model, He II is considered to be a combination of two components, one of which is superfluid with a density ρ_s and thus not subject to viscous forces and the other a normal liquid with a density ρ_n analogous to He I. In such a liquid (see [1], Ch. XVII, [9], Sect. 10 [19]) the density is $\rho = \rho_n + \rho_s$, such that as $T \to 0$, $\rho_n \to 0$ and all the liquid turns into He II. Conversely, when the temperature rises above the λ-point $\rho_s \to 0$, and all of the liquid is He I. Moreover, it is assumed that the superfluid and normal components move freely and without friction relative to one another. The crucial aspect of this model is that the motion of He II is characterized by two vectors of velocity, i.e., v_n, which is the velocity vector of the normal component and v_s, which is the velocity vector of the superfluid component. These concepts are sufficient to explain the results of the above experiments.

We begin with the viscosity paradox. In the experiments with torsional oscillations of a disc, the oscillations are stopped due to friction with the normal helium (hence $\eta_{He} \sim \eta_{HeI} \sim 10^{-6}$ Pa·s), the superfluid and normal components are not divided. In the capillary experiment only the superfluid component flows.

The mechanical caloric effect (Fig. 5.5b) arises because superfluid flow is not associated with heat transport. In a main, the superfluid component flows out of the vessel and there is no heat flow. Hence inside the vessel the temperature of the remaining liquid rises. In the experiment illustrated in Fig. 5.5a, the density ρ_n rises as the temperature is increased by the heating. This leads to a movement of the superfluid component which creates a thermostatic pressure due to the counterflow of superfluid component at the site of heating. As a result a fountain of helium emerges from the end of the tube. Finally, the wand of the torsional balance in the last of Kapitsa's experiment (Fig. 5.6) is affected by the normal component, and the superfluid component comprises the back flow.

E.L. Andronikashvili measured the densities ρ_s and ρ_n of the components using a stack of rotating metal discs in a vessel of liquid helium and suspended by an elastic thread. The idea behind this elegant experiment was that the normal component, which has viscosity, must be caused to move by the rotating discs and the system will possess more inertia the greater the mass of liquid. At the same time the superfluid component, which may not move because it has no viscosity, must therefore have a moment of inertia that is the same as that for the empty vessel [[25].

The most important aspect of the theory of superfluid liquids is that it predicts the propagation in liquid helium of waves called secondary sound (in 1945 V.P. Peshkov confirmed this experimentally)

In the linear approximation, the equations describing the propagation of sound in a superfluid liquid have the form [1]

$$\partial \rho/\partial t + \mathrm{div}\, \mathbf{j} = 0, \quad \mathbf{j} = \rho_n \mathbf{v}_n + \rho_s \mathbf{v}_s,$$

$$\partial(\rho S)/\partial t + \rho S\, \mathrm{div}\, \mathbf{v}_n = 0, \quad \partial \mathbf{j}/\partial t = -\nabla p, \quad \partial \mathbf{v}_s/\partial t = \nabla \mu.$$

Here μ is the chemical potential, which in the linear approximation satisfies $d\mu = -SdT + (1/\rho)dp$. Note that appearance of gradients in the equations of motion for the superfluid component reflects the potential of motion. By simply transforming the initial system it is possible to rewrite it in the form of two equations:

$$\partial^2\rho/\partial t^2 = \nabla^2 p, \quad \partial^2 S/\partial t^2 = (\rho_s S^2/\rho_n)\nabla^2 T.$$

At low temperatures these equations describe sound waves [9]. In fact, at low temperature it is possible to assume that the compressibility is caused by elastic forces between the molecules, i.e., the density ρ depends mainly on the pressure p and $\partial^2\rho/\partial t^2 = (\partial p/\partial \rho)_s^{-1} \partial^2 p/\partial t^2$. However we then get the equation

$$\partial^2 p/dt^2 - (\partial p/\partial \rho)_s \nabla^2 p = 0.$$

which describes sound waves in a normal liquid. For instance, by assuming that the solution has the form of a plane wave $p \sim \exp[i(\omega t - kx)]$ we find the dispersion law (5.12)[1]. In this case $v_n \approx v_s$ [1], i.e., both components in such a wave oscillate as a whole, and both masses move together at a velocity $\bar{v} = (\rho_n/\rho)v_n + (\rho_s/\rho)v_s$, which is the velocity of the center of mass. On the other hand, if it is assumed that at low temperatures the entropy S per unit mass is a significant function of temperature, then we set $S \approx S(T)$ and substitute $\partial^2 S/\partial t^2$ for $(\partial S/\partial T)_\rho \partial^2 T/\partial t^2$. This leads us to an equation for the waves called second sound:

$$\partial^2 T'/\partial t^2 - (S^2 \rho_s/\rho_n)(T/c_v)\nabla^2 T' = 0.$$

By setting the solution in a form of travelling waves, we find the dispersion law for second sound: $\omega^2 = S^2 \rho_s T(\rho_n c_v) k^2$, i.e., the velocity of the second sound is $v_{ph} = (S^2 \rho_s T/\rho_n c_v)^{1/2}$, where c_v is the specific heat capacity at constant volume. In such a wave $j \approx 0$ (the oscillation takes place at constant volume or pressure, moreover $c_v \sim c_\rho$), but then $v_n \approx (\rho_s/\rho_n), v_s$, i.e., the superfluid and

---

[1] Here and below F' is a small perturbation to the equilibrium values F of the variable F. The subscript zero will be omitted.

normal component oscillate out of phase. Hence there is no summed flow of material because the velocity ($\bar{v}$) of the center of mass of both components is zero (at the same time there is a relative motion between the superfluid and normal components). Recall, however, that the superfluid component does not transport heat. It is therefore understandable that second sound waves will be associated with temperature but not density oscillations (in this sense it is indicative that the variable in the wave equation for second sound is T'). Helium II is unique in that temperature oscillations can occur in it, i.e., reversible temperature perturbations as opposed to the irreversible propagation of temperature disturbances due to thermal conductivity which occurs in other substances. It should be noted that the two components of the liquid helium undergo compression and rarefaction separately. The compressions and rarefactions of the superfluid component, which as we have said, do not transfer entropy, accompany reversible increases and decreases in the temperature. The force opposing these changes, i.e., the restoring force, is associated with the gradient of the chemical potential (which arises because of changes in temperature without a change in pressure). It follows from the equation of motion for the superfluid component $\partial \mathbf{v}_s/\partial t = -\nabla\mu$ that the gradient of the chemical potential induces an acceleration of the superfluid component, which opposes the compressions and rarefactions.

To conclude, we note that the macroscopic two-liquid model, which is a classical model, cannot fully describe helium, which is a quantum liquid, i.e., a macroscopic material, with behavior that obeys quantum laws [11].

According to classical physics, ions in crystals at low temperatures (the simplest models are considered in Chapter 4) oscillate around their equilibrium positions (at $T = 0$ they are completely still), this being the reason for the regularity of solids. However, helium remains liquid at such low temperature (0-2K) that the wavelengths of the De Broglie waves, which define the thermal motion of the atoms in the liquid, are of the same order of magnitude as the distance between the atoms, i.e., only quantum effects are significant. Helium is not therefore have to solidify (remember that a quantum-mechanical oscillator, even in the ground state has an energy $\mathcal{E} = \hbar\omega/2$ and has null oscillations; see Chapter 1). This behavior of helium arises because the atoms only weakly interact and the energy of the null vibrations is relatively large. Crucial to the theory of quantum liquids is the concept of elementary perturbations or quasi particles (Chapter 4).

In 1947 L.D. Landau analyzed experimental data and suggested a dispersion law for quasi particles (the energy $\mathcal{E}$ as a function of momentum p), which is illustrated in Fig.

5.7. The initial straight section of the curve corresponds to quanta of sound, or phonons. At larger $p/\hbar$ the function $\mathcal{E}(p)$ reaches a maximum, after which the function declines and at a certain $p = p_0$ goes through a minimum, where $\mathcal{E}(p) = \Delta(p_0)$. Using small second order terms, the curve can be approximated around $p_0/\hbar$ by the function $\mathcal{E}(p) = \Delta + (p - p_0)^2/m^*$. The quasi particles corresponding to this region of momenta and energies $\mathcal{E} = \Delta + (p - p_0)^2/m^*$ are called rotons (m^* is the effective mass of the roton, and Δ is its minimum energy). At thermal equilibrium, disturbed phonons and rotons define the thermodynamic behavior of liquid helium because these quasi particles have energies closed to the minimum of the function $\mathcal{E} = \mathcal{E}(p)$ (phonons have energies close to $\mathcal{E} = 0$, and rotons have energies close to $\mathcal{E} = \Delta$). Both types of quasiparticles therefore describe different sections of the $\mathcal{E} = \mathcal{E}(p)$ curve, between which there is a continuous transition, i.e., both phonons and rotons belong to one physical object, viz., the quantum liquid He II. It was this energy spectrum which allowed L.D. Landau to explain the phenomenon of superfluidity ([9], [11]).

If a liquid flows in capillaries at a velocity v at $T = 0$ K, then viscosity is marked by a loss of kinetic energy of the liquid, and hence by a diminution of the liquid velocity. In a frame moving with the liquid, the helium is stationary and the capillary moves with velocity v. When there is viscosity, the helium in this system of coordinates must move, the movement beginning when elementary disturbances first appear. Suppose one quasiparticle with energy $\mathcal{E}(p)$ and momentum $\mathbf{p}$ has arisen. This means that the energy of the liquid E becomes equal to $\mathcal{E}(p)$ in the moving frame (in which the helium is at rest), while the momentum is $\mathbf{p}_0 = \mathbf{p}$. In the stationary frame (in which the capillary is at rest), mechanics yields the following for the relationship between energy and momentum

$$E = E_0 + \mathbf{p}_0 \mathbf{v} + Mv^2/2 , \quad \mathbf{p} = \mathbf{p}_0 + M\mathbf{v} ,$$

or

$$E = \mathcal{E}(p) + \mathbf{p}\mathbf{v} + Mv^2 + Mv^2/2 , \quad \mathbf{p} = \mathbf{p}_0 + m\mathbf{v} ,$$

where M is the mass of liquid, $Mv^2/2$ is the kinetic energy of the liquid before the disturbance. The change in energy due to the appearance of the quasiparticle $\mathcal{E}(p) + \mathbf{p}\mathbf{v}$ must be negative (the energy of the moving liquid must be smaller than the energy of the liquid at rest), i.e.,

$$\mathcal{E}(p) + \mathbf{p}\mathbf{v} < 0 . \tag{5.84}$$

where $T = 0$ K the quasiparticle appearing in the liquid is a phonon, i.e., $\mathcal{E} = v_{snd} p$ (Fig. 5.7), and (5.84) takes the form

$$v_{snd} p + \mathbf{pv} < 0.$$

Since $v_{snd} p$ may only be positive, the latter inequality is not satisfied when $v < v_{snd}$, i.e., the appearance of a phonon would be in violation of energy conservation. However if the liquid does not slow down, the helium flows through the capillary without drag, which leads to the appearance of superfluidity. (Decide for yourself what form the $\mathcal{E} = (p)$ curve would be if (5.84) was satisfied, for example for antiparallel **v** and **p**.) If $T \neq 0$ K, then there would already be a disturbance in the liquid. However the argument given above would remain valid and it would be necessary to consider the disturbance of another phonon. As in the preceding case, the appearance of a new phonon would violate the conservation law. However the existing quasiparticles will exchange energy with the walls of the capillary. They are what make up the normal viscosity component in the two-liquid model. The superfluid component - the other motion in a quantum liquid - does not have viscosity. A detailed analysis for the case $T \neq 0$ K is presented in section 23 [11]. We note that superfluidity is, in some sense, analogous to superconductivity: a charged electron liquid in a superconductor flows through the crystal lattice without friction (it does not exchange energy with the lattice and hence does not suffer drag).

5.6 Waves in a Plasma. Hydrodynamic Description

Main equations. A collection of freely moving, differently charged particles (an ionized gas) is called a plasma [12] if the Debye radius is small in comparison to the dimensions of the volume containing the gas.

Let us reconsider the physical meaning of the Debye radius. A plasma may be considered to be a mixture of three components - free electrons, positive ions, and neutral atoms or molecules. The quasineutrality of a plasma, i.e., the approximate equality of the densities of the electrons, is determined by electrostatic forces, which bind the negative and positive ions in the plasma. When a group of electrons is moved with respect to the ions, i.e., when the charges are divided, an electrostatic field is induced which seeks to re-establish the quasineutrality.

Suppose charges of one sign remain in a volume after a disturbance, this corresponding to a complete separation of charge. If the volumetric density of the charge $\rho = ne$ (n is the particle concentration and e is the particle charge), then the field in the volume satisfies the equation

WAVES WITH SMALL AMPLITUDES

$\operatorname{div} \mathbf{E} = 4\pi\rho$. Clearly we have $x \operatorname{div} \mathbf{E} \sim E/x \sim 4\pi n e$ and $E \sim 4\pi n e x$ for a region with linear dimensions, which corresponds to a change in the plasma potential in the area where the charges are divided by a value $V \sim Ex \sim 4\pi n e x^2$. If the potential difference V is large, then the charges are not separated: a strong field will expel particles with the same sign from the volume where the quasineutrality is disturbed and will draw in oppositely charged particles. If, however, a region occurs in the plasma that is so small that the field generated by an excess of one sort of particle is too weak to affect the particle motion significantly, then for given concentration and plasma temperature it is possible that the plasma's quasineutrality will be disturbed for $x < r_D$ (r_D is the characteristic linear dimension). We shall estimate r_D.

If in a region with linear dimensions of the order of r_D a complete charge separation occurs, then the potential energy of the charged particles will be of the order of the thermal energy of the particles, i.e.,

$$W_P = eV \sim 4\pi n e^2 r_D^2 \sim k_B T$$

(k_B is Boltzmann's constant, and T is the plasma temperature, which is taken at present to be the same for the electron and ion components). Hence,

$$r_D \sim [k_B T/4\pi n e^2]^{1/2} \sim (T/n)^{1/2} . \qquad (5.85)$$

This is called the screening radius. When such a point charge occurs in a plasma a strong electrostatic field, which is confined within a sphere of radius r_D (the Debye radius or length) is induced. Thus the Debye radius is a characteristic three-dimensional scale for a region in which the plasma's neutrality is disturbed, whereas the situation where $x \gg r_D$ is fulfilled is the one we are considering. The time t during which charge disturbance is conserved is proportional to r_D/v_e where v_e is the speed of the electrons (the fastest particles) and ia defined by $m_e v_e^2/2 \sim k_B T$ (m_e is the electron mass). Hence the characteristic time scale for plasma charge disturbance is

$$t \sim [k_B T/4\pi n e^2]^{1/2} [2k_B T/m_e]^{-1/2} \sim [m_e/4\pi n e^2]^{1/2} .$$

It is interesting that the time is independent on temperature. The frequency corresponding to this time is

$$\omega_p = [4\pi n e^2 m_e]^{1/2} \qquad (5.86)$$

and is called the plasma frequency.

Since a plasma may contain both singly and multiple charged ions, the electron and ion concentrations need not be equal. Moreover because the masses of electrons and ions are very different, the plasma generally is characterized by two temperatures – the electron temperature T_e and the ion temperature T_i. Only when average kinetic energy of the electrons and the ions are close can we talk about a plasma temperature T.

In order to describe the propagation of small amplitude waves in a plasma, it is convenient to use the two-liquid hydrodynamic model, in which the plasma is considered to be a mixture of electron and ion liquids.

The model works when the characteristic spatial scale is much larger than the mean free path and the characteristic temporal scale (the typical duration of the processes) t_p is much longer than the time τ between two collisions. As in conventional fluid mechanics it is sufficient to specify the velocity of a component $\mathbf{v}(x, y, z, t)$, its density $n(x, y, z, t)$, and the temperature $T(x, y, z, t)$ in order to give a full description.

The motion of a unit volume of the ion (subscript i) or electron (subscript e) component of the plasma obeys Newton's second law $n_{i,e} m_{i,e} \cdot d\mathbf{v}_{i,e}/dt = \sum \mathbf{F}_s$, where $\sum \mathbf{F}_s$ is the sum of the forces acting on the volume. Leaving out of our consideration the force due to gravity, let us look at these forces (first for the ion component). The main force is due to the pressure gradient and is equal to $-\nabla p_i$. As in conventional fluid dynamics there must be an equation of state relating the pressure, density, and temperature in order to close the system for the plasma. The pressure of each component of the plasma, given an isotropic distribution of charged particles, can be expressed, as for an ideal gas, by the equation of state $p_{i,e} = n_{i,e} k_B T_{i,e}$. Using this equation, we find that $(-\nabla p_i) = -\nabla n_i k_B T_i$. Because there is an electrostatic field in the plasma, the second force acting per unit volume of ions is the force due to the electrostatic field, which for singly charged ions is $-n_{i,e} e_{i,e} \nabla \phi$. The potential of the electrostatic field ϕ satisfies Poisson's equations

$$\Delta \phi = -4\pi e (n_i - n_e) , \qquad (5.87)$$

where $e_i = -e_e = e$. Because of the electrostatic field we have, in general, $\mathbf{v}_i \neq \mathbf{v}_e$, and hence there is a friction force $\mathbf{F}_{ei}$ between the components, the force being defined by

WAVES WITH SMALL AMPLITUDES   113

the momentum transferred per unit time by the electrons to the ions, with $\mathbf{F}_{ei} = -\mathbf{F}_{ie}$ [16].

Finally, if the plasma is in a magnetic field, then there is also a Lorentz force per unit volume, which is $-(1/c)e_{i,e}n_{i,e}[\mathbf{v}_{i,e}\mathbf{B}]$. Putting in the term for $\sum \mathbf{F}_i$ into the equation of motion per unit volume for ion liquid we get

$$\frac{\partial \mathbf{v}_i}{\partial t} + (\mathbf{v}_i \nabla)\mathbf{v}_i = -\frac{\nabla n_i k_B T_i}{m_i n_i} - \frac{e_i}{m_i}\nabla\phi + \frac{\mathbf{F}_{e,i}}{m_i n_i} - \frac{e_i}{cm_i}\left[\mathbf{v}_i \mathbf{B}_i\right] . \quad (5.88)$$

By analogy we obtain the following for the electron component

$$\frac{\partial \mathbf{v}_e}{\partial t} + (\mathbf{v}_e \nabla)\mathbf{v}_e = -\frac{\nabla n_e k_B T_e}{m_e n_e} - \frac{e_e}{m_e}\nabla\phi - \frac{\mathbf{F}_{e,i}}{m_e n_e} - \frac{e_e}{cm_e}\left[\mathbf{v}_e \mathbf{B}\right] . \quad (5.89)$$

Equations (5.88) and (5.89) are the Euler equations for two charged interpenetrating liquids, which interact together due to friction and via a self-consistent electrostatic field. If the plasma maintains quasineutrality and the ions are singly charged, then we have $n_i \approx n_e = n$. In this case it is possible to use a single liquid model of the fluid mechanics by adding (5.88) and (5.89). If, then, we neglect the Lorentz force, we obtain

$$m_i n\{\partial \mathbf{v}/\partial t + (\mathbf{v}\nabla)\mathbf{v}\} = -\nabla\{nk_B(T_i + T_e)\} , \quad (5.90)$$

where $\mathbf{v} = \mathbf{v}_i + (m_e/m_i)\mathbf{v}_e \approx \mathbf{v}_i$ (the terms associated with the "electrostatic friction" forces and friction due to collisions cancel each other out).

The electron and ion liquids must also obey continuity equations

$$\partial n_i/\partial t + \text{div } n_i \mathbf{v}_i = 0 , \quad (5.91)$$

$$\partial n_e/\partial t + \text{div } n_e \mathbf{v}_e = 0 . \quad (5.92)$$

We assume that ionization and recombination may be neglected.

Langmuir oscillations and waves in a plasma. We assume that all the electrons in a thin layer of a cold infinite noncolliding plasma ($T_e = T_i = 0$, $F_{ei} = F_{ie} = 0$) are gradually displaced to the right such that between planes 1 and 2 in Fig. 5.8a there are no electrons. The plasma ions will be considered to be immobile. To the right of plane 2 there will be an excess of charge and hence a restoring force $F_r = -eE_x$ will be induced because the charges are not compensated. We have already estimated E_x: if the electrons are moved by x', then $F_r \sim -4\pi n e^2 x'$. This force induces an acceleration in the electrons of $\ddot{x}' = -(4\pi n e^2/m_e)x$ and hence the motion of a group of displaced electrons can be described by the equation for harmonic oscillations with a plasma frequency of ω_p: $\ddot{x}' + \omega_p^2 x' = 0$. Such oscillations are called plasma or Langmuir oscillations in a cold noncolliding immobile plasma. We can describe them using (5.87)-(5.92). We assume that there is a zero magnetic field, that collisions may be neglected, that the ions do not oscillate and are homogeneously balanced by stationary phonons ($m_i \gg m_e$), and that the plasma is a one-dimensional stream of electrons moving at a velocity v_0 = const in the x-direction. We may also take into account the force associated with the pressure drop in the plasma, i.e., acoustic effects. Let as assume that initial disturbance has the form of a plane wave with the frequency ω and wave number k ($f' \sim \exp[i(\omega t - kx)]$). For small disturbances, the pressure in electron liquid is $p_e = p_0 + p'$, its concentration is $n = n_0 + n'$, and the velocity of the electron liquid is $v = v_0 + v'$, where all the $f' \ll f_0$ are perturbations and much less than the corresponding undisturbed variables. The pressure of the electron liquid can be given in the form $p_e(n_0 + n') = p_0 + m_e(\partial p_e/\partial \rho_0)n'$ ($\rho_0 = n_0 m_e$ is the density of the electron gas) and $\nabla p_e = m_e(\partial p_e/\partial \rho_0) \cdot (\partial n'/\partial x)$. Given these assumptions we obtain the following system from the equations for two-liquid plasma fluid mechanics (5.87), (5.88), (5.91), and (5.92):

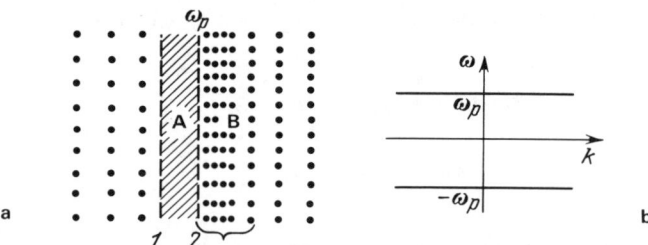

Fig. 5.8. An explanation of Langmuir plasma oscillations: a) all the electrons in a thin layer suddenly move to the right (A is the region of no electrons, and B is the region of excess electrons); b) dispersion curve $\Omega^2 = \Omega^2<Y>p<\backslash Y>$.

$$\frac{\partial v'}{\partial t} + v_0 \frac{\partial v'}{\partial x} = -\frac{1}{n_0} \frac{\partial p_e}{\partial \rho_0} \frac{\partial n'}{\partial x} + \frac{e}{m_e} \frac{\partial \phi'}{\partial x} = 0,$$

$$\frac{\partial^2 \phi'}{\partial x^2} = 4\pi e n', \quad \frac{\partial n'}{\partial t} + n_0 \frac{\partial v'}{\partial x} + v_0 \frac{\partial n'}{\partial x} = 0.$$

By substituting into these equations the values for v', n', $\phi' \sim \exp\{i(\omega t - kx)\}$ we can find the dispersion law for Langmuir waves from the condition that the resultant algebraic system is consistent

$$(\omega - kv_0)^2 = \omega_p^2 + k^2 \partial p_e / \partial \rho_0. \tag{5.93}$$

When $v_0 = 0$ this equation corresponds to the dispersion equation $\omega^2 = \omega_0^2 + k^2/LC$ for a chain of connected pendulums (Fig. 4.8).
Like (5.93) the equation was first obtained by Langmuir, who started from an analogy with acoustic waves in water (5.12). Here only the quantity $\partial p_e / \partial \rho_0$ remains unknown. In order to close the hydrodynamic equations, we must assume that the pressure of the electron liquid is isotropic and associated with the concentrations via an equation of state $p_e/n^\gamma = $ const. However $p_e = nk_B T_e$ and hence $\partial p_e / \partial \rho_0 = \gamma p_e / nm_e$ ($\rho_0 = m_e n$). It follows from kinetic theory [14] that $\gamma = 3$, i.e., $\partial p_e / \partial \rho_0 = 3k_B T_e / m_e$. The equation $p \sim n^3$ is an equation of state for a gas when there is one-dimensional adiabatic compression and it may be obtained from thermodynamics. Using these arguments we finally obtain from (5.93)

$$(\omega - kv_0)^2 = \omega_p^2 + (2k_B T_e / m_e) k^2. \tag{5.94}$$

The equation $p \sim n^3$ was obtained in [12] for a model in which the gas is in a medium with two parallel plane walls, the distance between which was slowly changing, using elegant qualitative considerations based on the conservation of the adiabatic invariant $v_\perp l = $ const, where $v_\perp$ is the component of the particle velocities perpendicular to the wall, and l the distance between the wall. Please consider this model on your own. The formula $v_\perp l = $ const is easily proved if reflections of the particles from the stationary wall are considered.

A graph of the dispersion equation for a median consisting of oscillators, corresponding to (5.94) when

$v_0 = 0$, is shown in Fig. 5.9. We shall draw in more detail on (5.93) for various special cases.

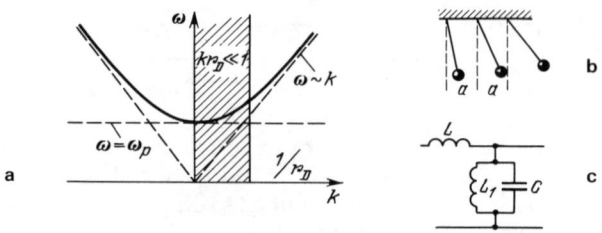

Fig. 5.9. a) Dispersion equation for a plasma that is a medium with dispersion at low frequencies. The bound on the k axes shows the limits for the validity of fluid mechanics theory $k \ll 1/r_D$. b) The mechanical and c) electrical analogs of waves in plasmas (5.94).

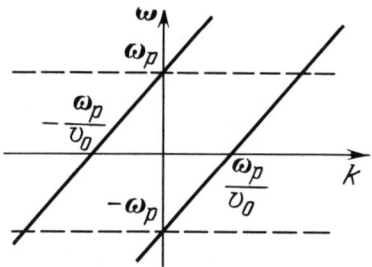

Fig. 5.10. The dispersion curves for a cold one-dimensional electron flow.

Plasma oscillations in a cold stationary plasma. The dispersion equation is obtained from (5.93) for $T_e = 0$ and $v_0 = 0$ and has the form $\omega^2 = \omega_p^2$, which we already know (Fig. 5.8b). In a cold plasma, Langmuir oscillations are not dispersed, and if the plasma is at rest they do not propagate because $v_{gr} = \partial\omega/\partial k = 0$. It should, however, be noted that the phase velocity is non-zero and equals $v_{ph} = \omega/k = \omega_p/k$ (k is the wave number of a plane wave disturbance).

Plasma oscillations in a one-dimensional cold flow $(v_0 \neq 0, T_e = 0)$. We find from (5.93) that

$$(\omega - kv_0)^2 = \omega_p^2 . \tag{5.95}$$

It can be seen that the waves of space charge, which are extensively used in VHF electronics [17], are a solution to (5.95), i.e., for slow waves with $k = \omega/v_0 - \omega_p/v_0$ and for fast waves with $k = \omega/v_0 + \omega_p/v_0$ (Fig. 5.10). Plasma oscillations in a one-dimensional cold stream are the Langmuir oscillations we have just looked at, and they transport electrons with a drift velocity of v_0. Moreover $v_{gr} = \partial\omega/\partial k = v_0$, and hence the waves of space charge are often called electrokinematic waves.

Plasma oscillations in a hot stationary plasma ($T_e \neq 0$, $v_0 = 0$.) We rewrite (5.93) for $v_0 = 0$ in the form

$$\omega^2 = \omega_p^2(1 + 3k^2 r_D^2) , \qquad (5.96)$$

where $r_D = (k_B T_e / m_e \omega_p^2)^{1/2}$ is the Debye radius (see (5.85)).

Dispersion equation (5.96) is only valid for long wavelength disturbances, when $kr_D \ll 1$ or $r_D \ll \lambda$. Electrons are moved during a period $2\pi/\omega$ a distance less than the wavelength; the compression must be adiabatic. Remember that we expanded the right-hand side of the equation of state into series and retained only one of the terms of the expansion. Hence dispersion equation (5.96) has a form analogous to the expansion about a small parameter (kr_D). Taking into account the final temperature of electrons in this approximation only yields a correction to the theory of cold plasmas. It is easy to see that $v_{gr} = 3\omega_p k r_D^2 / (1 + 3k^2 r_D^2)^{1/2}$, whence given ($kr_D \ll 1$), we have

$$v_{gr} \approx 3kr_D (k_B T_e / m_e)^{1/2} . \qquad (5.97)$$

The quantity $(k_B T_e / m_e)^{1/2}$ has the order of the thermal velocity of the electrons and thus the group velocity of the waves in a stationary hot plasma, as can be seen from (5.97), may be less than the thermal velocity. Thus, the waves transfer energy across a hot plasma, in contrast to the previous case, where the group velocity was simply equal to the drift velocity.

Using the kinetic theory, which is valid for any k, L.D. Landau noted that even when the friction forces were neglected, electron oscillations were damped (Landau damping). When $k > 1/r_D$ the damping is so large that there is no sense in drawing a dispersion curve for such values of k.

The reason for this effect is that if the velocity of the electrons is less than the phase velocity of the waves but close to it, then the electrons take energy from the waves and oscillations are damped. The more resonance particles there are, the greater the damping. If the distribution function for the plasma is rapidly and monotonically decreasing, then there will be more electrons lagging behind the wave (taking energy) than electrons overtaking the wave (giving energy). This is illustrated in Fig. 5.11.

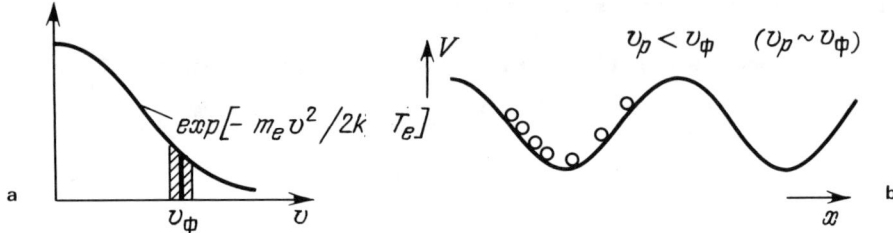

Fig. 5.11. An explanation of Landau damping: a) the shaded region corresponds to resonance electrons ($v_p \sim v_{ph}$); b) $v_p < v_{ph}$, but $v_p \sim v_{ph}$; here most of electrons are grouped on the accelerating slope of the potential hump of the plasma wave.

Ion-acoustic waves (ion sound). We will start from the equations for two-liquid fluid mechanics, assuming as before that $T_e \gg T_i$, that there is no magnetic field, and that collisions may be neglected. Moreover we shall neglect the inertia of the electrons in (5.89), i.e., the term $\partial v_e/\partial t + (v_e \nabla) v_e$. We therefore obtain the following system for a one-dimensional case from (5.87)–(5.91):

$$\frac{\partial v_i}{\partial t} = -\frac{e}{m_i}\frac{\partial \phi}{\partial x}, \quad 0 = -\frac{1}{n_e m_e}\frac{\partial}{\partial x}\left(n_e k_B T_e\right) + \frac{e}{m_e}\frac{\partial \phi}{\partial x},$$

$$\frac{\partial^2 \phi}{\partial x^2} = -4\pi e(n_i - n_e), \quad \frac{\partial n_i}{\partial t} + \frac{\partial}{\partial x}\left(n_i v_i\right) = 0.$$

Suppose the electrons have a constant temperature, i.e., T_e = const. Then it follows from the second equation in the linear approximation that $(k_B T_e/n_0)\partial n'_e/\partial x = e\,\partial \phi/\partial x$, where $n_e = n_0 + n'_e (n_0 \gg n'_e)$. Thus the first equation may be

rewritten in the form $\partial v'_i/\partial t = -(k_B T_e/m_i n_e)\partial n'_e/\partial x$. We can then find from the third equation $n'_i = n'_e - (1/4\pi e)\partial^2 \phi/\partial x^2$ or using the expression for $\partial \phi/\partial x$ that
$n'_i = n'_e - (k_B T_e/4\pi n_0 e^2)\partial^2 n'_e/\partial x^2$. Finally we can write the transformed system as:

$$\frac{\partial v'_i}{\partial t} + \frac{k_B T_e}{n_0 m_i}\frac{\partial n'_e}{\partial x} = 0, \quad \frac{\partial n'_i}{\partial t} + n_0 \frac{\partial v'_i}{\partial x} = 0, \quad (5.98)$$

$$n'_i = n'_e - \frac{k_B T_e}{4\pi n_0 e^2}\frac{\partial^2 n'_e}{\partial x^2}.$$

By comparing (4.39) for a long line, whose cell is shown in Fig. 4.7, with (5.98), it is easy to establish a direct link between them [15].

For clarity we give the quantities and parameters that are analogous between the long line and the plasma:
for a long line
$$u, \ I, \ I', \ L, \ C_1, \ C,$$
$$1/\sqrt{LC} = \omega_0, \quad 1/k_0^2 = c_1/c \sim (\Delta x)^2,$$
and for plasma
$$v_i, \ n_e, \ n_i, \ 1/n_0, \ m_i/4\pi e^2, \ m_i n_0/k_B T_e,$$
$$(4\pi n_0 e^2/m_i)^{1/2} = \omega_{pi}, \quad 1/k_0^2 = k_B T_e/4\pi n_0 e^2 = r_D^2.$$

We emphasize that size of the cell for the long line (Δx) corresponds to the Debye radius for the plasma.

Assuming that all the disturbances in the form of plane waves $\exp(i\omega t - ikx)$, we find the dispersion equation from (5.98) $\omega^2 = \omega_{pi}^2 k^2/(k_0^2 + k^2)$ (compare this with (4.33)), or

$$\omega^2 = c_s^2 k^2/(1 + k^2 r_D^2), \quad (5.99)$$

where $c_s = \sqrt{k_B T_e/m_i}$ is the velocity of ion sound. If $kr_D \ll 1$, then $\omega = c_s k$. As k increases, the frequency begin to grow more slowly than linearly, and the phase velocity of the wave begins to fall. Thus as $k \to \infty$ we have $v_{ph} \to 0$. The physical dispersion of ion sound arises because the ions oscillate against the background of electrons that are stationary on average: the pressure of the electrons compensating for the effect of the electrostatic field without displacing the electrons. Dispersion arises at high frequencies.

To illustrate the analogy with long lines (Chapter 4), we present the results of an experiment [15]. A diagram of the discharge tube used in the experiment is given in Fig. 5.12a. Standing waves are induced between the stationary grid and the anode and the probe is used to analyze the induced oscillations. Ion-sound waves with a frequency $f_{osc} \sim \sqrt{k_B T_e/m_i} \, 2/L$ were observed (L is the characteristic dimension of the plasma, for example, the tube length or distance between electrodes). The experimental results are given in Fig. 5.12b.

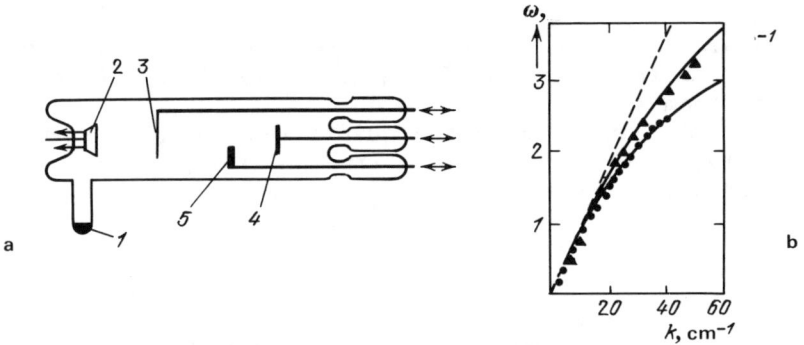

Fig. 5.12. a) Diagram of the discharge tube (the sign ⟷ denotes movable electrodes): 1) mercury; 2) cathode; 3) grid; 4) anode; 5) Langmuir probe. b) Calculated theoretical curves (solid lines) and experimental dispersion curves for two values of the discharge current: the circles are for $I_{ch} = 10$ mA; and the triangles are for $I_{ch} = 16$ mA; the dashed line is a straight line calculated using $\omega = \omega_0 k/k_0$ [15].

In our discussions of plasma, we had a mental picture of an ionized gas. In recent years solid-state plasma have been extensively investigated. For instance, a plasma of semiconductor or metal is defined as a set of stationary electrons and holes and ionized atoms associated with a crystal lattice. The collective oscillations in a solid-state plasma have much in common with the oscillations of a gas discharge plasma, which we have just considered [18-20].

CHAPTER 6

STABILITY AND INSTABILITY OF LINEAR SYSTEMS WITH
DISCRETE SPECTRA

6.1 General Notes and Definitions

The terms stability and instability are now so widely used that unless there are additional explanations it is almost always impossible to understand what is being discussed. The subject may be the stability of the system as a whole, the stability of its completely defined motion (trajectory or solution), or the stability of the equilibrium, etc. There are even different types of stability and instability. There may be stability in the large, which means with respect to arbitrary disturbances, or stability in the small, which is defined by the properties of the linearized equations. The qualifier to the word instability usually characterizes the physical mechanism by which the oscillation arises rather than the mathematical properties, i.e., dissipative instability, parametric instability, radiational instability etc.

We are concerned with the mechanism of instability and the study of the stability of motion in the small, i.e., using equations obtained by a series expansion of all the nonlinear equations about the solution we are interested in and by retaining only the linear terms (the linearization procedure we have already discussed). The most important stabilities we study are, first, the stability of the static position of the system, i.e., the equilibrium state of the linearized system with constant coefficients, and secondly, the stability of the periodic motions of the system, small deviations from which can be described by the linearized equations with periodic coefficients. We shall not, for the present, give an entirely strict definition of the stability of a linear system (and not of its solution): a dynamic system defined by a transfer coefficient $K(p)$ ($p = i\omega$) and subject to external influences Y is said to be stable if a small change in the external action leads to a small change in the motion of the system - to the coordinates X (these coordinates may be considered the output ones) (Fig. 6.1). To make this definition completely rigorous, it is necessary to define what is meant by a small disturbance, i.e., to define a distance measure between the solution being studied and the disturbed solution (mathematically speaking, a metric). The simplest method of defining a metric $d(x_{dstb}, x)$ is to take the difference between the absolute values of the coordinates: $|x_{dstb}(t) - x(t)| = d(x_{dstb}, x)$. We shall use this definition most often.

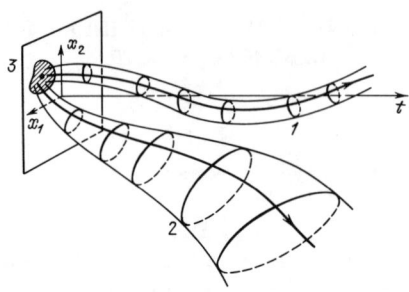

Fig. 6.1. Stability definitions: 1) the solution is stable; 2) the solution is unstable.

We now formulate various concepts of stability [1] for a system of the form

$$\dot{x}_i = f_i(x_1, x_2, \ldots, x_n) \quad (i = 1, 2, \ldots, n). \tag{6.1}$$

It is assumed that continuous derivatives $\partial f_i / \partial x_k$ $(i, k = 1, 2, \ldots, n)$ exist for (6.1) and that there is a solution

$$X_i(t) \quad (i = 1, 2, \ldots, n),$$

which when $t = t_0$ satisfies the initial conditions

$$X_i(t_0) = x_i^0 \quad (i = 1, 2, \ldots, n).$$

The solution $X_i(t)$ to (6.1) is called Lyapunov stable if for any $\varepsilon > 0$ there is a $\delta(\varepsilon) > 0$ such that for any solution $x_i(t)$ of (6.1) it follows from the inequalities

$$d_i(x_i(t_0), X_i(t_0)) = |x_i(t_0) - X_i(t)| < \delta(\varepsilon) \tag{6.2}$$
$$(i = 1, 2, \ldots, n)$$

that for all $t \geq t_0$ the following inequalities exist:

$$d_i(x_i(t), X_i(t)) = |x_i(t) - X_i(t)| < \varepsilon \qquad (6.3)$$
$$(i = 1, 2, \ldots, n)$$

In other words solutions close to the initial values remain close to them and that $t \geq t_0$. If for some very small $\delta(\varepsilon) > 0$ inequality (6.3) is not satisfied, if only for one $x_i(t)$, then the solution $X_i(t)$ is said to be unstable.

The concept of orbital stability is often used. It differs from Lyapunov stability in that only $d(\{x(t)\}, X(t)) < \varepsilon$ need follow from $d(x(t_0), X(t_0)) < \delta$, i.e., synchronicity between the disturbed and undisturbed trajectories is not required. Here $\{x(t)\}$ denotes the whole trajectory for $t > t_0$. It is only necessary that the disturbed solution (it may have a lag or lead) does not leave the bounds of the ε-neighborhood of the undisturbed solution. If as $t \to \infty$ the distance d between the disturbed and undisturbed solutions tends to zero $d \to 0$, then this stability is called asymptotic. If, however, for some $t > t_0$ we have $d \sim \exp(-\alpha t)$ ($\alpha > 0$), then the stability is called exponential.

Returning to the definition of system stability, it should be added that a system is stable in the small if its equilibrium state is Lyapunov stable; a system is stable in the large (or in a region) if its equilibrium state is stable for all finite regions, i.e., the sphere $\|x - X\| < R$.

A system is set to be absolutely stable if it has at least one equilibrium state which is asymptotically stable for all phase space; the system is globally asymptotically stable if for any trajectory it tends to some equilibrium state. Note that concepts associated with system stability are widely used in the theory of control and automatic regulation (see [2]).

We consider a very simple example of a system with one or two degrees of freedom — an oscillator with a small mass m, whose equation of motion is obtained from the equation of motion of a linear oscillator $m\ddot{x} + b\dot{x} + kx = 0$ (k and b are the elasticity and friction coefficients, respectively), if the mass is neglected. Then we have $b\dot{x} + kx = 0$ or

$$\dot{x} + ax = 0, \quad \text{where} \quad a = k/b. \qquad (6.4)$$

A solution to (6.4) satisfying the initial conditions $x(t_0) = x_0$ has the form $x = x_0 \exp[-a(t - t_0)]$. We wish to investigate the stability of the solution $X = x = 0$ using the definitions given above. If $k, b > 0$, then for $t \geq t_0$ we have $\exp[-a(t - t_0)] \leq 1$ and $|x| = |x_0|\exp[-\alpha(t - t_0)] < \varepsilon$

whence $|x_0| < \varepsilon$ for all $t \geq t_0$. The solution $X = 0$ is therefore stable for $a \geq 0$. When $a > 0$ we have $\lim_{t \to \infty} x(t) = \lim_{t \to \infty} x_0 \exp(-a(t - t_0)) = 0$ for all t_0 and x_0 and hence the solution $X = 0$ is exponentially stable.

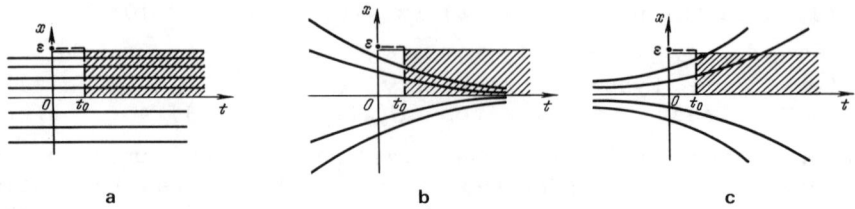

Fig. 6.2. A graphical interpretation of the stabilities and instabilities of solutions for the example of an oscillator with a small mass ($\dot{x} + ax = 0$): a) $a = 0$, the stability region is shaded; b) $a > 0$, the region of asymptotic stability in the whole is shaded; c) $a < 0$, the solutions leave the bounds of the ε-neighborhood and so is unstable.

When $a < 0$ the solution to (6.4) does not, for even very small $|x_0|$, obey the inequality $|x(t)| < \varepsilon$, and if t is large, then it tends to infinity for any $x_0 \neq 0$. Thus, the solution $X = 0$ is unstable for $a < 0$ (Fig. 6.2).

6.2 The Raus-Gurvits Criterion and Three-Dimensional Systems

For lumped systems with constant parameters, the deviation of the variables from equilibrium satisfies the equation

$$a_0 \frac{d^n x}{dt^n} + a_1 \frac{d^{n-1} x}{dt^{n-1}} + \ldots + a_{n-1} \frac{dx}{dt} + a_n x = 0 , \quad (6.5)$$

where all the a_n are real and $a_0 > 0$. It is necessary to investigate the stability of the solution to (6.5), i.e., $x = 0$. The equilibrium of the initial system is stable if as $t \to \infty$ we have $x \to 0$. We shall look for a solution to (6.5) in the form $x \sim \exp(pt)$ (p is a complex number). By substituting p into (6.5), we obtain the characteristic equation

$$\Delta(p) = a_0 p^n + a_1 p^{n-1} + \ldots + a_{n-1} p + a_n = 0 , \quad (6.6)$$

the roots of which define the character of the solution.

LINEAR SYSTEMS WITH DISCRETE SPECTRA

Table 6.1. Types of Equilibrium in a Three-Dimensional Phase Space.

Region in Parameter Space	Position of Roots in $p = p' + ip''$ Plane	Type of Equilibrium	Phase-Plane Diagram of the Equilibrium	Dimensionality of Stable and Unstable Manifolds
$p^3 + ap^2 + bp + c = 0$ $\quad \Delta = -a^2b^2 + 4b^3 + 4a^3c - 18abc + 27c^3 < 0$ (in this case the roots are real)				
$ab - c < 0$ $c < 0$ $b > 0$		unstable node		$\dim W^u = 3$ $\dim W^s = 0$
$ab - c > 0$ $c > 0$ $b > 0$		stable node		$\dim W^u = 0$ $\dim W^s = 3$
$ab - c < 0$ $c > 0 \quad b > 0$ or $c > 0 \quad b \leq 0$		saddle		$\dim W^u = 2$ $\dim W^s = 1$
$ab - c > 0$ $c < 0 \quad b > 0$ or $c < 0 \quad b \leq 0$		saddle		$\dim W^u = 1$ $\dim W^s = 2$
$p^3 + ap^2 + bp + c = 0$ $\quad \Delta = -a^2b^2 + 4b^3 + 4a^3c - 18abc + 27c^3 > 0$ (in this case there are always two complex conjugate roots)				
$ab - c < 0$ $c < 0$ $b > 0$		unstable focus		$\dim W^u = 3$ $\dim W^s = 0$
$ab - c > 0$ $c > 0$ $b > 0$		stable focus		$\dim W^u = 0$ $\dim W^s = 3$

Table 6.1. (Continued.)

Region in Parameter Space	Position of Roots in $p = p' + ip''$ Plane	Type of Equilibrium	Phase-Plane Diagram of the Equilibrium	Dimensionality of Stable and Unstable Manifolds
$ab - c < 0$ $c > 0 \quad b > 0$ or $c > 0 \quad b \leq 0$		saddle-focus		$\dim W^u = 1$ $\dim W^s = 2$
$ab - c > 0$ $c < 0 \quad b > 0$ or $c < 0 \quad b \leq 0$		saddle-focus		$\dim W^u = 2$ $\dim W^s = 1$
stable center		unstable node-focus		node-saddle

Equation (6.6) has n roots $p_m = \text{Re } p_m + i \text{Im } p_m$. Investigating the stability, therefore, reduces to a determination of where the roots lie in the complex plane p. If all the roots are located on the left-hand half-plane (to the left of the imaginary axis), then as t grows the deviation x will reduce as $\exp(-\text{Re } p_m t)$. Hence, the equilibrium will be exponentially stable. If even one root lies on the right-hand half-plane, then the equilibrium will be unstable. It is important that the positions of the roots are established without solving (6.6). The relation between the positions of the roots and the coefficients of the equation is a purely algebraic problem and there are a number of methods for assessing the real part of the roots of a characteristic equation from the coefficients of the polynomial [3,4]. Two extensively used and convenient methods are the Raus-Gurvits criterion and the D-partition method.

The Raus-Gurvits stability criterion relies on the fact that in order for all the roots of (6.6) to have negative

LINEAR SYSTEMS WITH DISCRETE SPECTRA   127

real parts (Re $p_m < 0$, i.e., all the roots are of the term $\Delta(p)$ lie to the left of the imaginary axis), it is necessary and sufficient for all the minors on the leading diagonal of the Gurvits matrix to be positive

$$D_n = \begin{pmatrix} a_1 & a_0 & 0 & 0 & \cdots & 0 \\ a_3 & a_2 & a_1 & a_0 & \cdots & 0 \\ a_5 & a_4 & a_3 & a_2 & \cdots & 0 \\ \cdots & \cdots & \cdots & \cdots & & \cdots \\ 0 & 0 & 0 & 0 & \cdots & a_n \end{pmatrix}. \quad (6.7)$$

The Gurvits matrix is structured such that the coefficients of the polynomial on the left-hand side of (6.6) lie along the leading diagonal (from a_1 to a_n) while the columns alternately contain either odd or even numbered coefficients (which include a_0). The remaining elements of the matrix (coefficients with indices less than 0 or larger than n) are zero. The minors of the leading diagonal of the Gurvits matrix take the form

$$\Delta_1 = a_1, \quad \Delta_2 = \begin{vmatrix} a_1 & a_0 \\ a_3 & a_2 \end{vmatrix}, \quad \Delta_3 = \begin{vmatrix} a_1 & a_0 & 0 \\ a_3 & a_2 & a_1 \\ a_5 & a_4 & a_3 \end{vmatrix}, \quad \ldots,$$

$$\Delta_n = \begin{vmatrix} a_1 & a_0 & 0 & 0 & \cdots & 0 \\ a_3 & a_2 & a_1 & a_0 & \cdots & 0 \\ a_5 & a_4 & a_3 & a_2 & \cdots & 0 \\ \cdots & \cdots & \cdots & \cdots & & \cdots \\ 0 & 0 & 0 & 0 & \cdots & a_n \end{vmatrix}.$$

Consequently the Raus-Gurvits stability criteria reduces to the requirement that

$$\Delta_1 > 0, \quad \Delta_2 > 0, \quad \ldots, \quad \Delta_n > 0. \quad (6.8)$$

We apply this criterion to an investigation of the roots of $p^2 + 2\gamma p + \omega_0^2 = 0$, which is the characteristic equation for a harmonic oscillator (1.1). Condition (6.8) reduces to a condition that the coefficients $\gamma > 0$ and $\omega_0^2 > 0$ are positive.

For a third order equation

$$p^3 + ap^2 + bp + c = 0 \qquad (6.9)$$

it is not sufficient that the coefficients are simply positive for the equilibrium to be stable. In fact, by writing out the Gurvits determinant, we find that the main minors are: $\Delta_1 = a$, $\Delta_2 = ab - c$, $\Delta_3 = c(ab - c)$. All the minors are positive if $ab > c$. If one of these conditions (the positiveness of the coefficient or $ab > c$) is not fulfilled, then the equilibrium is unstable. It can be seen from Table 6.1 that the character of the resulting instabilities mainly depends on the parameters.

The number of stable (unstable) roots is defined by the dimension of the stability W^s (instability W^u) manifold. This contains the trajectories near the equilibrium that are converging to it (diverging from it). If these manifolds are two-dimensional, then we see the usual stable (unstable) nodes or foci. Whether these manifolds contain nodes or foci depends on the sign of the discriminant

$$\Delta(a, b, c) = -a^2 b^2 + 4b^3 + 4a^3 c - 18abc + 27c^3 .$$

If $\Delta < 0$ then they are nodes and if $\Delta > 0$ they are foci.

6.3 The D-Partition Method

The Raus-Gurvits criterion method is not always convenient when establishing stability. For example, when n is large calculating the determinant is very cumbersome. Consequently, it is difficult to write out the stability condition in a general form. Moreover, if the system is unstable, then it is difficult to say how many roots have a positive real part, i.e., what the order of instability is. It would be better to have a criterion that is free from these drawbacks, which could be generalized for distributed systems (the left-hand side of the characteristic equation of which is not a polynomial, being in fact a quasipolynomial, i.e., a polynomial of order $\exp(\delta(p))$. In order to construct such a criterion it is useful to use the D-partition method. This is done as follows.

Let there be a parameter λ in the characteristic equation, i.e., $\Delta(\lambda, p) = 0$. We need to know how the order of stability changes as λ changes, i.e., what happens to the roots of the equation when they are moved across the p-plane. If when λ is changed the roots do not fall onto the imaginary axis, then from the point of view of stability nothing at all changes. If however one of the roots falls onto the imaginary axis, then this value of λ is critical because any farther small change in λ may lead to a change in the order of instability by one unit. We must relate changes in λ to whether or not the roots intersect the imaginary axis. Because the roots of a characteristic equation are complex, it is convenient to consider that λ is also complex. Suppose a root on the complex plane p intersects the imaginary axis. This will correspond on the λ complex plane to a transition of the parameter across some boundary separating regions with different orders of instability. By looking through all the values of p lying on the imaginary axis and comparing them to the values of λ, we construct in the λ-plane the D-partition boundary, i.e., the boundary separating the parameter plane into regions with different orders of instability.

In order to construct this boundary, we must relate the points on the p-plane with the points of the λ-plane, i.e., we must find the relation $\lambda = f(p)$ from the characteristic equation. If p changes from $-\infty$ to $+\infty$, then λ will describe a curve on the λ-plane. Moreover, it will do this in a certain direction. If we shade the region to the right of the imaginary axis, then on the curve lying in the λ-plane there must be shading to the right in the direction of motion. Thus we can see that a transition from the unshaded region to the shaded one increases the instability order by one. The p-plane is converted to the λ-plane by a conformal mapping. In order to construct this mapping we must be able to resolve the equation $\Delta(\lambda, p) = 0$ with respect to λ and it is necessary that $f(p)$ be continuous and differentiable, i.e., the function must be holomorphic.

Consider the simple example of $p^2 + p + \lambda = 0$. By resolving this equation with respect to λ, we find $\lambda = -p^2 - p$, whence when $p = i\omega$ we have $\lambda = \omega^2 - i\omega$. Consequently, Re $\lambda = \omega^2$, while Im $\lambda = -\omega$. Therefore, Re $\lambda = (\text{Im } \lambda)^2$. The boundary of the instability region is a parabola (Fig. 6.3). The stability region is inside the parabola and the region outside has an order of instability $D(\lambda) = 1$.

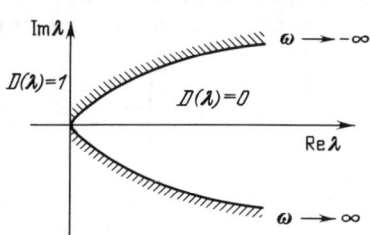

Fig. 6.3. Partitioning the λ-plane into regions with differing orders of instability for equation $p^2 + p + \lambda = 0$.

The D-partition method may be used even when the number

of roots in the characteristic equation is countable. As we saw in Chapter 4, this occurs for the spectrum of a resonator without radiation on the boundaries. If the resonator is one-dimensional, then the spectrum of wave numbers is always equidistant: $k = \pi n/l$ ($n = 1, 2, \ldots$). For a resonator with ideal mapping at the ends and $k = 2\pi n/l$ for a ring resonator. Because we have $D(\omega, k) = 0$, where k is now the actual number of modes ($k \sim n$) in the dispersion equation, it is not difficult to determine the stability boundary for a distributed system with a discrete spectrum by running through the n. A ring resonator is a simple example and we shall consider its stability with respect to wave disturbances propagating to the right. If there is no dispersion and no losses in the medium, then we can immediately obtain from the wave equation $u_t + v_o u_x = 0$ the frequency $\omega_n = v_o 2\pi n/l$. All the frequencies are real because a set of roots of the characteristic equation $D(p, n) = 0$ lies on the imaginary plane $p = i\omega$. Thus, the system is Lyapunov stable. If we have high-frequency losses, for example viscosity, then the equation of a travelling wave for the same medium takes the form $u_t + v_o u_x - \nu u_{xx} = 0$, and the characteristic equation is written in the form $\omega_n = (2\pi/l)(v_o + i\nu n 2\pi/l)n$. Now all the roots lie in the upper half-plane of the ω plane (or in the left half-plane of the p-plane), i.e., the stability is transient and becomes exponential.

We now introduce negative dissipation independent of the scale of the disturbance. For clarity, we shall consider that this non-equilibrium medium can be modeled by the chain shown in Fig. 6.4a. The equation of a travelling wave in such a medium is written as

$$u_t + v_o u_x - \nu u_{xx} - \gamma u = 0 . \qquad (6.10)$$

While searching for a solution $u(x, t) = v \exp(i\omega t - ikx)$ for the ring resonator, we obtain the characteristic equation

$$\omega_n = v_o \frac{2\pi}{l} n + i\nu \left[\left(\frac{2\pi}{l}\right)^2 n^2 - \frac{\gamma}{\nu} \right] .$$

The plane of the parameters γ/ν and l is partitioned in Fig. 6.4b into regions with different orders of the instability. Only static instability is possible in a short resonator $l < l_{sh} = 2\pi(\gamma/\nu)^{-1/2}$ because only one root Im $p = 0$ (i.e., Re $\omega = 0$) exists in the right-hand half-plane of the p-plane, this corresponding to an exponential curve in the spatially

LINEAR SYSTEMS WITH DISCRETE SPECTRA

homogeneous field. When $l > l_{sh}$ is increased, the instability order grows, although for any finite l the number of roots in the right-hand p-halfplane is always finite.

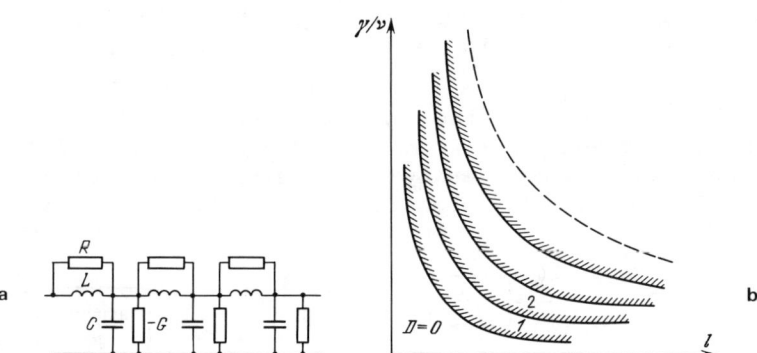

Fig. 6.4. a) Diagram of the chain corresponding to equation (6.10).
b) Partition of the plain of the parameters γ/ν, l into regions with varying orders of instability.

6.4 Stability of Non-Autonomous Systems

It is necessary to study for non-autonomous systems the stability of the motion caused by external forces. We shall do this for the example equation

$$a_0 \frac{d^n x}{dt^n} + a_1 \frac{d^{n-1} x}{dt^{n-1}} + \ldots + a_{n-1} \frac{dx}{dt} + a_n x = y(t) . \qquad (6.11)$$

Let us look at the motion with zero initial conditions

$$x(0) = dx(0)/dt = \ldots = d^{n-1}x(0)/dt^{n-1} = 0 . \qquad (6.12)$$

We take the Laplace transforms of (6.11) and (6.12) using the definition

$$F(p) = \int_0^\infty e^{-pt} f(t) dt , \qquad (6.13)$$

where $f(t)$ is the original function and $F(p)$ its transform. In transform space we have

$$X(p) = \mathcal{K}(p)Y(p) , \qquad (6.14)$$

where according to (6.13) $X(p) \leftarrow x(t)$, $Y(p) \leftarrow y(t)$,

$K(p) = 1/\Delta(p)$, $\Delta(p) = a_0 p^n + a_1 p^{n-1} + \ldots + a_n$ the ← sign denoting the relation between the transform and its inverse. Here $Y(p)$ is a vector whose components are the input variables, and $X(p)$ is a vector whose components contain the output variables. The function $K(p)$, which is the relation between the vectors, is called the transfer function. It depends on the coefficients $a_0, a_1, \ldots, a_n$, i.e., on the internal structure of the system. Equation (6.14) can be conveniently given in a form of a diagram (Fig. 6.5a). This is convenient when analyzing several linked systems (Fig. 6.5b), especially when there is feedback.

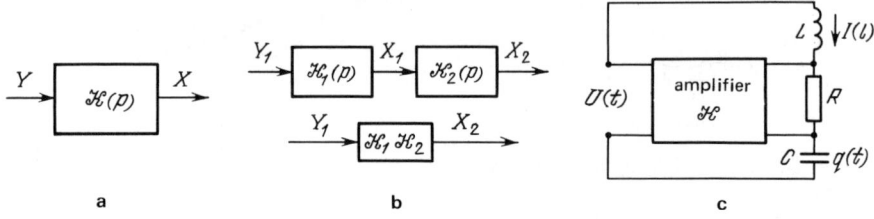

a   b   c

Fig. 6.5. a) A diagram explaining equation (6.14). b) Block diagram of two systems with the transfer coefficients $K_1(p)$ and $K_2(p)$ which are related. c) Diagram of an oscillating system with feedback $(U(t) = KRI(t))$: $K > 0$ is positive feedback; $K = 1$ is undamped oscillation; $K > 1$ is unstable; $K < 0$ is negative feedback.

Returning to the space of the original functions and using the correspondence relation between the original function and the product of the transform:

$$\int_0^t f_1(\tau) f_2(t - \tau) d\tau = f_1 * f_2 \rightarrow F_1 F_2 \,,$$

we find from (6.13) that

$$x(t) = \int_0^t k(\tau) y(t - \tau) d\tau \,, \quad \text{where } k(t) \rightarrow K(p) \,. \tag{6.15}$$

In order to find the inverse of the transform

$$K(p) = (a_0 p^n + a_1 p^{n-1} + \ldots + a_n)^{-1},$$

we break $K(p)$ into simple fractions. To do this we first

determine the zeros of the polynomial
$\Delta(p) = a_0 p^n + a_1 p^{n-1} + \ldots + a_n$. Suppose these are $\alpha_1, \alpha_2, \ldots, \alpha_n$ than it is clear that

$$\Delta(p) = a_0 (p - \alpha_1)(p - \alpha_2) \ldots (p - \alpha_n). \qquad (6.16)$$

Suppose that all the α_i are different. In this case the function is

$$\mathcal{K}(p) = \frac{1}{\Delta(p)} = \frac{A_1}{p - \alpha_1} + \frac{A_2}{p - \alpha_2} + \ldots + \frac{A_n}{p - \alpha_n} \qquad (6.17)$$

and this must have simple poles at $p = \alpha_1$, $p = \alpha_2, \ldots, p = \alpha_n$. We multiply (6.17) by $(p - \alpha_1)$ and take the limit as $p \to \alpha_1$, i.e., $\lim_{p \to \alpha_1} \dfrac{p - \alpha_1}{\Delta(p)} = A_1$. However, $A(\alpha_1) = 0$ and so

$$A_1 = \lim_{p \to \alpha_1} \frac{p - \alpha_1}{\Delta(p)} = \lim_{p \to \alpha_1} \left(\frac{\Delta(p) - \Delta(\alpha_1)}{p - \alpha_1} \right)^{-1} = \left(\frac{\partial \Delta}{\partial \alpha_1} \right)^{-1} = \frac{1}{\Delta'(\alpha_1)}.$$

We find the other coefficients in a similar way and so obtain [5]:

$$\mathcal{K}(p) = \frac{1}{\Delta(p)} = \sum_{i=1}^{n} \frac{1}{\Delta'(\alpha_i)} \frac{1}{p - \alpha_i}. \qquad (6.18)$$

Hence

$$k(t) = \sum_{i=1}^{n} \frac{\exp(\alpha_i t)}{\Delta'(\alpha_i)}. \qquad (6.19)$$

By substituting (6.19) into (6.15), we finally get

$$x(t) = \int_0^t \sum_{i=1}^{n} \frac{\exp(\alpha_i \tau)}{\Delta'(\alpha_i)} y(t - \tau) d\tau. \qquad (6.20)$$

It is easy to see that the solution $x(t)$ is bounded if all the exponents have negative real parts. Consequently, all the roots of the characteristic polynomial corresponding to an autonomous system must lie to the left of the imaginary axis. Therefore, when investigating the stability of a non-autonomous system, it is possible to use the same criteria as for an autonomous system.

6.5 Instability Mechanisms

No doubt everyone has witnessed the increased sound in a hall with a public address system, which arises when a microphone is placed too close to the loudspeaker or when there is too much amplification. The increasing sound is evidence of self- energizing in the system consisting of the microphone, amplifier and loudspeaker, i.e., its instability. The culprit of this is positive feedback. Indeed this mechanism is at the heart of most generators. Let us look in detail at an example, namely the diagram in Fig. 6.5c. The feedback in this system is proportional to the current and the amplification potential is $U(t) = \mathcal{K}\, RI(t)$, where $\mathcal{K}$ is the voltage amplification coefficient. Thus the circuit is described by the equation

$$LdI(t)/dt + RI(t) + C^{-1}q(t) = \mathcal{K}\, RI(t) \ .$$

This can easily be transformed into the equation for a discharge through a capacitor:

$$\ddot{q} + \frac{\omega_0}{Q}\dot{q} + \omega_0^2 q = \mathcal{K}\frac{\omega_0}{Q}\dot{q} \ ,$$

or (6.21)

$$\ddot{q} + \frac{\omega_0}{Q^*}\dot{q} + \omega_0^2 q = 0 \ ,$$

where $Q^* = Q/(1 - \mathcal{K})$, $Q = \omega_0 L/R$ is the circuit's Q-factor, and $\omega_0^2 = 1/LC$ is the natural frequency of the circuit.

The effect of the feedback is completely determined by a modified Q-factor Q^*. If $\mathcal{K} > 0$, then there is positive feedback: the value of Q^* is greater than Q, which is equivalent to an diminution of loss in the circuit. When $\mathcal{K} = 1$ the loss is completely compensated by the positive feedback, and when $\mathcal{K} > 1$ instability arises in the system, i.e., an oscillation exponentially grows over time. When there is negative feedback $\mathcal{K} < 0$, the loss in a circuit is increased and the oscillation is exponentially damped.

Although (6.21) was obtained for a concrete circuit, it describes any harmonic oscillator with damping and feedback proportional to velocity. When the feedback is not too large and $\mathcal{K} > 1$ the instability on the phase plane (6.21) corresponds to an unstable focus (Fig. 1.7, where the plane of the parameters $2\gamma = \omega_0/Q$, ω_0^2 is partitioned into areas with different types of equilibrium). This sort of instability occurs in oscillators for which the friction (resistance, conductivity) as a function of velocity (current, potential) has a falling section (Fig. 6.6).

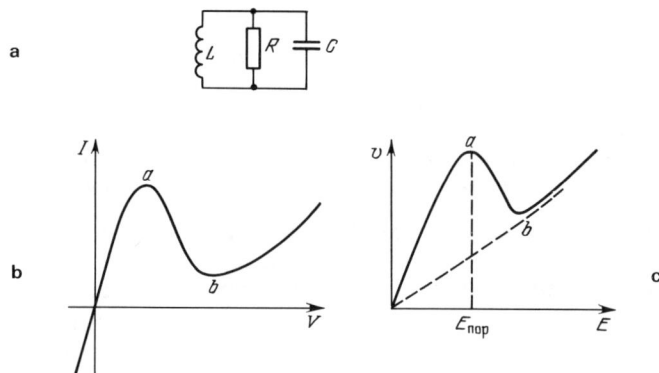

Fig. 6.6. a) Oscillating circuit with an element R which has negative feedback. b) A voltage current diagram for R when it is a tunneling diode. c) The average drift velocity of the electrons as a function of the potential of the electrostatic field for a Gann diode.

We shall look at the formation of the ab section in the characteristic $v = v(E)$ (Fig. 6.6c) or $j = j(E)$ (falling section in Fig. 6.6b) for a Gann diode [6]. The structure of the energy zones in a gallium arsenide (GaAs) n-type semiconductor is shown in Fig. 6.7. This substance is widely used for Gann diodes. The conduction zone in GaAs has one main lower valley and several upper troughs, which in the (100) direction. Electrons in the main valley have a mobility μ_1, which is much larger than the mobilities μ_2 of the electrons in the upper troughs.

When the potential of the electrostatic field in a semiconductor is slight, almost all the electrons are in the lower valley, and so the electron concentration in the upper troughs is $n_2 = 0$ and the overall concentration n is n_1 in the lower valley. The current density through the semiconductor is $j_1 en\mu_1 E$ (Fig. 6.7b). When the potential of the field increases enough for some of the electrons to obtain more energy than $\Delta \mathcal{E}$, they move to the region with lower mobility, i.e., to the upper trough. These transfers are at a threshold value of the field $E = E_{th}$. At $E = E_{max}$

all the electrons transfer to the upper troughs and the current density becomes $j_2 = en\mu_2 E$. If $\mu_1 E_{th} > \mu_2 E_{max}$, then $j_1 > j_2$, i.e., when E changes from E_{th} to E_{max} the current

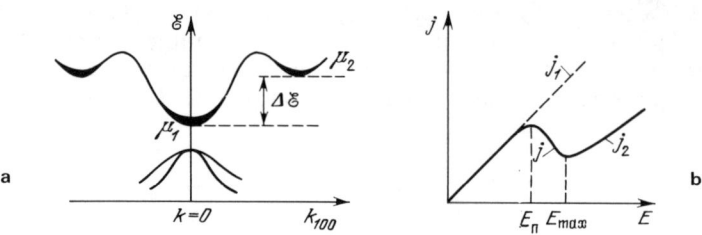

Fig. 6.7. a) Qualitative structure in the energy zones of GaAs: $\mu_1 \approx 8000$ cm^2/(V·s), n_1; $\mu_2 \approx 100-200$ cm^2/(V·s), n_2. b) An explanation of the mechanism of negative differential resistance in GaAs using the model of valley-to-valley transitions for the electrons: $j = e(\mu_1 n_1 + \mu_2 n_2)E$; $j_1 = en\mu_1 E$; $j_2 = en\mu_2 E$.

density reduces as the field increases and it is described by $j = e(\mu_1 n_1 + \mu_2 n_2)E$. Thus for changes in field strength of $E_{th} < E < E_{max}$ there is a section in the $j = j(E)$ curve where $\partial j/\partial E < 0$, i.e., a section with a negative differential conductivity (a negative differential resistance). Because $\partial j/\partial E = \partial [e(\mu_1 n_1 + \mu_2 n_2)E]/\partial E$ it follows from the condition $\partial j/\partial E < 0$ that

$$\partial n_2/\partial E > n_1/E .\qquad(6.22)$$

To derive (6.22) we used the fact that $\partial n_1/\partial E = -\partial n_2/\partial E$ and neglected terms proportional to μ_2 because $\mu_1 \gg \mu_2$. Equation (6.22) means that for the E changes where there is a negative differential resistance, small changes in E cause many electrons to transfer from the lower valley to the upper trough (a change in electron concentration).

This qualitative assessment of the formation of negative resistance is valid if the electrons transfer from the lower zone to the upper one for the whole volume of the semiconductor evenly and simultaneously. In reality, the situation is more complicated because the appearance of a section on the current-voltage diagram with negative resistance leads to instability in the current flow through the semiconductor and moving areas of stronger field, domains, are formed. There are, therefore, nonlinear effects,

LINEAR SYSTEMS WITH DISCRETE SPECTRA

which are considered in the second half of the book.

Unconservative instability in a system with one degree of freedom, examples of which are unstable foci or unstable nodes, has not to be associated with friction or viscosity. For example, in Volterra's ecological model (Chapter 1), which describes changes in the populations in the predator-prey system,

$$\dot{N}_{pred} = \gamma_{pred} N_{pred} (N_{prey} - \nu_{pred}) ,$$

$$\dot{N}_{prey} = \gamma_{prey} N_{prey} (1 - N_{pred} - \nu_{prey}) ,$$

(6.23)

where N_{pred} and N_{prey} are the numbers of predator and prey, respectively, and γ_{pred} and γ_{prey} are parameters for the reproduction rates, the parameters ν_{pred} and ν_{prey} define the rate at which predator and prey exit the sphere of influence (this may be mortality for mammals or fish, or the rate of flow in a cultivator for microorganisms such as bacteria or viruses). If we apply Mono's [7] empirical finding that the predator reproduction rate is proportional to the prey population if the prey population is small, and weakly depends on a prey population if it is large, then instead of (6.23) we can use [7] (taking $\nu_{pred} = \nu_{prey} = \nu$)

$$\dot{N}_{pred} = N_{pred} N_{prey} / (r + N_{prey}) - \nu N_{pred} ,$$

$$\dot{N}_{prey} = N_{prey} (1 - \nu_{prey}) - N_{pred} N_{prey} / (r + N_{prey})$$

(6.24)

where r is a constant. These kinetic equations have two equilibriums, with $N^0_{pred} = N^0_{prey} = 0$ and $N_{pred} = r$, $N^0_{prey} = r\nu/(1 - \nu)$. It is obvious that the second equilibrium only makes sense for $0 < \nu < 1$. Let us determine for what types of equilibrium they are valid. The linearized system for a disturbance n_{pred} and n_{prey} ($N_{pred} = N^0_{pred} + n_{pred}$, $N_{prey} = N^0_{prey} + n_{prey}$) has the form

$$\dot{n}_{pred} = \frac{\left(N^0_{pred}n_{prey} + N^0_{prey}n_{pred}\right)}{r + N^0_{prey}} - \frac{N^0_{pred}N^0_{prey}}{(r + N^0_{prey})^2} \cdot n_{prey}$$

$$- \nu n_{pred},$$

$$\dot{n}_{prey} = n_{prey}(1-\nu) - \frac{\left(N^0_{pred}n_{prey} + N^0_{prey}n_{pred}\right)}{r + N^0_{prey}}$$

$$- \frac{N^0_{pred}N^0_{prey}}{\left(r + N^0_{prey}\right)^2} \cdot n_{prey}.$$

(6.25)

Substituting into this the coordinates of the second equilibrium (the first equilibrium for this model is not of interest because as $N_{pred} \to 0$ Mono's law is invalid), we find

$$\ddot{n}_{prey} - \nu(1-\nu)\dot{n}_{prey} + \nu(1-\nu)^2 n_{prey} = 0. \quad (6.26)$$

Because $0 < \nu < 1$ the equilibrium we are interested in is always an unstable focus (see the parameter plain in Fig. 1.7.). In essence, the instability here is also associated with the action of positive feedback, i.e., as the prey population increases the rate of increase of the predator population also rises, which is reflected in Mono's law.

The mechanisms for instability that we have so far examined more or less reduce to one, namely the effect of negative friction (resistance, conductivity, viscosity etc.). In fact indeed both the existence of positive feedback and the inclusion into an *LC*-circuit of an element with a falling current-voltage characteristic lead to the appearance in the oscillator equation of the same term, i.e., *h dx/dt*, where *h* < 0 (negative losses). When the positive losses are large enough they overwhelm this instability. These are very simple mechanisms for instability and quite common, especially for radio electricians. However, they do not exhaust all of the variety of oscillational instabilities; there are oscillational instabilities for which the incorporation into the system of positive losses (friction, resistance etc.) not only does not overwhelm the instability, indeed it sometimes increases it.

We have already encountered an example of instability which is in no way related to negative dissipation. This is the unbounded secular growth of an oscillation in an oscillator without friction, that is being affected by a resonating harmonic influence.[1] When there are no such disturbances the oscillator oscillates with a fixed amplitude, and a very small disturbance causes the oscillation to grow to a very large value (up to infinity as $t \to \infty$). The mechanism of this instability is very simple. It is a periodic disturbance in phase with the oscillation of the oscillator, as the result of which there is a swing. Increasing oscillations in a Hamiltonian system (i.e., a system without dissipation) due to the resonance extraction of energy from a source is also possible when the source is not oscillational. A sufficient condition for this is the presence in the system of a few degrees of freedom, for example (modes interacting between themselves). This sort of instability is, in particular, the reason for the flapping of an aircraft's wings, which are growing longitudinal flexing oscillations.

Assume that a wing is an oscillational system with two degrees of freedom, one of which corresponds to flexing (x_1) oscillations and the other (x_2) to circular oscillations. We can write the equation for the wing in the absence of aerodynamic forces [8] as

$$I_1 \ddot{x}_1 + I_{12} \ddot{x}_2 + k_1 x_1 = 0 , \quad I_{12} \ddot{x}_1 + I_2 \ddot{x}_2 + k_2 x_2 = 0 , \qquad (6.27)$$

where I_1 and I_2 are the generalized moments of inertia for flexing and rotating, respectively, k_1 and k_2 characterize the wing's elasticity, and I_{12} is the moment of inertia, which defines the inertial link between the flexing and rotating modes, the degrees of freedom of the oscillation. The magnitude of this relation depends on the distance between the center of mass of the wing and the center of rotation (Fig. 6.8). When these centers coincide, the link is zero and (6.27) turns into a system of two independent oscillators with frequencies $n_1^2 = k_1/I_1$, $n_2^2 = k_2/I_2$. When there is a link, the normal frequencies of the longitudinal flexing vibrations of the wing are given by the characteristic

---

[1] The simplest example of instability in a system without friction is a sphere on a hill. In phase space this instability, which is associated with "negative energy" of an oscillator, corresponds to a saddle type equilibrium.

equation

$$(1 - I_{12}^2/I_1 I_2)\omega^4 - (n_1^2 + n_2^2)\omega^2 + n_1^2 n_2^2 = 0 .$$

We know that normal oscillations lie outside the partial frequencies, because the link extends the frequencies. For a wing in flight, (6.27) is supplemented by integral aerodynamic forces (for the lifting force and forces tending to turn the wing):

$$\ddot{x}_1 + n_1^2 x_1 + (I_{12}/I_1)\ddot{x}_2 - b_1 v^2 x_2 = 0 ,$$

$$(I_{12}/I_2)\ddot{x}_1 + \ddot{x}_2 + n_2^2 x_2 - b_2 v^2 x_2 = 0 ,$$

(6.28)

where v is a dimensional parameter characterizing the velocity of flight, and b_1 and b_2 are positive constants that depends on the geometry of the wing [8]. All the dissipative effects will be neglected, and the system is Hamiltonian. However the equilibrium $x_1 = x_2 = 0$ may, even so, be unstable if the aircraft's velocity is too large. The instability boundary is determined by equating the discriminant to zero

$$\left[\frac{n_1}{n_2} + \frac{n_2}{n_1} + \left(b_1 \frac{I_{12}}{I_2} - b_2\right)v^2\right]^2 - 4\left(1 - \frac{I_{12}^2}{I_1 I_2}\right)\left(1 - b_2 \frac{n_1}{n_2} v^2\right) = 0.$$

(6.29)

We leave the reader to determine the critical speed at which flapping is induced. A concrete numerical value of the parameters may, for instance, be found in [8]. The reason for the instability that leads to flapping is that in their special link between the flexing and rotating oscillations in a flow of air. The phase relations that arise between these modes are such that an oscillating wing takes energy from the oncoming air flow. The introduction of dissipation may also make this phase relations more optimal with respect to the extraction of energy from the flow and thus increase the stability.

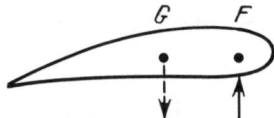

Fig. 6.8. Schematic drawing of a wing, showing the non-coincidence of the centers of mass G and rotation F.

CHAPTER 7

STABILITY OF DISTRIBUTED SYSTEMS WITH CONTINUOUS SPECTRA

7.1 General Remarks

When we investigate the stability of a bounded distributed system or resonator, the problem is made more difficult than the corresponding lumped system by the fact that the spectrum of the complex fundamental frequencies is countable. By going through all the possible spatial disturbances, i.e., all the wave numbers k_n acceptable given the boundary conditions, we can by determining all the roots of the characteristic equation $D(\omega, k_n) = 0$, completely determine the stability. Obviously there will be difficulties, but these are technical.

If the system is semibounded or infinite, then it is not clear how, in general, to define stability and further thought is necessary. In fact, if we are now to consider the stability of disturbances in the region of space that interests us, we must look at the evolution of the spatial localized perturbation, a problem with the boundary conditions:

$$u(x, t) = \frac{1}{2\pi} \sum_s \int_{-\infty}^{\infty} u_s(k, 0) \exp[i\omega_s(k)t - ikx] \, dk, \quad (7.1)$$

where $u_s(k,0)$ is the spatial spectrum of the initial disturbance, and the summation is over all normal waves. Concerning the behavior of the disturbance at a given point or localized area, it should be noted that exponential growth in time of separate k-components of the spatial spectrum does not guarantee the growth in time of the disturbance at that point or region. In fact, the disturbance may simply die away in the region or leave it completely, while growing over time. This sort of travelling or convective instability is observed, for example, in various shear flows (for instance, heated streams, Fig. 7.1), and in various electronic systems, such as a travelling wave tube or a plasma penetrated by an electron beam.

A growing disturbance that does not die away in a given region, i.e., at every point in the region where the disturbance grows, is a true (in our old understanding) instability. This instability is called absolute[1]. The formal definitions, therefore, most be such that if

$$\lim_{t \to \infty} u(x, t) \to \infty \quad (x \in (x_1, x_2)) ,\qquad(7.2)$$

where $u(x, t)$ is a disturbance (x_1, x_2 are the bounds of the region that are of interest and contain the instability), then the instability is absolute; if, however,

$$\lim_{t \to \infty} u(x, t) \to 0 \quad (x \in (x_1, x_2)) ,\qquad(7.3)$$

then the instability is convective.

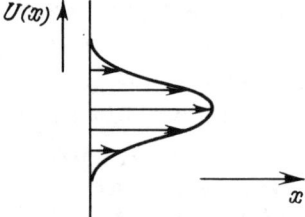

Fig. 7.1. The velocity profile in a heated jet.

It is natural that the form of the instability depends on the choice of coordinates. If we move with a disturbance that is travelling and growing in time, then in this new system of coordinates the instability is not convective but absolute. Conversely, if a system has an absolute instability and we transfer to new variables $t_{instab} = t$, $x_{instab} = x - v_0 t$, where v_0 exceeds the largest velocity at which the disturbance propagates (such a transformation is not always possible. For example, there is no reason to go to coordinate system that is moving faster than light), then the instability stops being absolute and becomes convective.

The problem of distinguishing between absolute and convective instabilities is closely related to another, perhaps, even more important problem for applications, namely the identification of amplification and filtering in

---

[1] L.D. Landau and E.M. Lifshits [18] were the first to distinguish between absolute and convective instabilities in their analysis of hydrodynamic instabilities.

semibounded systems disturbed by lumped sources. We shall clarify this problem in more detail.

Suppose at the boundary $x = 0$ of a medium described by a dispersion equation $D(\omega,k) = 0$ a signal is received. For simplicity we shall assume that it is a radio signal with a carrier frequency ω_0. We further assume that the roots of the equation $D(\omega_0,k) = 0$ are complex, and that there are roots with Im $k < 0$ and Im $k > 0$. As the signal propagates through the medium along the x-axis, it would seem that the signal should grow along x because the solution has the form

$$\exp(\text{Im } kx) \exp(i\omega_0 t) \exp(-i \text{ Re } kx) \text{ for Im } k > 0.$$

In general, this is not correct. For example, if we try to induce an oscillation with a frequency $\omega < \omega_0$ in a chain of linked pendulums, (Fig. 7.2a, the system's dispersion diagram is given in Fig. 7.2b), we do not get any amplification of the oscillation along the x-axis, indeed it is exponentially damped (Fig. 7.2c). No oscillation is induced, and above this critical frequency there is filtering even though for $\omega < \omega_0$ there is a root to the equation $D(\omega_0,k) = 0$ with Im $k > 0$. The reason for this is that the existence of roots for $D(\omega_0,k) = 0$ in the upper half-plane of the complex k-plane does not in itself mean that there is amplification. A wave corresponding to such a root may propagate to the left (Re $k < 0$) and hence it will be damped in the direction in which it propagates (Fig. 7.2c). In contrast to the analogous problem concerning the instability of a sinusoidal solution in time (for which $t > 0$ always), here both directions for the change in the variable (x) have sense.

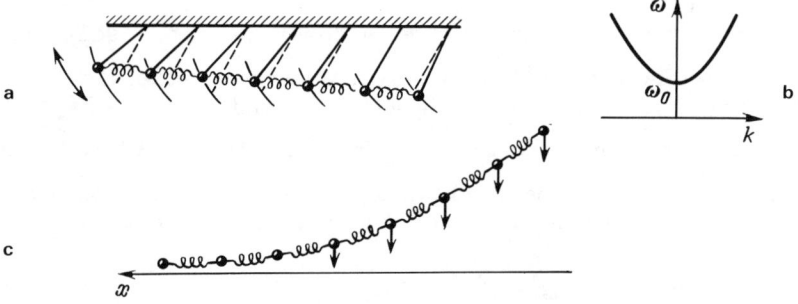

Fig. 7.2. a) Oscillation (in the plane perpendicular to the Figure) in a chain of linked pendulums. b) Dispersion diagram for this oscillating system. c) Damping of the oscillation in the direction in which it propagates [15].

In this chapter we shall consider various examples of unstable and amplifying media and comparatively simple criteria which can be used to distinguish amplification from filtering and to define which stability occurs in the system, absolute or convective.

7.2 Examples of Instability

Jean's instability. Let us consider the stability of a spatially stationary, homogeneously distributed gravitating gas using the equations of fluid mechanics [1]. By linearizing these equations against the background of the steady state solution:

$$
\begin{aligned}
&a) \quad \partial \rho/\partial t + \mathbf{v}\nabla\rho = -\rho\nabla\mathbf{v} , \\
&b) \quad \partial \mathbf{v}/\partial t + (\mathbf{v}\nabla)\mathbf{v} = -\rho^{-1}\nabla p - \nabla\Phi , \\
&c) \quad \Delta\Phi = 4\pi G \rho ,
\end{aligned}
\qquad (7.4)
$$

where ρ is the density, $\mathbf{v}$ is the velocity, p is the pressure, Φ is the potential of the gravitational field, and G is the gravitational constant, we obtain the following wave equation for one-dimensional disturbances to the density:

$$\partial^2 \rho'/\partial t^2 - c^2 \partial^2 \rho'/\partial x^2 - 4\pi G \rho_0 \rho' = 0 . \qquad (7.5)$$

Here $c^2 = dp/d\rho$ is the square of the isothermal velocity of sound. This equation describes the evolution of a disturbance against the background of a stationary solution to (7.4) in a form $\rho = \rho_0$, $\mathbf{v} = \mathbf{v}_0 = 0$, $\Phi = \Phi_0 = $ const. At the same time when $\rho = \rho_0$ we find that $\Phi = \Phi_0 = $ const is not a solution of Poisson's equation (7.4c). However when the stability of the stationary solution with $\Phi_0 = \phi(x)$ is analyzed close to the center of the bounded region, $\Phi_0 = $ const is considered a good approximation [1]. The following dispersion equation follows from (7.5) for disturbances of the form $\rho' = \rho_0 \exp(i\omega t - ikx)$

$$\omega^2 = k^2 c^2 - 4\pi G \rho_0 . \qquad (7.6)$$

It can be seen from here that when $k^2 < (4\pi G/c^2)\rho_0$ a homogeneous distribution of density is unstable: $\omega^2 < 0$. At a nonlinear stage in a process this leads to the appearance of gravitational "bubbles" (they are one-dimensional for us) with a spatial scale of $\lambda > \lambda_{cr} = \sqrt{\pi c^2/G\rho_0}$. The greatest increment corresponds to $\lambda \to \infty$ and is equal to Im $\omega_\infty = 2\sqrt{2\pi G \rho_0}$. The form of the dispersion curves of (7.6) is shown in Fig. 7.3a. Note that the dispersion equation (7.6) simultaneously describes the wave disturbances for the system of linked pendulums (in the long wave approximation), except that in contrast to Fig. 7.2 in this case we are considering the stability of the stationary state, in which all the pendulums are standing upside-down (Fig. 7.3b).

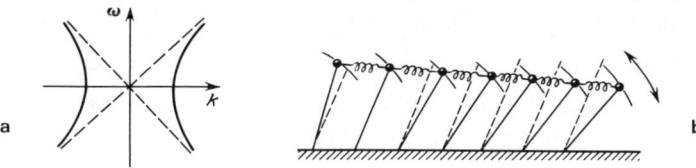

Fig. 7.3. a) Dispersion curves of (7.6). b) Oscillations (in a plane perpendicular to the Figure) in a system of linked pendulums as described by (7.6).

There is, therefore, instability in system (7.5). We shall leave a consideration of whether this instability is absolute or convective to the next section.

Turing instability. In 1952 Turing considered a model for the kinetics of a chemical reaction with an account of diffusion. In this model instability appears leading to the appearance of spatial structures. Turing's model and others like it have, therefore, aroused great interest as models for the structures that arise in biological systems [2-5]. We shall consider the stability of the steady state using a simple Turing model, which describes a reaction between two species with concentrations X_1 and X_2 in a one-dimensional reactor:

$$\frac{\partial X_1}{\partial t} = f_1(X_1, X_2) + D_1 \frac{\partial^2 X_1}{\partial x^2}, \quad \frac{\partial X_2}{\partial t} = f_2(X_1, X_2) + D_2 \frac{\partial^2 X_2}{\partial x^2}.$$

(7.7)

Here D_1 and D_2 are the different constants for one-dimensional diffusion along the x-axis.

We associate system (7.7) with a concrete system of chemical reactions:

$$A + B \rightleftarrows D + \mathcal{E},$$

$$A \underset{k_{-1}}{\overset{k_1}{\rightleftarrows}} X_1, \quad 2X_1 + X_2 \underset{k_{-2}}{\overset{k_2}{\rightleftarrows}} 3X_1,$$

$$B + X_1 \underset{k_{-3}}{\overset{k_3}{\rightleftarrows}} D + X_2, \quad X_1 \underset{k_{-4}}{\overset{k_4}{\rightleftarrows}} \mathcal{E}.$$

For simplicity we shall assume that the kinetic coefficients are $k_1 = k_2 = k_3 = k_4 = 1$ and $k_{-1} = k_{-2} = k_{-3} = k_{-4} = 0$. Thus the system of the corresponding kinetic equations together with the terms for one-dimensional diffusion in the x-direction has the form

$$\partial X_1/\partial t = A + X_1^2 X_2 - BX_1 - X_1 + D_1 \partial^2 X_1/\partial x^2,$$

$$\partial x_2/\partial t = BX_1 - X_1^2 X_2 + D_2 \partial^2 X_2/\partial x^2.$$
(7.8)

The model described by (7.8), was suggested by Prigogine and Lefev and is called the brusselator. It is the basic model for describing processes in chemical kinetics.

The system's (7.8) spatially homogeneous steady state (i.e., when $\partial/\partial t = \partial^2/\partial x^2 = 0$) is

$$X_1^0 = A, \quad X_2^0 = B. \tag{7.9}$$

In order to investigate these systems' stability we find an equation for small deviations x_1' and x_2' from (7.0). By assuming that $X_1 = X_1^0 + x'$ and $X_2 = X_2^0 + x_2'$ and linearizing (7.8), we arrive at

$$\partial x_1'/\partial t - (B - 1)x_1' - A^2 x_2' = D_1 \partial^2 x_1'/\partial x^2,$$

$$\partial x_2'/\partial t + A^2 x_2' + Bx_1' = D_2 \partial^2 x_2'/\partial x^2.$$
(7.10)

The solution to (7.10) will be solved in the form of concentration waves

$$x_1', x_2' \sim \exp(pt - ikx), \tag{7.11}$$

where $-ip = \omega$ is an unknown circular frequency, and k is an unknown wave number. By substituting (7.11) into (7.10), we find the characteristic equation

$$p^2 - \theta p + \Delta = 0 , \quad (7.12)$$

where

$$\theta = - \{A^2 + 1 - B + k^2(D_1 + D_2)\} ,$$

$$\Delta = A^2 - (B - 1)D_2 k^2 + A^2 D_1 k^2 + D_1 D_2 k^4 . \quad (7.13)$$

Suppose $D_1 = D_2 = 0$. If we are interested in the stability of the steady state over time, we must determine the position of the roots of $p^2 - \theta_1 p + \Delta_1 = 0$ with $\theta_1 = -\{A^2 + 1 - B\}$ and $\Delta_1 = A^2$ in the complex p-plane. A system without diffusion is thus stable when

$$\Delta_1 = A^2 > 0, \quad \theta_1 = -\{A^2 + 1 - B\} < 0 \quad (7.14)$$

We need to know whether diffusion can turn the stable state of (7.9) in the homogeneous model into an unstable state.

It follows from (7.12) that the system will be unstable when $\Delta < 0$, whence using (7.13) we obtain the condition

$$\Delta = D_1 D_2 k^4 + [A^2 - (B - 1)D_2 + AD_1]k^2 + A^2 < 0 . \quad (7.15)$$

In order for this inequality to be fulfilled, k^2 must be in the interval with boundaries k_1^2 and k_2^2 are defined by the equality $\Delta = 0$, i.e., in the interval where

$$k_{1,2}^2 = (2D_1 D_2)^{-1} \Big\{ - [A^2 D_1 - (B - 1)D_2]$$

$$\pm \sqrt{[A^2 D_1 - (B - 1)D_2]^2 - 4D_1 D_2} \Big\} . \quad (7.16)$$

Note that now $D_1 \cdot D_2 \neq 0$ since we allow for diffusion.

We may therefore say that the appearance in the reactor of diffusion indeed transforms a stable state to instability. It is noticeable that this instability is very selective: disturbances that are periodic in space grow with a spatial period in the bounded interval[2].

----

[2] Here we must allow for the boundedness of the system's dimensions [5].

3. Here are two more examples, which illustrate the operation of a distributed microwave amplifier (a travelling-wave tube TWT) and a distributed microwave generator (backward-wave tube BWT). We looked at a travelling wave tube in Chapter 4 in a discussion of spatial resonance (Fig. 4.24). There we mentioned that in order to describe amplification correctly it was necessary to add the equation $MI = \text{const } E$ (using the notation of section 4.4, M is an operator), which describes the inverse influence of the wave-carrying system on the electron beam and describes the grouping of electrons into clusters, to the equation for the disturbance of a wave-carrying system without losses by the current to the electron beam :

$$\frac{\partial E}{\partial x} + i \frac{\omega}{v_{ph}} E = -\frac{1}{2}\left(\frac{\omega}{v_{ph}}\right)^2 KI . \qquad (7.17)$$

Equation (7.17) was obtained by assuming that all the variables change over time as $\exp(i\omega t)$. Moreover, ω is a real number because a travelling wave tube is an amplifier, in which the signal, whose frequency is given by the external signal generator, grows exponentially along the length of the tube.

Suppose the electron beam is described by the equations of fluid mechanics. We shall assume that the beam fills all space and that the motion is one-dimensional, i.e., nothing changes in directions perpendicular to the direction of motion (in microwave electronics this is called the model of an infinitely wide beam). Thus three equations are sufficient to describe this charged liquid (we neglect collisions of particles, i.e., viscosity), namely, an Euler equation for velocity,

$$\frac{\partial v}{\partial t} + v \frac{\partial v}{\partial x} = \frac{e}{m}\left(E + E_{spch}\right) , \qquad (7.18)$$

a continuity equation

$$\frac{\partial}{\partial x}(\rho v) + \frac{\partial \rho}{\partial t} = 0 \qquad (7.19)$$

and a generalized Poisson equation, which relates the gradient of the electrostatic field due to the body charge with the density of the electron liquid's body charge

$$\partial E_{spch}/\partial x = 4\pi\rho . \qquad (7.20)$$

The electron beam is assumed to be ionically

compensated, i.e., this medium of charged particles is electrically neutral as a whole.

Since we are interested in stability, it is sufficient to consider linearized equations, assuming $v = v_0 + v'$ and $\rho = \rho_0 + \rho'$, and the current density $\rho v = j_0 + j'$ ($j_0 = \rho_0 v_0$), where v_0, ρ_0, and j_0 are the constant components of the variables, and v', ρ', and j' are small perturbations of these variables (each perturbation is much smaller than the corresponding constant variable). The linearized equations (7.18)-(7.20) take the form

$$\frac{\partial v'}{\partial t} + v_0 \frac{\partial v'}{\partial x} = \frac{e}{m}\left(E + E_{spch}\right), \quad \frac{\partial j'}{\partial x} = -\frac{\partial \rho'}{\partial t}, \quad (7.21)$$

or because $j' = v_0 \rho' + \rho_0 v'$ we have

$$\frac{\partial j'}{\partial t} + v_0 \frac{\partial v'}{\partial x} = \rho_0 \frac{\partial v'}{\partial t}, \quad (7.22)$$

$$\frac{\partial E'_{spch}}{\partial x} = 4\pi \rho'. \quad (7.23)$$

Assuming that all the variables change over time as $\exp(i\omega t)$ and by introducing the operator $\hat{\mathcal{L}} = i\omega + v_0 \partial/\partial x$, we can rewrite (7.21)-(7.23) as

$$\hat{\mathcal{L}} v' = \frac{e}{m} E + \frac{e}{m} E'_{spch}, \quad \hat{\mathcal{L}} j' = i\omega \rho_0 v', \quad (7.24)$$

$$E'_{spch} = -\frac{4\pi}{i\omega} j'.$$

When deriving (7.24) we used (7.23) and the continuity equation in the form $\partial j'/\partial x = -i\omega \rho$.

By excluding v' and E'_{spch} from (7.24), we obtain

$$\hat{\mathcal{L}}^2 j' + \omega_p^2 j' = (i\omega \rho_0 e/m) E. \quad (7.25)$$

The simplest method of transforming from an infinitely wide electron beam to a beam with infinite cross-section S is to introduce a reduced plasma frequency $\omega_q = R\omega_p$ instead of the plasma frequency $\omega_p = \sqrt{4\pi e \rho_0/m}$, where R is a reduction coefficient ($0 < R \le 1$), which takes into account the effect of the surrounding walls on the beam [6].

Thus for a current $i' = j'S$, which is grouped in the beam due to the effect of the wave-carrying system's field we have for (7.25)

$$\frac{\partial^2 i'}{\partial x^2} + 2i \frac{\omega}{v_0} \frac{\partial i'}{\partial x} - \left(\frac{\omega^2}{v_0^2} - \frac{\omega_q^2}{v_0^2} \right) i' = i \frac{\omega}{v_0} \frac{I_0}{2V_0} E , \quad (7.26)$$

where $I_0 = \rho_0 v_0 S$ is the constant beam current, $V_0 = \sqrt{v_0^2 m/2e}$ is the beam's accelerating voltage. If (7.17) and (7.26) are to be self-consistent and assuming that i' and E change in space as $\exp(-ikx)$, where k is a wave number, then we arrive at a dispersion equation

$$(\omega - kv_{ph})(\omega - kv_0 - \omega_q)(\omega - kv_0 + \omega_q) = \omega^3 C^3 , \quad (7.27)$$

where $C^3 = (I_0 K/4V_0)(v_0/v_{ph})^2$, and C is known from the theory of microwave amplification [7]. It is not difficult to see from (7.27) that when a wave from the wave-carrying system $k = \omega/v_{ph}$ and two waves from the beam, a fast wave of the space charge $(k = (\omega - \omega_q)/v_0)$ and a slow wave of the space charge $(k = (\omega + \omega_q)/v_0)$ interact. A necessary condition for amplification in space is that the wave number is complex when the frequency ω is real, and because $E \sim \exp(-ikx)$ for waves propagating to the right, there will be instability in space if and only if Im $k > 0$.

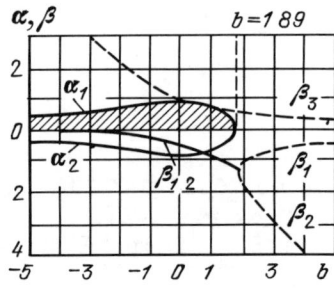

Fig. 7.4. α and β versus the asynchronicity between the beam and a cold wave; α_1 corresponds to a wave growing with distance [8].

A solution to (7.27) is shown in Figure 7.4 for $\omega_q = 0$ in the form of $\alpha = (\text{Im } k)/k_e C$ and $\beta = (k_e - \text{Re } k)/k_e C$

($k_e = \omega/v_0$) as functions of the asynchronicity $b = (k_0 - k_e)/k_e C$ ($k_0 = \omega/v_{ph}$) between the beam and a cold wave. It is assumed, that damping and the force of the space charge have a negligible effect on the interaction. It is easy to see that when $b = 0$ ($v_0 = v_{ph}$) there is a maximum increment $(\text{Im } k)_{max} = (\sqrt{3}/2)k_e C$ and instability region is bounded by $b = 3\sqrt[3]{2}/2 \approx 1.89$.

In the mid 1950's Piddington [9] began to argue that the theory covering wave tubes and twin-beam tubes (about which we shall have more to say in this Chapter) was incorrect. He maintained that the spatial increase of waves as predicted by the theory was incorrect and that the error was an incorrect treatment of the dispersion equation. Piddington showed that sometimes waves exponentially decaying along the x-axis may sometimes be taken by mistake as amplified waves. However he, himself, erred in his conclusion by deciding that the case of complex k with real ω always corresponded to filtering.

We shall look at one more example, the backward-wave tube. In this tube the electron beam moves through an artificial medium, in which waves may propagate with a longitudinal electrostatic field. The dispersion of this medium is such that the phase velocity of the waves at some frequency Ω is equal to the electron velocity and the group velocity is negative, i.e.,

$$v_{ph}(\Omega) = v_0, \quad v_{gr}(\Omega) < 0. \quad (7.28)$$

In real devices the artificial medium with the required properties is a periodic electrodynamic structure, a delay system. Because of (7.28), when the electron beam interacts with a wave in the system distributed feedback arises, that is these small wave disturbances propagating at v_{gr} travel upstream and thus connect the input and output of the system. It is thus possible to have either amplification (regeneration) or self-excitation in the tube. In electronics a backward-wave tube is mainly used to generate monochromatic microwave oscillations (a diagram of a backward-wave tube is given in Fig. 23.6 [10-11]).

It is easy to show that the dispersion equation of the system consisting of the electron wave and the reverse electromagnetic wave has the form

$$(\omega - kv_{ph})(\omega - kv_0 - \omega_q)(\omega - kv_0 + \omega_q) = -\omega^3 C^3, \quad (7.29)$$

i.e., it differs from (7.27) only by the sign of the

right-hand side. If we are interested in a self-exciting system, then it is ω and not k that is unknown. Since we are interested in generation, it follows that we are interested in instability over time and hence we wish to know what sense the complex values of k has in this case. Usually this is found by solving equations like (7.17) (it is necessary to change the sign on the right-hand side in this equation for a backward-wave tube) and (7.6) simultaneously for initial and boundary conditions corresponding to the physics of the situation.

It follows from (7.27) that the field E and, the group current i' may be described by three waves:

$$E = \sum_{i=1}^{3} E_i(0)e^{-ik_i x}, \quad i' = \sum_{i=1}^{3} i'_i(0)e^{-ik_i x}. \tag{7.30}$$

The unknown amplitudes $E_i(0)$ and $i'_i(0)$ are determined for a travelling wave tube from the initial conditions ($x = 0$)

$$\sum_{i=1}^{3} E_i(0) = E(0), \quad \sum_{i=1}^{3} i'_i(0) = 0, \quad \sum_{i=1}^{3} \frac{\partial i'_i(0)}{\partial x} = 0. \tag{7.31}$$

where $E(0)$ is the amplitude of the input signal, the second condition means that the beam is not grouped at the input, and the third condition means that the beam is not velocity modulated at the input. Thus it is possible to find the distribution of field along the length of the interaction space. It follows from the solution that at long wavelengths a wave with Im $k > 0$, which also determines the amplification factor of the travelling tube, dominates. For example, when $b = 0$, the amplification is

$$G = \frac{E(1)}{E(0)} \sim \exp\left(\frac{\sqrt{3}}{2} 2\pi C \frac{1}{\lambda}\right) \sim \exp\left(\frac{\sqrt{3}}{2} 2\pi CN\right),$$

where $\lambda = 2\pi/k_e$, and $N = 1/\lambda$ is the number of wavelengths that fit into the interaction space.

In the case of a backward-wave tube generator we must solve the boundary value problem in order to determine the conditions under which an oscillation is induced (Im $\omega = 0$), assuming that $i'(0) = \partial i'(0)/\partial x$ and $E(1) = 0$ (no input signal). The following start-up parameters were obtained at which an oscillation will be induced: $b_{st} = 1.522$, $(CN)_{st} = 0.314$. When v_0 does not differ much from v_{ph} and $C \ll 1$, the solution to (7.29) may be found in the form $k = (\omega/v)(1 + iC\delta)$, which leads to the equation

$\delta^2(\delta + ib) = -i$. When $b = b_{st}$ the roots of this equation are $\delta_1 = 0.725 + 0.151\ i$, $\delta_2 = -0.725 + 0.151\ i$, and $\delta_3 = -1.822\ i$ [11]. It is clear that a wave with $\text{Im}\ k_1 > 0$ does not play the same role it does in a travelling wave tube, and the field is defined by the superposition of all three waves because initially the boundary condition $E(l) = 0$ is not fulfilled. However, a clear difficulty arises using this approach due to the necessity of solving the boundary value problem. At the same time, it would be desirable not to have to solve a problem with initial, and what is more boundary, conditions, but to restrict our consideration to unbounded systems, i.e., analyze the dispersion equation and use it to study stability.

7.3 Absolute and Convective Instability. The Characteristics Method

It is in general exceedingly difficult to determine the behavior of an arbitrary disturbance (whether it is carried away in the x-direction or is widened embracing new regions in the $+x$- or $-x$-directions) by only using the dispersion equation of a system and without analyzing a concrete solution like (7.1). However, for a wide class of distributed systems, and particularly those described by hyperbolic partial differential equations, this can be done relatively simply (note that hyperbolic equations also describe oscillations in systems of linked pendulums (Figs. 7.2 and 7.3) and inviscid gravitating gases, and many other important systems). For such systems it is only necessary to determine on the (x,t-plane) the borders of the region in which the disturbance propagates (Fig. 7.5), the borders corresponding to the system's characteristics and having a maximum and minimum slope. We already know the simplest example of a hyperbolic equation; this is the ordinary wave equation $u_{tt} - a^2 u_{xx} = 0$. Here there are two families of characteristics: $x - at = C_1$ and $x + at = C_2$. The first family corresponds to disturbances propagating to the right and the second family to disturbances propagating to the left. Because in this case the system is linear, an arbitrary disturbance (which is the superposition of the two families) will widen in the $+x$- and $-x$-directions. Thus if we "organize" instability in such a system (formally this can be done by adding a term $-b^2 u_t$ to the left-hand side of the equation), then the instability will be absolute. The region in which the disturbance propagates encompasses both half-spaces (i.e., to the left and to the right of the initial region along the x-axis (Fig. 7.5)). Thus, instability in a homogeneous gravitating gas (Jean's instability) and instability in a backward-wave tube generator are absolute.

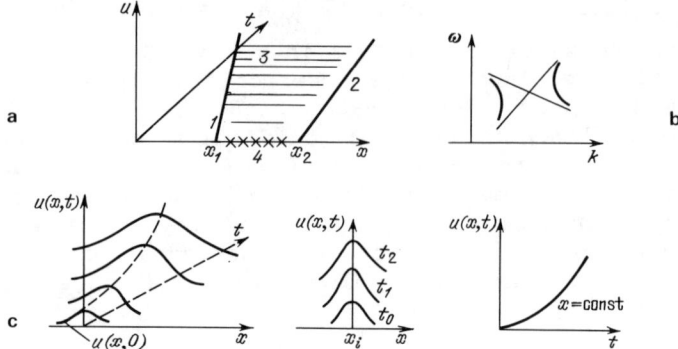

Fig. 7.5. a, b) Relationship between the characteristics of hyperbolic systems (the x, t-plane) and the asymptotes of the corresponding dispersion equations (the ω, k-plane) for absolute instability in a twin-beam system:
1, 2) characteristics of different families; 3) region in which the disturbance propagates; 4) region of the initial disturbance.
c) Figures explaining the development in the system of an absolute instability.

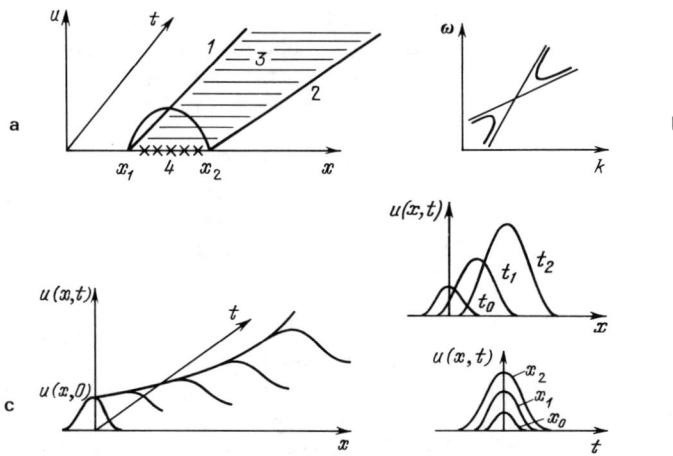

Fig. 7.6. a, b) Link between the characteristics of the hyperbolic system (the x, t-plane) and the asymptotes of the corresponding dispersion equations (the ω, k-plane) for convective instability for a twin-beam system. 1-4) As in Fig. 7.5. c) Figures explaining the development of convective instabilities in the system.

The characteristics of a hyperbolic system turn out to be related to the asymptotes of the dispersion curves of the corresponding linearized equations. The characteristics and asymptotes have the same slopes on the (x, t)-plane and (ω, k)-plane, respectively. Because of this it is possible to say for hyperbolic systems, for which the number of asymptotes with finite slopes is the same as a number of normal waves, whether the instability is absolute or convective from the form of the dispersion curves. If the slopes of the asymptotes of the dispersion curves have opposite signs, then the instability is absolute (Fig. 7.5). If, however, they are directed in the same direction, then the instability is convective (Fig. 7.6). In the first case the region in which the disturbance propagates will be as in Fig. 7.5a and in the second case as in Fig. 7.6a.

We now cover the elements of the theory of characteristics [12,13]. We write the system of initial equations in the form

$$\frac{\partial u_i}{\partial t} + \sum_{k=1}^{n} a_{ik}(u) \frac{\partial u_k}{\partial x} + b_i(u) = 0 \quad (i = 1, 2, \ldots, n), \quad (7.32)$$

where u_i is a variable describing our system and $a_{ik}(u)$ and $b_i(u)$ are nonlinear functions of $u_1, \ldots, u_n$. An equation like (7.32) is usually called quasilinear. It does not contain nonlinear functions of the derivatives. Any of the properties of a characteristic may serve as a definition of one. We will call the lines in the (x,t)-plane bounding the influence region characteristics. If a disturbance is given on some arc AB in the (x,t)-plane, then it will affect the solution $u_i(x, t)$ of (7.32) only in the region bounded by the characteristics passing through A and B. Because a characteristic divides the disturbed region from the undisturbed, it is impossible to use (7.32) to determine uniquely the derivatives $\partial u_i/\partial n$ normal to the characteristic by specifying all the quantities u_i along the characteristic, (i.e., only the $\partial u_i/\partial s$ are known). We can use this to find an equation for the characteristics. We use V to denote the slope of the characteristic with respect to the t-axis and express $\partial u_i/\partial x$ and $\partial u_i/\partial t$ in terms of $\partial u_i/\partial s$ and $\partial u_i/\partial n$, i.e.,

$$\partial u_i/\partial t = (V^2 + 1)^{-1} \partial u_i/\partial s - V(V^2 - 1)^{-1} \partial u_i/\partial n,$$

$$\partial u_i/\partial x = V(V^2 + 1)^{-1}\partial u_i/\partial s + (V^2 + 1)^{-1}\partial u_i/\partial n .$$

After substituting these derivatives into (7.32) we have

$$\sum_{k=1}^{n} (a_{ik} - V\delta_{ik})\partial u_k/\partial n = -\sum_{k=1}^{n} (Va_{ik} + \delta_{ik})\partial u_k/\partial s - b_i(u) .$$

(7.33)

This is a linear inhomogeneous system in $\partial u_k/\partial n$ with a known right-hand side. To make it impossible to determine $\partial u_k/\partial n$ from this equation, it is necessary to set the determinant to zero, i.e.,

$$\mathrm{Det}(a_{ik} - V\delta_{ik}) = 0 \qquad (7.34)$$

(δ_{ik} is the Kronecker delta). This is the equation we require for a characteristic. Because this is an n-order polynomial in V, we find the slope of the n families of characteristics. If the system is linear and a_{ik} is independent of u, then the characteristics are straight lines in the x, t)-plane, whose slopes are V_l, where V_l ($l = 1, 2, \ldots, n$) is a root of (7.34).

The linearized system (7.32) is described by the dispersion equation

$$\mathrm{Det}\left(a_{ik} - \frac{\omega}{k}\delta_{ik} - \frac{1}{k}b_{ik}\right) = 0 \qquad \left(b_{ik} = \left.\frac{\partial b_i}{\partial u_k}\right|_{u=u_0}\right) . \qquad (7.35)$$

A comparison between (7.35) and (7.34) shows that in the limit as $k \to \infty$ the slope of the dispersion curves corresponds to the slope of the characteristics.

Let us determine the form of the instability in a system of two interpenetrating electron beams that are moving along the x-axis using a criterion based on the estimate of where the asymptotes are placed. The asymptotes' dispersion characteristics are given in Fig. 7.8 for colliding beams and in Fig. 7.10c they are given for beams that are travelling in the same direction. In the first case, the slopes of the asymptotes have opposite signs, and consequently the instability in this system is absolute, whereas the instability in the second system is convective.

7.4 Waves in Flows. Electron Beams. Helmholtz Instability

Suppose we have two interpenetrating charged liquids (for instance, two electron or ion beams), the interaction between which is determined by a common longitudinal electrostatic field of a space charge E_{spch}. As we did when we studied a travelling-wave tube, we shall assume the medium is conservative, and neglect the friction for this medium (viscosity). The flows are infinitely wide and move either together in the x-direction (codirectional beams), or in opposite directions (counterdirectional beams) with different absolute constant velocities $|v_{01}|$ and $|v_{02}|$.

This theoretical model corresponds to twin-beam tubes, which have been well studied in microwave electronics [7,8,14]. In experimental setups two cathodes are used, and the difference between the potentials ensures a difference between the velocities of the electron beams. The cathode construction is chosen so as to provide well interpenetrating beams (for example, one cathode construction involves two plane spirals, one inside the other, such that the electrons emitted by one cathode passes between the spirals of the second cathode, thus providing good mixing).

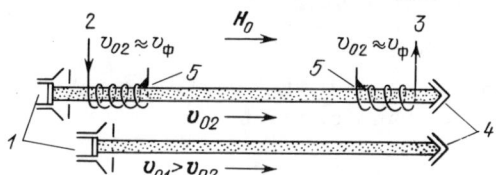

Fig. 7.7. Diagram of a twin-beam amplifier: *1)* electron beams; *2, 3)* The input and output devices; *4)* collectors; *5)* matching loads; H_0 focusing magnetic field. The beams are shown apart.

In order to introduce the signal that is to be amplified onto one or both beams, the usual practice is to use sections of spiral with a high-frequency electrostatic field to modulate the electrons. The velocity of one of these beams is chosen to be close to the phase velocity of the wave v_{ph} in the spiral since then the modulation of the beam by the input signal is efficient. Because the beam amplitude increases exponentially with distance, the signal induced in the output spiral has a much larger amplitude than the signal input to the tube (Fig. 7.7). Initially a twin-beam tube seemed to be very promising, especially for millimeter waves, because it combined a long interaction with the absence of a delay system. However, it turned out, that higher frequencies required smaller differences between the beam velocities and increased current densities. The scope for bringing the beam velocities close together is limited be the electron velocity spread, which can be characterized by a velocity distribution function for the electrons. Clearly when $|v_{01}| - |v_{02}|$ is comparable to the spread in velocities, the two beams are practically indistinguishable. Because of this, twin-beam amplifiers are not used in microwave electronics, however they have became the standard example for the theory of wave instability [15-18].

Let us consider further two ionically balanced electron beams which are described by the linearized equations of fluid mechanics

$$\frac{\partial v'_1}{\partial t} + v_{01}\frac{\partial v'_1}{\partial x} = -\frac{e}{m}E'_{spch}, \quad \frac{\partial v'_2}{\partial t} + v_{02}\frac{\partial v'_2}{\partial x} = -\frac{e}{m}E'_{spch},$$

$$\frac{\partial \rho'_1}{\partial t} + v_{01}\frac{\partial \rho'_2}{\partial x} + \rho_{01}\frac{\partial v'_1}{\partial x} = 0, \quad \frac{\partial \rho'_2}{\partial t} + v_{02}\frac{\partial \rho'_2}{\partial x} + \rho_{02}\frac{\partial v'_2}{\partial x} = 0,$$

$$\frac{\partial E'_{spch}}{\partial x} = 4\pi(\rho'_1 + \rho'_2). \tag{7.36}$$

If we assume that all the variables change in time as $exp(i\omega t)$, (7.36) is transformed to

$$E'_{spch} = -\frac{4\pi}{i\omega}\left(j'_1 + j'_2\right),$$

$$\frac{\partial^2 j'_1}{\partial x^2} + 2i\frac{\omega}{v_{01}}\frac{\partial j'_1}{\partial x} - \left(\frac{\omega}{v_{01}}\right)^2 j'_1 = i\omega\frac{1}{4\pi}\frac{\omega_{p1}^2}{v_{01}^2}E'_{spch},$$

$$\tag{7.37}$$

$$\frac{\partial^2 j'_2}{\partial x^2} + 2i\frac{\omega}{v_{02}}\frac{\partial j'_2}{\partial x} - \left(\frac{\omega}{v_{02}}\right)^2 j'_2 = i\omega\frac{1}{4\pi}\frac{\omega_{p2}^2}{v_{02}^2}E'_{spch},$$

$$\omega_{p1}^2 = 4\pi\frac{e}{m}\rho_{01},$$

$$\omega_{p2}^2 = 4\pi\frac{e}{m}\rho_{02}.$$

This system corresponds to a self-consistent model of a disturbance in an electron wave-guide by an electron beam. The first equation describes the disturbance to the electron wave-guide by the beams, and the other two describe the grouping of the electron beams due to the summed field of the space charge of the two electron beams.

This approach allows us to explain the physical mechanism of a twin-beam amplifier with codirected beams by analogy with a travelling-wave tube.

The input device modulates the velocity and density of a slow electron beam and this causes a cloud with a periodic electron structure to form in space, i.e., alternating regions of dense and less dense electrons. This situation, as was shown in Chapter 5, corresponds to the propagation in the beam of two waves of space charge, i.e., a fast wave and a slow wave, whose phase velocities are $v_{ph.f,s} = v_{02}/(1 \mp \omega_q/\omega)$. Thus, the modulated beam in the twin-beam system has an analogous role to the delay system in the travelling-wave tube. The second, fast beam ($v_{02} < v_{01}$) interacts with the longitudinal component of the slow wave in the first beam. Thus, as was the case for a travelling-wave tube, when the velocity v_{01} of the second beam is appropriately chosen, the second beam will give off energy to the high frequency field, as the result of which there is amplification of the input signal. By excluding E'_{spch} from (7.37), we finally arrive at

$$\frac{\partial^2 j'_1}{\partial x^2} + 2i \frac{\omega}{v_{01}} \frac{\partial j'_1}{\partial x} - \left[\left(\frac{\omega}{v_{01}}\right)^2 - \left(\frac{\omega_{p1}}{v_{01}}\right)^2\right] j'_1 = -\left(\frac{\omega_{p1}}{v_{01}}\right)^2 j'_2,$$

(7.38)

$$\frac{\partial^2 j'_2}{\partial x^2} + 2i \frac{\omega}{v_{02}} \frac{\partial j'_2}{\partial x} - \left[\left(\frac{\omega}{v_{02}}\right)^2 - \left(\frac{\omega_{p2}}{v_{02}}\right)^2\right] j'_2 = -\left(\frac{\omega_{p2}}{v_{02}}\right)^2 j'_1.$$

This system of equations (7.38) admits of the solution $j'_1 = j'_2 = 0$, when the beams move without interacting with each other. In order to found out whether this movement is stable we shall search for a solution to (7.38) in the form $j'_{1,2} = \Psi_{1,2} \exp(-ikx)$.

By substituting it into (7.38) we obtain a system of linear algebraic equations which define the distribution coefficients $\Psi_{1,2}$ (the vector $\Psi(\Psi_1, \Psi_2)$ is also called the polarization vector). Equating the determinant of this system to zero yields a dispersion equation for the situation

$$D(\omega, k) = \omega_{p1}^2(\omega - kv_{02})^2 + \omega_{p2}^2(\omega - kv_{01})^2$$
$$- (\omega - kv_{01})^2(\omega - kv_{02})^2 = 0 . \quad (7.39)$$

Note that the coefficients of the equations are real although, at the same time, the roots of the equation (ω or k) may be complex. We shall now look in detail at various special cases.

Suppose that the beams are identical, but in opposite directions, i.e.,

$$\rho_{01} = \rho_{02} = \rho_0 \quad (\omega_{p1} = \omega_{p2} = \omega_p), \quad v_{01} = -v_{02} = v_0 . \quad (7.40)$$

Using (7.40) dispersion equation (7.39) takes the form

$$D(\omega, k) = (\omega - kv_0)^2(\omega + kv_0)^2 - \omega_p^2(\omega - k_0)^2 - \omega_p^2(\omega + kv_0)^2 = 0$$

or

$$\omega^2 = (k^2 v_0^2 + \omega_p^2) \pm \sqrt{4\omega_p^2 k^2 v_0^2 + \omega_p^4} . \quad (7.41)$$

It can be seen from (7.41) that ω may be complex if $(k^2 v_0^2 + \omega_p^2) < \omega_p \sqrt{4k^2 v_0^2 + \omega_p^2}$, i.e., when

$$k < \sqrt{2}\omega_p/v_0 . \quad (7.42)$$

In follows from (7.42) that $\lambda_p < \sqrt{2}\lambda$ ($k = 2\pi/\lambda$, $\omega_p/v_0 = 2\pi/\lambda_p$), i.e., only long-wave disturbances are unstable. We emphasize that k is real here. The dispersion characteristics defined by (7.41) are shown in Fig. 7.8.

In order to understand the Figure, let us go through a chain of transitions: one beam to two noninteracting beams to two interacting beams (Fig. 7.9). It follows from (7.39) (we know this from Chapter 5) that there are two waves of space charge, a fast and a slow one, in one disturbed electron beam. If $v_{02} = -v_0 = 0$, $\omega_{p2} = 0$, $\omega_{p1} = \omega_p$, then we have from (7.39) $(\omega - kv_0)^2 = \omega_p^2$, $\omega = kv_0 \pm \omega_p$, and when $v_{01} = v_0 = 0$ we have $\omega_{p1} = 0$, $\omega_{p2} = \omega_p$ - $(\omega + kv_0)^2 = \omega_p^2$, $\omega = -kv_0 \pm \omega_p$.
It follows from analysis of Fig. 7.9c and by comparing it with Fig. 7.9a and b that curve 1 of the dispersion characteristic of interacting beams corresponds to the slow wave, and curve 2 corresponds to the fast wave. It can be seen from Fig. 7.8 that instability is impossible for fast

waves, because any real value of the wave number k for fast waves corresponds to a real value of the frequency ω.

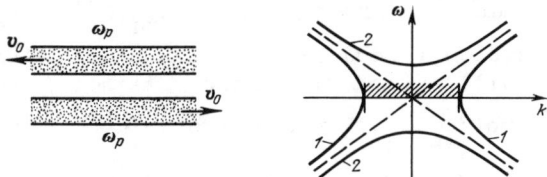

Fig. 7.8. Model of two identical opposite beams and the dispersion characteristics defined by (7.41). When $\omega = 0$ we have $k = \pm \sqrt{2}\, \omega_p/v_0$ and when $k = 0$ we have $\omega = \pm \sqrt{2}\, \omega_p$; at large k and small ω_p we have $\omega \approx \pm kv_0$. The band of real k for which there are complex ω is shaded.

For slow waves in the band of wave numbers $|k| < \sqrt{2}\, \omega_p/v_0$ (in Fig. 7.8 this band is shaded) the frequency ω will be complex and when $\mathrm{Im}\,\omega < 0$ the disturbance will grow over time.

Thus there will be instability in the conservative system we are analyzing. This is in itself a fascinating fact. The energy needed to maintain this instability is taken from the "nonwave" motion of the evenly moving flows. Whereas in the case of codirected streams we made an analogy with a travelling-wave tube, here there is an analogy with a backward-wave tube. In fact, there is feedback in the system because one of the beams is directed in the opposite direction to the other. If the screen is taken out of the scheme shown in Fig. 7.9b, then the upper beam may be considered to be an electron delaying system with inverse waves (the energy is transferred at a velocity $-v_0$). The slow wave that arises in such a transmission line will interact with the electron beam moving at velocity v_0. At certain currents, the system will be self-exciting and no input device will be necessary. The oscillations are caused by fluctuations in the density of the volume charge in the beam. We shall now investigate identical codirected beams, i.e., the case when $\rho_{01} = \rho_{02} = \rho_0$ with $\omega_{p1} = \omega_{p2} = \omega_p$, $v_{01}v_{02} > 0$. Under these conditions we obtain from (7.40)

$$\omega_p^2[(\omega - kv_{01})^{-2} + (\omega - kv_{02})^{-2}] = 1. \qquad (7.43)$$

The dispersion characteristic of this system is illustrated in Fig. 7.10c. The qualitative form of its curves can easily be obtained by going from the case of independent beams, one of which is stationary ($v_{02} = 0$, $v_{01} > 0$) to the case where they interact, and then to the case when codirected beams interact, when $v_{01} > 0$, $v_{02} > 0$. The dispersion equation (7.43) has four roots as before, and each disturbance will contain four terms, for example $v'_{1,2} = \sum_{m=1}^{4} \Psi_{1,2m} \exp(i\omega_m t - ik_m x)$. Two of these (curve 1 of the dispersion characteristic in Fig. 7.10c) do not grow over time, but the other two (curve 2) may grow because a real k in the shaded region corresponds to complex values of ω. However, the instability here is different to the case we encountered with oppositely directed beams. Because the beams are moving in the same direction, a disturbance will be carried together with the beam, i.e., at some point in space the disturbance may be damped.

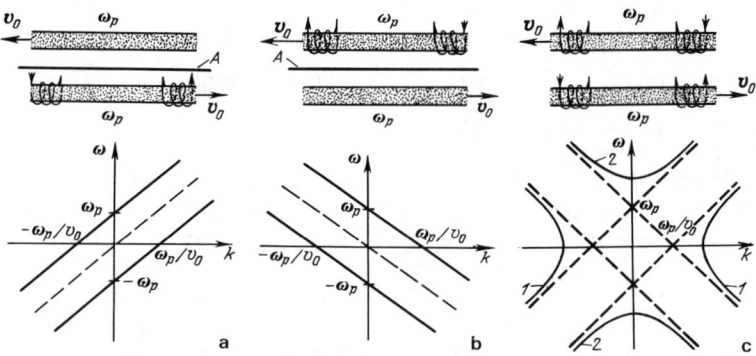

Fig. 7.9. a, b) Models of two beams and the dispersion characteristics of one beam modulated at the input of the device (the other beam is separated from the first by a screen (A)). c) Model of two interacting beams and the dispersion characteristics. The dashed lines indicate their dispersion characteristics for noninteracting beams.

We solve (7.43) for $k = k(\omega)$ assuming that ω is real. By introducing [7] a semi-difference velocity $\delta = (v_{01} - v_{02})/2$, an average velocity $v_{av} = (v_{01} + v_{02})/2$, and a wave number $k = \omega/v_{av} + i\gamma$, we can rewrite (7.43) as:

$$\left(i\frac{\gamma v_{01}}{\omega_p} + \frac{\omega \delta}{v_{av} \omega_p}\right)^{-2} + \left(i\frac{\gamma v_{02}}{\omega_p} - \frac{\omega \delta}{v_{av} \omega_p}\right)^{-2} = 1 . \qquad (7.44)$$

If δ is small in comparison to v_{01} and v_{02}, then we can consider that $\gamma v_{01}/\omega_p \approx \gamma v_{02}/\omega_p \approx \gamma v_{av}/\omega_p$. Whence we find from (7.44) that

$$\left(\frac{\gamma v_{av}}{\omega_p} \right)_{1-4} = \pm \left[\left(\frac{\omega \delta}{v_{av} \omega_p} \right)^2 + 1 \pm \left[\left(\frac{\omega \delta}{\omega_p v_{av}} \right)^2 + 1 \right]^{1/2} \right]^{1/2}. \quad (7.45)$$

It follows from an analysis of (7.44) that the maximum value of the increment in the growth of the wave occurs at $\text{Im } k = \gamma_{max} = \omega_p/2v_{av}$ where it is $\omega \delta/\omega_p v_{av} = \sqrt{3}/2$. When $\omega \delta/\omega_p v_{av} \geq \sqrt{2}$, the value $\gamma v_{av}/\omega_p$ becomes purely imaginary and all four waves are constant in amplitude. Thus, harmonic disturbances grow along the x-direction.

We shall now analyze Helmholtz's instability [19].[1] When the interaction between flows of liquid is being considered, it is usual to consider the problem in two dimensions because the velocity of the flow must depend both on the longitudinal coordinate x and on the transverse coordinate y (Fig. 7.11a). However, in the special case when the boundary across which the flows interact can be considered to be unwetted, the problem can be reduced to one dimension.

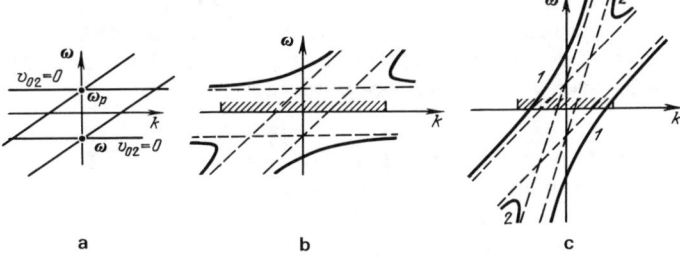

Fig. 7.10. Dispersion characteristics of two a) non-interacting and b) interacting electron beams, one of which is stationary ($v_{02} = 0$) and c) two interacting codirected beams. The shaded areas indicate where real k correspond to complex values of ω.

We shall assume that both liquid layers slip past each other at constant velocities v_{01} and v_{02}, the section of surface of the velocity jump is plane, and the density of the liquids are constant and equal to ρ_{01} and ρ_{02} because the liquids do not mix (Fig. 7.11a). Suppose a weak disturbance y' arises in the interface itself, in the liquid velocity $\mathbf{v}'$,

---

[1] The instability at an interface between two moving liquids when $\rho_{01} \neq \rho_{02}$ is called a Kelvin--Helmholtz instability.

and in the liquid pressure p'. Let, moreover, y', $\mathbf{v}'$, and p' be proportional to $\exp(i\omega t - ikx)$. For an incompressible liquid we have the following on one side of the surface jump from Euler's equation and the continuity equation (Chapter 5):

$$\partial v'_1/\partial t + v_{01} \partial v'_1/\partial x = -\rho_{01}^{-1} \text{grad } p'_1, \quad \text{div } \mathbf{v}'_1 = 0 \qquad (7.46)$$

(in the first equation it is assumed that the velocity in the x-direction is constant).

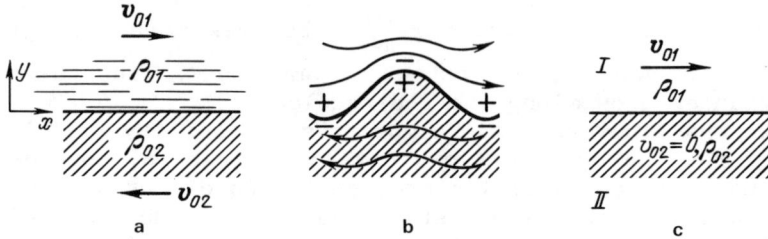

Fig. 7.11. Helmholtz instability [19]: a) no disturbance at the interface and two layers of liquid slip at the interface in opposite directions; b) the interface is disturbed and the form of the current lines is sketched and the pressure distribution across the disturbed surface of the tangential velocity jump is given; c) initial model for analyzing I) surface winds and II) stationary water.

Applying the operator div to both parts of (7.46) and using the liquid's incompressibility, we obtain

$$\partial^2 p'_1/\partial y^2 + \partial^2 p'_1/\partial x^2 = 0. \qquad (7.47)$$

The solution of (7.47) is naturally found in the form

$$p'_1 = p'_1(y)\exp(i\omega t - ikx). \qquad (7.48)$$

Whence for a liquid filling the space below the interface $y > 0$, we find from (7.47) and (7.48)

$$p'_1(y) = Ae^{-ky}e^{i(\omega t - kx)}. \qquad (7.49)$$

We use $y' = y'(x, t)$ to denote the shift in the interface and so for the transverse velocity component v'_y at the interface itself we have

$$v'_{y1} = dy'_1/dt = i(\omega - kv_{01})y'_1. \qquad (7.50)$$

We can find from Euler's equation for the v'_{y1}-components of velocity and using (7.50) a link between a pressure p'_1 and the shift in the interface y'_1:

$$p'_1 = -(\omega - kv_{01})^2 \rho_{01} y'_1/k . \qquad (7.51)$$

It is clear that the pressure p'_2 on the outside of the interface, for which $y < 0$, will be expressed in a form similar to (7.51), the only difference being the opposite sign:

$$p'_2 = (\omega + kv_{02})^2 \rho_{02} y'_2/k . \qquad (7.52)$$

It was assumed in (7.52) that $v_{02} < 0$. The pressures at the interface must be equal and so the dispersion equation will have the following form

$$(\omega - kv_{01})^2 = -\rho_{02}/\rho_{01} (\omega + kv_{02})^2 , \qquad (7.53)$$

$$\omega_{1,2} = k(\rho_{01} + \rho_{02})^{-1} \left\{ \left(\rho_{01} v_{01} - \rho_{02} v_{02} \right) \right.$$

$$\left. \pm i \left(v_{01} + v_{02} \right) \sqrt{\rho_{01} \rho_{02}} \right\} . \qquad (7.54)$$

It was in (7.54) that we found the frequency must be complex and always fulfilling the condition Im $\omega < 0$ for real k. This is Helmholtz's instability, i.e., an absolute instability. The instability's mechanism is quite easy to explain from Bernoulli's equation $v^2 + 2p/\rho = $ const. If a disturbance arises at the interface, say, the liquid below the interface is elevated, then the current lines are shifted. Where the current lines are more dense, a transverse pressure gradient arises which amplifies the disturbances (Fig. 7.11b and equations (7.51) and (7.52)). It is interesting that Rayleigh proposed this mechanism as an explanation for the flapping of sails and flags in the wind. However in reality, another mechanism associated with the breaking away of vortices occurs.

7.5 Amplification and Filtering. Separation Criteria

From the point of view of physics it would seem obvious

that a system in which there is convective instability could be used to amplify signals. Thus, if a dispersion equation $D(\omega, k) = 0$ for real ω has complex solutions for k and the asymptotes of the dispersion curves have slopes with the same sign (Fig. 7.6b and 7.10c), then there will be amplification in the system. In terms of phase-plane diagrams this means that the area of propagation lies on the same side of the $x = 0$ boundary as the signal. The inverse case, i.e., when the asymptotes have slopes of different signs, corresponds to filtering.

We shall only use such a simple criterion for separating amplification from filtering for hyperbolic systems. For more general types of system, more complicated criteria exist [15-18, 20-22], one of which is Briggs's [18] criterion, which we shall now present. When solving the dispersion equation $D(\omega, k) = 0$, we shall assume that ω is complex with Im ω < 0. We shall determine whether a complex solution for k corresponds to amplification or filtering as follows: if as Im $\omega \to \infty$ the sign of Im k changes, then we have amplification, and if the sign does not change, we have filtering.

In other words, there will be amplification in the system if it is sensitive to a fall in the signal of a time, and then will be filtering if the system does not sense this fall (a wave simply does not penetrate into the medium, such as is the case for collisionless plasmas, where $kc = \pm i\sqrt{\omega_p^2 - \omega^2}$ when $\omega < \omega_p$). The physics of this criterion is associated with the causality principle. If we assume that a system is disturbed by a source whose signal changes over time as $\exp(i \operatorname{Re} \omega \cdot t) \exp(- \operatorname{Im} \omega \cdot t)$ and that Im $\omega \to -\infty$, then all the waves must be damped with distance from the source because of the finite speed at which the disturbance can propagate. Consequently, when a wave is amplified with real ω, the sign of Im k must be changed as Im $\omega \to -\infty$, i.e., the wave must be damped with time in the same direction in which it is amplified when Im $\omega = 0$.

We conclude this chapter by presenting another two examples of distributed amplifiers. One of these [25] is the acoustic amplifier invented by C. Bell. In this amplifier a thin jet of water is directed against a small rubber diaphragm connected to the sound indicator - a trumpet. Waves propagating in the stream of water induce vibrations in the diaphragm, which are transformed into sound waves at the exit of the trumpet. The existence of waves that grow with distance can be proved as follows. A metronome or musical box is placed close to the nozzle from which the water jet is ejected [25], this in modern terminology being the input device. The sound signal nearing the trumpet is amplified so much that it is loud enough to be heard in a lecture theater.

A simple theory for amplification is presented in [25], which is close in form to the theory of Helmholtz instability. Consider a cylindrical, laminar, homogeneously moving stream of an incompressible liquid with density ρ_0, which can be described by Euler's hydrodynamic equation for radial v_r and longitudinal v_z velocity components; the disturbance in the azimuthal component ϕ is neglected. Given that the initial disturbance induces variation of the form $\exp(i\omega t - ikz)$, where ω is real, the linearized equations of motion have the form

$$(\omega - kv_0)v'_z = kp'/\rho_0 , \qquad (7.55)$$

$$(\omega - kv_0)v'_r = -(i/\pi_0)\partial p'/\partial r , \qquad (7.56)$$

where v_0 is the constant velocity of liquid in the z-direction. By combining the condition that the liquid is incompressible and (7.55) and (7.56), we obtain a differential equation div $\mathbf{v}' = 0$ for p', which has the solution

$$p' = AI_0(kr)\exp(i\omega t - ikz) , \qquad (7.57)$$

where A is a constant, $I_0(kr)$ is a modified type-one, zero-order Bessel function. The disturbance causes the liquid interface to curve, and it was shown in [25] that this leads to the following expression for the variable pressure at the boundary:

$$p' = S\left(k^2 - \frac{1}{r^2}\right)r' = \frac{\left(S/\rho_0 r^2\right)\left(k^2 r^2 - 1\right)}{(\omega - kv_0)^2}\frac{\partial p'}{\partial r} , \qquad (5.58)$$

where S is the surface tension and $v'_r = i(\omega - kv_0)r'$. Using (7.57) and (7.58), we arrive at the dispersion equation

$$(\omega - kv_0)^2 = \frac{S}{\rho_0 r^2}\left[k^2 r^2 - 1\right]k\frac{I_1(kr)}{I_0(kr)} . \qquad (7.59)$$

If we assume that $k = \omega/v_0 + \delta$ ($|\delta| \ll \omega/v_0$) and substitute k for ω/v_0 into the right-hand side of (7.59), then we obtain

$$k = \frac{\omega}{v_0} \pm i \left(\frac{S}{\rho_0 r^3 v_0^2}\right)^{1/2} \frac{1}{20 \log e} F\left(\frac{\omega r}{v_0}\right). \qquad (7.60)$$

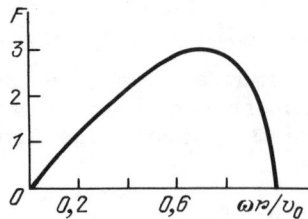

Fig. 7.12. Graph of the function $F(\omega r/v_0)$. For low frequencies this function (which means a maximum amplification coefficient per unit length) is proportional to frequency, which attains a maximum at $\omega r/v_0 = 0.7$ and falling to zero at a frequency $f_{max} = v_0/2\pi r$ [25].

The form of the $F(\omega r/v_0)$ function is shown in Fig. 7.12. For growing waves the coefficient of maximum amplification (in decibels per unit length) is

$$G_{max} = (S/\rho_0 r^3 v_0^2)^{1/2} F(\omega r/v_0). \qquad (7.61)$$

Following [25], we estimate G_{max} assuming that the diameter of the flow is 0.1 cm, and the frequency corresponding to the condition $\omega r/v_0 = 1$ (Fig. 7.12) is 5000 Hz. From these data the velocity of the flow must be 1570 cm/s (this is equivalent to a 12.6-m head of water). Then using the graph in Fig. 7.12 we obtain $G_{max} \sim 1.43$ dB/cm ($S \sim 73$ dyn/cm, ρ_0 g/cm..) for a frequency around 3500 Hz.

A similar type of instability is found in electronics for tube beams in longitudinal magnetic fields, which compensate for the Coulomb repulsion forces in the bulk charge [26]. Several photographs from [26] are given in Fig. 7.13 which illustrate the evolution of this instability in drift space. The instability of hollow beams is close to the instability of thin charged ribbons in crossed electrostatic or magnetic fields, for which there is a simple qualitative explanation [7]. If we use a system of coordinates in the tube beam problem that moves along the magnetic field at a static velocity for the electron flow, then the motion of the electrons will be the same as in a beam in crossed fields, thus being perpendicular to both the electrostatic and magnetic fields. It is interesting that a number of well-known hydrodynamic theorems about the stability of various plane-parallel flows are also valid for electron beams in cross fields with arbitrary distributions of density in the cross section (for example, there is an analog to Rayleigh's theorem, that an electron beam at a kink in the velocity profile must be unstable).

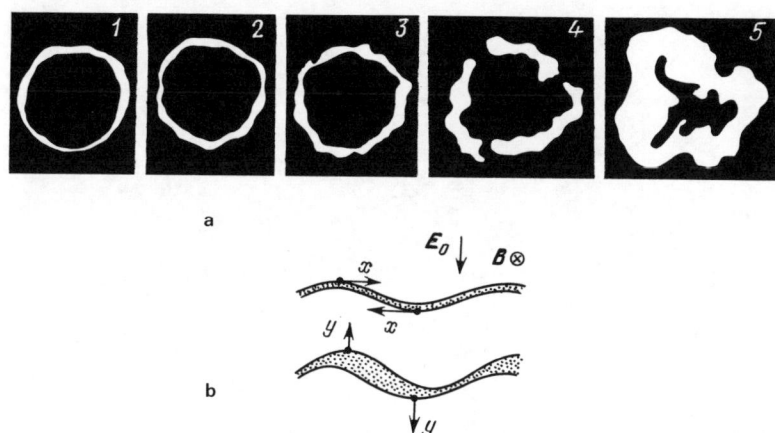

Fig. 7.13. a) The evolution of an instability in a drifting cylindrical electron beam through a longitudinal magnetic field. The photographs [26] are cross-sections of the beam at different positions in the drift space. Larger photograph numbers indicate longer drift distances. b) Illustration of the photographs for a thin ribbon in a crossed field. Local rises in the charge density cause kinks in the ribbon; it becomes unstable, and the initial disturbance grows (see [7], Chapter 5).

CHAPTER 8

PROPAGATION VELOCITY OF WAVES

8.1 Various Introductions of the Concept of Group Velocity

We have already used the concepts of phase and group velocity for waves as they are well-known from the course in general physics. In simple situations no additional explanation would be necessary. However when the medium is active and contains variable parameters, the propagation velocity of waves requires extra consideration [1-4].

If a change in a function that describes a wave process can be represented in the form $u(x, t) = u_0 \text{Re}\{\exp[i(\omega t - kx)]\}$, where $u_0 = $ const, then such a monochromatic wave propagates at the velocity

$$v_{ph} = \omega/k . \qquad (8.1)$$

This is the phase velocity of the wave, and it defines the velocity of separate troughs, peaks or wave nodes $u(x, t)$. If we introduce a phase $\phi = \omega t - kx$, which is linear in the independent variables, then $\phi = $ const for an observer moving at a velocity v_{ph}. In fact, $d\phi/dt = \partial\phi/\partial t + (dx/dt)\partial\phi/\partial x = 0$, when $dt/dx = v_{ph}$, because by definition we have $\partial\phi/\partial t = \omega$, and $\partial\phi/\partial x = -k$. However, it is obvious that the signal cannot be transmitted using a monochromatic wave because of its homogeneity in space and in time (it must exist over all time t from $-\infty$ to $+\infty$ and along the whole x-axis from $-\infty$ to $+\infty$). Clearly there are no such waves in nature: every wave process has both a beginning and end, i.e., a real signal always has a finite frequency width and propagates in general at a rate not equal to v_{ph}. Suppose we now change the amplitude or phase of the wave in some way so that it can transmit information. Let us take the concrete case in which at the initial moment in time $t = 0$ the wave is given by the spatial distribution $u = \text{Re}\{f(x)\exp(-ik_0 x)\}$, with $f(x)$ changing slowly with respect to $\exp(-ik_0 x)$. It would be expected that the wave will propagate with constant amplitude $u = \mathrm{Re}\{f(x)\exp(i\omega_0 t - ik_0 x)\}$, i.e., at a velocity $v = \omega_0/k_0$ (compare (8.1)). However, in media with dispersion it is not so. In fact, we represent $f(x)$ in the form of a Fourier integral

$$f(x) \sim \int_{-\infty}^{\infty} g(k)e^{-ikx}dk , \quad \text{where } g(k) \sim \int_{-\infty}^{\infty} f(x)e^{ikx}dx .$$

Whence

$$u(x, 0) \sim e^{-ik_0 x} \int_{-\infty}^{\infty} g(k)e^{-ikx}dk \sim \int_{-\infty}^{\infty} g(k)e^{-i(k+k_0)x}dk .$$

Note that our integral is a continuous set of waves with constant amplitude existing along the whole of the x-axis from $-\infty$ to $+\infty$. Then for groups of waves (a wave packet) we have

$$u(x, t) \sim \int_{-\infty}^{\infty} g(k)\exp[i\omega t - i(k + k_0)x]dk . \qquad (8.2)$$

In a dispersing medium $\omega = \omega(k)$. The slowness of the change in $f(x)$ with respect to $\exp(-ik_0 x)$ means that $g(k)$ is nonzero only when $k \ll k_0$, and whence the function $\omega(k)$ may be expanded into series and truncated to the first two terms:

$$\omega = \omega_0 + (\partial\omega/\partial k)_0 k . \qquad (8.3)$$

By substituting (8.3) into (8.2) we obtain

$$u(x, t) \sim \exp(i\omega_0 t - ik_0 x) \int_{-\infty}^{\infty} g(k)\exp\{-ik[x - (\partial\omega/\partial k)_0 t]\}dk$$

or

$$u(x, t) = f[x - (\partial\omega/\partial k)_0 t]\exp(i\omega_0 t - ik_0 t) .$$

If we consider $f[x - (\partial\omega/\partial k)_0 t]$ to be the amplitude of a wave that is changing, the wave's phase velocity being $v_{ph} = \omega_0/k_0$, then the amplitude change propagates with a group velocity[1].

---

[1] Hamilton [9] was the first to introduce the concept of the propagation velocity of a group of waves.

$$v_{gr} = (\partial\omega/\partial k)_{k=k_0} . \qquad (8.4)$$

We shall include one more term in the expansion of $\omega(k)$ than is in (8.3), i.e.,

$$\omega = \omega_0 + (\partial\omega/\partial k)_0 k + (\partial^2\omega/\partial k^2)_0 k^2/2 .$$

In order to neglect the additional phase $(\partial^2\omega/\partial k^2)_0 (k^2 t/2)$ in the exponent in (8.2), the inequality $|(\partial^2\omega/\partial k^2)(k^2 t/2)| \ll 1$ most be fulfilled, this inequality being rewritten in the form

$$|(\pi/v_{gr})(\partial v_{gr}/\partial\lambda)_{\lambda_0} \Delta| \ll 1 .$$

Here we introduce the distance $\Delta = v_{gr} t$ the amplitude covers in time t, and use the fact that $k = 2\pi/\lambda$, where λ is the wavelength. The distance Δ is the scale of which our consideration is valid; it is greater the smaller $(\partial v_{gr}/\partial\lambda)\lambda_0$ is. And so, the group velocity is a parameter of motion of a wave packet in a dispersing medium if the packet also maintains its form and dimension, i.e., in a distance of the order Δ. In a certain sense, the packet in this case is like a particle in classical mechanics, and the group velocity of the whole packet is like the particle velocity.

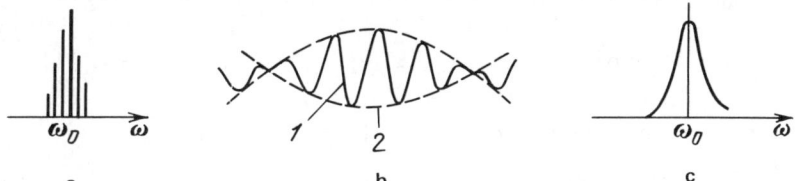

a b c

Fig. 8.1. a) A narrow discrete spectrum containing all the components close to ω_0; b) a packet of waves bounded by a modulation envelope (2), which carries all the information about the signal, in contrast to the high-frequency carrier (1); c) example of a continuous signal spectrum.

Let us consider another way of introducing the concept of group velocity. We shall analyze the propagation of a signal with a discrete frequency spectrum (Fig. 8.1a), i.e.

$$u = \text{Re}\{u_0 \exp(i\omega_0 t - ik_0 x) + u_1 \exp(i\omega_1 t - ik_1 x)$$

$$+ u_2 \exp(i\omega_2 t - ik_2 x) + ...\} .$$

We represent this superposition on monochromatic waves with

frequencies ω_0, ω_1, ω_2, ... in the form

$$u = \text{Re}\{F \exp(i\omega_0 t - ik_0 x)\}, \qquad (8.5)$$

where

$$F = u_0 + u_1 \exp[i(\omega_1 - \omega_0)t - i(k_1 - k_0)x]$$
$$+ u_2 \exp[i(\omega_2 - \omega_0)t - i(k_2 - k_0)x] + \ldots \qquad (8.6)$$

The function $F(x, t)$ is called the complex envelope of a high-frequency signal in space and of the time [2]. The reason for this name is simple to understand if we introduce $F = A \exp(i\phi)$. We then have form (8.5) $u = A \cos(\omega_0 t - k_0 x + \phi)$, which is a quasiharmonic wave (A is the envelope, $(\omega_0 t - k_0 x)$ is the high-frequency phase, and ϕ is the slowly changing phase). If the signal's spectrum is narrow (all the spectrum's components are concentrated around ω_0), then all differences of the form $\omega_n - \omega_0$ and $k_n - k_0$ ($n = 1, 2, \ldots$) are slow. Consequently, the function F in (8.5) will change more slowly than $\exp(i\omega_0 t - ik_0 x)$. The exponential factor corresponds to the propagation of the monochromatic waves with a frequency ω_0, which is called the carrier frequency. We rewrite (8.6) in the form

$$F = u_0 + u_1 \exp\{-i(k_1 - k_0)[x - (\omega_1 - \omega_0)t/(k_1 - k_0)]\}$$
$$+ u_2 \exp\{-i(k_2 - k_0)[x - (\omega_2 - \omega_0)t/(k_2 - k_0)]\} + \ldots \qquad (8.7)$$

For a narrow spectrum it is possible to assume that

$$(\omega_1 - \omega_2)/(k_1 - k_2) = (\omega_2 - \omega_0)/(k_2 - k_0) = \ldots = d\omega/dk = v_{gr}$$

(the equality becomes more exact the narrower the spectrum), and hence

$$F = u_0 + u_1 \exp\{-i(k_1 - k_0)(x - v_{gr})\}$$
$$+ u_2 \exp\{-i(k_2 - k_0)(x - v_{gr}t)\} + \ldots = F(x - v_{gr}t). \qquad (8.8)$$

This argument allows us to define the group velocity as the propagation velocity of a signal envelope (Fig. 8.1b). If the link between ω and k in the dispersion equation is linear and homogeneous, then $d\omega/dk = \omega/k = v_{gr} = v_{ph}$ and the wave packet propagates in the same way as an individual monochromatic wave (this is the attribute that characterizes a medium without dispersion).

For a signal with a continuous spectrum that fills a narrow interval around some fixed frequency $\omega = \omega_0$ (Fig. 8.1c), equation (8.8) remains the same [2]. Obviously, the concept of group velocity remains valid as before in this approach, only so long as the packet is not distorted, i.e., for relatively slow periods of time and for signals with narrow boundaries.

We now introduce the concept of group velocity using more general considerations for waves which are modulated in a quasiharmonically smooth way both by amplitude and by frequency, i.e., they have the form $u(x, t)\exp[i\Psi(x, t)]$, where Ψ is a rapidly oscillating phase (we may also consider a wide k-packet for which the change in k is of the order of k itself). The instantaneous frequencies and wave number are defined by derivatives of the phase as

$$\omega(x, t) = \partial\Psi/\partial t , \quad k(x, t) = - \partial\Psi/\partial x \qquad (8.9)$$

and clearly satisfy

$$\partial k/\partial t + \partial\omega/\partial x = 0 . \qquad (8.10)$$

If we expand Ψ into series about some point (x_0, t_0), then ω and k coincide with the local frequency and wave number in the traditional definition, when the characteristic scale of the changes in ω and k are much larger than $1/\omega$ and $1/k$. That is, we assume that over an interval of space much larger than the modulation period but smaller than the characteristic scale of its changes, the local frequency is close to the frequency of a sinusoidal wave with the given local value of k. Hence ω and k are related by the dispersion equation $\omega = \omega(k)$. By using this in (8.10), we obtain

$$\partial k/\partial t + (\partial\omega/\partial k)\partial k/\partial x = 0 \text{ or } \partial k/\partial t + v_{gr}(k)\partial k/\partial x = 0 , (8.11)$$

where $v_{gr}(k) = \partial\omega/\partial k$. In this way it is possible to give yet another definition that is important for the understanding of the kinematics of wave motion, that is, the group velocity $v_{gr}(k)$ is the velocity at which a disturbance in the wave number k propagates. Equation (8.11) for k is a hyperbolic nonlinear equation even when the initial problem is linear. It follows from this equation that k is constant along curves in the (x, t)-plane for which $dx/dt = v_{gr}$. Whence it follows that $v_{gr} = $ const, i.e., the curves are straight lines defined by

$$x - v_{gr}t = \text{const} \qquad (8.12)$$

(see Fig. 8.2).

It is clear that instead of (8.11) we may use

$$\partial\omega/\partial t + v_{gr}(\omega)\partial\omega/\partial x = 0, \qquad (8.13)$$

which is also nonlinear. Hence, when considering dispersion we may discuss "frequency nonlinearity". The left-hand side of (8.13) is $d\omega/dt$ taken along the line $dx/dt = v_{gr}(\omega)$ in the (x, t)-plane, i.e., (8.13) means that $\omega = \text{const}$ along this line. This means that $v_{gr}(\omega) = \text{const}$ along the curve $t - x/v_{gr}(\omega) = \zeta(\omega)$, where $\zeta = \text{const}$ for a given ω. The function $\zeta(\omega)$ is defined by the modulus of the frequency at $x = 0$ and hence the overall solution to (8.13) has the form

$$\omega = \Omega[t - x/v_{gr}(\omega)], \qquad (8.14)$$

where Ω is an arbitrary function inverse to $\zeta(\omega)$. The solution (8.14) will be discussed in great detail in the second half of this book when we consider the theory of simple waves whose behavior is determined by the fact that each point of the profile of a simple wave moves at the velocity $v(\omega)$, which is constant but different for different ω. Therefore it is possible to represent the wave as a set of independent groups each moving at its own velocity.

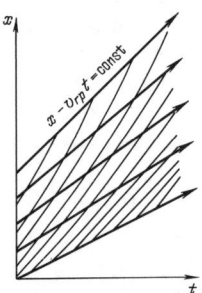

Fig. 8.2. The behavior of a group of waves on the (x, t)-plane. Waves with a wave length $\lambda = 2\pi/k = \text{const}$ are found in the bold-face straight lines along which $x - v_{gr}t = \text{const}$. The trajectories of the wave peaks arising from nothing and dissipating at the front are shown by the thin lines [3].

Clearly, depending on the frequency modulations these groups may move apart or move together, overtaking one another and moving apart again. If we construct the curves on the (x, t)-plane then it is possible to obtain, for example, a focus, i.e., a point at which two or three groups come together and then move apart again (Fig. 8.3). The solution will therefore not be unique (in a flow of non-interacting particles). This is a clear analogy with the behavior of the light in ordinary geometric optics. For example, it is shown in [4] that this analogy is not coincidental, and for a dispersing medium it is natural to discuss the approximation of space-time geometric optics.

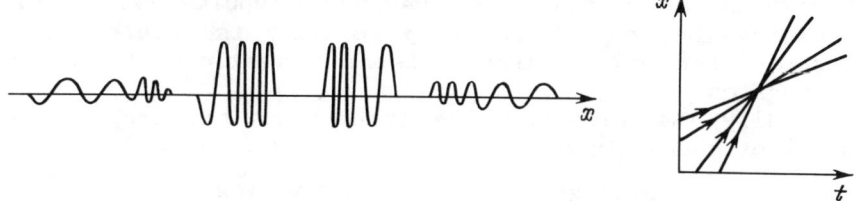

Fig. 8.3. The compression and subsequent rarefaction of a frequency-modulated wave and the corresponding space-time diagram.

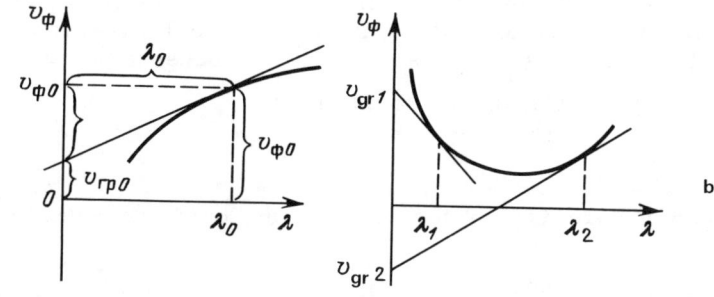

Fig. 8.4. A graphical definition of v_{gr}: a) for $v_{gr} > 0$; b) $v_{gr} > 0$ for $\lambda = \lambda_1$ and $v_{gr} < 0$ for $\lambda = \lambda_2$. The projection on the ordinate axis of the tangent to the curve $v_{ph}(\lambda)$ at the point $\lambda = \lambda_0$ is equal to $|v_{ph0} - \lambda_0 (\partial v_{ph}/\partial \lambda)_0|$.

If we replace ω by $v_{ph} k$ and use $k = 2\pi/\lambda$, then we obtain Rayleigh's formula from the definition $v_{gr} = \partial\omega/\partial k$, i.e.,

$$v_{gr} = v_{ph} - \lambda \partial v_{ph}/\partial \lambda . \qquad (8.15)$$

It can be seen from (8.15) that the group velocity may be either positive or negative and either larger or smaller than the phase. A simple way of determining v_{gr} from the curve

$v_{ph} = v_{ph}(\lambda)$ was suggested by Ehrenfest. This method can easily be understood from Fig. 8.4. Examples of real waves in which v_{gr} and v_{ph} are in opposite directions are the inverse electromagnetic waves or inverse spatial harmonics of electromagnetic waves which propagate in the delay systems used in amplifiers and generators like travelling wave tubes.

We did not touch on the question of the velocity propagation of short impulses in dispersing media. A modern presentation of this problem is given in [5]. We shall only emphasize that for short impulses, and impulses with wide band-widths the concept of group velocity is indeterminate because the form of the impulse becomes greatly distorted as it propagates.

Finally, we introduce the concept of the propagation velocity of energy in a medium:

$$V = \frac{\text{average density of energy flow}}{\text{average energy density}}.$$

M. A. Leontovich has shown that if there is no absorption in the medium and no rotation of the polarization plane, V coincides with v_{gr}. Clearly, it is necessary, as before, that the packet's spectrum is narrow.

8.2 Group Velocity of Waves in Some Continuous Media

1. For gravity waves in deep water $\omega = \sqrt{gk}$ (Chapter 5), i.e., $\partial\omega/\partial k = (1/2)(g/k)^{1/2}$, but $v_{ph} = \omega/k = (g/k)^{1/2}$ and hence,

$$v_{gr} = (1/2)v_{ph}. \qquad (8.16)$$

It can then be seen from (8.16) that for long waves the phase and group velocities may be larger than the velocity of light c in vacuum. However no signal may propagate at a velocity greater than c. (Mandel'shtam [1] proved this statement by analyzing the propagation of a signal in two inertial systems moving at a constant relative velocity V_0).[2] In our example

---

[2] Prove this for yourself, starting from the formula $t' = [t - (V_0 x)/c^2]/\sqrt{1 - (V/c)^2}$, which links the time origins t and t' in these systems, but if the signal's

the paradox is easily explained, the dispersion law and hence the formula for v_{ph}, was derived for an incompressible liquid. This assumption itself is in contradiction with the theory of relativity.

2. It follows from (5.58) that for capillary waves $\omega = k^{3/2}(\sigma/\rho)^{1/2}$ and $v_{gr} = d\omega/dk = (3/2)(\omega/k) = (3/2)v_{ph}$, i.e., the propagation velocity of the energy v_{gr} of the capillary waves is greater than the velocity v_{ph} of the peak (for a given wavelength).

3. We noted in Chapter 5 that in a one-dimensional "cold" flow of electrons, Langmuir oscillations are transferred by the electrons with a drift velocity v_0, i.e., $v_{gr} = v_0$. Moreover, it was established that in a stationary "hot" plasma a wave transfers energy at a velocity $v_{gr} \approx 3kr_D\sqrt{kT_e/m_e}$, which is much smaller than the thermal velocity. Let us consider now the propagation of a transverse plane wave across the ionosphere, which consists of stationary free electrons.

In contrast to Chapter 5 we shall analyze the oscillations in a plasma using electrodynamics. Starting from Maxwell's equations and the continuity equation and assuming that all the variables change in the form $\exp(i\omega t - ikr)$, we find [6]:

$$k^2 E - k(kE) - (\omega^2/c^2)D = 0 . \qquad (8.17)$$

Now, using the equations of motion for a charged particles, the vector of the electrostatic induction **D** may be expressed in terms of the vector of the potential of the electrostatic field **E**: $D = \hat{\varepsilon}E$, where $\hat{\varepsilon}$ is the tensor of the dielectric permittivity of the medium. Substituting this into the expression for **D** in (8.17), we obtain a system of linear homogeneous equations because $D_\alpha = \sum_\beta \varepsilon_{\alpha\beta} E_\beta$, where $\varepsilon_{\alpha\beta}$ is a matrix. The dispersion equation [6] follows from the condition that this system of equations is consistent:

$$\text{Det} \left| k^2 \delta_{\alpha\beta} - k_\alpha k_\beta - (\omega^2/c^2)\varepsilon_{\alpha\beta} \right| = 0 , \qquad (8.18)$$

---

velocity in the unshaded system is greater than c, then in the shaded system the causality principle is broken, i.e. the signal arrives earlier than it left.

where δ is the Kronecker delta.

For an isotropic plasma without a magnetic field $\mathbf{D} = \varepsilon \mathbf{E}$ and for longitudinal waves we find from (8.17) that

$$\varepsilon = 0, \qquad (8.19)$$

and for cross-sectional waves

$$\varepsilon_\perp = k^2 c^2 / \omega^2. \qquad (8.20)$$

We can express $\varepsilon_\perp$ from the equation of motion for electrons $\partial \mathbf{v}_\perp / \partial t = (e/m_e) \mathbf{E}_\perp$ and hence $\mathbf{v}_\perp = (e/im_e\omega)\mathbf{E}_\perp$, and so we have for the current density $\mathbf{j}_\perp = e\rho_0 \mathbf{v}_\perp = -i\omega_{pe}^2 \mathbf{E}_\perp / 4\pi\omega$. By definition $\mathbf{D} = \varepsilon_\perp \mathbf{E}_\perp = \mathbf{E}_\perp + 4\pi \mathbf{j}_\perp / i\omega$. Thus,

$$\varepsilon_\perp = 1 - \omega_{pe}^2 / \omega^2. \qquad (8.21)$$

Equating (8.20) and (8.21), we obtain

$$\omega^2 = \omega_{pe}^2 + k^2 c^2. \qquad (8.22)$$

And hence, in the ionosphere $v_{ph} = c/\sqrt{1 - \omega_{pe}^2/\omega^2}$ and $v_{gr} = c\sqrt{1 - \omega_{pe}^2/\omega^2}$, i.e., v_{ph} is always larger than c, $v_{gr} < c$, and $v_{ph} v_{gr} = c^2$.

The latter relationship is not as general as is often thought. For instance, for various transmission lines used in technology and microwave electronics, the link between v_{ph} and v_{gr} changes form [10], i.e.,

$$v_{ph} v_{gr} = \frac{c^2}{\varepsilon\mu} \left(1 \mp \frac{\tau}{k} \frac{d\tau}{dk}\right)^{-1}, \qquad (8.23)$$

where $\tau^2 = k^2 - \beta^2$, k is the wave number in the medium, β is the phase constant for the transmission line, and ε, μ are the relative dielectric and magnetic permittivities of the medium; the upper sign corresponds to $k > \beta$ (fast waves when $\varepsilon = \mu = 1$), and the bottom sign corresponds to $K < \beta$ (slow waves; $\tau^2 = \beta^2 - k^2$). It follows from (8.23) that $v_{ph} v_{gr} = c^2$ only if $\varepsilon = \mu = 1$ and $d\tau/dk = 0$. For instance, $d\tau/dk = 0$ for metal wave carriers without losses and a homogeneous

PROPAGATION VELOCITY OF WAVES   181

dielectric filling, i.e., $v_{ph}v_{gr} = c^2/\varepsilon\mu$. Typically, we have v_{gr} and v_{ph} for damping systems, and these are much smaller than the velocity of light in the medium.

4. Let us now look at internal waves in a stratified liquid (Chapter 5). Let the medium be the unbounded and Vyasyalya's frequency be N = const. Then (5.62), viz., $\sin \theta = \mu(\omega/N)$, is valid. To be specific, let us assume that the wave propagates in the positive z-direction, i.e., $\mu = +1$. We use (5.62) as the definition of group velocity or use (5.61), which is the same: $\omega = N\xi/k = N[k_x^2 + k_y^2]^{1/2}/[k_x^2 + k_y^2 + k_z^2]^{1/2}$, assuming that the velocity $\mathbf{v}_{gr} = \mathbf{x}_0 \partial\omega/\partial k_x + \mathbf{y}_0 \partial\omega/\partial k_y + \mathbf{z}_0 \partial\omega/\partial k_z$, where $\mathbf{x}_0$, $\mathbf{y}_0$, and $\mathbf{z}_0$ are the corresponding unit vectors [3]. After differentiation and simple transformations, we obtain [7]

$$\mathbf{v}_{gr} = (Nk_z/k^3)[k_z\xi/\xi - \xi\nabla z] . \qquad (8.24)$$

It is not difficult to see from (8.24) that $(\mathbf{v}_{gr}\mathbf{k}) = 0$, i.e., $\mathbf{v}_{gr}$ is directed perpendicular to $\mathbf{k}$ (Fig. 8.5). When $q \to 0$, $s \to 0$ and $\omega \gg \Omega$, we have from (5.37)

$$\partial V(z)/\partial z + (\xi^2/\omega^2\rho_{00})\mathcal{P} = 0, \quad \text{and hence when}$$

$V(z) = V_0\exp\{i(\omega t - k_z z)\}$ we find for the pressure $p = [ik_z\omega^2\rho_{00}/\xi^2]\exp\{i(\omega t - \mathbf{kr})\}$. Consequently, $\nabla p = -i\mathbf{k}p$ is codirectional with $\mathbf{k}$. It follows from (5.30) and (5.31) that for internal waves we have $V_{0x} = k_x/\omega$ and $V_{0y} = k_y/\omega$. Using (5.19) and these relations for particle velocities we obtain

$$\mathbf{v} = (-k_z\xi/\xi^2 + \nabla z)v_z , \quad v_z = -iv_0\exp(i\omega t - i\mathbf{kr}) . \qquad (8.25)$$

However we immediately have from (8.25) that $(\mathbf{vk}) = 0$, i.e., the particles move along trajectories perpendicular to $\mathbf{k}$ in the plane containing the vector $\mathbf{k}$ and z-axis. By using the expression for the pressure p and particle velocity $\mathbf{v}$ it is easy to find from (8.25) the following relation for the density of the energy flow $\mathbf{P} = (1/4)(p\mathbf{v}^* + cc)$ averaged over time:

$$\mathbf{P} = (1/2)\rho_{00}v_0^2 N^2 \mathbf{v}_{gr} . \qquad (8.26)$$

By analogy we obtain for the average density of energy

$$W = (1/2)\rho_{00} vv^*$$

$$W = (1/2)\rho_{00} v_0^2 N^2 \ . \tag{8.27}$$

It follows from (8.26) that the energy flow is codirectional with the group velocity vector (Fig. 8.5) and the propagation velocity of the energy in a medium $V = P/W$ is exactly equal to the group velocity.

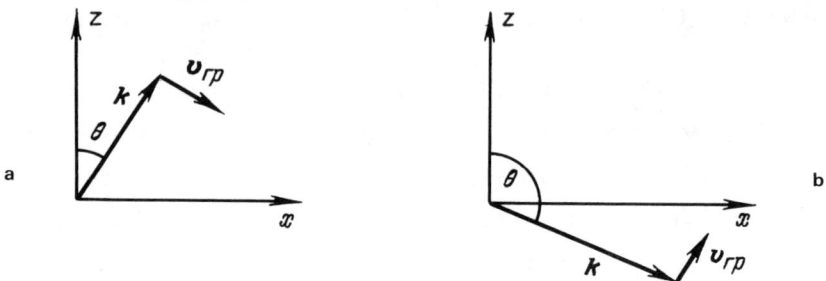

Fig. 8.5. Defining the group velocity of internal waves in a stratified liquid [7]: a) the wave moves upwards and the energy flow is downwards ($k_y = 0$, $\mu = 1$); b) the wave moves downwards and the energy flow is downwards ($k_y = 0$, $\mu = -1$).

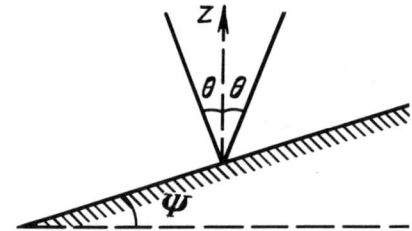

Fig. 8.6. On the reflection of a wave from an inclined floor.

Prove for yourselves that if the floor is sloping then the following is obeyed for wave reflection: the angle of incidence is equal to the angle of reflection, but with respect to the vertical to the plane of the floor and not the normal to the plane of the floor (Fig. 8.6). In order for the reflected wave to cancel the component of the incident wave's velocity that is perpendicular to the boundary, it is necessary that

$$k_{inc}\sin(\theta - \Psi) = k_{ref}\sin(\theta + \Psi) \ ,$$

when $\mathbf{k}_{inc}$ and the normal to the boundary lie in the plane of Fig. 8.6 (prove this!). An interesting conclusion must follow from this: the wavelength must change on reflection. There is no contradiction here because any wave of a given frequency may have any wavelength (Chapter 5).

It is easy to show from (5.83) that for Rossby waves when $k_y = 0$ the phase and group velocities of the waves have opposite directions. In general, when $k_y \neq 0$, we have

$$\partial\omega/\partial k_x = (\beta/\xi^2)\left(k_x^2 - k_y^2\right)/\left(k_x^2 + k_y^2\right) ,$$

$$\partial\omega/\partial k_y = (\beta/\xi^2) 2 k_x k_y / \left(k_x^2 + k_y^2\right)$$

and for the group velocity we have

$$\mathbf{v}_{gr} = (\beta/\xi^2)(\mathbf{x}_0 \cos 2\gamma - \mathbf{y}_0 \sin 2\gamma) ,$$

i.e., $\mathbf{v}_{gr}$ is directed from the end of the vector $\boldsymbol{\xi}$ to the center of the circle given by (5.83), with angle γ being between $\boldsymbol{\xi}$ and k_x-axis.

These examples show that for a medium with anisotropic dispersion, i.e., for a medium with a dispersion equation $\omega = \omega(k_x, k_y, k_z)$, the group velocity vector behaves quite strangely. It seems quite clear that from the point of view of wave kinematics, the concept of group velocity may be generalized to multidimensional systems. Without going into detail [3, 8], we shall write out the main equations. Let the vector $\mathbf{x}$ have coordinates x_1, x_2 and x_3 in a modulated wave $\mathbf{u}(\mathbf{x},t) \exp[i\Psi(\mathbf{x},t)]$. We define

$$\partial\Psi/\partial t = \omega , \quad \partial\Psi/\partial x_i = -k_i \quad (i = 1, 2, 3) , \tag{8.28}$$

where k_1, k_2 and k_3 are the components of the wave vector. The dispersion equation has the form $\omega = \omega(k_1, k_2, k_3)$ or

$$\partial\Psi/\partial t = \omega(-\partial\Psi/\partial x_1, -\partial\Psi/\partial x_2, -\partial\Psi/\partial x_3) . \tag{8.29}$$

By differentiating (8.29) with respect to x_i and using the definitions in (8.28), we obtain the three-dimensional analog of (8.11) in the form

$$\partial k_i/\partial t + U_j \partial k_i/\partial x_j = 0 , \tag{8.30}$$

where $U_j = \partial\omega/\partial k_j$ is a component of the group velocity vector. If $dx_j/dt = U_j$, then the k_i component of the wave vector is constant, and there is motion at constant velocity U_j along the straight line $x_j - U_j t$ = const. It is shown in [3] that the group velocity for sinusoidal waves $U_j = \partial \omega / \partial k_j$ coincides with the energy propagation velocity (for internal waves we have seen this already) in any homogeneous anisotropic systems.

A more complicated case involves the propagation of waves in nonhomogeneous, unsteady-state dispersing media, where $\omega = \omega(\mathbf{k}, \mathbf{x}, t)$. In this case the group velocity behaves like the beam velocity. We shall not touch on this point or any other more complicated topic, referring the reader instead to [3, 8, 11].

CHAPTER 9

ENERGY AND MOMENTUM OF WAVES

9.1 Equation for the Transport of the Average Energy Density by Wave Packets in Dispersing Media

Waves, like any moving object, transport energy as they propagate. The amount of transported energy depends on the nature of the wave: it may be quite significant in sea waves, which are able to move huge boulders during storms, or comparatively small, for instance electromagnetic waves of light, reaching the Earth from the Sun (power per square of surface of ~ 1 kW/m^2). Like moving particles, waves have momentum. However the existence of wave momentum should not cause doubt, although it is less noticeable than wave energy. For example, the light pressure of the Sun's radiation on the Earth orbit is very small, $p = 4.5 \cdot 10^{-7}$ Pa [1, 2].

We shall obtain an equation in this Chapter to describe the transfer of energy and momentum of waves in dispersing media [3-6].

When deriving the energy transport equation we shall proceed as in the derivation of the equation for the evolution of the wave vector (Chapter 8), that is we shall decline from using the Fourier integral and start, instead, from the Klein-Gordon equation with constant coefficients [3]:

$$u_{tt} - V^2 u_{xx} + \beta^2 u = 0 . \qquad (9.1)$$

By multiplying both sides of (9.1) by u_t we obtain

$$\frac{\partial}{\partial t} \left(\frac{1}{2} u_t^2 + \frac{1}{2} \beta^2 u^2 \right) - V^2 u_t u_{xx} = 0 . \qquad (9.2)$$

By adding the term $V^2 u_x u_{xt} = \frac{\partial}{\partial t} \left(\frac{1}{2} u_x^2 \right)$ to the left hand side of (9.2) and taking it away, we find that

$$- V^2 u_t u_{xx} - V^2 u_x u_{xt} = - V^2 \frac{\partial}{\partial t} (u_t u_x) .$$

These transformations yield an equation for the conservation of energy in the form

$$\frac{\partial}{\partial t} \left[\frac{1}{2} u_t^2 + \frac{1}{2} V^2 u_x^2 + \frac{1}{2} \beta^2 u^2 \right] + \frac{\partial}{\partial x} (- V^2 u_x u_t) = 0 , \qquad (9.3)$$

where $\left[\frac{1}{2} u_t^2 + \frac{1}{2} V^2 u_x^2 + \frac{1}{2} \beta^2 u^2 \right]$ is the energy density and $(-V^2 u_x u_t)$ is the energy flow.

Let us consider a group of waves (or as is often said a wave packet) which is changing slowly in space and over time. For such a wave group we have

$$u \sim \mathrm{Re}\, (A e^{i\Psi}) = a \cos (\Psi + \phi) , \qquad (9.4)$$

where $a = |A|$ and $\phi = \arg A$. Using (9.4), we calculate the energy density and the density of the energy flow. It is clear that

$$u_t \sim - i\omega a \sin (\Psi + \phi) + a_t \cos (\Psi + \phi) - \phi_t a \sin (\Psi + \phi) ,$$

where $u_t^2/2 \sim \omega^2 a^2 \sin^2 (\Psi + \phi)$ because the slow changes in a and ϕ means that terms containing a_t and ϕ_t can be neglected. In this approximation, it is easy to calculate the remaining terms in the energy density, which finally yields

$$\left[\frac{1}{2} u_t^2 + \frac{1}{2} V^2 u_x^2 + \frac{1}{2} \beta^2 u^2 \right] \sim \frac{1}{2} (\omega^2 + V^2 k^2) a^2 \sin^2 (\Psi + \phi)$$
$$+ \frac{1}{2} \beta^2 a^2 \cos^2 (\Psi + \phi) , \qquad (9.5)$$

in this we assumed that $\partial\Psi/\partial t = \omega$ and $\partial\Psi/\partial x = - k$. By analogy we have for the density of the energy flow

$$V^2 u_x u_t \sim V^2 \omega k a^2 \sin^2 (\Psi + \phi) . \qquad (9.6)$$

If we use an equation containing higher order derivatives instead of (9.1), then clearly when it is worked out using (9.4) additional terms containing derivatives with respect to ω and k will appear. However, because we are considering a slowly changing wave packet, ω and k also change slowly and these terms may be neglected. Let us look at the values of

ENERGY AND MOMENTUM OF WAVES

(9.5) and (9.6) averaged over the period. This is justifiable because we are interested in significant (average) changes in ω, k and a, and not in small oscillations. Thus for the averaged values of the energy density and energy flow density we obtain for the assumptions we have made the equation

$$\mathcal{E} = (\omega^2 + V^2 k^2) a^2/4 + \beta^2 a^2/4 , \qquad (9.7)$$

$$S = V^2 \omega k a^2/2 . \qquad (9.8)$$

The dispersion equation follows from (9.1), i.e.,

$$\omega^2 = \beta^2 + V^2 k^2 . \qquad (9.9)$$

Using (9.9), the equations (9.7) and (9.8) take the final form

$$\mathcal{E} = (\beta^2 + V^2 k^2) a^2/2 , \qquad (9.10)$$

$$S = V^2 \omega k a^2/2 . \qquad (9.11)$$

By definition $v_{gr} = d\omega/dk$, and hence we attain from (9.9)

$$v_{gr} = V^2 k / \sqrt{V^2 k^2 + \beta^2} . \qquad (9.12)$$

We find from (9.10)-(9.12) and using (9.9) that

$$V_{gr} = \mathcal{E}/S . \qquad (9.13)$$

The generality of this expression was already been noted in Chapter 8.

Returning now to (9.3) and using (9.13) as a justification, we can assume that the conservation of the average energy density is expressed by the differential equation

$$\frac{\partial \mathcal{E}}{\partial t} + \frac{\partial}{\partial x} \left(v_{gr} \mathcal{E} \right) = 0 . \qquad (9.14)$$

It is shown in [3] that this equation corresponds to the situation in which the total energy between the two straight lines $x - v_{gr1,2} t = \text{const}$ on the (x, t)-plane remains constant. To prove this, consider the expression for energy

$$\langle \mathcal{E}(t) \rangle = \int_{x_1(t)}^{x_2(t)} \mathcal{E} \, dx , \qquad (9.15)$$

where x_1 and x_2 are points moving with velocities $v_{gr}(k_1)$ and $v_{gr}(k_2)$. It is clear that

$$\frac{d\langle \mathcal{E}(t) \rangle}{dt} = \int_{x_1(t)}^{x_2(t)} \frac{\partial \mathcal{E}}{\partial t} \, dx + v_{gr}(k_2)\mathcal{E}_2 - v_{gr}(k_1)\mathcal{E}_1 , \qquad (9.16)$$

and that this value is zero, as follows from (9.14). It is no less obvious that (9.16) turns into (9.14) at the limit as $x_2 - x_1 \to 0$.

The expression for the averaged energy density may be written in the form $\mathcal{E} = F(k)a^2$. By substituting this expression into (9.14) we obtain

$$F(k)\left\{ \frac{da^2}{\partial t} + \frac{\partial}{\partial x}\left(v_{gr} a^2\right) \right\} + \frac{\partial F}{\partial k} a^2 \left\{ \frac{\partial k}{\partial t} + v_{gr} \frac{\partial k}{\partial x} \right\} = 0 .$$

However, as was shown in Chapter 8, we have $\partial k/\partial t + v_{gr} \partial k/\partial x = 0$, and hence

$$\frac{\partial a^2}{\partial t} + \frac{\partial}{\partial x}\left(v_{gr} a^2\right) = 0 . \qquad (9.17)$$

Expressions like (9.14) and (9.17) can easily be extended to the multidimensional case. Such a generalization has done in [3] for the Klein--Gordon equation and the equation $u_{tt} - V^2 \nabla^2 u_{tt}$. The equation for the transport of the averaged energy density by a wave packet in a medium with the given dispersion takes the form

$$\frac{\partial \mathcal{E}}{\partial t} + \frac{\partial}{\partial x_j}\left[\left(v_{gr}\right)_j \mathcal{E}\right] = 0 \quad \text{or} \quad \frac{\partial a^2}{\partial t} + \frac{\partial}{\partial x_j}\left[\left(v_{gr}\right)_j a^2\right] = 0 . \qquad (9.18)$$

These results are somewhat unsatisfactory because they were obtained for concrete equations. Wysem has shown [3] that the "averaged variation principle" is valid for the functions $a(\mathbf{x}, t)$ and $\Psi(\mathbf{x}, t)$, the application of which results in (9.18).

9.2 Density of the Energy of an Electromagnetic Wave in a Medium with Dispersion

The well-known expression $S = (c/4\pi)[EH]$ for the density of electromagnetic energy is valid even in a medium with dispersion [5, 6]. A no less well-known equation follows from Maxwell's equations, i.e.,

$$\text{div } S = (1/4\pi)[E\partial D/\partial t + H\partial B/\partial t] , \qquad (9.19)$$

where E and D, and H and B are the field strength and induction of the electrostatic and magnetic fields respectively.

If there is no dispersion, and if the permittivities ε and μ are real constants, then (9.19) expresses the change in the density of the electromagnetic energy per unit volume $\mathcal{E} = (1/8\pi)(\varepsilon E^2 + \mu H^2)$, i.e., $\partial \mathcal{E}/\partial t + \text{div } S = 0$. When there is dissipation the energy density of the thermal losses is defined by the imaginary parts of ε and μ, i.e.,

$$Q = \frac{\omega}{4\pi}\left(\text{Im } \varepsilon <E^2> + \text{Im } \mu <H^2>\right) , \quad \frac{\partial \mathcal{E}}{\partial t} + \text{div } S + Q = 0 ,$$

where the <> indicate averaging over time.

We find $\mathcal{E}$ following [5]. Consider a narrow wave packet consisting of monochromatic components with frequencies close to some ω_0, i.e., a narrow packet with bound width $\Delta\omega \ll \omega_0$:

$$E = E_0(t)e^{i\omega_0 t} , \quad \text{Re } E = \frac{1}{2}\left[E_0(t)e^{i\omega_0 t} + E_0^*(t)e^{-i\omega_0 t}\right] ,$$

$$H = H_0(t)e^{i\omega_0 t} , \quad \text{Re } H = \frac{1}{2}\left[H_0(t)e^{i\omega_0 t} + H_0^*(t)e^{-i\omega_0 t}\right]$$

(there are similar expressions for D and B), where $E_0(t)$ and $H_0(t)$ are functions of time that change slowly in comparison to $\exp(i\omega_0 t)$. We substitute the expression for the real parts of the field strength E and H and the expressions for D and B into (9.19), and then average the result over the period $2\pi/\omega_0$. It is clear that the rapidly changing terms like $E_0(\partial D_0/\partial t)\exp(2i\omega_0 t)$ and $E_0^*(\partial D_0^*/\partial t)\exp(-2i\omega_0 t)$ will disappear during the averaging, and only terms like the following will remain:

$$M = (1/16\pi)[E\partial D^*/\partial t + E^*\partial D/\partial t]$$

(we do all the transformations only with the first terms on the right-hand side of (9.19)). We now express the derivative

$\partial D/\partial t$ in form $\hat{f}E$ where the operator $\hat{f} = (\partial/\partial t)\hat{\varepsilon}$. Suppose this operator acts on $E = E_0 \exp(i\omega_0 t)$. Clearly, if $E_0 = $ const (the field is purely harmonic), then $\hat{f}E = i\omega_0 \varepsilon(\omega_0) E$ or $\hat{f}E = f(\omega_0)E$, where $f(\omega) = i\omega\varepsilon(\omega)$. We expand the $E_0(t)$ function into a Fourier integral, which corresponds to representing it as a group of monochromatic components $E_{0\omega}\exp[i(\omega - \omega_0)t]$ with $E_{0\omega} = $ const:

$$E_0(t) \sim \int_{-\infty}^{\infty} E_{0\omega} \exp[i(\omega - \omega_0)t] d(\omega - \omega_0) .$$

Because $E_0(t)$ is a function of time that is slowly changing, only components for which $\Delta\omega = |\omega - \omega_0| \ll \omega_0$ is true will enter the integral. This allows us to write the following relation:

$$\hat{f}E_{0\omega} e^{i(\omega_0+\Delta\omega)t} = f(\omega_0 + \Delta\omega) E_{0\omega} e^{i(\omega_0+\Delta\omega)t} \approx f(\omega_0) E_{0\omega} e^{i(\omega_0+\Delta\omega)t}$$
$$+ \Delta\omega \frac{df(\omega_0)}{d\omega_0} E_{0\omega} e^{i(\omega_0+\Delta\omega)t} .$$
(9.20)

It is easy to see that

$$\partial E_0/\partial t \sim \int_{-\infty}^{\infty} i\Delta\omega E_{0\omega} e^{i\Delta\omega t} d(\Delta\omega) .$$
(9.21)

We integrate (9.20) with respect to $\Delta\omega$ from $-\infty$ to $+\infty$, which corresponds to an inverse Fourier transform. Using (9.21), we find

$$\hat{f}e^{i\omega_0 t} \int_{-\infty}^{\infty} E_{0\omega} e^{i\Delta\omega t} d(\Delta\omega) = f(\omega_0) e^{i\omega_0 t} \int_{-\infty}^{\infty} E_{0\omega} e^{i\Delta\omega t} d(\Delta\omega)$$

$$- i \frac{df(\omega_0)}{d\omega_0} e^{i\omega_0 t} \int_{-\infty}^{\infty} i\Delta\omega E_{0\omega} e^{i\Delta\omega t} d(\Delta\omega)$$

$$= f(\omega_0) E_0 e^{i\omega_0 t} - i \frac{df(\omega_0)}{d\omega_0} \frac{\partial E_0}{\partial t} e^{i\omega_0 t} .$$

Omitting the zero subscript in ω_0, we obtain

$$\frac{\partial D}{\partial t} = i\omega\varepsilon(\omega)E + \frac{d(\omega\varepsilon)}{d\omega}\frac{\partial E_0}{\partial t}e^{i\omega_0 t}. \qquad (9.22)$$

Remember that $\varepsilon(\omega) = \operatorname{Re}\varepsilon(\omega) + i\operatorname{Im}\varepsilon(\omega) = \varepsilon'(\omega) + i\varepsilon''(\omega)$. The frequency range in which $\varepsilon''(\omega)$ is smaller than $\varepsilon'(\omega)$ is called the transparency region of the medium (by analogy with magnetic permeability). In such regions it is possible to assume that $\varepsilon''(\omega) = 0$, so that $Q = 0$. If we now assume the loss is $\varepsilon(\omega) = \varepsilon'(\omega) = \varepsilon^*(\omega)$, we have a relation for M, i.e.,

$$M = \frac{1}{16\pi}\frac{d(\omega\varepsilon)}{d\omega}\left(E_0^*\frac{\partial E_0}{\partial t} + E_0\frac{\partial E_0^*}{\partial t}\right)$$

$$= \frac{1}{16\pi}\frac{d(\omega\varepsilon)}{d\omega}\frac{d}{dt}\left(E_0^*E_0\right) = \frac{1}{16\pi}\frac{d(\omega\varepsilon)}{d\omega}\frac{d}{dt}\left(EE^*\right).$$

Because all the mathematical arguments are similar for a magnetic field, we can write the expression for the averaged energy density:

$$\langle\mathcal{E}(t)\rangle = \langle\mathcal{E}_{el}\rangle + \langle\mathcal{E}_m\rangle = \frac{1}{16\pi}\left[\frac{d(\omega\varepsilon)}{d\omega}\langle|E|^2\rangle + \frac{d(\omega\mu)}{d\omega}\langle|H|^2\rangle\right]. \qquad (9.23)$$

We shall show another simple method for obtaining the energy relations in media with time and spatial dispersion, a method based on the use of the dispersion equation for the system [7,8]. Consider a one-dimensional wave $v' = \operatorname{Re}\{ve^{i(\omega t - kx)}\}$, where v', is, for example, the rate of a disturbance in an electron beam. Suppose the wave velocity is disturbed by an external wave $F' = \operatorname{Re}\{F\exp i(\omega t - kx)\}$ (for example, the longitudinal electric component of a travelling electromagnetic wave), which determines the values of ω and k. The amplitudes v and F are defined such that the average power over the period of the interaction between the disturbance and the external wave is proportional to $(F'v'^*)$. If v and F are linearly related by the relation $D(\omega, k)v = -iF$, where $D(\omega, k)$ is an analytical function of ω and k, then we have two equations: the energy per unit length averaged over the period

$$\langle\mathcal{E}\rangle = (\partial D/\partial\omega)vv^*/4 \qquad (9.24)$$

and the energy flow per unit length averaged over the period

$$<S> = (\partial D/\partial k)vv^*/4 .\qquad(9.25)$$

In the absence of an external interaction we have $D(\omega, k) = 0$ and $v_{gr} = d\omega/dk = (\partial D/\partial k)(\partial D/\partial \omega)^{-1} = <S>/<\mathcal{E}>$, where the complete differential is taken along the whole of the dispersion curve.

We would leave the reader to prove the very useful formulae (9.24) and (9.25). As an example of the application we consider waves of space charge in an electron beam, starting from the equation for the density of the group current j' given a disturbance to the flow by an external travelling electromagnetic wave with a longitudinal electrical component of the field E' (Chapter 7). Assuming that all the variables change over time as $\exp(i\omega t)$, this equation has the form

$$\frac{\partial^2 j'}{\partial x^2} + 2i\frac{\omega}{v_0}\frac{\partial j'}{\partial x} - \left[\left(\frac{\omega}{v_0}\right)^2 - \left(\frac{\omega_q}{v_0}\right)^2\right]j' = \frac{i\omega\omega_p^2}{4\pi v_0^2}E' ,\qquad(9.26)$$

where v_0 is the beams constant velocity, and $\omega_q = R(\omega, k)\omega_p$. If $E' \sim \exp(i\omega t - ikx)$, then we have from (9.26)

$$(4\pi/\omega S)\left[(\omega - kv_0)^2/\omega_p^2 - R^2(\omega, k)\right]j'S = -iE' ,\qquad(9.27)$$

i.e.,

$$D(\omega, k) = \left[(\omega - kv_0)^2/\omega_p^2 - R^2(\omega, k)\right]\frac{4\pi}{\omega S} ,\qquad(9.28)$$

where S is the beam's cross-sectional area. The form of $D(\omega, k)$ is defined such that the average power over the period of the interaction between the electron beam and the external travelling wave is $(1/2)\text{Re}(E' j'^* S)$. When the external disturbance is taken away, we have $D(\omega, k) = 0$ and $(\omega - kv_0)^2 = R(\omega, k)\omega_p^2$, i.e., two waves of space charge in a drifting beam: a fast wave ($\omega - v_0 k = R\omega_p$) and a slow wave ($\omega - v_0 k = -R\omega_p$). We find from (9.24), (9.25) and (9.28) that

$$<\mathcal{E}> = \frac{2\pi j' j'^* S}{\omega\omega_p}\left[\frac{\omega - kv_0}{\omega_p} - R\omega_p\frac{\partial R}{\partial \omega}\right] ,\qquad(9.29)$$

$$\langle S \rangle = \frac{2\pi j' j'^* S v_0}{\omega \omega_p} \left[\frac{\omega - k v_0}{\omega_p} + R \frac{\omega_p}{v_0} \frac{\partial R}{\partial k} \right]. \quad (9.30)$$

If the beam is infinitely wide and $R = 1$, then $v_{gr} = \langle S \rangle / \langle \mathcal{E} \rangle = v_0$ and the energy transfer is only associated with the kinematic motion of beam. However for a beam of a finite thickness we have

$$v_{gr} = v_0 \left[\frac{\omega - k v_0}{\omega_p} + R \frac{\omega_p}{v_0} \frac{\partial R}{\partial k} \right] \left[\frac{\omega - v_0 k}{\omega_p} - R \omega_p \frac{\partial R}{\partial \omega} \right]^{-1},$$

i.e., the energy propagation is determined both by the kinematics of the beam and by the second term in the square brackets, which are of electromagnetic origins. Equations (9.29) and (9.30) are also valid for the relativistic case, if in the definition of ω_p we use the longitudinal relativistic mass. The equations are useful in the noise theory of electron beams. It is interesting that when $\partial R / \partial \omega = \partial R / \partial k = 0$ we have the following for the fast (subscript f) and slow (subscript s) waves of space charge from (9.29) and (9.30):

$$\langle \mathcal{E}_{f,s} \rangle = \pm \frac{2\pi S}{\omega \omega_p} j' j'^*, \quad \langle S_{f,s} \rangle = \pm v_0 \langle \mathcal{E}_{f,s} \rangle, \quad (9.31)$$

i.e., the fast wave has a positive energy and the slow wave has a negative energy. We have devoted the next chapter to waves with negative energies.

9.3 Momentum of a Wave Packet

Suppose in a medium which is moving relative to the observer at a velocity $|V| \ll c$ (c is the velocity of light) a wave packet is propagating. Its energy in the coordinates frame moving at a velocity V is equal to $\mathcal{E}_v$, and in the stationary reference frame the energy is $\mathcal{E}_v^0 \neq \mathcal{E}_v$. In our arguments we shall use the fact [4] that when $|V| \ll c$ we have Galilean physical invariance: the laws governing changes in state of physical systems are independent of the inertial frame of reference (for mechanics this means that Newton's equations are invariant with respect to Galilean transformations). First we shall consider whether $\mathcal{E}_v^0$ and $\mathcal{E}_v$ are related. To do this we consider a particle of mass m

moving with respect to the observer at a velocity $v_0 = V + v$ apart from the wave packet. The quantity v is the relative velocity of motion. The extra kinetic energy of the particle is

$$\mathcal{E}^0 = mv_0^2/2 = mV^2/2 + mvV + mv^2/2 \ . \tag{9.32}$$

Because the particle's momentum is $p = mv$ and $\mathcal{E} = mv^2/2$ is the energy in the reference frame moving at a velocity V, we have $\mathcal{E}^0 = \mathcal{E} + pV$, omitting the constant $mV^2/2$. We shall further assume that the particle and wave packet exchange energy and momentum. The following relation between energy and momentum in the moving medium is a consequence of Galilean invariance:

$$\mathcal{E}_v^0 = \mathcal{E}_v + PV \ . \tag{9.33}$$

The structure of (9.33) arises from the fact that it must coincide exactly with the equation written above for the particle. The wave and the particle interact most effectively when the conditions for spatial resonance are fulfilled, i.e., when the particle's velocity v equals the wave's phase velocity v_{ph}. This condition is most conveniently written in the form of the condition for Cherenkov radiation $\omega - kv = 0$. Interaction with the wave changes (reduces) the particle's energy $\Delta\mathcal{E} = \Delta(mv^2/2) = mv\Delta v = v\Delta p$ which is associated with changes in its momentum. We must write a similar relation for the wave packet as a consequence of Galilean invariance. If it is assumed that the resultant change in energy $\Delta\mathcal{E}_v$ and momentum ΔP of the wave packet is proportional to the square of the amplitude, then $\Delta\mathcal{E}_v$ and ΔP are proportional, i.e., when there is spatial resonance we have $\mathcal{E}_v = vP$. The momentum P is directed along the k vector because the velocity component of the particle perpendicular to k may be arbitrary. Thus it follows from the condition $\omega = kv$ that $P = (k/\omega)\mathcal{E}_v$, whence it can be seen that $(v_{ph}P) = \mathcal{E}_v$ (the wave's phase velocity is a ratio of its energy to its momentum). If we introduce the amplitude of the wave by the expression $\mathcal{E}_v = \omega|a^2| = \omega N$, where N is the number of waves in the packet with a given wave number k [4], then $P = kN$. By using the last two expressions for $\mathcal{E}_v$ and P in (9.33), we find that $\mathcal{E}_v^0 = \omega N + kV = \omega_0 N$, where $\omega_0 = \omega + kV$ is a Doppler frequency.

CHAPTER 10

WAVES WITH NEGATIVE ENERGY. LINKED WAVES

10.1 General Notes

In the preceding chapter we discovered that the energy density and the energy flow density of the slow space-charge wave in an electron beam are negative (see (9.31) and (9.32)). At first sight this seems to contradict some general principles. For example, some energy must be spent in disturbing the electromagnetic wave packet in a medium with dissipation, and hence when energy ceases being "pumped" from the outside, the dissipation existing in the dispersing system (even if it is small) must transform all the energy

$$<W(t)> = \frac{1}{16\pi}\left[\frac{d(\omega\varepsilon)}{d\omega}<|E|^2> + \frac{d(\omega\mu)}{d\omega}<|H|^2>\right]$$

(Chapter 9) into heat. Since entropy must increase, the heat must be released and not absorbed, and hence we obtain [1]

$$<W(t)> > 0, \quad d(\omega\varepsilon)/d\omega > 0, \quad d(\omega\mu)/d\omega > 0. \quad (10.1)$$

However, this is only true for equilibrium media. Equation (10.1) need not be fulfilled in nonequilibrium media; and it is in such media that disturbances and wave propagation may occur with negative energy. The physical meaning of this will be made clear below.

The nonequilibrium may arise for a variety of reasons [2], e.g., unbalanced flows in particular directions, external fields, density gradients, and temperature gradients. There are some very well-known examples of nonequilibrial media: electron beams, interacting with the fields of a braking system (travelling-wave tubes, backward-wave tubes), and plasma with velocity distribution functions that have many peaks for the charged particles (a special case is the interaction between an electron beam and a plasma), and media with negative conductivity or viscosity (tunneling and Gann semiconductors), and boundary layers and other types of shear flow in hydrodynamics. We shall now try to explain why waves with negative energy arise in such systems and why waves with particular signs are disturbed and what the results of an interaction between linked waves and energies with various signs are.

10.2 Waves with Positive and Negative Energies

The concept of a wave with a negative energy first appeared in microwave electronics in the form of the Tchu theorem on kinetic power [21]. Tchu demonstrated that in a slow space-charge wave in an electron beam the wave was related to a negative kinetic power. The following very important step in understanding waves with negative energies was made by P.A. Sterrok [3] who showed that in a medium moving at a velocity u, the energies of the fast and slow waves, which are seen to be stationary by the observer, are expressed by

$$W_f = W_0(1 + u/v_{ph}) , \quad W_{sl} = W_0(1 - u/v_{ph}) , \qquad (10.2)$$

where v_{ph} and $(-v_{ph})$ are the velocities of the waves in the moving reference frame, and W_0 is the energy in this system, without making any assumptions about the nature of the waves. It can be seen from (10.2) that when $v_{ph} < u$ the value of W_{sl} is negative whilst the group velocity of both waves is positive. A simpler, though lest strict derivation of (10.2) than in [3] was presented in [4]. Later waves with negative energies were widely discussed both in the periodical [5] and book literature [6,7,18,22].

Physically, a wave with negative energy is one such that the total energy of the system consisting of the wave and the medium is decreased as the wave's amplitude increases. Besides waves in nonequilibrium media, longitudinal electrostatic waves also possess negative energies if their spectra lay in regions of the medium's anomalous dispersion $d\varepsilon/d\omega < 0$, for which the energy density is $<W_{el}>$ = $(\omega/16\pi)(d\varepsilon/d\omega)<E>^2 < 0$. We shall clarify the meaning of the term negative energy using the example that we know already of the distribution of waves of space charge in a drifting electron beam. The linearized equations for the problem are in the notation we have used before

$$\partial v'/\partial t + v_0 \partial v'/\partial x - (e/m)E_{spch} = 0 , \qquad (10.3)$$

$$\partial \rho'/\partial t + \rho_0 \partial v'/\partial x + v_0 \partial \rho'/\partial t = 0 , \qquad (10.4)$$

$$\partial E_{spch}/\partial x = 4\pi \rho' . \qquad (10.5)$$

WAVES WITH NEGATIVE ENERGY. LINKED WAVES 197

Suppose all the variables change as $f' \sim \exp(i\omega t - ikx)$. Then for the system (10.3)-(10.5) to be consistent it follows that $(\omega - kv_0)^2 = \omega_p^2 = (4\pi e \rho_0)/m$ and $\omega - kv_0 = \omega_p$ correspond to a fast wave of space charge and $\omega - kv_0 = -\omega_p$ corresponds to a slow wave. We find from (10.3) and (10.5) that

$$v' = (e/m)E_{spch}/i(\omega - kv_0) \text{ and } E_{spch} = -4\pi\rho'/ik,$$

and hence we have $v' = (4\pi e/m)\rho'/k(\omega - kv_0)$ or

$$\frac{v'}{v_0} = \frac{\omega_p^2}{kv_0(\omega - kv_0)} \frac{\rho'}{\rho_0}. \tag{10.6}$$

Since $(\omega - kv_0) = \pm \omega_p$ and $kv_0 = \omega \mp \omega_p$ we find from (10.6) a relation linking the variable velocity components and the densities of the volume charge for the slow and fast waves, i.e., respectively

$$\frac{v'_{sl}}{v_0} = -\frac{\omega_p}{\omega + \omega_p} \frac{\rho'_{sl}}{\rho_0}, \quad \frac{v'_f}{v_0} = \frac{\omega_p}{\omega - \omega_p} \frac{\rho'_f}{\rho_0}. \tag{10.7}$$

It can be seen from (10.7) that in a slow wave the velocity disturbance is out of phase to the density disturbance (the minus sign in the first formula in (10.7)), whilst in the fast wave it is in phase with the density disturbance (the plus sign in the second formula in (10.7)). Equation (10.7) can be simplified further if $\omega_p \ll \omega$ (as is typical, for example, for vacuum microwave electronics [8]), whence

$$\frac{v'_{sl}}{v_0} = -\frac{\omega_p}{\omega} \frac{\rho'_{sl}}{\rho_0}, \quad \frac{v'_f}{v_0} = \frac{\omega_p}{\omega} \frac{\rho'_f}{\rho_0}. \tag{10.8}$$

These expressions (10.7) (or (10.8)) immediately explain why the energy is positive in the fast wave and negative in the slow wave. In fact, it follows from (10.8) that for a fast wave in a region where disturbances lead to increases in density ρ'_f, the velocity of a particle is greater than v_0, and the in regions where the density decreases, the electron velocity is smaller than v_0. Therefore when a fast wave is excited in a stream electrons are accelerated faster than v_0, and the resultant kinetic energy carried by the beam is greater than the energy of the undisturbed beam. If a slow wave is excited in region where compressions have formed (increases in ρ'_{sl}), the electron velocity is, by contrast, slower than v_0 and larger than v_0 where rarefactions occur (decreases in ρ'_{sl}). As the result, when a slow wave is excited in an electron beam, we get electrons that are slower than v_0 and the energy carried with the beam is smaller than the energy of a beam without waves.

Because the dispersion equation for the system has the form

$$\varepsilon(\omega, k) = 1 - \omega_p^2/(\omega - kv_0)^2 = 0, \qquad (10.9)$$

it can be seen that on the curves $\omega - kv_0 = -\omega_p$, which correspond to the slow wave, we have

$$\partial\varepsilon/\partial\omega = 2\omega_p^2/(\omega - kv_0)^3 = -2/\omega_p < 0, \qquad (10.10)$$

i.e., the energy of this wave is negative. At the same time for a fast wave we have $\partial\varepsilon/\partial\omega = 2/\omega_p > 0$ and the wave's energy is positive.

It is clear that in contrast to the slow wave, a wave with positive energy is one such that a growth in its amplitude corresponds to an increase in the total energy of the system consisting of the medium and the wave.

Let us try to obtain an expression for the density of the energy in the electron beam proceeding directly from the one-dimensional equation of motion for the beam in a longitudinal electron field $\partial v/\partial t + v\partial v/\partial x = (e/m)E_x$, the equation $j = \rho v$ for the current density, and the one-dimensional equation of continuity $\partial j/\partial x + \partial \rho/\partial t = 0$. Let us consider, as was done in [9], the product $E_x j$. Using equation of motion and the formula for j, we find

$$E_x j = (m/e)[\partial v/\partial t + v\partial v/\partial x]\rho v. \qquad (10.11)$$

Equation (10.11) and the continuity equation take the form

$$\partial W_d/\partial t + \partial S_d/\partial x - E_x j = 0, \qquad (10.12)$$

where the density of the kinetic energy is

$$W_d = (m/2e)\rho v^2 \qquad (10.13)$$

and the density of the kinetic energy of the flow is

$$S_d = (m/2e)\rho v^3, \qquad (10.14)$$

where $S_d/W_d = v$, i.e., the full velocity of the beam.

When deriving (10.12)-(10.14) we made no assumption about the smallness of the disturbances. If, however, we assume that $v = v_0 + v'$, $\rho = \rho_0 + \rho'$, $j = j_0$ $j' = j_0 + v_0\rho' + \rho_0 v'$ (each perturbation is much smaller than the constant value of the variable), then by retaining second order terms in (10.13) we obtain

$$W_d = (m/2e)[\rho_0(v_0^2 + 2v_0v' + v'^2) + (v_0^2 + 2v_0v')\rho'] \qquad (10.15)$$

We calculate the kinetic energy density over the period for a drifting beam assuming that $E_x = E_{spch}$ and that the beam is locally disturbed at the input by a high frequency signal with frequency ω, which maintains itself, i.e., a wave of space charge propagates in the beam, the wave being of the form

$$E_{spch} = E_{spch}^0 \exp\{i\omega t - i(k + \omega_p/v_0)x\}$$
$$+ E_{spch}^0 \exp\{i\omega t - i(k - \omega_p/v_0)x\},$$

where E_{spch}^0 is determined by the initial disturbance. Using (10.15) we have

$$<W_d> = \frac{1}{2\pi}\int_0^{2\pi} \frac{m}{2e}\rho_0 v_0^2 d(\omega t) + \frac{1}{2\pi}\int_0^{2\pi} \frac{m}{e}\rho_0 v_0 v' d(\omega t)$$

$$+ \frac{1}{2\pi}\int_0^{2\pi} \frac{m}{2e}\rho_0 v'^2 d(\omega t) + \frac{1}{2\pi}\int_0^{2\pi} \frac{m}{2e}\rho' v_0^2 d(\omega t)$$

$$+ \frac{1}{2\pi}\int_0^{2\pi} \frac{m}{e}\rho' v_0 v' d(\omega t).$$

The first integral is the energy density of an undisturbed

beam which we denote $\langle W_d^0 \rangle$. Because v' and ρ' are superpositions of harmonic terms (waves of space charge), the second and fourth integrals are zero. Thus we are interested in

$$\delta \langle W_d \rangle = \langle W_d \rangle - \langle W_d^0 \rangle =$$

$$= \frac{m\rho_0}{4\pi e} \int_0^{2\pi} v'^2 d(\omega t) + \frac{mv_0}{2\pi e} \int_0^{2\pi} v'\rho' d(\omega t). \quad (10.17)$$

Since $v' = (e/m)E_{spch}/i(\omega - kv_0)$, and v' and ρ' for $\omega_p \ll \omega$ are related by (10.8) then by calculating the integrals in (10.16) we have the following for a fast and a slow wave of space charge, respectively:

$$\delta \langle W \rangle_{df} \approx \frac{(E^0_{spch})^2}{8\pi} \frac{\omega}{\omega_p} > 0,$$

$$\delta \langle W \rangle_{dsl} \approx - \frac{(E^0_{spch})^2}{8\pi} \frac{\omega}{\omega_p} < 0.$$

(10.18)

The approximation sign appears because we ignored terms (10.18) that we obtained from the first integral. This is valid when $\omega_p \ll \omega$ because this term is a factor ω/ω_p less than the second term.

Note that if we take $E_{spch} = -4\pi j'/i\omega$ in (9.31) and consequently $j'j'^* = \omega^2 E^2_{spch}/16\pi^2$, then we arrive at (10.18).

Thus the formula $\langle W \rangle_{el} = (\omega/16\pi)(d\epsilon/d\omega)|E_{spch}|^2$ leads to the same result when $d\epsilon/d\omega = \pm 2/\omega_p$. When $\omega_p \ll \omega$ we have the following from both formulae in (10.2):

$$W_{f,sl}/W_0 = \pm \omega/\omega_p, \quad (10.19)$$

because $u = v_0$, $v_{ph} = \pm v_0 \omega_p/\omega$.

To sum up, waves have negative energy when the velocity and density disturbances are out of phase. It would appear that this explanation for the occurrence of waves with negative energies is quite general, as it applies both to electronics and to many situations in fluid mechanics, in which compressibility is fundamental. Such a simple interpretation of the physical sense of waves with the negative energy is not always possible for incompressible liquids. For instance, if the flow is density stratified [20], then the explanation is only valid if a density disturbance is taken to mean a disturbance to the density gradient. In flows without stratification, boundary layer flows for example, we must consider velocity waves and pressure waves [19], [2].

Concerning the conditions that must be satisfied in order for waves with negative energies to arise in a medium, it is clear that the slow wave must be able to give up some of its energy either to the medium or to other waves. We shall illustrate this using the example of a resistive amplifier [8] (Fig. 10.1). The electron beam is modulated in the input device and passed through a dielectric tube whose inner surface is covered in an absorbing layer. The beam induces a variable charge in the tube. The field created by the induced charge in turn acts on the electron beam and changes the varying component of the beam current.

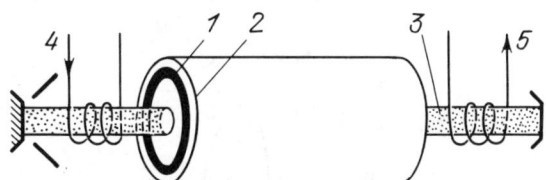

Fig. 10.1. Illustrative diagram of a resistive amplifier: 1) resistive layer; 2) dielectric tube; 3) electron beam; 4 and 5) input and output devices.

After passing through the tube, the beam is fed into the output device. The input action excites two waves of space charge in the beam, the fields of the space charge causing charges to move in the resistive walls. This in turn leads to Joule losses of energy from the wave. The losses affect the fast and slow waves differently. The fast waves (the wave with positive energy) dies away, whilst the slow wave grows increasing in amplitude as it gives off energy to the medium.

Experimental proof that the slow space-charge wave grows in a resistive amplifier is illustrated in Fig. 10.2. The above can easily be confirmed using the simplest theory, the linearized equations of which are

$$\frac{\partial^2 j'}{\partial x^2} + 2i \frac{\omega}{v_0} \frac{\partial j'}{\partial x} - \left(\frac{\omega}{v_0}\right)^2 j' = \frac{i\omega \rho_0}{2V_0} E, \qquad (10.20)$$

$$E = -\frac{4\pi(j' + j_{cd})}{i\omega} = -\frac{4\pi j'}{i\omega} - \frac{4\pi\sigma E}{i\omega}, \qquad (10.21)$$

where j_{cd} is the current density in the absorbing coating, and σ is the covering's conductivity. Assuming that the process has a wave nature (j', $E \sim \exp(i\omega t - ikx)$) we obtain from the following dispersion equation from the condition that (10.20) and (10.21) are consistent:

$$\omega_p^2/(\omega - kv_0)^2 + i\, 4\pi\sigma/\omega = 1 . \qquad (10.22)$$

We rewrite (10.22) in the form

$$(\omega - kv_0 - \omega_p)(\omega - kv_0 + \omega_p) = i(4\pi\sigma/\omega)(\omega - kv_0)^2 . \qquad (10.23)$$

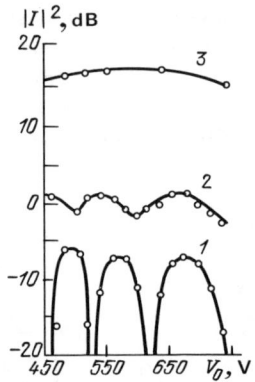

Fig. 10.2. Square of the relative grouped current versus acceleration voltage [10]: 1) dielectric medium replaced by a metal surface (two waves of space charge with constant amplitude are excited); 2) the beam moves in a resistive medium, but the beam current is small (growing and decaying waves are excited); 3) the current is large (the growing slow wave predominates).

Assuming, furthermore, that $\omega - kv_0 \approx -\omega_p$, which corresponds to a disturbance with a slow-charge wave, we obtain $\omega - kv_0 - \omega_p \approx -2\omega_p$ and (10.23) becomes $\omega - kv_0 \approx -\omega_p - i \cdot 2\pi\sigma\omega_p/\omega$, whence

$$k \approx (\omega/v_0 + \omega_p/v_0) + i2\pi\sigma\omega/\omega_p v_0 , \qquad (10.24)$$

i.e., $Re\, k$ is the slow wave's phase constant and $Im\, k = 2\pi\sigma\omega/\omega_p v_0 > 0$. Thus the slow wave grows as it propagates:

WAVES WITH NEGATIVE ENERGY. LINKED WAVES

$$j', E \sim \exp\{-i(\omega/v_0 + \omega_p/v_0)x + 2\pi\sigma\omega x/\omega_p v_0\}.$$

Similar arguments show that the fast space-charge wave dies away (please, do the calculations yourself).

The energy of transverse electromagnetic waves may be negative, for example in a medium of two-level particles. In this case we have

$$\varepsilon = 1 - \omega_0^2 N_{12}/(\omega^2 - \omega_{12}^2 + 2i\gamma\omega_{12}), \qquad (10.25)$$

where ω_{12} is the transition frequency, $\omega_0^2 = 4\pi e^2 Nd/m$ (d characterises the relationship between the particle and the field, and N is the particle concentration), $N_{12} = (n_1 - n_2)/n_2$, and n_1 and n_2 are the populations of the lower and upper levels [11]. The energy of a wave at frequency ω, where $\omega - \omega_{12} \gg \gamma$, is approximately

$$\partial(\omega^2\varepsilon)/\partial\omega = 2\omega[1 + \omega_{12}^2 N_{12}\omega_0^2/(\omega^2 - \omega_0^2)] \qquad (10.26)$$

and it may be negative if the medium is inverted, i.e., the upper level has a greater population than the lower level. According to (10.26) a wave for $\omega - \omega_{12} < \omega_0\sqrt{n_2 - n_1}/2$ will have a negative energy. It is clear that the interaction between waves with negative energy and waves with positive energy must lead to instability. Both waves will grow in amplitude.

10.3 Coupled Waves. Synchronicity.
Normal and Anomalous Doppler Effects

The linear equations for coupled waves can formally be obtained from the equations for coupled oscillations (for the two coupled waves from (2.21)) if time is replaced by a coordinate and the frequency is replaced by the propagation constant. The analog of the oscillator's energy will be the power transferred along the wave-carrying medium. We shall restrict ourselves to a weak link, where the phenomenological derivation of the equations for coupled waves is elementary. In the absence of a link we have

$$da_1/dx = -ik_1 a_1, \quad da_2/dx = -ik_2 a_2, \qquad (10.27)$$

where a_1 and a_2 are normalized such that $|a_1|^2$ and $|a_2|^2$ are the power fluxes carried by the waves, and k_1 and k_2 are the waves' propagation constants. If we now couple the waves, but consider the coupling to be weak (k_1 and k_2 remain the same as they were in the absence of a link, while c_{12} and c_{21} are small in comparison to k_1 and k_2), than we get

$$da_1/dx = -ik_1 a_1 + c_{12} a_2, \quad da_2/dx = -ik_2 a_2 + c_{21} a_1 \quad (10.28)$$

where c_{12} and c_{21} are the coefficients of the link between the waves (compare (10.28) and (2.21)). We assume further that any damping of the wave can be neglected, i.e., in the absence of the link k_1 and k_2 are real values, that $k_{1,2} = \omega/v_{1,2}$, and that $v_{1,2}$ are the phase velocities of the uncoupled waves. The overall average power for a weak link is approximately the sum of the powers in the uncoupled systems, i.e.,

$$P = 2\{\pm |a_1|^2 \pm |a_2|^2\} \approx \text{const}, \quad dP/dx = 0 . \quad (10.29)$$

The plus and minus signs correspond to waves with positive and negative energies.

When one wave is of space charge in an electron beam and the other is an electromagnetic wave in a damping system, (10.29) is the mathematical formulation of Tchu's theorem in the microwave theory of electronic devices with extended interaction in the two-wave approximation [12,13]. For instance, Tchu's theorem on the kinetic power for a drifting electron beam has the form $P = 2\{|a_f|^2 - |a_{sl}|^2\} = \text{const}$.

Since $|a|^2 = aa^*$, it follows from the second expression in (10.29) that

$$\frac{dP}{dx} = \pm \left\{ a_1 \frac{da_1^*}{dx} + a_1^* \frac{da_1}{dx} \right\} \pm \left\{ a_2 \frac{da_2^*}{dx} + a_2^* \frac{da_2}{dx} \right\} = 0 . \quad (10.30)$$

By substituting (10.28) and its complex conjugate equation into (10.30), we find

$$(\pm c_{12}^* \pm c_{21}) a_1 a_2^* + (\pm c_{12} \pm c_{21}^*) a_1^* a_2 = 0 . \quad (10.31)$$

Equation (10.31) is valid for any a_1 and a_2, and hence

$$c_{21} = \pm c_{12}^* \quad \text{when} \quad P = \pm |a_1|^2 - |a_2|^2 , \quad (10.32)$$

$$c_{21} = \mp c_{12}^* \quad \text{when } P = \pm |a_1|^2 + |a_2|^2. \tag{10.33}$$

Assuming that a_1, $a_2 \sim \exp(i\omega t - ikx)$, using (10.32) and (10.33), and given the condition that (10.28) is consistent, we obtain the following dispersion equations:
for codirectional flows of power (the same signs for $|a_1|^2$ and $|a_2|^2$ in the first expression in (10.29)

$$(k - \omega/v_1)(k - \omega/v_2) = |c_{12}|^2, \tag{10.34}$$

for counterdirectional flows of power (different signs for $|a_1|^2$ and $|a_2|^2$)

$$(k - \omega/v_1)(k - \omega/v_2) = -|c_{12}|^2. \tag{10.35}$$

In a system without losses, four types of linkage between interacting waves are possible (see Table 10.1, which is taken from [23]). The dispersion diagrams of the uncoupled waves are shown by dashed lines, the solid lines corresponding to the possible linkages.

We shall illustrate the table by a concrete example from microwave electronics using the theory for the interaction between a straight electron beam and a running electromagnetic wave (Chapter 7). We return to the equation with the disturbance of a wave-carrying system by the current of the electron beam

$$\frac{\partial E}{\partial x} + i\frac{\omega}{v_{ph}} E = -\frac{1}{2}\left(\frac{\omega}{v_{ph}}\right)^2 KI \tag{10.36}$$

and the equation for current concentrated in the beam due to the effect of the field of the wave-carrying system

$$\frac{\partial^2 I}{\partial x^2} + 2i\frac{\omega}{v_0}\frac{\partial I}{\partial x} - \left(\frac{\omega^2}{v_0^2} - \frac{\omega_q^2}{v_0^2}\right) I = i\frac{\omega}{v_0}\frac{I_0}{2V_0} E. \tag{10.37}$$

If we have $I(0) = 0$ and $(\partial I/\partial x)_{x=0} = 0$, then (10.37) may be rewritten as

$$I(x) = \frac{I_0 k_e}{4V_0 k_q} \int_0^x E(\xi)\exp[-i(k_e-k_q)(x-\xi)]d\xi$$

$$- \frac{I_0 k_e}{4V_0 k_q} \int_0^x E(\xi)\exp[-i(k_e+k_q)(x-\xi)]d\xi \qquad (10.38)$$

$$= I_s(x) + I_f(x)$$

where $k_e = \omega/v_0$ and $k_q = \omega_q/v_0$, and the indices s and f correspond to the integrals associated with the slow and fast space-charge waves. Two first order equations may be written instead of (10.37) or (10.38), i.e.,

$$\partial I_{sl}/\partial x + i(k_e + k_q)I_{sl} = -(k_e I_0/4V_0 k_q)E(x), \qquad (10.39)$$

$$\partial I_f/\partial x + i(k_e - k_q)I_f = (k_e I_0/4V_0 k_q)E(x). \qquad (10.40)$$

The condition for resonance interaction is

$$k_e + k_q = k_0 = \omega/v_{ph} \quad \text{or} \quad v_{ph} = v_0/(1 + \omega_q/\omega) \qquad (10.41)$$

and it corresponds to the condition for synchronicity between an electromagnetic wave in a damping system and a fast space-charge wave in the beam; and the conditions

$$k_e - k_q = k_0 = \omega/v_{ph} \quad \text{or} \quad v_{ph} = v_0/(1 - \omega_q/\omega) \qquad (10.42)$$

correspond to the condition for synchronicity between a wave in a damping system and the slow space-charge wave. When (10.41) or (10.42) are fulfilled in the system (10.36), (10.39) and (10.40) it is possible to have only two and not three equations. In this case, the following interactions are possible (Table 10.1) with the corresponding devices:

1. Interaction between the fast space-charge wave (a wave with positive energy) and the forward wave in the wave carrying structure (a wave with positive energy); the group velocity is in the same direction; a travelling-wave tube is the suppressor; at certain values of the constant beam current and accelerating potential for a given frequency the input signal is completely damped.
2. Interaction between a space-charge wave and a backward-wave in a wave carrying system (both waves have positive energy, but the group velocities are in opposite

directions); a backward-wave tube is the suppressor; the input signal can only be completely damped if the interaction space is infinitely long.

3. Interaction between the slow space-charge wave (a wave with negative energy) and the forward wave in the system (a wave with positive energy); the group velocities are in the same direction; a travelling-wave tube is the amplifier.

4. Interaction between the slow space-charge and the reverse wave in the system; the group velocities are in opposite directions; a backward-wave tube is the generator.

Let us consider, by a way of example, case 1, which corresponds to the equation system

$$\frac{\partial E}{\partial x} + ik_0 E = -\frac{1}{2}k_0^2 KI_f, \quad \frac{\partial I_f}{\partial x} + i(k_e - k_q)I_f = \frac{k_e I_0}{4V_0 k_q} E$$

and dispersion equation

$$(k - k_0)(k - k_e + k_q) = (k_e C)^3 / 2k_q .$$

TABLE 10.1. Dispersion Characteristics of Weakly Coupled Waves

$\omega - \kappa$ Diagram	Properties of Coupled-Wave System	Examples of Coupled-Wave Systems
	κ is a real for all ω and vice versa	The coupling of two waves with positive energy or two waves with negative energy
	ω is real for κ; κ is complex for all real ω; the waves damp in space; no stability	ditto
	κ is complex for real ω and vice versa; κ takes values corresponding to amplification for real ω; convective instability	Couple between a wave with positive energy and one with negative energy; group velocities of waves in same direction
	κ is real for all real ω; ω is complex for all real κ; absolute instability	Couple between a wave with positive energy and one with negative energy; group velocities of waves in opposite directions

We have following from the condition for perfect synchronicity between the waves $k_0 = k_e - k_q$:

$$k_{1,2} = k_0 \pm k_e C/2(QC)^{1/4}, \qquad (10.43)$$

where $k_q/k_e C = 2(QC)^{1/2}$.

The initial conditions $E(0) = E_{in}$ and $I_f(0) = 0$ given that $I_f \sim E/(k - k_e + k_q)$ (see (10.41))

and $E(x) = \sum_{m=1}^{2} E_m(0) e^{-ik_m x}$ may be rewritten as

$$E_1(0) + E_2(0) = E_{in},$$
$$E_1(0)/(k_1 - k_0) + E_2(0)/(k_2 - k_0) = 0. \qquad (10.44)$$

It follows from (10.44) using (10.43) that

$$E(x) = E_{in} e^{-i(k_e - k_q)x} \cos\frac{k_e Cx}{2(QC)^{1/4}},$$

i.e., there is periodic transfer of energy between the interacting waves and the condition for complete suppression of the input signal is $CN_{sup} = (2n + 1)(QC)^{1/4}/2$. If we make the substitutions $a_1 = E/k_0\sqrt{2K}$ and $a_2 = I_f\sqrt{2Y_c}$ where $Y_c = I_0 k_e/2V_0 k_q$ is the wave conductivity of the beam, then we have $|C_{12}|^2 = k_e^2 C^2/4(QC)^{1/2}$ in (10.34), and because $\omega/v_1 = \omega/v_2 = k_0 = k_e - k_q$ we find that (10.42) and (10.34) coincide [13]. Note that the case of two interacting beams corresponds in microwave electronics to the case of large space charge ($4QC \sim 1$), where a wave in a transmission line may not be simultaneously close in speed to both of the space-charge waves.

An elegant experiment demonstrating growing oscillation in two coupled wave systems was constructed by K. Katler [14]. He built a mechanical generator with a running wave. The transmission line was made from a set of transverse heavy strips mounted along a steel wire (Fig. 10.3). When one of the strips was turned by a small angle and then released, a slow torsion wave propagated along the line due to the torsion in the wire (its velocity depended on the resistance of the wire to torsion and the rotational inertia of the heavy transverse strips). In order for the wave systems to be able to move with respect to each other (there were two), each transmission line was mounted on the periphery of a bicycle wheel and linked into a ring (the wheels were able to rotate independently about a common axis). Small cylindrical magnets were fixed to the ends of each transverse plate (Fig. 10.3a). The magnets were positioned such that the strips attracted each other. The wave interactions in the system could best be seen when the wheels were rotated in opposite directions. Initially they rotated independently, but at a certain velocity a small random disturbance would lead to an oscillation that grows. Initially two waves were excited on the rings; when the rotation slowed these oscillations disappeared but another oscillation having three waves around the ring was induced simultaneously; the character of the oscillations changed continuously with frequency (Fig. 10.4). In Katler's experiment the interaction was stopped when seven waves were induced around the wheel ring.

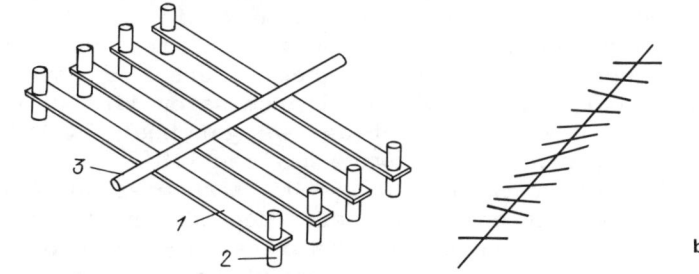

Fig. 10.3. a) Transmission line consisting of a set of rigid transverse plates (1) with small steel magnets (2) attached to a wire (3).
b) Illustration of the propagation of twisting waves along the transmission line [4].

Fig. 10.4. Photographs demonstrating the operation of a mechanical generator of travelling waves [14]. The wheel speed is such that a) four waves fit around the rim; b) five waves fit around the rim; c) six or seven wave fit around the rim.

Up to now we have mainly used space-charge waves and the resonance interactions with electromagnetic waves in wave-carrying structures as examples. There is, however, another class of fundamental waves in electron beams moving in longitudinal focus in magnetic fields and in crossed electrostatic and magnetostatic fields, viz., cyclotron waves [12, 13]. These waves propagate in electron flux in the form of high frequency disturbances of the transverse flux velocity with phase velocities

$$v_{ph,s,f} = v_0/(1 \pm \omega_c/\omega) , \qquad (10.45)$$

where $\omega_c = (e/m)B$ is the cyclotron frequency, B is the magnetic induction, and v_0 is the constant component of the longitudinal velocity of the electrons, the indices s and f (+ and - in (10.45)) correspond to the slow and fast cyclotron electron waves. Cyclotron waves are plane-polarized and perpendicular to the constant magnetic field. The fast cyclotron wave has positive energy and the slow one has negative energy [12, 13]. When there is synchronous interaction between a wave in a transmission line with one of the cyclotron waves in the beam, we get what is known as cyclotron resonance in microwave devices [13].

As we have seen for waves in transmission lines and electron beams, the energy transfer depends on the properties of the interacting waves. It would be nice to know whether the result of the interaction could be predicted and the nature of the energy transfer determined before, say, determining the sign of the wave energies. It turns out that this is possible if we use the theory of irradiation at superlight velocities in a medium [15-17].[1]

If a radiator (charged particle, electric dipole, etc.) is moving in a medium with a refractive index n, then a consequence of the Doppler effect in a system of coordinates associated with the stationary medium is that radiation will have a frequency [15,16]

$$\omega(\theta) = \omega_0/|1 - \beta n \cos \theta| , \qquad (10.46)$$

where ω_0 is the frequency of the radiation in the system of coordinates in which the radiator is at rest; $\beta = v_0/c$; v_0 is the velocity of the radiator; c is the velocity of light; and

---

[1] Cyclotron resonance in microwave devices with crossed fields are considered in [13] (pp. 486-489). A wider consideration of this subject can be found in [5] (pp. 489-494) and [23].

WAVES WITH NEGATIVE ENERGY. LINKED WAVES

θ is the angle between v_0 and the direction of observation. When $\beta n < 1$ the Doppler effect is called normal and when $\beta n > 1$ it is called anomalous [17] (the Doppler effect in a refracting medium is considered in detail in [24]). It is important that the nature of the anomalous Doppler effect does not change when the field is kept within narrow channels or grooves in a medium or concentrated close to a boundary [15-17]. The radiation associated with the normal Doppler effect leads to a damping of the field and the radiation associated with anomalous Doppler effect leads to amplification of the field. When a retarded electromagnetic wave ($v_{ph} = c/n$) interacts with an electron beam moving at a velocity v_0 in a straight line, i. e., oscillators vibrating with a reduced plasma frequency ω_q, (10.46) takes the form

$$\omega(\theta)|_{\theta=0} = \omega(0) = \omega = \omega_q / |1 - v_0/v_{ph}| . \qquad (10.47)$$

When $\beta n < 1$ (the normal Doppler effect, $v_0 < v_{ph}$) we have from (10.47) $\omega = \omega_q/(1 - v_0/v_{ph})$ or $v_{ph} = v_0/(1 - \omega_q/\omega)$, is the same as the synchronicity condition (10.42) of a wave in a wave-carrying system with a fast space-charge wave. The oscillation is damped, which in electronics means that when the synchronicity condition (10.42) is fulfilled, the electrons are grouped in the accelerating phase of the high frequency field and take energy from the waves.

When $\beta n > 1$ ($v_0 > v_{ph}$) we have $v_{ph} = v_0/(1 + \omega_q/\omega)$, which is the same as (10.41) and corresponds to synchronicity between the wave in the transmission line and the slow spacer-charge wave. In this case the electrons are grouped in the retarding phase of the field (the radiation associated with the anomalous Doppler effect drives the oscillation) and when (10.41) is satisfied we would expect amplification or generation of the oscillation. Thus there is a physical analogy between the induced normal Doppler effect and synchronicity between an electromagnetic wave and the electron wave with positive energy (the fast wave) and an analogy between the induced anomalous Doppler effect and synchronicity between an electromagnetic wave and the wave with negative energy (the slow wave). We should emphasize that it is only valid to apply the analogy to microwave devices for the two-beam approximation (condition (10.41) or (10.42) for the fast space-charge approximation, or condition (10.45) for the cyclotron resonance regime), when the electromagnetic wave interacts with electron oscillators, whose fundamental frequency is ω_q or ω_c (the oscillating properties appearing when there is a high-frequency field).

In the synchronous regimes typical for microwave electronic devices with long interactions, for which $v_o \approx v_{ph}$, both "electron" waves "work" and we have induced Cherenkov radiation.

CHAPTER 11

PARAMETRIC SYSTEMS AND PARAMETRIC INSTABILITY

1.1 General Comments

Systems whose parameters change over time or in space are usually called parametric.

The simplest mechanical parametric system is a simple pendulum whose length changes over time $l = l(t)$ or whose fulcrum moves. The electrical analog of such a system is an oscillating circuit whose capacitance changes over time $C = C(t)$. A mathematical analysis of these parametric systems yields ordinary differential equations whose coefficients depend on time.

Problems involving the propagation of waves in media with periodically changing parameters is often encountered in physics and technology. These problems occur when studying the propagation of waves in stratified media, the motion of electrons in the field of a crystal's ion lattice, in light transmitted through a medium in which a sound wave has been induced, etc. The parameters of the medium may change both over the time and in space. If they change synchronously in time and all points in space or only in space, then the mathematical analysis reduces to an investigation of ordinary differential equations with coefficients that depend on time or coordinates.

We shall distinguish between two classes of problem corresponding to the two classes of parametric systems.

1. Resonance parametric systems. Such a system has a characteristic time for the changes in the parameters of the same order as the characteristic time for the changes in the system's variables. For example, if the frequency of a harmonic oscillator, which can be described by the equation $\ddot{x} + \omega_0^2 x = 0$, depends on time ($\omega_0 = \omega_0(t)$) and the period τ_{ω_0} for the change in the parameter ω_0 is of the same order as $t_{chr} \sim 2\pi/\omega_0$, then the oscillator belongs to the class of resonance parametric systems.

2. Nonresonance parametric systems. These involve situations when the parameters change much more rapidly or

much more slowly than the characteristic time for the change in the variables: $\tau_{\omega_0} \ll t_{chr}$ or $\tau_{\omega_0} \gg t_{chr}$. We include in this class systems for which the parameter changes periodically and even systems for which the formal resonance condition $n\omega_{parametric} = m\omega_{natural}$ is fulfilled, but for which the numbers n or m are large (the situation which we call resonance corresponds to small integer values of n and m). An example is the motion of an electron in an atom if an external high-frequency field is imposed.

In this chapter we shall consider resonance parametric systems and systems with rapidly changing parameters. Effects associated with slowly changing parameters are left to the next chapter.

11.2 Parametric Resonance. Floquet's (Blokh's) Theorem. Mathieu's Equation

The classical example of parametric resonance is a swing. Everyone knows that it is easy to make a swing move if the impulse is given at the top of its trajectory and thus by moving it center of mass twice in a period it is possible to increase the effective length of the swing. A natural model of a swing is the mathematical pendulum whose length changes as $l = l_0[1 + \mu(a/l_0)\cos \omega_p t]$ (Fig. 11.1a). The equation of motion has the form $\ddot{x} + (g/l_0)[1 + \mu(a/l_0)\cos \omega_p t]^{-1} x = 0$. If $\mu a \ll l_0$, then writing g/l_0 as ω_0^2 we obtain Mathieu's equation [1, 2]:

$$\ddot{x} + \omega_0^2[1 - \mu(a/l_0)\cos \omega_p t]x = 0 . \tag{11.1}$$

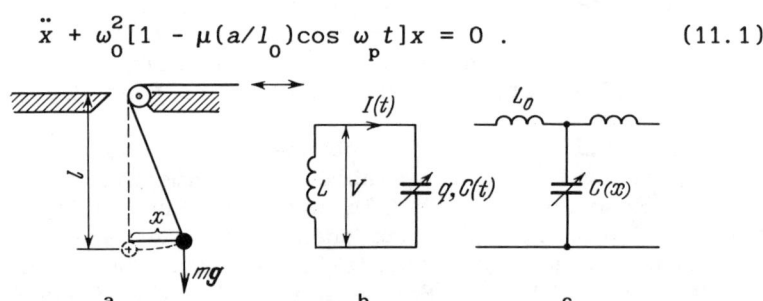

Fig. 11.1. Examples of simple parametric systems: a) pendulum with a length changing over time; b) oscillating circuit with changing capacitance; c) A long line whose capacitance periodically changes with coordinates.

The electrical analogy of this pendulum is an oscillating circuit with a changing capacitance

$C(t) = C_0[1 + \mu(C_1/C_0)\cos \omega_p t]$ (Fig. 11.1b) or
$C(x) = C_0\{1 - \mu(C_1/C_0)\cos q_1 x\}$ (Fig. 11.1c). The capacitance may be changed mechanically, for example by a motor moving a capacitor plate. In order for the amplitude of the oscillations to grow, it is necessary to introduce energy into the circuit to do work against the force of the electrostatic field in the capacitor. This means that the plate must be removed when the charge on the capacitor is at a maximum and moved back when the charge on the capacitor has reached zero. The corresponding equation for the oscillation is $L\ddot{q} + (1/C(t))q = 0$, where $C(t) = (S\varepsilon)/4\pi d(t)$, and S is the area of the capacitor plate, ε is the dielectric permittivity of the material filling the capacitor, and $d(t)$ is the distance between the plates, which varies over time. If $d(t) = d_0\{1 + \mu(\delta/d_0)\cos \omega_p t\}$, than the equation for the oscillation in the charge takes the form

$$\ddot{q} + \omega_0^2\left\{1 + \mu \frac{\delta}{d_0} \cos \omega_p t\right\} q = 0 ,$$

(11.2)

$$\omega_0 = \frac{1}{\sqrt{LC_0}} , \quad C_0 = \frac{S\varepsilon}{4\pi d_0} .$$

The same equation, with the exception that t is replaced by x, is used to analyze the propagation of waves in a medium whose parameters periodically depend on the coordinates. One possible realization of such a medium is shown in Fig. 11.1c. We have chosen a long line whose capacitance periodically depends on the coordinates. A similar medium is described by the telegrapher's equations $\partial I/\partial x = -C(x)\partial U/\partial t$, $\partial U/\partial x = -L_0 \partial I/\partial t$ which lead to the wave equation $\partial^2 U/\partial x^2 - L_0 C(x)\partial^2 U/\partial t^2 = 0$.

We look for the solution the wave equation in the form $U(x, t) = V(x)\exp(i\omega t)$, and also let

$$C(x) = C_0\{1 - \mu(C_1/C_0) \cos q_1 x\}, \quad v_0^2 = 1/L_0 C_0.$$

Then we obtain the following equation for $V(x)$:

$$d^2 V(x)/dx^2 + (\omega/v_0)^2[1 - \mu(C_1/C_0)\cos q_1 x]V(x) = 0 . \quad (11.3)$$

We shall restrict ourselves to considering parametric systems with one degree of freedom, which can be described by a general equation, Hill's equation:

$$\ddot{x} + \omega^2(t)x = 0 ,\qquad(11.4)$$

where $\omega^2(t)$ is a periodic function of time.

The questions we wish to answer are: Is it possible for there to be instability in parametric systems? If so under what conditions does instability arise? What are the boundaries of the instability region?

We now present some of the theory of differential equations with periodic coefficients, in particular Floquet's theorem, which determines the structure of the solution to a system of linear differential equations with periodic coefficients. The general form of the theorem is formulated thus: a system with n degrees of freedom described by a differential equation of order $2n$ with periodic coefficients of period T has $2n$ linearly independent solutions which form a fundamental system. Each of these solutions has the form $x_i(t) = \Phi_i(t)\exp(\lambda_i t)$ where $\Phi_i(t)$ is a periodic function with period T. The term $\exp(\lambda_i t)$ is called a Lyapunov exponent and λ_i is called a Lyapunov characteristic index. The $\phi_i(t)$ are called Floquet functions. We shall explain Floquet's theorem for a second order system, i.e., for (11.4).

We choose arbitrarily two special, linearly independent solutions $x_1(t)$ and $x_2(t)$ to (11.4). Because of the periodicity of the coefficients in (11.4) we have $\omega^2(t + T) = \omega^2(t)$, and hence the functions $x_1(t + T)$ and $x_2(t + T)$ will also be solutions to (11.4). Like any solution, they can be expressed in terms of the fundamental system as follows:

$$x_1(t + T) = \alpha_1 x_1(t) + \beta_1 x_2(t) ,$$

$$x_2(t + T) = \alpha_2 x_1(t) + \beta_2 x_2(t) .$$

Solutions x_1 and x_2 may always be chosen such that $\beta_1 = \alpha_2 = 0$ (we leave the reader to prove this independently). This means that

$$x_1(t + T) = s_1 x_1(t) , \qquad x_2(t + T) = s_2 x_2(t) ,\qquad(11.5)$$

i.e., the solution repeats itself each period multiplied by a constant factor s_1 (or s_2). It is clear that $\ddot{x}_1 + \omega^2(t)x_1 = 0$ and $\ddot{x}_2 + \omega^2(t)x_2 = 0$. By multiplying the first of these by x_2 and the second by x_1 and taking one from another, we obtain

$$\ddot{x}_1 x_2 - \ddot{x}_2 x_1 = \frac{d}{dt}(\dot{x}_1 x_2 - \dot{x}_2 x_1) = 0,$$

i.e., $\dot{x}_1 x_2 - \dot{x}_2 x_1 = $ const. Consequently

$$\dot{x}_1(t)x_2(t) - \dot{x}_2(t)x_1(t) = \dot{x}_1(t+T)x_2(t+T)x_1(t+T),$$

which using (11.5) yields an equation for the relationship between s_1 and s_2 in the form

$$s_1 s_2 = 1. \tag{11.6}$$

We introduce a new constant λ_i (usually complex) in the relationship $s_i = \exp(\lambda_i T)$ ($i = 1, 2$). It then follows from (11.6) that $\lambda_1 = -\lambda_2 = \lambda$. We also introduce a new function $\Phi_i(t) = x_i(t)\exp(-\lambda_i t)$. It is easy to show that if (11.5) is fulfilled then the function $\Phi_i(t)$ is periodic with period T. Consequently, solutions $x_1(t)$ and $x_2(t)$ have the form

$$x_1(t) = \Phi_1(t)\exp(\lambda_1 t), \quad x_2(t) = \Phi_2(t)\exp(\lambda_2 t),$$

and the general solution to (11.4) may be written as

$$x(t) = c_1 e^{\lambda t}\Phi_1(t) + c_2 e^{-\lambda t}\Phi_2(t). \tag{11.7}$$

If Re $\lambda \neq 0$ then one of the terms on the right-hand side of (11.7) will grow over time and $x(t)$ will increase, i.e., instability is possible in the system.

The growth of an oscillation in a parametric system is called parametric resonance [2]. In order to determine under which conditions parametric resonance arises, we make the function $\omega^2(t)$ in (11.4) a little more concrete.

Let $\omega^2(t) = \omega_0^2[1 - \mu b \cos \omega_p t]$, which transforms (11.4) into Mathieu's equation, i.e.,

$$\ddot{x} + \omega_0^2[1 - \mu b \cos \omega_p t]x = 0. \tag{11.8}$$

We have seen how one-dimensional formulations for wave propagation reduce to Mathieu's equation as well. There is a generalization of Floquet's theorem (in the three-dimensional case) for the propagation of waves in three-dimensional period structures. This is called Blokh's theorem [1, 3].

For an arbitrary μ the solution to (11.8) can be expressed in terms of special functions (called Mathieu's functions) which have been tabulated and whose properties are well understood. Let us attempt to solve a problem using the simple function assuming that $\mu b \ll 1$. When $\mu = 0$, the solution to (11.8) is well known. It is hoped that when $\mu \neq 0$ but small, the solution will only differ a little from the known one, and a correction can be added by a recurrence method, i.e., every approximation is determined from the previous one. Thus, we use the theory of perturbations for the solution to (11.8) [21] starting from a knowledge of the solution when $\mu = 0$; $x(t) = x_0(t) = A \cos(\omega_0 t + \phi)$. We shall look for a solution to (11.8) when $0 < \mu \ll 1$ in the form

$$x(t) = x_0(t) + \mu W^{(1)}(t) + \mu^2 W^{(2)}(t) + \ldots + \mu^n W^{(n)}(t). \quad (11.9)$$

A solution in the form (11.9) only has sense when the correction $W^{(i)}$ to the zero approximation $x_0(t)$ does not grow over time. Substituting (11.9) into (11.8) and grouping terms with the same order of μ together yields an equation for $W^{(i)}$ ($1 \leq i \leq n$), i.e.,

$$\ddot{x}_0 + \omega_0^2 x + \mu\{\ddot{W}^{(1)}(t) + \omega_0^2 W^{(1)}(t) - x_0 \omega_0^2 b \cos \omega_p t\}$$
$$+ \mu^2 \{\ddot{W}^{(2)}(t) + \omega_0^2 W^{(2)}(t) - W^{(1)}(t)\omega_0^2 b \cos \omega_p t\} + \ldots \quad (11.10)$$
$$+ \mu^n \{\ddot{W}^{(n)}(t) + \omega_0^2 W^{(n)}(t) - W^{(n-1)}(t)\omega_0^2 b \cos \omega_p t\} \equiv 0$$

All the terms in (11.10) have different order of magnitude and cannot cancel each other out. Thus in order for it to be an identity each of the expressions in the brackets must be zero. Thus we have obtained a recurrence system of equations to find the i-th approximation. It can be seen from (11.10) that each of the equations is the equation of a harmonic oscillator acted on by an external force in the form of a collection of harmonics. For example, the correction to the first approximation has the form

$$\ddot{W}^{(1)}(t) + \omega_0^2 W^{(1)}(t) = \omega_0^2 Ab \cos(\omega_0 t + \phi)\cos \omega_p t$$

or

$$\ddot{W}^{(1)}(t) + \omega_0^2 W^{(1)}(t) = (\omega_0^2 Ab/2)\{\cos[(\omega_0 - \omega_p)t + \phi]$$
$$+ \cos[(\omega_0 + \omega_p)t + \phi]\}.$$

The driving force on the right-hand side of this expression

contains two harmonic components at most, those for the frequencies $(\omega_0 - \omega_p)$ and $(\omega_0 + \omega_p)$. If $W^{(1)}$ is not to grow over time, then it is necessary that these harmonics do not resonate with the oscillation at a frequency ω_0, i.e., it is necessary that $\omega_0 - \omega_p \neq \omega_0$ and $\omega_p \neq 2\omega_0$. However we are interested in the case of resonance, remember the swing. When there is resonance, $W^{(1)}$ grows linearly over time and hence a solution of the form $x(t) = x_0(t) + \mu W^{(1)}(t)$ only has sense for durations of time of the order of several periods. How can the solution be corrected to be of use during resonance?

Note that the terms which grow are those terms in the correction $W^{(1)}(t)$ which have the form of the main part of the solution $x_0(t) \sim A \cos(\omega_0 t + \phi)$. Indeed, in the resonance case $\omega_{forced} \approx \omega_0$ and $W^{(1)}(t) \sim t \cos(\omega_0 t + \phi)$. However, in the solution $x(t)$, the term $W^{(1)}(t)$ is multiplied by the small parameter μ, i.e., the term $W^{(1)}(t)$ is as before sinusoidal in form locally over time $x(t)$, and has frequency ω_0, in spite of the secular growth. However, over long periods the amplitude and phase of the solution may differ a great deal from their initial values. This indicates a way out of the situation. If the amplitude and phase of the main part of the solution are not considered constants and rather slowly changing functions of time, i.e., $A = A(\mu t)$ and $\phi = \phi(\mu t)$ such that their change takes into account the resonance terms $W^{(1)}(t)$ (i.e., the secular part of $W_c^{(1)}(t)$ is summed over each period with $x_0(t)$), then the correction will be of the same order as μ and over long periods of time the resonance components will not exist because of the difference in $(W^{(1)} - W_c^{(1)})$ [20].

Similar summations of the resonance components in different orders of the theory of perturbation with the main part of the solution is the main idea behind majority of methods using small parameters, including those for nonlinear systems.

We now return to our main problem, and consider the resonance case $\omega_p = 2\omega_0 + \mu\delta$, where $\mu\delta = -\delta'$ is a small disturbance. The solution to (11.8), which using the expression for ω_p has the form

$$\ddot{x} + \omega_0^2 x = \mu \omega_0^2 b \cos[(2\omega_0 + \delta')t]x , \qquad (11.11)$$

will be

$$x(t) = A(\mu t)\cos[(\omega_0 + \delta'/2)t] + B(\mu t)\sin[(\omega_0 + \delta'/2)t] + \mu W^{(1)}$$
(11.12)

where $A(\mu t)$ and $B(\mu t)$ are functions of time that change more slowly than cos or sin. No terms with frequencies differing from $(\omega_0 + \delta'/2)$ by more than $n(2\omega_0 + \delta')$, where n is an integer, were included in the main part of (11.12) because they are many more orders of magnitude smaller than μ than is necessary for a first approximation. Once again the functions $A(\mu t)$ and $B(\mu t)$ are defined such that the correction $W^{(1)}$ does not grow. By substituting (11.12) into (11.11) and equating the coefficients of μ at the first order (we assume that $\dot{A} \sim \mu A$ and $\dot{B} \sim \mu B$), we obtain an equation for $W^{(1)}$, i.e.,

$$\ddot{W}^{(1)} + \omega_0^2 W^{(1)} = \omega_0 [2\dot{A} + \delta'B - (\mu b \omega_0/2)B]\sin[(\omega_0 + \delta'/2)t]$$
$$+ \omega_0 [-2\dot{B} + \delta'A + (\mu b \omega_0/2)A]\cos[(\omega_0 + \delta'/2)t].$$

By now using the freedom of choice of $A(\mu t)$ and $B(\mu t)$ we require that this equation has no resonance terms on the right-hand side, i.e., $\dot{A}$ and $\dot{B}$ are defined by

$$\dot{A} = -(\mu/2)(\delta - \omega_0 b/2)B, \quad \dot{B} = -(\mu/2)(\delta + \omega_0 b/2)A.$$

These are the equations for the slowly changing amplitudes that we want.

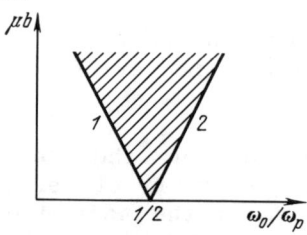

Fig. 11.2. Boundaries of the zone of parametric instability corresponding to the main resonance zone (shaded):
1) $\mu b/4 = 1 - 2\omega_0/\omega_p$; 2) $\mu b/4 = 2\omega_0/\omega_p - 1$.
In constructing the Figure we used the relationship $\omega_0/\omega_p = \omega_0/(2\omega_0 + \delta) \approx 1/2 - \delta/4\omega_0$, when $\delta \sim \mu$.

The solution of this linear system of equations is sought, as usual, in the form $A, B \sim \exp(\lambda t)$. Discarding the trivial solution, we obtain the characteristic equation to determine λ, i.e.,

$$\lambda^2 = -(\mu^2/4)(\delta^2 - \omega_0^2 b^2/4) .\qquad(11.13)$$

When the disturbance is small we have $-\omega_0 b/2 < \delta < \omega_0 b/2$, and the amplitudes A and B will grow; the system has parametric instability. The inequality defines the zone of the main resonance, whose boundaries are illustrated in Fig. 11.2.

If there is a loss $\sim\mu\nu$ in the system, then the main instability zone will be determined by the inequality

$$-\sqrt{\omega_0^2 b^2/4 - 4\nu^2} < \delta < \sqrt{\omega_0^2 b^2/4 - 4\nu^2} ,$$

where ν is the damping decrease. It follows from this that when there is dissipation even at exact resonance ($\delta = 0$) a finite depth to the parameters modulation is necessary for instability to arise, the threshold value being $b = 4\nu/\omega_0$ (Fig. 11.3a).

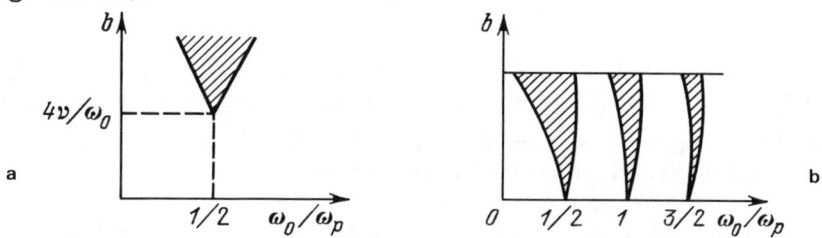

Fig. 11.3. The behavior of the instability region as described by the asymptotic solution of Mathieu's equation: a) the appearance of the excitation threshold of parametric oscillations, which arise as a result of damping; b) the narrowing of the instability regions as the zone number increases.

It is not difficult to determine the boundaries of the main instability zone and to do so with quite adequate accuracy, that is to the order of μ. The main part of the solution to (11.1) in this case is naturally found in a form

$$x(t) = A \cos[(\omega_0 + \delta'/2)t] + B \sin[(\omega_0 + \delta'/2)t]$$
$$+ \mu A_1 \cos[3(\omega_0 + \delta'/2)t]$$
$$+ \mu B_1 \sin[3(\omega_0 + \delta'/2)t] .$$

Since we have $\dot{A} = \dot{B} = \dot{A}_1 = \dot{B}_1 = 0$ on the boundaries of instability zone we can find the boundaries of the instability region after substituting $x(t)$ in this form into (11.1) to the accuracy of the terms of order μ, e.g.,

$$\delta = \pm w_0 b/2 - \mu\omega_0 b^2/32 .$$

We suggest the reader derives (11.13), referring in the case of difficulty to [4] (problem 1, p. 107).

It can be seen that parametric resonance must occur for any $\omega_p \approx 2\omega_0/n$, when n is an integer. This includes the case for $n = 2$. It is qualitatively clear that in order to push a swing the impulse need only be given once a period, but it will be necessary to push harder to obtain the previous result.

Let us try to find the solution to (11.11) when $\omega_p = \omega_0 + \mu\delta$, i.e., in the second parametric resonance zone. The solution is given in a form

$$x(t) = A(\mu t)\cos[(\omega_0 + \delta')t] + B(\mu t)\sin[(\omega_0 + \delta')t]$$

$$+ \mu W^{(1)}(t) + \mu^2 W^{(2)}(t) .$$

In order to eliminate the resonance terms from the right-hand sides of the equations of the different approximations (i.e., from both equations for $W^{(1)}$ and for $W^{(2)}$) the derivatives $\dot A$ and $\dot B$ are divided into two terms, i.e.,

$$\dot A = \mu F_1(A, B) + \mu^2 F_2(A, B), \qquad \dot B = \mu \phi_1(A, B) + \mu^2 \phi_2(A, B) .$$

where F_1 and ϕ_1 are defined by the condition that $W^{(1)}(t)$ does not increase, and F_2 and ϕ_2 by the condition that $W^{(2)}(t)$ does not increase [20]. Repeating the above operations (but changing $\dot A$ and $\dot B$ for F_1 and ϕ_1), it is not difficult to see that in the first approximation in μ there is no parametric instability. By calculating the first solution $W^{(1)}$ at the frequency $\approx 2\omega_0$ and substituting it together with F_2 and ϕ_2 into the right-hand side of the equation for $W^{(2)}$, which is obtained from (11.10), we obtain

$$\ddot W^{(2)} + \omega_0^2 W^{(2)} = [2\omega_0 F_2 + \delta^2 B]\sin\omega_p t$$

$$+ [-2\omega_0 \phi_2 + \delta^2 A]\cos\omega_p t \qquad (11.14)$$

$$+ W^{(1)}(2\omega_p t)\omega_0^2 b\cos\omega_p t .$$

Here $W^{(1)}(2\omega_p t)$ is the forced solution of the equation for $W^{(1)}$ at the frequency of the second harmonic. By calculating this quantity and determining (from the condition that the terms with frequency ω_0 are absent from the right-hand side of (11.14)) F_2 and ϕ_2, we obtain the desired equations for $A(t)$ and $B(t)$. We leave the reader to complete the algebra independently. We shall only provide the equation for the boundaries of the second instability zone for the special case $\dot{A} = \dot{B} = 0$, i.e.,

$$-5\mu b^2 \omega_0/24 < \delta < \mu b^2 \omega_0/24$$

([4], problem 2, p. 108). It can be seen that the bandwidth of the second zone of parametric instability is an order of magnitude narrower than the first zone ($\delta \sim \mu$). The parametric resonance zone will for increasing n get narrower by μ^n (Fig. 11.3b). The instability increment in these zones will be correspondingly smaller.

When the modulation depth of the parameter is large the right-hand side of $\ddot{x} + \omega_0^2 x = x\omega_0^2 b \cos \omega_p t$ is no longer small and the asymptotic method of solution is not applicable. It then becomes necessary to use tables or to solve Mathieu's equation numerically.

In this section we shall generalize the theory of coupled oscillations, which was presented briefly in Chapter 2, to the case when the coupling parameter changes over the time (parametric linkage). As happens when two oscillations with different frequencies are mixed in a non-linear component, frequency mixing occurs when even one of the system's parameters changes over the time.

We introduce this generalization [5] using an example of a parametric oscillation system with constant capacitance that includes a parallel capacitance $C(t)$. The system is called a degenerate two-frequency system, and it is illustrated in Fig. 11.4. The charge on the capacitors connected in parallel, C_1 and $C(t)$, is determined from $q = [C_1 + C(t)]V$, and so the equation for the current flowing

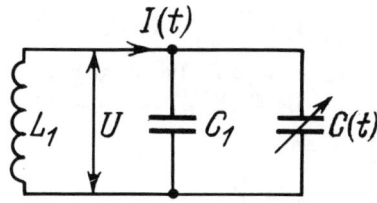

Fig. 11.4. Circuit of a degenerate two-frequency system.

through the inductive coil has the form

$$\frac{d}{dt}\left\{\left[C_1 + C(t)\right]V\right\} = I , \qquad (11.15)$$

and the voltage across the coil is $dI/dt = V/L_1$. We shall assume that

$$C(t) = C_0 + C_p(t) = C_0\{1 + (\Delta C/C_0)\cos(2\omega_1 + \phi)\} , \qquad (11.16)$$

$\omega_1 = [L_1(C_1 + C_0)]^{-1/2}$, $2\omega_1 = \omega_p$ is the pumping frequency, and ϕ is the pumping phase, which determines a phase shift for the pumping relative to the phase change in the charge on the capacitor.

We rewrite (11.15) in the form

$$\frac{dV}{dt} = \frac{I}{C_{11}} - \frac{d}{dt}\left[\frac{C_p(t)}{C_{11}} V\right] , \qquad (11.17)$$

$C_{11} = C_0 + C_1$. We multiply (11.17) by $\pm i\omega_1 C_{11}$ and dividing by (11.15) we obtain

$$\frac{d}{dt}\left(I \pm i\omega_1 C_{11} V\right) = -i\omega_1\left(\mp I + \frac{V}{i\omega_1 L_1}\right) \mp \frac{d}{dt}\left\{i\omega_1 C_p(t)V\right\}. \qquad (11.18)$$

It seems natural, ignoring the fact that C_p is a function of time, to introduce into (11.18) normal oscillations, as was done in Chapter 2. The amplitudes of these oscillations, as can be seen from the structure of (11.18), can be determined easily from the relationships

$$a = (\sqrt{L_1}/2)(I + i\omega_1 C_{11} V) , \quad a^* = (\sqrt{L_1}/2)(1 - i\omega_1 C_{11} V) . \qquad (11.9)$$

It is clear, here, that $a - a^* = i\omega_1 C_{11} \sqrt{L_1} V$. Moreover, we determine ω_1 to be $\omega_1 = 1/\sqrt{L_1 C_{11}}$. We use the definition of ω_1 to transform the first term on the right-hand side of (11.18) to the form $i\omega_1[\mp I - i\omega_1 C_{11} V]$. By using (11.19) equation (11.18) may be written as

$$\frac{da}{dt} = i\omega_1 a - \frac{d}{dt}\left[\frac{C_p(t)}{2C_{11}}(a - a^*)\right] , \qquad (11.20)$$

$$\frac{da^*}{dt} = -i\omega_1 a^* + \frac{d}{dt}\left[\frac{C_p(t)}{2C_{11}}(a - a^*)\right]. \quad (11.21)$$

It can be seen that as $\Delta C \to 0$ ($C_p \to 0$) equations (11.20) and (11.21) correspond to the equations of the normal oscillations of oscillator, which as we have said, can be represented by two oppositely rotating vectors. When $\Delta C \neq 0$, the oscillations become parametrically coupled (the pumping is associated with the normal oscillations). As in Chapter 2, we restrict ourselves to the case of weak linkages, assuming that $\Delta C/2C_{11} \ll 1$. Under this condition the oscillation can be considered close to normal. Moreover, it is possible to set $a(t) = A(t)\exp(i\omega_1 t)$, where $A(t)$ is a function that changes more slowly than $\exp(i\omega_1 t)$, i.e., $dA/dt \ll i\omega_1 A$. To see at which frequencies components appear on the right-hand side of (11.21) due to which we get
$C_p(t) = (\Delta C/2)\{\exp[i(2\omega_1 t + \phi)] + \exp[-(2\omega_1 t + \phi)]\}$,
we look at (11.20) as an example. Clearly the following terms appear on the right-hand side of (11.20):

$$\frac{\Delta C}{2C_{11}}\left\{A(t)\exp[i(3\omega_1 t + \phi)] - A^*(t)\exp[i(\omega_1 t + \phi)]\right.$$

$$+ A(t)\exp[-i(\omega_1 t + \phi)]$$

$$\left. - A^*(t)\exp[-i(3\omega_1 t + \phi)]\right\}.$$

It is assumed that Q-factor of the circuit is large, then components such as ($\pm 3\omega_1$) can be neglected. Moreover, it is also possible to discard the term $A(t)\exp[-i(\omega_1 t + \phi)]$ (think for yourselves why this is possible). The analogous terms may be eliminated from the right-hand side of (11.21) too. As the result we get

$$da/dt = i\omega_1 a + c_{12}\exp(2i\omega_1 t)a^*,$$
$$da^*/dt = -i\omega_1 a^* + c_{21}\exp(-2i\omega_1 t)a, \quad (11.22)$$

where $c_{12} = c_{21}^* = i\omega_1(\Delta C/4C_{11})\exp(i\phi)$.

In deriving (11.22) we used the fact that

$$\frac{d}{dt}\left\{A^*(t)\exp i(\omega_1 t + \phi)\right\} = \frac{dA^*}{dt}\exp i(\omega_1 t + \phi)$$

$$+ i\omega_1 A^*(t)\exp i(\omega_1 t + \phi)$$

$$\approx i\omega_1 A^*(t)\exp i(\omega_1 t + \phi)$$

$$= i\omega_1 \exp(2i\omega_1 t) a^* ,$$

because $dA^*/dt \ll i\omega_1 A^*$ (analogously

$$\frac{d}{dt}\left(\exp(-2i\omega_1 t)a\right) \approx -i\omega_1 \exp(-2i\omega_1 t)a).$$

We change the variables in (11.22) to $A(t)$. It is easy to see that $da/dt = \left(\frac{da}{dt} + i\omega_1 t\right)\exp(i\omega t)$ and hence $dA/dt = c_{12} A^*$, $dA/dt = c_{21} A$. We obtain a system with constant coefficients using the assumption that $\Delta C \ll C_{11}$ and by neglecting the harmonics. If $A(t) \sim \exp(\lambda t)$, then

$$\lambda^2 = \omega_1^2 (\Delta C/4C_{11})^2 \quad \text{or} \quad \lambda = \pm \omega_1 (\Delta C/4C_{11}) . \quad (11.23)$$

It follows from (11.23) that there will be rising and decaying oscillations. It is easy to see that (11.23) is the same as the equation we obtained above for the first approximation when solving the problem by the method of perturbations if, in the latter, we set $\delta = 0$.

If we set the initial conditions $a(0)$ and $a^*(0)$, then it is easy to find the full solution of the problem, as was done in [5]. The main feature of the solution is the strong dependence of the phase shift between the pumping ϕ and the change in the charge on the capacitor. Suppose, for example, that at the initial moment in time $t = 0$ we have $V(0) = V_{max}$, $I(0) = 0$. In this case $C_p(0) = \Delta C \cos \phi$ (ϕ is the phase shift between the pumping and the initial voltage across the capacitor). If, moreover, $\phi = \pi/2$, and $C_p(t) = \Delta C \sin(2\omega_1 t)$, then as was shown in [5] we have

$$a(t) = (i/2)\sqrt{C_{11}}\, V_{max} \exp\{i\omega_1 t + \omega_1 (\Delta C/4C_{11}) t\},$$

$$a^*(t) = -(i/2)\sqrt{C_{11}}\, V_{max} \exp\{-i\omega_1 t + \omega_1 (\Delta C/4C_{11}) t\}.$$

Thus, when $\phi = \pi/2$ there is parametric instability, the solution grows exponentially in time. At the same time when $\phi = -\pi/2$ the solution exponentially decays. Find for yourself how the phase shift between the pumping and the potential must be chosen for arbitrary initial conditions in order to obtain an increasing solution (note that it is convenient to look for a $(\phi - 2\theta)$, where $\theta = \arctan\left[\sqrt{C_{11}}V(0)/\sqrt{L_1}I(0)\right]$ is the oscillation phase a).

The power associated with oscillations a and a^* and the pumping power obey the Manley-Rowe relations (the conservation of energy for an oscillating system together with a pumping source), i.e.,

$$\frac{P_p}{2\omega_1} = \frac{P_{\omega_1}}{\omega_1} = \frac{P_{-\omega_1}}{\omega_1}, \quad P_{\omega_1}(t) = \frac{d}{dt}|a(t)|^2,$$

$$P_{-\omega_1}(t) = \frac{d}{dt}|a^*(t)|^2.$$

This relationship is easily interpreted as the power of the pumping source evenly distributed between the normal oscillations a and a^* (a power may not be assigned to one oscillation). It is obvious that the above is only valid for the pumping frequency $\omega_p = 2\omega_1$.

The Manley-Rowe relations [6] for a model of nonlinear capacitance, which is associated with an equivalent external circuit containing generators with frequency ω_1, ω_2, $\omega_1 + \omega_2$, $\omega_1 - \omega_2$, ..., $m\omega_1 + n\omega_2$ (m and n are integrals, and ω_1 and ω_2 are incomparable frequencies), active conductors, and ideal filters (filters have zero resistance at the generator frequency and infinite resistance at all other frequencies) takes the form

$$\sum_{m=0}^{\infty}\sum_{n=-\infty}^{\infty} \frac{mP_{m,n}}{m\omega_1 + n\omega_2} = 0, \quad \sum_{m=-\infty}^{\infty}\sum_{n=0}^{\infty} \frac{nP_{m,n}}{m\omega_1 + n\omega_2} = 0,$$

where $P_{m,n} = P_{-m,-n}$ is the average power entering the nonlinear capacitor at frequencies $\pm |m\omega_1 + n\omega_2|$. Although our general consideration will be for nonlinear capacitors and not those changing over time, it can be shown that these concepts are equivalent [5].

11.3 Waves in Periodic Structures.
The Mathieu Zone and the Brillouin Diagram

When analyzing waves in media with periodically changing parameters we use (11.3). We have already stated that the formal difference between (11.3) and (11.1) or (11.2) is only that the variable V is a function of coordinates and not of time. However, the physical sense of the solution to (11.3) is completely different to that for, say, (11.1). Indeed, it might be expected that a wave would be amplified simply because it is propagating in a periodically inhomogeneous medium. However this is not so because there is nowhere from which to take energy for such an amplification. Nevertheless, it follows from the formal analogy between the equations that the solutions will exponentially increase with coordinate:

$$V(x) = A_1 e^{\lambda x} \sin k_0 x + A_2 e^{-\lambda x} \cos k_0 x .$$

This means that our medium allows waves to propagate into opposite direction, the forward and return waves.

When we are looking for instability over time we will only be interested in a solution corresponding to the positive characteristic index λ, (i.e., - $\exp(\lambda t)$; we consider parameter values inside the Mathieu zone). Now it is necessary to select one of the two solution terms. We are helped here by the physical reasoning that at equilibrium (if only for an inhomogeneous medium) the forward and return waves may not grow at the same time. Therefore only $A_1 = 0$ may be correct. The forward and return waves must therefore exponentially decay in the x direction.

Thus, if the wave number of the wave is inside the Mathieu zone, then the wave will be nondistributing, i.e., it is an opaque zone. Outside an opaque zone the characteristic index λ is purely imaginary, i.e., a wave with a corresponding k_0 will propagate (true, it will be spatially modulated).

Thus waves in periodic nonhomogeneous media may only propagate under certain conditions. When $k_0 = q_1/2$, for example, when the wavelength of an incident wave Λ_0 is twice the characteristic scale of the homogeneity of the medium, viz., the lattice wavelength Λ_1, a wave will not propagate

($\lambda_0 = 2\lambda_1$ is called a Bragg condition for reflection from a periodic structure). A physical explanation is quite simple: a return wave appears due to resonance reflection ($\Lambda_0 = 2\Lambda_1$) from even the smallest inhomogeneity. True, this wave is weak but because of the resonance effect along the x-coordinate it builds up and a standing wave appears, i.e., at a certain length all of the energy of the incident wave goes into the reflected wave. When $\Lambda_0 = 2\Lambda_1$ (or close to this resonance region) the forward and return waves are strongly associated. The following opaque zones correspond to waves scattered on specially harmonic inhomogeneities. Inside these zones we have $k_0 \sim nq_1/2$.

If the depth of modulation of the parameter characterizing the periodic medium is not small, then in general a wave in the medium is described [8] by

$$d^2\Psi/dx^2 + f(x)\Psi = 0 , \qquad (11.24)$$

$$f(x) = f(x + 2\pi/K) = \sum_{n=0}^{\infty} a_n \cos(nKx) , \qquad (11.25)$$

where the a_n are the coefficients of the Fourier series of the periodic function $f(x)$ with period $2\pi/K$ ($\Lambda = 2\pi/K$ is the structure period). Hill functions, of which the Mathieu functions are special cases (when only a_0 and a_1 are nonzero), are solutions to (11.24). Note that the function $f(x)$ in (11.25) may be odd. The solution to (11.24) may be sought for in the form

$$\Psi(x) = A(x)\exp(-ikx) + B(x)\exp(ikx) ,$$

where $A(x)$ and $B(x)$ are periodic functions with period $2\pi/K$ and k (the analog of λ) is a characteristic indicator depending on the coefficients a_n. By expanding $A(x)$ and $B(x)$ into Fourier series, we obtain

$$\Psi = \sum_{n=-\infty}^{\infty} \left\{ A_n \exp[-i(k+nK)x] + B_n \exp[i(k+nK)x] \right\} .$$

Each n corresponds to a spatially harmonic wave, and the quantity $k_n = k + nK$ has the meaning of a wave number for these harmonics. Note that the spatial harmonic cannot be excited independently. By substituting this solution into (11.24) it is possible to obtain a dispersion equation for k as a function of the coefficients a_n [1,8]. If we have $a_0 \neq 0$

and $a_1 \neq 0$ and all the remaining $a_n = 0$ in (11.25) then we have from (11.25)

$$d^2\Psi/dx^2 + (a_0 + a_1 \cos Kx)\Psi = 0 ,\qquad(11.26)$$

i.e., we arrive at an equation like Mathieu's, the stability diagram of which is given in Fig. 11.5a. The shaded region in the diagram is the nontransmission (instability) region, in which $k = mK/2 + i\alpha$, ($m = 0, 1, 2 \ldots$), α is real, and $mK/2 = |k|$ is the value of $|k|$ at the region's boundaries. A straight line passing alternately through zones of transmission and nontransmission corresponds to a constant modulation depth ($a_1/K^2 = $ const) and constant signal frequency $K = $ const as (a_0/K^2) is changed in Fig. 11.5a. When all the $a_n \neq 0$, the stability diagram changes somewhat (see Fig. 11.5b). The borders of the regions intersect, i.e., the zones of nontransmission are changed. Note that the behavior of nontransmission zones is often interpreted to be the result of the existence of distributed feedback arising as a wave propagates due to reflections of elements of the periodic structure following one another. Brillouin diagrams, which are graphical representations of the dispersion equation, are more often used in the theory of periodic structures than diagrams of the Mathieu zones, to which they are related. We look at this diagram using the example of an infinite medium with a weak periodic inhomogeneity. If a wave is propagating in a homogeneous linear medium, then $\omega = \pm v_0 k$ (Fig. 11.6a). When there is a weak (infinitesimal) periodic inhomogeneity in the medium, then a spatial harmonic, a wave with dispersion equation $k_n = \pm v_0 k + nK$ ($n = 0, \pm 1, \pm 2, \ldots$) will rise. When the disturbance is infinitesimal the harmonics do not interact (Fig. 11.6b). As the disturbance grows, strong coupling arises between the harmonics at the points where the dispersion characteristics in Fig. 11.6b intersect (Fig. 11.6c), and as the result nontransmission zones arise. At the borders of these zones the interacting harmonics have oppositely signed group velocities (Chapter 8). The function $\omega = v_0 k_0 + (\mu c_1/4c_0)v_0 k_0$ is graphed in Fig. 11.6d for dispersion equation (11.24), which illustrates the formation of nontransmission zones. The regions on the (ω, k)-plane corresponding to real ω and k, i.e., the transmission zones, are called Brillouin zones. We leave the reader to work out for himself exactly how the stability diagrams (Mathieu zones) are related to the Brillouin diagrams [1,7].

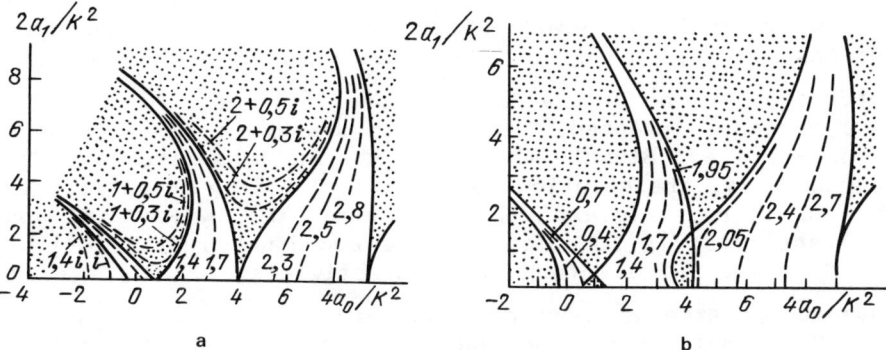

Fig. 11.5. a, b) Stability diagrams for the (11.26) and (11.25) taken from [8]. The instability regions are shaded with dots.

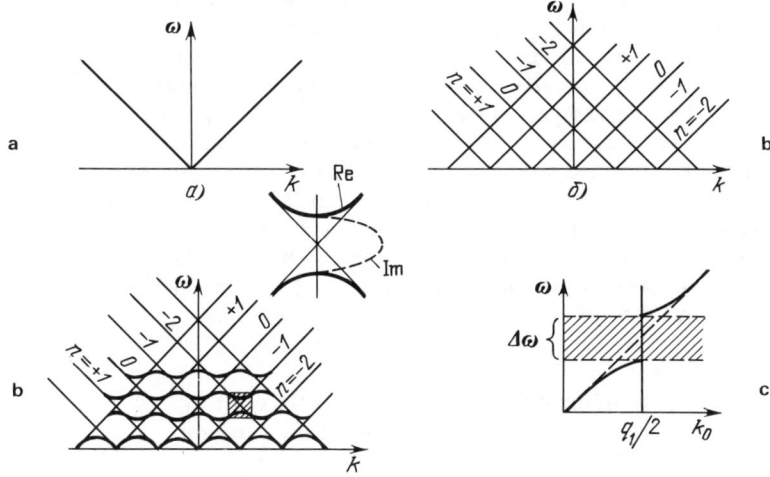

Fig. 11.6. Construction of Brillouin diagrams for a periodically disturbed infinite medium: a) for a homogeneous medium; b) for a medium with an infinitely small periodic inhomogeneity leading to the appearance of noninteracting spatial harmonics; c) a finite disturbance (the solid line) and the harmonics are strongly linked; d) the behavior of the nontransmission zones for a system described by equation $\omega = v_0 k_0 \pm (\mu c_1/4c_0) v_0 k_0$.

Another situation that leads to an equation like (11.25) is the motion of an electron in the field of a crystal's ionic lattice. Electron density waves are described by the Schrodinger equation with a periodic potential, i.e.,

$$(\hbar/2m)\nabla^2\Psi + [E - V(\mathbf{r})]\Psi = 0 , \qquad (11.27)$$

where E is the total energy, and $V(\mathbf{r})$ is the potential energy, which is a periodic function of the coordinates (the periods d_1, d_2, and d_3 of the change along each axes are determined by the crystal's structure ([1], Sec. 40)). As we have already mentioned, there is an analog to the Floquet's theorem , i.e., Blokh's theorem, for equations like (11.27). Consequently, the solution to (11.27) should be sought for in the form $\Psi(r) = A(r)\exp(\lambda r) + B(r) \exp(-\lambda r)$, where $A(r)$ and $B(r)$ are periodic functions in the coordinates with periods d_1, d_2 and d_3. The methods for investigating (11.27) are in essence the same as those discussed above.

Note that there is an unusually wide circle of problems requiring analysis of waves in periodic structures. Its interest is mostly associated with technological problems. By way of example, we indicated the creation of new types of dampening systems for electronic microwave devices [9], periodically loaded antennae for travelling waves [10], and transformers and filters for bulk and surface acoustic waves [11, 12]. The analysis of periodic structures is also required in biology, where it is mainly associated with processes in complex insect eyes (the multilayered chitin covering of the horse-fly; the eyestalk of the nut butterfly, which consists of periodic disks in eye wave guide; part of the eyestalk of the thickhead, which is a round wave guide with a corrugated surface) [8]. Many other interesting examples of waves in passive and active periodic structures can be found in [8].

11.4 Motion in a Rapidly Oscillating Field. Kapitsa's Pendulum. Free Electron Lasers

Up to now, we have, in our study of the behavior of systems with slow parameters, we have restricted ourselves to parametric resonance, i.e., to the special case when the frequency of the change in the system's parameter is of the same order as the fundamental frequency ($\omega_p \approx 2\omega_0/n$, n is a small number). We have seen that it is possible to have an exponential instability. We look now to see what happens if the parameter changes very much faster than the fundamental frequency of the system, i.e., $\omega_p \gg \omega_0$.

Let as consider a nonlinear oscillator on which a force that is periodic in x is acting:

$$\ddot{x} + f(x) = F(x)\cos \Omega t . \qquad (11.28)$$

Here $\Omega \gg 1/T$, where $T = 2\pi/\omega_0$ is the characteristic period of the motion of the autonomous system along its trajectory. We shall look for a solution to (11.28) as the sum $X(t) + \mu\chi(t)$

of slowly and rapidly oscillating parts, where $X(t)$ and $\chi(t)$ change, respectively, with characteristic times $T \sim 2\pi/\omega_0$ and $\tau \sim 2\pi/\Omega$, while $\mu \sim \omega_0/\Omega \ll 1$. This form of solution is physically justifiable because an oscillator can only weakly respond to the rapid external fluctuations due to inertia. By substituting this solution into (11.28) and taking into account that $f(X + \mu\chi) \approx f(X) + \mu\chi(\partial f/\partial x)_X$,

$$F(X + \mu\chi) \approx F(X) + (\partial F/\partial x)_X,$$

we obtain the equation

$$\ddot{X} + \mu\ddot{\chi} = -f(X) - \mu\chi(\partial f/\partial x)_X + [F(X) + \mu\chi(\partial F/\partial x)_X]\cos \Omega t . \tag{11.29}$$

This equation contains fluctuating and slowly changing terms. These can easily been sorted out by averaging (11.29) over the period $\tau = 2\pi/\Omega$. The result is two coupled equations, i.e.,

$$\ddot{X} \approx -f(X) + \langle\mu\chi(\partial F/\partial x)_X\cos \Omega t\rangle ,$$

$$\mu\ddot{\chi} = -\mu\chi(\partial f/\partial x)_X + F(X)\cos \Omega t .$$

Since the term $\mu\ddot{\chi}$ in the last equation is of the order $\mu\Omega^2\chi \sim \omega_0\Omega\chi$, and hence is not small, while the term $\mu\chi(\partial f/\partial x)_X$ is small, the equation can be integrated immediately. As the result we get

$$\chi = -(F(X)/\mu\Omega^2)\cos \Omega t \tag{11.30}$$

Note that when integrating over a rapid time the function $F(X)$ may be considered constant. By substituting (11.30) into the equation for X, we obtain

$$\ddot{X} + f(X) = -\langle(F(X)/\Omega^2)(\partial F/\partial x)_X\cos \Omega t\rangle,$$

whence we finally get

$$\ddot{X} + f(X) + \frac{1}{2}\frac{F(X)}{\Omega^2}\left(\frac{\partial F}{\partial x}\right)_X = 0 .$$

We have obtained a very important result, one that is intuitively very unexpected. Instead of vibrating insignificantly under the action of the rapid external fluctuations, maintaining its average motion along the trajectory that the autonomous analog would follow, our new effective oscillator behaves very differently. A significant term proportional to the square of the amplitude of the external fluctuations has been added to the restoring force.

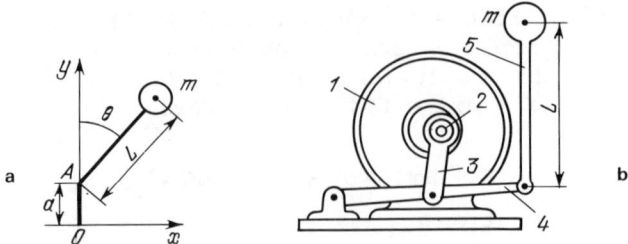

Fig. 11.7. A pendulum with a vibrating suspension [13]:
a) the theoretical model; a mathematical pendulum of length L and mass m freely rotating about its suspension point A, which vibrates along the axes (about point O) with frequency ω and amplitude a.
b) Diagram of a device to test Kapitsa's pendulum: an electric motor (1) from a sewing-machine (4000-6000 rpm) drives an eccentric axis (2) to which a connecting shaft (3) is fixed; the other end of the connecting shaft drives a lever (4), one end of which rotates about a stationary support, and the other end is fastened to the shaft of a pendulum (5) (L ~ 15 cm) such that the pendulum can swing freely (a ~ 3-4 mm).

This result was first obtained in 1951 by P.L. Kapitsa and applied to a pendulum with a rapidly vibrating fulcrum [13,14]. The theoretical model of Kapitsa's pendulum and a diagram of the device for testing it are shown in Fig. 11.7. The equation of motion of a Kapitsa's pendulum has the form

$$mL^2\ddot{\theta} = M_\theta - mLa\omega^2 \sin \omega t \sin \theta , \qquad (11.31)$$

where M_θ is the moment of the external forces (when the moment of the external forces is created by gravity, we have $M_\theta = mgL \sin \theta$) [13] assuming that $\theta(t) = \phi(t) + \beta(t)$ $\phi(t)$ and $\beta(t)$ have the same sense as $X(t)$ and $\mu\chi(t)$, respectively), we get the following for the quantities averaged of a time $\tau = 2\pi/\omega$:

$$\langle\theta(t)\rangle \approx \phi , \quad \langle\beta(t)\rangle = 0 ,$$

i.e., as in the previous case, we exclude motion from the equations by averaging the angle β, and the angle θ is replaced by φ, which characterizes the position of the pendulum around which small vibrations take place. The effect of the vibrating suspension point on the oscillations of the pendulum are in this approximation very simple: a vibration situation sets in which a force couple tends to orient the pendulum in the direction of the fulcrum's vibration, i.e., along the y-axis. This situation is expressed thus:

$$<M> = - (ma^2\omega^2/4)\sin 2\phi . \qquad (11.32)$$

It does not depend from the length of the pendulum and is proportional to the square of the amplitude of the fulcrum's vibration. The equation of motion (11.31) together with (11.32) can now be represented as $mL^2\ddot\phi = M_{eff}$, where $M_{eff} = M_\phi - (ma^2\omega^2/4)\sin 2\phi$ (M_ϕ is obtained from M_θ by substituting the angle θ for ϕ). In a gravity field we have $M_\phi = mgL \sin \phi$, i.e., the pendulum's equilibria, which are defined as $M_{eff} = 0$, includes the trivial one $\phi = 0$, which corresponds to the pendulum being upside down. In order for this equilibrium to be stable, it is necessary that $(dM_{eff}/d\phi)_{\phi=0} < 0$, whence we get a stability condition $a^2\omega^2 > 2gL$. When this condition is satisfied, the vertical position of Kapitsa's pendulum (Fig. 11.7a) is stable. Experimentally this was reported as follows: "When the device was switched on the pendulum began to behave as if it was being acted upon by a force directed along the axis of the fulcrum's vibration. Because the frequency of vibration of the fulcrum was large the pendulum seemed to the eye to be blurred and the vibration was not noticeable. Therefore the stability situation was very surprising. If the pendulum was pushed to the side it began to swing like a normal pendulum. These oscillations decayed and the pendulum returned to the vertical" [13].

"If the device was twisted so that the pendulum vibrated in the horizontal plane, then the effect of gravity on the motion was excluded. If the pendulum was touched very carefully and moved to one side, then one's finger feels the pressure caused by the vibrational moment and it easy to be convinced that its greatest value occurs when the angle of turn is 45°." [14]. When the pendulum is in the usual stable position, the vibration of the fulcrum leads to a decrease in the period of the pendulum's oscillation. This means that any vertical oscillation affecting the pendulum with a period smaller than the pendulum's period will always accelerate the motion of the pendulum (this was demonstrated on a double pendulum [14]).

The theory was generalized for three-dimensional motion in electromagnetic fields [15]. It has been suggested that the motion of electrons in weakly homogeneous variable fields be used to create a microwave laser [16]. Recently, a similar approach has been successfully applied to the theory of free-electron lasers. These work by either irradiating electrons into a periodic static field (the ubitron) or scattering the waves using beams of relativistic electrons (scatteron) [17,18]. Diagrams of these lasers are shown in Fig. 11.8.

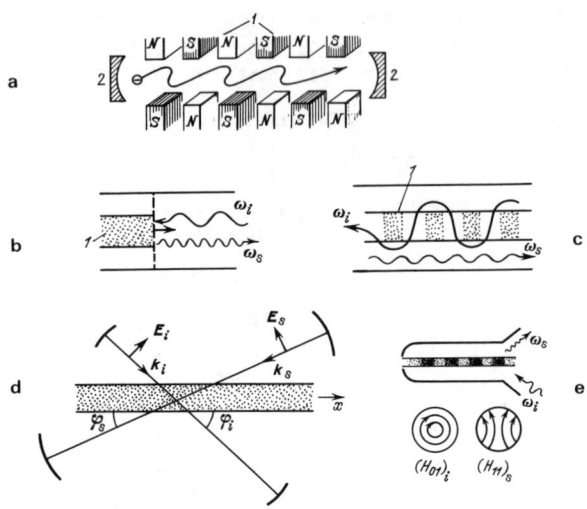

Fig. 11.8. Diagrams of free-electron lasers.
a) The ubitron laser. This is pumped by a periodic magnetic system (1) and the beam's current spectrum contains harmonics whose velocities are greater than that of light and which interact with the field in the resonator (2).
b) The scatteron is a laser with mirror reflection from the rapidly moving leading front of the electron beam (1).
c) The scatteron with the pumping wave scattered at density disturbances (1) caused by a combined wave with frequency $\omega_i - \omega_s$. This scattering leads to the appearance of the signal. The pumping (subscript i) and signal (subscript s) may exist as different types of oscillation in the electrodynamic structure.
d, e) diagram of models of the scatteron used in the theory.

The simplest theory applicable to Fig. 11.8d is presented in [19]. Moving uniformly in the x-direction (the direction of the electron beam) at the phase velocity of the combined wave $v_{ph} = \Omega/k_x = (\omega_s - \omega_i)/(k_s \cos \phi_s + k_i \cos \phi_i)$ in the inertial reference frame K', the fields of both waves acting on the beam

$$E = \text{Re}\{E_i \exp i(\omega_i t - k_i r) + E_s \exp i(\omega_s t - k_i r)\}$$

have one frequency, with $E' = \text{Re}\{\hat{E}(r), \exp i\omega t\}$. The averaged motion of an electron, whose velocity is $v' \ll c$, is determined in this coordinate system by a force (Miller's force [15])

$$F' = -(e^2/4m\omega^2)\nabla' |\hat{E}|^2.$$

When E_i and E_s are constant this force is determined only by the combination wave. The expression for F' is valid when focusing electron fields are not present. Using this approach, the physical processes in the scatteron may be interpreted as the radiation of particles when there is spatial resonance with the wave beats: $v \approx v_{ph} \approx (\omega_s - \omega_i)/\kappa_x$, $\kappa = k_s - k_i$ (v is the electron velocity). Because the action of the combined wave is similar to the action of the normal wave, the mechanism for the induced scattering is qualitatively as follows. When two waves with frequencies ω_i and ω_s and amplitudes E_i and E_s act on an electron beam, a periodic force with the frequency $\omega_s - \omega_i$ and amplitude $E_i E_s$ begins to act on the electrons. Given that $v \approx v_{ph}$ the combined wave acting on the beam causes the beam to group. The electron density then changes in amplitude proportional to $(E_i E_s)/(\omega_i - \omega_s)^2$ with frequency $\omega_s - \omega_i$. Since the electric permittivity of the electron beam and its reflective index are determined by the electron density, changes in the density also cause changes in these parameters. The pumping wave (E_i, ω_i) is then scattered on these disturbances in the refractive index. However, the frequency difference $\omega_s - \omega_i$ and the pumping frequency ω_i are superimposed, which leads to a signal wave with frequency ω_s. Because of the synchronicity condition ($v \approx v_{ph}$), the grouped electron beam strengthens the field of the combination wave, much as happens in a travelling wave tube, and therefore the theory of the scatteron is similar to the theory of the travelling wave tube (Chapter 7), the high frequency electrostatic field in the travelling wave tube theory being changed for the effective field of the combined wave.

CHAPTER 12

ADIABATIC INVARIANTS.
PROPAGATION OF WAVES IN INHOMOGENEOUS MEDIA

12.1 The Wentsel-Kramers-Brillouin (VCB) Approximation and Adiabatic Invariants

We saw in Fig. 11.3b that if the frequency ω_p of the change in the system's parameter is much smaller than the fundamental frequency ω_0 ($\omega_p \ll \omega_0$), then there is practically no instability. The instability zone becomes narrower as the ratio ω_0/ω_p increases. This case of a very slow, adiabatic, change in the parameter (of which an example is the oscillation of a pendulum whose length is changing slowly) is very interesting for a discussion of oscillations and waves, and at the same time it is often encountered in practice.

We take as our basic model an oscillator with a slowly changing frequency. Its equation has the form

$$\ddot{x} + \omega_0^2(t)x = 0 . \qquad (12.1)$$

Here the characteristic time T of the change in the parameter (frequency ω_0) is large: $T \gg 2\pi/\omega_0$. We introduce a slow time $\tau = t/T$. The equation (12.1) may be rewritten as:

$$x'' + T^2\omega_0^2(\tau)x = 0$$

(here and later the primes indicate differentiation with respect to the slow time). We change the variables

$$x(\tau) = \exp\left(\int_0^\tau y d\tau\right) . \qquad (12.2)$$

It is clear that $x'(\tau) = xy$, $x''(\tau) = xy^2 + xy'$, and hence in place of (12.1) we obtain $x\left[y^2 + y' + T^2\omega_0^2(\tau)\right] = 0$, which when $x \neq 0$ is the same as Riccatti's equation, i.e.,

$$y^2 + y' + T^2\omega_0^2(\tau) = 0 .\qquad(12.3)$$

Thus in place of a linear second order equation (12.1) we have obtained a first-order nonlinear equation. However, this nonlinear equation is simpler to investigate.

Since the change in the parameter is slow, we shall look for an approximate solution to (12.3) in the form of an asymptotic expansion

$$y = Ty_0 + y_1 + T^{-1}y_2 + \ldots T^{-(n-1)}y_n + \ldots ,\qquad(12.4)$$

where the small parameter will be $(1/T)^1$. By substituting (12.4) into (12.3) we obtain

$$T^2 y_0^2 + 2Ty_0 y_1 + y_0^2 + Ty_0' + y_1' + T^2\omega_0^2(\tau) + \ldots = 0 .$$

Separating the terms by degree yields

$$y_0 = \pm i\omega_0(\tau) , \quad y_1 = - y_0'/2y_0 = -(1/2)(\ln \omega_0)' .\qquad(12.5)$$

We shall use only the first two terms in the expansion (12.4). In this approximation, using (12.5) we find from (12.2) that

$$x(\tau) = \exp\left\{\int_0^\tau (Ty_0 + y_1)d\tau\right\}$$

$$= A \exp\left\{-(1/2)\ln \omega_0(\tau) + iT\int_0^\tau \omega_0(\tau)d\tau\right\} + cc ,$$

where A is a constant, and cc means the complex-conjugate of the term corresponding to the second sign (minus) in y_0.

The final approximate solution is written in the form

$$x(\tau) = (A/\sqrt{\omega_0(\tau)})e^{i\theta} + c.c. ,\qquad(12.6)$$

-----

[1] Various asymptotic methods for solving equation with variable coefficients, containing both small and large parameters, are presented in [1].

where $\theta = T\int_0^\tau \omega_0(\tau)d\tau$ is the full phase. In this way, the solution corresponds to an oscillator with a changing amplitude and frequency. The essential result is that the amplitude of these oscillations decays and grows slowly, that is adiabatically, because of the slow change in ω_0. The solution (12.6) is called the Wentsel-Kramers-Brillouin (WKB) solution [2,12]. It was first obtained as a solution of Schrodinger's equation for waves propagating in a weakly nonhomogeneous medium.

We shall try to sort out the physical meaning of this solution. Let us calculate the energy of our oscillator with its slowly changing frequency. We know that $\mathcal{H} = (\dot{x}^2 + \omega_0^2 x^2)/2$. We have $x(t) = (A/\sqrt{\omega_0(t)})\cos\theta$, where $d\theta/\delta t = \omega_0(t)$ and $\dot{x}(t) \approx A\sqrt{\omega_0(t)}\sin\theta$ (the term containing $\dot{\omega}(t)$ is very small and therefore we neglect it). Thus $\mathcal{H} = A^2\omega_0(t)/2$, where A = const, or

$$\mathcal{H}/\omega_0(t) = \text{const} , \qquad (12.7)$$

i.e., the ratio of the energy stored by the oscillator to its frequency, when the parameter is changing slowly, is conserved over time. A quantity conserved during a slow change in the parameters of a dynamic system is called an adiabatic invariant [3,4,16].

A very important conclusion follows from (12.7): in a slowly changing field it is possible to change considerably (e.g., increase) the energy of the oscillator, i.e., to use the oscillator for amplification. To illustrate why the quantity $\mathcal{H}/\omega_0(t)$ is conserved, a quantum analogy might help, i.e., the description of an oscillator as a quasiparticle. The energy of an oscillator is $\mathcal{H} = \hbar\omega_0 N$, where $\hbar\omega_0$ is the energy of an elementary oscillation of a quantum or quasiparticle, and N is the number of quasiparticles or quanta. When the change in the parameter is slow, the number of quanta, clearly, may not change since they cannot be merged, i.e., the number of quanta is an adiabatic invariant. The total energy of the oscillator changes due to changes in the energy of the quanta, or quasiparticles. Thus, the sense of the adiabatic invariant in (12.7) is quite transparent.

When there is resonant parametric amplification, the situation is quite different. The energy of the oscillating system grows because of an increase in the number of quanta, the energy of each quantum not changing. Remember, for example, that the energy of one pumping quantum is approximately $2\hbar\omega_0$ for the main resonance while a signal oscillation has a fundamental oscillator frequency of ω_0, i.e., the energy of one pumping quantum is equal to that of two quanta for the signal oscillator ($2\hbar\omega_0 = \hbar\omega_0 + \hbar\omega_0$). In other words, one pumping quantum is broken into two signal quanta, due to which the oscillation's total energy increases with frequency ω_0.

It can be shown quite easily that if the system is partitioned into n normal oscillators, then there must be n independent adiabatic invariants.

Let is consider another elegant method for obtaining the adiabatic invariant, this one based on the application of the approximate direct variation method [5] which is close to Whitham's method [6]. We consider (12.1) to be the Euler equation for the variational problem, i.e., to be the condition for the functional I to be stationary. Since the Euler equation for $I = \int_a^b F(x, y, y')dx$ has the form [7]

$$F'_y(x, y, y') - \frac{d}{dx} F'_{y'}(x, y, y') = 0,$$

the functional corresponding to (12.1) is written as

$$I = \int_0^t \frac{1}{2}\left(\dot{x}^2 - \omega^2 x^2\right)dt . \qquad (12.8)$$

The reader is offered the opportunity to verify for himself the validity of (12.8). We assume, further, that

$$x(t) = X(t)T[\Omega(t)] , \qquad (12.9)$$

where $X(t)$ and $\Omega(t)$ are functions that change slowly over time, and $T[\Omega(t)]$ is a periodic function such that

$$T(\Omega + 2\pi) = T(\Omega) , \qquad (12.10)$$

$$\langle T \rangle_\Omega = \frac{1}{2\pi} \int_0^{2\pi} T(\Omega)d\Omega = 0 , \qquad (12.11)$$

$$\langle T^2 \rangle_\Omega = 1 . \qquad (12.12)$$

It can be shown that (12.10) and (12.12) do not reduce the generality of the solution because X and Ω are yet undefined.

ADIABATIC INVARIANTS 243

At the same time (12.11) defines the character of the solution. By using (12.9) to find $\dot{x}$, we can write an equation for the averaged functional (12.8), viz.

$$<I>_\Omega = \frac{1}{2} \int_0^t \left\{ <\dot{X}^2 T^2>_\Omega + \left<2 X \dot{X} \dot{\Omega} \left(T \frac{dT}{d\Omega}\right)\right>_\Omega \right.$$

$$\left. + \left<X^2 \left(\frac{dT}{d\Omega}\right)^2 \dot{\Omega}^2\right>_\Omega - <\omega^2 X^2 T^2>_\Omega \right\} dt .$$

Since X, T, and ω_0 change slowly over time it can be shown that $<\dot{X}^2 T^2>_\Omega \approx \dot{X}^2$ (using (12.12)), that $<2X\dot{X}\dot{\Omega}(T\ dT/d\Omega)>_\Omega \approx 0$ (using (12.11), and that $<X^2(dT/d\Omega)^2 \dot{\Omega}^2>_\Omega \approx \alpha X^2 \dot{\Omega}^2$, where

$$\alpha = \frac{1}{2\pi} \int_0^{2\pi} (dT/d\Omega)^2 d\Omega , \quad -\omega^2 <X^2 T^2>_\Omega \approx -\omega^2 X^2$$

(using (12.12)). Thus we obtain a new functional

$$I \approx <I>_\Omega \approx \frac{1}{2} \int_0^t [\dot{X}^2 + (\alpha \dot{\Omega}^2 - \omega^2) X^2] dt ,$$

which contains yet another dependent variable $\dot{\Omega}$. Remember, that if $I = \int_a^b F(x, y, y', z, z') dx$, then the stationary condition [7] is

$$\frac{\partial F}{\partial y} - \frac{d}{dx}\frac{\partial F}{\partial y'} = 0 , \quad \frac{\partial F}{\partial z} - \frac{d}{\partial z}\frac{\partial F}{\partial z'} = 0 .$$

By varying I, we obtain two Euler equations

$$\partial X \rightarrow X(\alpha \dot{\Omega}^2 - \omega^2) - \ddot{X} = 0 , \qquad (12.13)$$

$$\delta \Omega \rightarrow \frac{d}{dt} \alpha X^2 \dot{\Omega} = 0 . \qquad (12.14)$$

It is rational to choose the test function for T (see (12.1)) in the form $T = T_1 \sin[\Omega(t)]$ where $T_1 = \text{const}$. Then $\alpha = T_1^2/2$

and, by assuming that $T_1 = \sqrt{2}$ we have

$$T = \sqrt{2} \sin \Omega, \quad \alpha = 1. \tag{12.15}$$

Because X changes slowly over time, it follows from (12.13) that $X(\alpha\Omega^2 - \omega^2) \approx 0$ and using (12.15) that

$$\Omega = \int_0^t \omega(t)dt. \tag{12.16}$$

We find from (12.14) that

$$X^2\dot{\Omega} = X^2\omega(t) = \text{const}.$$

We have finally regained the adiabatic invariant, and the solution to (12.9) has the form

$$x(t) = \frac{\text{const}}{\sqrt{\omega(t)}} \sin \int_0^t \omega(t)dt. \tag{12.17}$$

This direct variational method is interesting because it can be used for both linear and nonlinear waves [5].

12.2 Equivalence Between a Rotor and an Oscillator

As an example of a rotor, let us consider an electron moving in a homogeneous constant magnetic field. Under arbitrary initial conditions the electron will move in a spiral whose axis lies along the magnetic field. We are interested in the special case when the electron's initial velocity has no component along the field and it rotates in a circle in a plane perpendicular to the field with a cyclotron frequency $\omega = (e/m)B$. Now suppose the magnetic field $B(t)$ is directed along the z-axis and changes slowly over the cyclotron period, $T = 2\pi/\omega$. The alternating magnetic field induces an electric field $\mathbf{E} = -[\mathbf{z}_0\mathbf{r}](dB/dt)/2$ (the formula is written for a system of units in which the velocity of light is $c = 1$, and $\mathbf{z}_0$ is the unit vector in the z-direction). The equation of motion $\ddot{\mathbf{r}} = (e/m)([\dot{\mathbf{r}}\mathbf{B}] + \mathbf{E})$ using the expressions for ω and $\mathbf{E}$ takes the form

$$\frac{d^2\mathbf{r}}{dt^2} + \left[\mathbf{z}_0 \frac{d\mathbf{r}}{dt}\right]\omega(t) + \frac{1}{2}[\mathbf{z}_0\mathbf{r}]\frac{d\omega}{dt} = 0. \tag{12.18}$$

ADIABATIC INVARIANTS

When projected on the x and y-axes we obtain instead of (12.18) a fourth-order system of equations, i.e.,

$$\ddot{x} - \omega(t)\dot{y} - \dot{\omega}(t)y/2 = 0 , \qquad (12.19)$$

$$\ddot{y} + \omega(t)\dot{x} - \dot{\omega}(t)x/2 = 0 . \qquad (12.20)$$

It would seem that there should be two independent adiabatic invariants in this system. We shall show, however, that in reality this system for a rotor is equivalent to that for oscillator, and has only one invariant. By multiplying (12.20) by i, adding it to (12.19) and making the substitution of the complex number $\xi = x + iy$, we obtain in place of (12.19) and (12.20) one complex equation, i.e.,

$$\ddot{\xi} + i\omega(t)\dot{\xi} + (i/2)\dot{\omega}(t)\xi = 0 . \qquad (12.21)$$

We now make the substitution in (12.21) of

$$\xi = u \exp\left\{-(i/2)\int_0^t \omega(\tau)d\tau\right\} = u \exp(\theta) ,$$

which yields

$$\dot{\xi} = (\dot{u} - i\omega u/2)\exp(\theta)$$

and

$$\ddot{\xi} = [\ddot{u} - i\omega\dot{u} - (1/4)\omega^2 u - (1/2)i\dot{\omega}u]\exp(\theta).$$

By using these expressions in (12.21), we arrive at the equation for a harmonic oscillator, whose fundamental frequency is the Larmor frequency, i.e.,

$$\ddot{u} + \omega_L^2(t)u = 0 . \qquad (12.22)$$

Equation (12.22) differs from (12.1), whose solution we have found, only by the fact that u is now complex. However, because the same independent equations are obtained for $Re\ u$ and $Im\ u$ the change introduces nothing new. Hence,

$$u(t) = (A/\sqrt{\omega(t)}) \exp\left[i \int_0^t \omega(t)dt\right].$$

To find out what the adiabatic invariant is now, we write $\dot{u}\dot{u}^* = \omega_L(t)|A|^2$. It is clear that the invariant is $(\dot{u}\dot{u}^*)/\omega_L(t) = $ const. Concerning the physical meaning of this invariant, it is easy to show that $\dot{u}\dot{u}^* = \dot{\xi}\dot{\xi}^*$ if $\dot{u}/u = \dot{u}^*/u^*$.

The latter is satisfied, i.e., u and u^* change over time in the same way, because (12.22) is an equation with real coefficients. However, we have $\dot{\xi}\dot{\xi}^* = \dot{x}^2 + \dot{y}^2 = v_\perp^2$, where $v_\perp$ is the velocity of the electron's transverse rotation. Thus, $v_\perp^2/\omega_L(t) = $ const or $\mathcal{E}_{kin}/B(t) = $ const, i.e., the electron's kinetic energy changes in proportion to the amplitude of the magnetic fields. The quantity $uu^*\omega_L(t) = |A|^2$ is also an adiabatic invariant, whence $(x^2 + y^2)\omega_L(t) = $ const. However $v_\perp^2 = (x^2 + y^2)\omega_L^2$, i.e., this is the invariant we know. We have arrived at the interesting conclusion, that is the energy of an electron-oscillator in a slowly changing magnetic field may be greatly changed. For example, the electron oscillator may be given a continuous high frequency electric field. This will occur if the quasistatic component of the field changes slowly over time.

12.3 Propagation of Waves in Inhomogeneous Media. The Approximation of Geometric Optics

Waves may propagate in inhomogeneous media, i.e., media whose properties change over time, in a large variety of ways. However, the mathematical problem concerning the propagation of an harmonic wave in a inhomogeneous medium may in most cases, be reduced to finding the solution of Helmholtz's equation

$$\nabla^2 f + k^2(x, y, z)f = 0 \qquad (12.23)$$

where f is a scalar function. Clearly the solution to (12.23) mainly depends on the choice of function $k^2(x, y, z)$.

The simplest problem is when k^2 depends on only one coordinate, for example, Cartesian coordinate x, which corresponds to a layer inhomogeneous medium. In certain approximations, the atmosphere and ionosphere of the Earth, ocean water, the earth's core, and optic fibers are such media.

In the general case the propagation of a plane wave in a medium whose properties depend on x is described by

$$\partial^2 u/\partial x^2 - [\varepsilon(x)/v_{ph}^2]\partial^2 u/\partial t^2 = 0, \qquad (12.24)$$

where $\varepsilon(x)$ is the function characterizing the medium's properties (for electromagnetic waves the dielectric permittivity) and smoothly changing along x, and v_{ph} is the phase velocity of the waves (the physical nature of the waves does not concern us) in a homogeneous medium.

We shall be interested in the steady state propagation of a monochromatic wave, i.e., we shall consider that

$$v(x, t) = u(x)e^{i\omega t} + \text{c.c.} , \qquad (12.25)$$

and that the amplitude of the wave does not depend on time. This means, for example, that when the wave decays as $\exp(i\omega t - ikx)$ at the boundary of a medium, we do not have to wait long for a steady state to become established in the medium. In order to solve an equation like (12.25) we can represent (12.24) as

$$\partial^2 u/\partial x^2 + k_0^2 \varepsilon(x) u = 0 , \qquad (12.26)$$

where $k_0^2 = \omega^2/v_{ph}^2$. The smoothness of the medium's inhomogeneity means that over a wavelength $\lambda = 2\pi/k_0$ the value $\varepsilon(x)$ does not change very much, i.e., $\lambda d\varepsilon(x)/dx \ll \varepsilon(x)$. Equation (12.26) is similar to equation (12.1) or (12.22), which we have investigated with the exception that the changes in amplitude occur along the coordinate, and not over time. By making the substitution $u(x) = \exp\left\{\int_0^x y dx\right\}$, we again get Riccatti's equation $y^2 + y' + k_0^2 \varepsilon(x) = 0$, where k_0 is a large parameter. We look for a solution in the form $y(x) = k_0 y_0(x) + y_1(x) + (1/k_0) y_2(x) + \ldots$ and for the zeroth and first approximations we get $y_0 = \pm i\sqrt{\varepsilon(x)}$, and $y_1 = -\frac{1}{2}\frac{d}{dx}\left(\ln\sqrt{\varepsilon(x)}\right)$. Both signs of y_0 now have clear physical meaning, they correspond to the forward and return waves. The solution to (12.26) also has the form of a WKB solution

$$u(x) = \frac{A}{\sqrt[4]{\varepsilon}} \exp\left(-ik_0 \int_0^x \sqrt{\varepsilon}\, dx\right) + \frac{B}{\sqrt[4]{\varepsilon}} \exp\left(ik_0 \int_0^x \sqrt{\varepsilon}\, dx\right) ,$$

$$(12.27)$$

which corresponds to the geometric optics approximation. Here A and B are constants, i.e., in this approximation there is no scattering or transformation of one wave into another wave despite the inhomogeneity of the medium. If, for example, there is no return wave, then one does not appear, and if there was one, then its amplitude is unchanged. The return waves being independent, a forward wave propagates, is deformed, but does not interact with the return ones. This occurs because $\varepsilon(x)$ changes smoothly and the reflected wave is exponentially small.

It should be noted that many physical problems lead to an equation like (12.26). We shall list some of those relating to microwave electronics.

When analyzing slipping, the instability on an electron beam drifting through crossed electrostatic and magnetostatic fields, the model usually used is of electrons without high frequency disturbances and any density moving in a straight line with a transverse velocity gradient $dv_e/dy = \omega_p^2/\omega_c = \omega_c r^2$, where ω_p and ω_c are the plasma and cyclotron frequencies.

In an analysis of the high frequency wave processes occurring in such a model it is assumed that all the variables change as $\exp(i\omega t - ikx)$ over time and in the direction the wave propagates (along the x-coordinates). The equation for the dependent variable F, which is related to the y-component of the velocity by the formula $F(y) = [v_y/v_e(y)]\sqrt{1 - s^2(y)}$, has the form

$$d^2F/ds^2 - r^{-4}Q(s)F = 0,$$

where

$$Q(s) = 1 + 2r^2/(s^2 - 1) + 3r^4/(s^2 - 1)^2, \quad s = -[\omega - y)]/\omega_c$$

(in place of the dimensionless coordinate). The solution of this problem and finding the correction to the VCB-approximation is considered in [9], for example.

In microwave electronics the solution to (12.26) is also used for the propagation of waves of space charge in an accelerated electron beam [10].

Since the geometric optics approximation is very important for the solution of many physical problems, we shall dwell on the theory for the propagation of waves in a medium whose properties change slowly in the propagation direction. In doing so we shall keep to the traditional form of presentation[2]. This will yield to a better understanding of the physical meaning of this approximation.

We assume that a wave's amplitude and propagation direction only change significantly over distances L many

---

[2] Various books are devoted to a systematic presentation of the methods of geometric optics and the application to wave processes in nonhomogeneous media, e.g., [8], and Chapter VII in [17]. The monographs [13-16, 18] are also for a study of methods.

times larger than the wavelength λ. In this case, it is possible to partition space into sections $l \ll L$ ($\lambda_3 \ll l$), within which the wave may be considered to be planar[3] while the medium is considered homogeneous. As a result of this partitioning, we obtain surfaces (wave surfaces) on which the wave's phase at a given moment in time is constant, and we determine the wave's propagation direction at each point as the direction of the normal to the wave surface at that point. Usually, the definition of a ray is also introduced, i.e., the line whose tangent is the direction of the wave's propagation at each point (the definition is valid for isotropic media, to which we shall restrict ourselves). This allows us to reduce the consideration of the propagation of a wave to the propagation of a ray and to use the geometric optics approximation. Thus, geometric optics is an abstraction from the wave nature of the ray and imposes the following conditions on the dimensions of the homogeneous sections:

$$L \gg l \gg \lambda . \qquad (12.28)$$

We shall derive the main equation of geometric optics, which is called the characteristic function equation.

Let the field of a monochromatic wave be described by the function

$$f(\mathbf{r}, t) = f_0(\mathbf{r}) \exp\{i[\omega t - k_0 \Psi(\mathbf{r})]\} , \qquad (12.29)$$

where $f_0(\mathbf{r})$ and $\Psi(\mathbf{r})$ are real functions and $\mathbf{r}$ is the radius vector of the current point. In the case of the planar wave $f_0(\mathbf{r})$ is constant at the surface of the front, which is defined by the equation $\Psi(\mathbf{r}) = $ const. We shall assume that $f_0(\mathbf{r})$ and grad $\Psi(\mathbf{r})$ change noticeably over a distance $L \gg \lambda$, i.e.,

$$|\text{grad } f_0(\mathbf{r})| \ll k_0 f_0(\mathbf{r}) , \qquad (12.30)$$

$$|\text{grad } \Psi(\mathbf{r})| \ll k_0 \Psi(\mathbf{r}) . \qquad (12.31)$$

In other words, we assume that the properties of the medium change slowly over a distance of the order of a wavelength.

We substitute the function given by (12.29) into

----

[3] Note that the direction of a planar wave is constant and the same as the normal to the equal-phase plane; in the case of a homogeneous wave the equal-phase plane and the equal-amplitude plane are parallel.

(12.23), assuming further that $k^2(\mathbf{r}) = k_0^2 n^2(\mathbf{r})$, where $n(\mathbf{r})$ is the refractive index of the inhomogeneous medium. After some simple manipulations, we obtain

$$(\nabla^2 f_0)/(k_0^2 f_0) - (2i\nabla\Psi\nabla f_0)/k_0 f_0$$

$$- (i\nabla^2\Psi)/k_0 - [(\nabla\Psi)^2 - k^2/k_0^2] = 0 . \qquad (12.32)$$

The terms in (12.32) have various orders of magnitude. Starting from the fact that f_0 and $\nabla\Psi$ change over a distance of order L, we find that

$$(\nabla^2 f_0)/(k_0^2 f_0) \sim \lambda^2/L^2 , \quad (2i\nabla\Psi\nabla f_0)/(k_0 f_0) \text{ and } (i\nabla^2\Psi)/k_0 \sim \lambda/L,$$

and the last term is independent of (λ/L). We neglect the first terms in (12.32) and after separating them, equate the real and imaginary parts of the equation to zero. We obtain

$$(\nabla \Psi)^2 = k^2/k_0^2 = n^2(\mathbf{r}) , \qquad (12.33)$$

$$f_0 \nabla^2\Psi + 2\nabla\Psi\nabla f_0 = 0 . \qquad (12.34)$$

Equation (12.33) is called the characteristic function equation because it determines the phase (characteristic function). Equation (12.34) relates the amplitude and the phase of a wave and is called the transport equation. The propagation of a wave is approximately described by (12.33) and (12.34) when

$$|\nabla^2 f_0| \ll k_0 |f_0 \nabla^2\Psi| , \qquad (12.35)$$

$$|\nabla^2 f_0| \ll 2k_0 |\nabla f_0 \nabla\Psi| , \qquad (12.36)$$

i.e., when the discarded term is smaller than each of the two terms with the next order of magnitude smaller and which have been retained in the equations. Inequalities (12.35) and (12.36) are qualitative criteria for the applicability of the geometric optics approximation.

A more correct method for obtaining equations like (12.33) and (12.34) is to neglect the small terms in the solutions rather than those in the equation, i.e., a method close to the one used to obtain (12.5). We look for a solution $f(\mathbf{r})$ in the form of a series in powers of $(1/k_0)$, i.e.,

ADIABATIC INVARIANTS 251

$$f(r) = [f_0(r) + f_1(r)/k_0 + f_2(r)/k_0^2 + \ldots]\exp[-ik_0\Psi(r)] .$$
(12.37)

By substituting the expansion of (12.37) into (12.23) and equating terms with the same order of magnitude, we obtain[4]

$$\left[(\nabla\Psi)^2 - k^2/k_0^2\right]f_0 = 0 ,$$
(12.38)

$$f_0\nabla^2\Psi + 2\nabla f_0 \nabla\Psi = 0 ,$$
(12.39)

$$f_1\nabla^2\Psi + 2\nabla f_1 \nabla\Psi = i\nabla^2 f_0$$
(12.40)

Equations (12.38) and (12.39) are the same as the characteristic function and transport equations. By determining Ψ and f_0, we find f_1 etc.

The characteristic function equation is considered to be the main equation of geometric optic, and is a nonlinear, first order, partial differential equation:

$$(\partial\Psi/\partial x)^2 + (\partial\Psi/\partial y)^2 + (\partial\Psi/\partial z)^2 = n^2(x, y, z) .$$
(12.41)

We introduce the notation $\partial\Psi/\partial x = P_x$, $\partial\Psi/\partial y = P_y$, $\partial\Psi/\partial z = P_z$ when $d\Psi = P_x dx + P_y dy + P_z dz$ and (12.41) is turned into a system of ordinary differential equations, viz.,

$$dx/P_x = dy/P_y = dz/P_z = d\Psi/[2(P_x^2 + P_y^2 + P_z^2)]$$
(12.42)

$$= dP_x/(\partial n^2/\partial x) = dP_y/(\partial n^2/\partial y) = dP_z/(\partial n^2/\partial z) = ds/2n ,$$

where ds is an element of the ray's trajectory and s is introduced as an independent variable whose meaning will become clear. By equating each element in (12.42) to the following, we obtain:

---

[4] Equation (12.38) corresponds to the terms of order $\sim k_0^2$, (12.39) corresponds to terms of order $\sim k_0^1$, and (12.40) corresponds to terms of order $\sim k_0^0$.

$$dx/ds = P_x/n, \quad dy/ds = P_y/n, \quad dz/ds = P_z/n,$$

$$dP_x/ds = (\partial n^2/\partial x)/2n, \quad dP_y/ds = (\partial n^2/\partial y)/2n, \quad (12.43)$$

$$dP_z/ds = (\partial n^2/\partial z)/2n, \quad d\Psi/ds = n.$$

We introduce the new variables $l_x = P_x/n$, $l_y = P_y/n$, $l_z = P_z/n$, $l_x^2 + l_y^2 + l_z^2 = 1$, which are directed cosines of the ray ($dx/ds = l_x$, $dy/ds = l_y$, $dz/ds = l_z$). Using these, it is easy to show that the following equation follows from (12.43):

$$\frac{d}{ds}\left(n\frac{d\mathbf{r}}{ds}\right) = \text{grad } n. \qquad (12.44)$$

Because $|l| = 1$, we obtain from $d\mathbf{r}/ds = \mathbf{l}$ that $(ds)^2 = (dx)^2 + (dy)^2 + (dz)^2$. It can be seen from the latter relation that s is the length of the curve $\mathbf{r}(s)$ and $\mathbf{l}$ is a unit vector tangent to $\mathbf{r}(s)$.

When n = const, i.e., in a homogeneous medium, equation (12.44) changes into equation $d^2\mathbf{r}/ds^2 = 0$. By integrating the later we obtain the equation of a straight line for $\mathbf{r}$, i.e., $\mathbf{r} = \mathbf{a}s + \mathbf{b}$, which is obvious since a ray in a homogeneous medium is a straight line. In general, when $n = n(\mathbf{r})$, equation (12.44) together with the boundary conditions given for the direction of the ray at $\mathbf{r} = \mathbf{r}_0$ allow us to find the trajectory of the ray $\mathbf{r}(s)$. When the ray's trajectory has been found, the characteristic function (or phase) may be determined from $d\Psi/ds = n$ in the form of a curvilinear integral along the ray's trajectory, i.e.,

$$\Psi = \int_A^B n[\mathbf{r}(s)]ds. \qquad (12.45)$$

The ray $\mathbf{r}(s)$ is orthogonal to the surface Ψ = const. Since a wave in geometric optics is considered to be a bundle of rays, the change in the intensity along a ray can be found using the transport equation (12.34), which is more convenient when written in another form. By multiplying (12.34) by f_0 and using $2f_0\nabla f_0 = \nabla f_0^2$ we find an equivalent to the transport equation, thus:

$$\nabla(f_0^2 \nabla\Psi) = 0. \qquad (12.46)$$

ADIABATIC INVARIANTS

However grad $\Psi = n\mathbf{l}$, and so it follows from (12.46) that

$$\mathrm{div}(f_0^2 \mathbf{n l}) = 0 . \tag{12.47}$$

Let us consider a surface $\Psi' = \mathrm{const}$ and separate out a small area $d\sigma_1$ bounded by bundle of rays for which $f = f'$. We continue these rays until they intersect another surface $\Psi'' = \mathrm{const}$, on which the rays are bounded by another surface $d\sigma_2$, where $f = f''$. We integrate (12.47) over the volume inside this ray tube, whence by Gauss's theorem we have

$$\int_V \mathrm{div}\left(f_0^2 \mathbf{n l}\right) = \oint_\Sigma nf_0^2 \mathbf{l m}\, d\sigma = 0 , \tag{12.48}$$

where $\mathbf{m}$ is the unit vector of the external normal to the surface, such that at the end of the tube's surface we have $\mathbf{l m} = 0$, and on the surface $d\sigma_1$ is the scalar product $\mathbf{l m} = -1$ and $\mathbf{l m} = 1$ at the surface $d\sigma_2$. Thus, inside the ray tube we have

$$n_1(f')^2 d\sigma_1 = n_2(f'')^2 d\sigma_2 = nf_0^2 d\sigma = \mathrm{const} = f_{00}^2 , \tag{12.49}$$

where $d\sigma$ is the current tube cross section, nf_0^2 is proportional to the flow's energy density, and $nf_0^2 d\sigma$ is proportional to the energy carried by the wave along the ray tube. We have for the intensity from (12.49)

$$f_0^2 = (f_{00}^2)/(n d\sigma) = (f')^2 (n_1 d\sigma_1)/(n d\sigma) . \tag{12.50}$$

It was shown above that in a homogeneous medium a ray propagates in a straight line. What then is the wave's intensity in this case? We single out a wave surface from the ray element being analyzed $d\sigma$, as shown in Fig. 12.1. The ray MN intersects the wave surface at the point O. The surface in general has two different radii of curvature, whose centers O_1 and O_2 lay on MN. Let ab and cd be elements of the two main circles of curvature passing through point O. The centers of these circles will, then, lay at the points O_1 and O_2. The lengths of ab and cd are proportional, respectively, to the radii $R_1 = O_1O$ and $R_2 = O_2O$, and the area of the element of wave surface is $d\sigma \sim R_1 R_2$. Thus we find from (12.50) that

$$f_0^2 \approx \text{const}/(R_1 R_2) \ . \qquad (12.51)$$

It is possible to conclude from (12.50) that the intensity along a ray is a function of the distance from the centers of curvature of the wave surfaces (given points of the ray). If both radii of curvature are the same, then

$$f_0^2 \approx \text{const}/R^2 \ , \qquad (12.52)$$

and the wave's field is

$$f = (\text{const} \cdot e^{-ikR})/R \ . \qquad (12.53)$$

It follows from (12.52) and (12.53) that the bundle of rays diverges from a point source or converges to a point, whilst the wave surfaces are concentric spheres. As $R_1 \to 0$ or $R_2 \to 0$ (to the centers of curvature of the wave's surfaces), the intensity turns to infinity. Let us consider all the possible waves in a bundle bearing in mind this property. This consideration will lead to the conclusion that the intensity of a wave turns to infinity at two surfaces, which are the geometric loci of all the centers of curvature of the wave surfaces. These surfaces are called caustics. They are the geometric envelopes of a system of rays[5], i.e., in terms of geometric optics the field behind a caustic is zero and a ray cannot penetrate it. In the case of rays with spherical wave fronts, both caustic surfaces merge at one point, the focus.

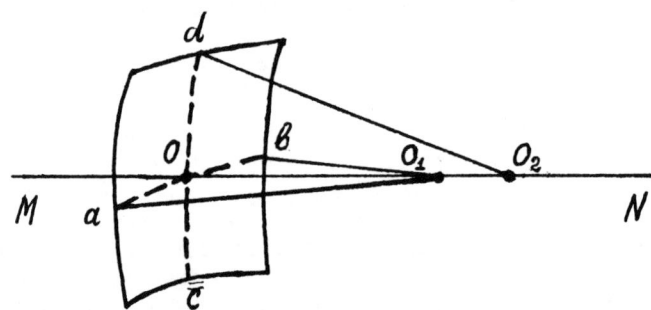

Fig. 12.1. Element of a wave surface used to calculate the intensity of a wave in a homogeneous medium.

---

[5]The latter statement follows from the properties of a geometric locus of the centers of curvature of a family of surfaces: rays, being the normals to the wave surfaces, are tangent to caustics.

12.4 The Propagation of Waves in a Plane-Layer Medium in the Geometric Optics Approximation

When studying the propagation of electromagnetic waves in an isotropic plasma, radio waves through the Earth atmosphere, acoustic waves in a liquid, a waveguide with irregular filling, or the Earth's core, etc., a widely used model is that of a nonhomogeneous medium whose properties only change in one direction, i.e., n $n(x)$ when $x > 0$ and $n = n_0 = $ const when $x \leq 0$. Here and below we shall not be considering random fluctuations in the properties of a medium. We consider a wave, whose plane of incidence is the xz-plane, while the wave vector is at an angle θ to the x-axis. The characteristic function equation for a plane-layer medium has the form

$$(\partial \Psi/\partial x)^2 + (\partial \Psi/\partial z)^2 = n^2(x) . \qquad (12.54)$$

We find from (12.42) using $n = n(x)$, that

$$P_z = a = \text{const} , \qquad (12.55)$$

$$d\Psi/dx = P_x = \sqrt{n^2(x) - a^2} . \qquad (12.56)$$

Since $l_z = P_z/n$, then given $x \leq 0$, when $n = n_0$, we have $P_z = nl_z = n_0 \sin \theta_0$ and $P_x^2 = n^2 - n_0^2 \sin \theta$. Thus we have, by integrating, from (12.56)

$$\Psi = n_0 z \sin \theta_0 \pm \int_0^x \sqrt{n^2(x) - n_0^2 \sin^2 \theta_0} \, dx . \qquad (12.57)$$

The sign before the root is determined from the direction in which the ray propagates: a minus corresponds to propagation in the positive x-direction, and a plus to a propagation in the negative x-direction. Using (12.57) it is easy to obtain the equation for the ray's trajectory from (12.42), i.e.,

$$z = \int_0^x \frac{n_0 \sin \theta_0 \, dx}{\sqrt{n^2(x) - n_0^2 \sin \theta_0}} . \qquad (12.58)$$

Equation (12.44) leads to

$$n(x) \sin \theta(x) = \text{const} = n_0 \sin \theta_0 , \qquad (12.59)$$

which is a generalization of Snell's law for a plane layer medium. It can be seen from (12.59) that when $dn/dx > 0$ the angle θ reduce with height, and when $dn/dx < 0$ the angle θ grows, i.e., in a plane-layer medium a ray diverges, in other words we have reflection. It is easy to understand that at a height x_0, determined from $n(x_0) = n_0 \sin \theta_0$, the ray will reverse, which is analogous to complete internal reflection..

We determine the wave's amplitude from the transport equation (12.34), rewriting $\nabla \Psi$ and $\nabla^2 \Psi$ for convenience. Using (12.59) we have

$$\nabla \Psi = \mathbf{z}_0 n_0 \sin \theta_0 + \mathbf{x}_0 n(x) \cos \theta , \qquad (12.60)$$

$$\nabla^2 \Psi = \frac{d}{dx} \left(n \cos \theta \right) , \qquad (12.61)$$

and, finally,

$$f_0 \frac{d}{dx} \left(n \cos \theta \right) + 2 n_0 \sin \theta_0 \frac{\partial f_0}{\partial z} + 2n \cos \theta \frac{\partial f_0}{\partial z} = 0 . \qquad (12.62)$$

The characteristic system of ordinary differential equations corresponding to (12.62) has the form

$$dz/(2n_0 \sin \theta_0) = dx/(2n \cos \theta) = df_0 \bigg/ \left[f_0 \frac{d}{dx} (n \cos \theta) \right] .$$

Whence

$$f_0 = f_{00}/\sqrt{n \cos \theta} = f_{00}/\sqrt[4]{n^2(x) - n_0^2 \sin^2 \theta_0} . \qquad (12.63)$$

The general solution (12.29) of the wave equation may be written using (12.63) as

$$f = \frac{f_{00}}{\sqrt[4]{n^2(x) - n_0^2 \sin^2 \theta_0}} \exp \left\{ i \left[\omega t - k_0 z n_0 \sin \theta_0 \mp k_0 \int_0^x \sqrt{n^2(x) - n_0^2 \sin^2 \theta_0} \, dx \right] \right\} . \qquad (12.64)$$

The two signs in (12.64) correspond to two waves propagating in the positive and negative x-directions. Both waves propagate independently, i.e., in the geometric optics approximation there is no partial reflection of waves from an

ADIABATIC INVARIANTS

inhomogeneous medium[6]. When $n^2 = n_0^2 \sin^2\theta_0$, i.e., in the region where the ray turns, the wave's amplitude tends to infinity, and consequently, (12.64) is invalid.

As has been shown, when a wave propagates in an inhomogeneous medium foci and caustics may form. If the equation for the family of waves at an angle θ_0 from some point with coordinates $(0,x)$ has the form

$$r = r(\theta_0, x), \qquad (12.65)$$

where θ_0 is a parameter, then the equation of the caustic - the envelope of the family of waves - is found [17] by excluding θ_0 from (12.65) and from

$$\partial r(\theta_0, x)/\partial \theta_0 = 0. \qquad (12.66)$$

Clearly the geometric optics approximation is inapplicable in regions close to caustics.

Let us look at inequalities (12.35) and (12.36), which are the conditions for the applicability of geometric optics in a case of a plane layer medium. We determine the values of $\nabla^2 \Psi$ and $\nabla^2 f_0$ using (12.57) and (12.63). After some simple transformations, we find

$$\left| \frac{d}{dx}(n \cos \theta) \right| \Big/ (n \cos \theta)^2 \ll k_0, \qquad (12.67)$$

$$\left| \frac{d^2}{dx^2}(n \cos \theta) \right| \Big/ (n \cos \theta) \ll k_0 \left| \frac{d}{dx}(n \cos \theta) \right|. \qquad (12.68)$$

Inequality (12.68) is a restriction on grad $(n \cos \theta)$ and is less significant than (12.67). To see the meaning behind (12.67) let $\cos \theta = 1$, whence

$$(\lambda/2\pi)|dn/dx|/n \ll 1, \qquad (12.69)$$

where $\lambda = \lambda_0/n$ is the wave length in the medium. It follows from (12.69) that the geometric optics approximation is valid when the properties of the medium change slowly over distances of the order of a wavelength. As $n \to 0$, the wavelength in the medium turns to infinity $\lambda \to \infty$, and the changes in the medium's properties, even for small

---

[6] Thus, the solution, as might be expected, has a form analogous to a WKB solution (12.27).

derivatives $|dn/dx|$, will be large over distances of the order of a wavelength. Thus, inequality (12.69) is violated. Clearly, it is also violated when $|dn/dx|$ is large.

When the incident wave is vertical in a nonhomogeneous medium, zero refractive index of a medium is a condition for reflection of the wave from the medium. This condition is fulfilled, for example, for a plasma in the radio zone [18]. At the same time, as was shown in [18], sharp gradients in n (the derivative $|dn/dx|$ is large) may only occur in thin plasmas sporadically. In the absence of absorption we have the following for a plasma with concentration $N(x)$ and frequency ω

$$n^2(x) = 1 - [4\pi e^2 N(x)]/(m\omega^2) . \qquad (12.70)$$

At large $N(x)$ or small ω, the square of the refractive index falls away to zero. In the case of vertical incidence, we have $n = 0$ and due to (12.70) we have at the reflection point

$$N(x) = (m\omega^2)/4e^2 = 1.24 \times 10^{-8} f^2 \text{ (Hz)} . \qquad (12.71)$$

This formula is one of the main relations used to interpret ionospheric data and the results of investigations by radio astronomy of the Sun's atmosphere (see [18]).

Where the geometric optics approximation is no longer valid, it is necessary either to find a method to go beyond the boundaries of the approximation or to find an exact solution of the equation describing wave propagation in a nonhomogeneous medium. The exact solution to such an equation

$$\nabla^2 f + k_0^2 \varepsilon(x) = 0 \qquad (12.72)$$

may be found for various special cases of $\varepsilon = \varepsilon(x)$, e.g., linear, parabolic, exponential functions. We shall look for a solution to (12.72) close to the turning point because further away from it the solution (12.64) "works" well. We assume that the medium is a planar layer (its properties do not change in the z-direction and consequently $f \sim \exp(-iaz)$, where $a = \text{const} = k_0 \sin \theta_0$ at $n_0 = 1$), and we consider a wave the plane of whose propagation lies in the xz-plane. Equation (12.70) then takes the form

$$d^2 f/dx^2 + k_0^2 [\varepsilon(x) - \sin^2\theta_0] f = 0 . \qquad (12.73)$$

The solution must be found for $\{k_0^2[\varepsilon(x) - \sin^2\theta_0]\}$. When the properties of the medium change smoothly in a neighborhood of the point where the ray turns, it can be considered that $\varepsilon(x)$ changes linearly, i.e., $\varepsilon(x) = 1 + (d\varepsilon/dx)x = 1 - x/x_1$, where

x_1 is the distance from the start of the inhomogeneous layer to the reflection region. In this case (12.73) may be written as

$$d^2f/dx^2 + k_0^2[\cos^2\theta_0 - x/x_1]f = 0 . \qquad (12.74)$$

We change the variables in (12.74) $\xi = (k_0 x_1)^{2/3}[\cos^2\theta_0 - x/x_1]$, which leads to

$$d^2f/d\xi^2 + \xi f = 0 , \qquad (12.75)$$

whose solution is well known and has the form

$$f = f_0 \Phi(\xi) , \qquad (12.76)$$

where $\Phi(\xi) = (1/\sqrt{\pi})\int_0^\infty \cos[t^3/3 + t\xi]dt$ is the Airy function.

The solution to (12.75) may be represented in terms of Hankel functions of order 1/3 (see [17]).
When $\xi > 0$ we have

$$f(\xi) = f_0\sqrt{\pi/3}\ \xi^{1/2}\left\{H_{1/3}^{(1)}\left[(2/3)\xi^{3/2}\right]\exp\left[(i2\pi)/3\right]\right.$$

$$\left. + H_{1/3}^{(2)}\left[(2/3)\xi^{3/2}\right]\exp\left[(-i2\pi)/3\right]\right\} , \qquad (12.77)$$

and when $\xi < 0$ we have

$$f(\xi) = f_0 \xi^{1/2}\sqrt{\pi/3}\ \exp(-i2\pi/3)H_{1/3}^{(2)}\left[-i(2/3)|\xi|^{3/2}\right] . \qquad (12.78)$$

If $\xi \gg 1$, then it is natural to use the asymptotic expression for the Hankel functions and large values of the argument, which yields

$$H_{1/3}^{(1)}\left[(2/3)\xi^{3/2}\right] = \sqrt{3/(\pi\xi^{3/2})}\exp\left\{i\left[(2/3)\xi^{3/2} - (5/12)\pi\right]\right\}, \qquad (12.79)$$

$$H_{1/3}^{(2)}\left[(2/3)\xi^{3/2}\right] = \sqrt{3/(\pi\xi^{3/2})}\exp\left\{i\left[-2/3\xi^{3/2} + (5/12)\pi\right]\right\}. \qquad (12.80)$$

By substituting (12.79) and (12.80) into (12.77) and (12.78), we find for $\xi > 0$

$$f(\xi) = 2f_0 \xi^{-1/4} \cos\left[(2/3)\xi^{3/2} + \pi/4\right], \qquad (12.81)$$

and for $\xi < 0$

$$f(\xi) = f_0 \xi^{-1/4} \exp\left[- (2/3)|\xi|^{3/2}\right]. \qquad (12.82)$$

Returning to the variable x, we finally obtain from (12.81) and (12.82) for $(\cos^2\theta_0 - x/x_1) > 0$ the equation

$$f(x) = \frac{f_0 \cdot (k_0 x_1)^{-1/6}}{\sqrt[4]{\cos^2\theta_0 - x/x_1}} \left\{ \exp\left[- ik_0 \int_0^x \sqrt{\cos^2\theta_0 - x/x_1}\, dx + i\pi/4\right] \right.$$

$$\left. + \exp\left[ik_0 \int_0^x \sqrt{\cos^2\theta_0 - x/x_1}\, dx - i\pi/4\right] \right\}, \qquad (12.83)$$

and for $(\cos^2\theta_0 - x/x_1) < 0$ the equation

$$f(x) = \frac{f_0 \cdot (k_0 x_1)^{-1/6}}{\sqrt[4]{|\cos^2\theta_0 - x/x_1|}} \exp\left\{- k_0 \int_0^x \sqrt{|\cos^2\theta_0 - x/x_1|}\, dx\right\}. \qquad (12.84)$$

It is easy to see that (12.83) only differs from solution (12.64), which was obtained in the geometric optics approximation, by a constant correction in the phase of the incident and reflected waves. Moreover, as distinct from (12.64), there is no explicit dependence on z and t in (12.83) and (12.84).

The field $f(x, z)\exp(i\omega t)$ when $(\cos^2\theta_0 - x/x_1) > 0$ is a standing wave. The wave's amplitude increases as the reflection region is approached, but remaining finite everywhere. There is no wave process when $(\cos^2\theta_0 - x/x_1) < 0$ and only an exponentially damped field enters this region. When $(\cos^2\theta_0 - x/x_1) = 0$ there is complete reflection of the wave (the incident and reflected waves are twisted in phase by $\pi/2$ after reflection). This process is illustrated in Fig. 12.2, which is taken from [17].

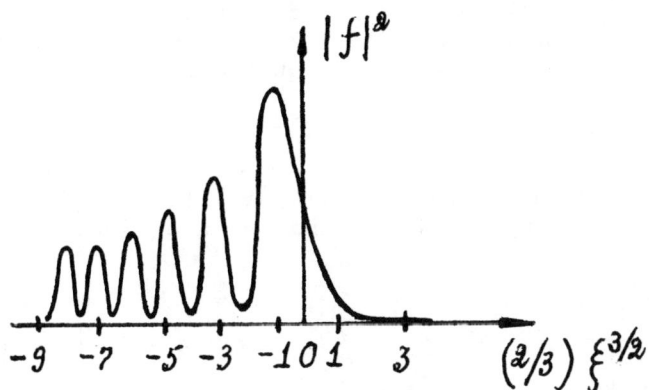

Fig. 12.2. The function $|f|^2$ versus $(2/3)\,\xi^{3/2}$ which illustrates the way the field amplitude changes close to the reflection zone in a nonhomogeneous medium.

The field $f(x, z)\exp(i\omega t)$ when $(\cos^2\theta_0 - x/x_1) > 0$ is a standing wave. The wave's amplitude increases as the reflection region is approached, but remaining finite everywhere. There is no wave process when $(\cos^2\theta_0 - x/x_1) < 0$ and only an exponentially damped field enters this region. When $(\cos^2\theta_0 - x/x_1) = 0$ there is complete reflection of the wave (the incident and reflected waves are twisted in phase by $\pi/2$ after reflection). This process is illustrated in Fig. 12.2, which is taken from [17].

12.5. Linear Wave Interaction in an Inhomogeneous Medium.

It was noted several times above that in the geometric optics approximation there is no partial reflection of waves on an inhomogeneous medium, i.e., the waves propagate independently. There are several ways to get a reflection in the solution. One is to find the correction for the next approximation to the WKB solution due to which the return wave is coupled. However, other ways are possible, for example, the Van der Pol method. We shall use this approach for a concrete example, that is, the transition layer problem. We formulate the problem thus: let there be a layer of width l in which the medium's properties change smoothly. A wave with amplitude $A(0) = A_0$ is incident on the boundary $x = 0$ of the layer. The amplitude of the return (reflected) wave at the boundary $x = l$ is zero. The problem is to find the amplitude $B(x)$ of the wave that arises due to reflection from smooth inhomogeneities, i.e., the amplitude of the wave propagating in the negative x-direction. We introduce a new variable $V = du/dx$ and write (12.24) more conveniently in the form of a system of two-order equations:

$$du/dx = V, \quad dV/dx = -k_0^2 \varepsilon(x) u. \qquad (12.85)$$

The solution to system (12.85) will be sought in the form (12.27) assuming that A and B are functions of the coordinates

$$v = \frac{A(x)}{\sqrt[4]{\varepsilon(x)}} e^{-i\varphi} + \frac{B(x)}{\sqrt[4]{\varepsilon(x)}} e^{i\varphi}, \quad \text{where } \varphi = k_0 \int_0^x \sqrt{\varepsilon(x)}\, dx.$$

Since we have introduced to new variables A and B in place of the one variable u, one of the relations linking these new variables can be arbitrarily discarded. We require for convenience that

$$\frac{A'(x)}{\sqrt[4]{\varepsilon(x)}} e^{-i\varphi} - \frac{A(x)\varepsilon'(x)}{4\sqrt[4]{\varepsilon(x)}} e^{-i\varphi}$$

$$+ \frac{B'(x)}{\sqrt[4]{\varepsilon(x)}} e^{i\varphi} - \frac{B(x)\varepsilon'(x)}{4\sqrt[4]{\varepsilon(x)}} e^{i\varphi} = 0. \qquad (12.86)$$

Whence $V = -ik_0 \varepsilon^{1/4} A \exp(-i\varphi) + ik_0 \varepsilon^{1/4} B \exp(i\varphi)$. By substituting this expression into the second equation of system (12.85), we find

$$-ik_0 \varepsilon^{1/4} A' e^{-i\varphi} - ik_0 \frac{\varepsilon' A e^{-i\varphi}}{4\varepsilon^{3/4}} +$$

$$+ ik_0 \varepsilon^{1/4} B' e^{i\varphi} + ik_0 \frac{\varepsilon' B e^{i\varphi}}{4\varepsilon^{3/4}} = 0.$$

Combining this equation with condition (12.86) yields the following system of equations:

$$(4\varepsilon A' - \varepsilon' A) e^{-i\varphi} + (4\varepsilon B' - \varepsilon' B) e^{i\varphi} = 0,$$

$$(4\varepsilon A' + \varepsilon' A) e^{-i\varphi} - (4\varepsilon B' + \varepsilon' B) e^{i\varphi} = 0.$$

By solving this system for the derivatives A' and B', we obtain

$$A' = \frac{\varepsilon' B}{4\varepsilon} e^{2i\varphi} , \quad B' = \frac{\varepsilon' A}{4\varepsilon} e^{-2i\varphi} . \qquad (12.87)$$

Equation (12.87) is exact: we have only made a substitution of variables, from u and V to A and B. However, because the inhomogeneity is weak, ε' is smaller than ε and consequently, A and B change slowly. Therefore, in order to solve (12.87) we can used the method of subsequent approximations, assuming at the zeroth approximation that $\varepsilon' = 0$ and $A = A_0$.

By substituting $A = A_0$ into the second equation of (12.87), we obtain

$$B(x) = A_0 \int_0^x \frac{\varepsilon'}{4\varepsilon} \exp\left(-2ik_0 \int_0^x \sqrt{\varepsilon} \, dx\right) dx .$$

Including the corrections for the first-order approximation gives

$$A(x) = A_0 + \int_0^x \frac{\varepsilon'}{4\varepsilon} B(x) \exp\left(2ik_0 \int_0^x \sqrt{\varepsilon} \, dx\right) dx .$$

This is already outside the bounds of geometric optics because the waves interact: their amplitudes are coupled.

Let us consider another example of linear interaction between waves, one which is important in microwave electronics. In Chapter 7 we discussed a distributive amplifier, the travelling wave tube, and a distributed generator, the standing wave tube. One of the main advantages of the travelling wave tube, making it the main device for satellite communications, is that it carries a large amplification factor of a wide spectrum of amplifiable frequencies (an octave and more). A serious defect in the operation of the amplifier is the excitation of parasite self-excited oscillations on the return wave (the physics of the self-excited oscillations is the same as in the standing wave generator). A popular method of countering parasitic self-excitation is to increase the starting current necessary for oscillations to begin. This can be achieved by smoothly changing the geometric parameters of the retarding system along the length of the interaction space, i.e., by smoothly changing the phase velocity of the return wave. A simple formulation yields a linear interaction between the slow space-charge wave in the electron beam (see Chapter 10) and the return electromagnetic wave, whose phase velocity changes smoothly in a direction of the electron motion [11][1]. The equation for the induction of a longitudinal component to the potential of the electrostatic field E of this return wave by the grouped current I_M, which is caused by the excitement in the electron beam of a slow space-charge wave, has the form

$$dE/dx + [ik_0(x) - (1/k_0(x))(dk_0(x)/dx)]E = (1/2)k_0^2(x)KI_M(x), \quad (12.88)$$

where $k_0(x) = \omega/v(x) = k_{00}\xi(x)$, $k_{00} = 2\pi/\lambda_0$. For the second equation we use (10.39), i.e.,

$$dI_M/dx + i(k_e + k_q)I_M = -(k_e I_0/4V_0 k_q)E(x). \quad (12.89)$$

We shall now use the notation of Chapter 10. The boundary conditions for the system of (12.88) and (12.89) are

$$I_M(0) = 0 \; ; \; E(0) = E_0, \quad (12.90)$$

where E_0 is the unknown amplitude of the high-frequency electric field's potential of the inverse wave.

By excluding $E(x)$ from (12.88) and (12.89), we obtain

$$d^2 I_M/dx^2 + \left\{ i[k_e + k_q + k_0(x)] - (1/k_0(x))(dk_0(x)/dx) \right\} dI_M/dx$$

$$+ \left\{ i(k_e + k_q)[ik_0(x) - (1/k_0(x))(dk_0(x)/dx)] \right.$$

$$\left. + [C^2 k_0^2(x)]/4\sqrt{QC} \right\} I_M = 0. \quad (12.91)$$

We look for a solution to (12.91) in the form

$$I_M(x) = v(x)\exp\left\{ \frac{1}{2}\int \left\{ [k_e + k_q + k_0(x))^2 + (2i/k_0(x))(k_e + k_q \right.\right.$$

$$+ k_0(x))(dk_0(x)/dx)$$

$$\left.\left. - (dk_0(x)/dx)^2/k_0^2(x) \right\}^{1/2} \right\}. \quad (12.92)$$

---

[1] The method now presented was suggested by A.Yu. Dmitriev.

Then after substituting (12.92) into (12.91) we obtain an equation for $v(x)$, i.e.,

$$\frac{d^2v}{dx^2} + \left\{ \frac{c^2 k_0^2(x)}{4\sqrt{QC}} + \frac{[k_0(x) - k_e - k_q]^2}{4} - \frac{i}{2} \frac{k_e + k_q}{k_0(x)} \frac{dk_0(x)}{dx} \right.$$

$$\left. - \frac{3}{4k_0^2(x)} \left(\frac{dk_0(x)}{dx}\right)^2 + \left(\frac{d^2 k_0(x)}{dx^2}\right) \middle/ k_0^2(x) \right\} v = 0 . \qquad (12.93)$$

Let $k_0(x)$ change such that the last two terms ($\sim (dk_0(x)/dx)^2$ and $\sim d^2k_0(x)/dx^2$) in the curly brackets can be neglected, and such that in the third term it can be considered that $\frac{k_c + k_q}{k_0(x)} \frac{dk_0(x)}{dx} \approx \frac{dk_0(x)}{dx}$. Then instead of (12.93) we have

$$\frac{d^2v}{dx^2} + F(x)v(x) = 0 ,$$

where

$$F(x) = \frac{c^2 k_0^2(x)}{4\sqrt{QC}} + \frac{[k_0(x) - k_e - k_q]^2}{4} - \frac{i}{2} \frac{dk_0(x)}{dx} . \qquad (12.94)$$

If now the condition for the applicability of geometric optics $\frac{\lambda_0}{2\pi} \left|\frac{dF(x)}{dx}\right| \ll F(x)$ is fulfilled for $F(x)$, then using the first of the boundary conditions in (12.90), we obtain

$$I_M(x) = \frac{2iA}{\sqrt[4]{F(x)}} e^{-i \int_0^x \frac{[k_0(\xi) + k_e + k_q]}{2} d\xi} \cdot \sin \int_0^x \sqrt{F(\xi)} \, d\xi ,$$

(12.95)

where A is an arbitrary constant. By substituting (12.95) into (12.89), we find the following expression for the distribution of the high-frequency electrostatic field along the length of the interaction space:

$$E(x) = -\frac{2k\sqrt{QC}}{c^2} A e^{-i\int_0^x \frac{k_0(\xi) + k_e + k_q}{2} d\xi} \cdot \left\{ 2iF^{1/4}(x)\cos\int_0^x \sqrt{F(\xi)}\, d\xi \right.$$

$$\left. + \left[(k_0(x) - k_e - k_q)F^{1/4}(x) - \frac{iF^{-5/4}(x)\, dF(x)}{2\, dx}\right] \sin\int_0^x \sqrt{F(\xi)}\, d\xi \right\}$$

(12.96)

The arbitrary constant A may be determined from the second boundary condition in (12.90), but this is not necessary because the condition for self-excited oscillations in the return wave

$$E(1) = 0 \qquad (12.97)$$

is such that all constants decay (1 is the length of the interaction space).

In order to obtain a simple analytic solution, we shall assume that

$$F(x) \approx \frac{c^2 k_0^2(x)}{4\sqrt{QC}}. \qquad (12.98)$$

Using (12.98) and (12.96), we find from the self-exciting condition (12.97) that

$$\frac{Ck_{00}}{2\sqrt[4]{QC}} \int_0^1 \xi(x)\, dx = \frac{\pi}{2}, \qquad (12.99)$$

$$k_{00}\xi(x) = k_0 + k_q \qquad (12.100)$$

In the simplest case encountered in practice (as in a plane-layer medium), i.e., $\xi(x) = 1 + \alpha x$, we find from the starting conditions (12.99) and (12.100) that

$$\frac{Ck_{00}1}{2\sqrt[4]{QC}}\left(1 + \frac{\alpha 1}{2}\right) = \frac{\pi}{2}, \qquad (12.101)$$

$$k_e + k_q - k_{00} = \alpha k_{00} 1. \qquad (12.102)$$

If the parameters of the standing-wave tube are the same as a corresponding tube in which the phase velocity of the return wave is unchanged along the interaction space, then it follows from (12.101) and (12.102) that the ratio of the starting currents will be

$$I_{v_\Phi}(x) \Big/ I_{v_\Phi} = [1 + (\alpha l)/2]^{-2} . \qquad (12.103)$$

In a general case for (12.99) and (12.100), we have

$$I_{v_\Phi}(x) \Big/ I_{v_\Phi} = l^2 \left[\int_0^l \xi(x) dx \right]^{-1} . \qquad (12.104)$$

We may conclude from (12.103) and (12.104) that when the gradient of $k_0(x)$ (or $\alpha l < 0$) is negative, the starting current is increased, which yields a method for increasing the stability of a travelling-wave tube amplifier against self-excitation in the return wave.

Two very simple examples of waves with linear interactions have been considered. The number of similar examples could be increased, there are thousands of them, and they come from very different areas of physics, e.g., hydrodynamics, plasma physics, electrodynamics, acoustics, and in recent years the physics of liquid crystals, ferrodielectrics, waveguides, planar waveguides. The authors of a review [19] consider that linear interaction between waves is now the most important problem in the linear theory of oscillations and waves[2]. Investigations in this area began in the 1950's with studies of the propagation of waves in ionospheric plasma (see [18, 20, 21]) and investigation of irregular waveguides for microwaves and acoustic waves in layered media (for example [22, 13]). Generally, the phenomenon of linear interaction between waves (the linear transformation of modes) arises if the geometric amplitude of a wave may be changed nonadiabatically as the wave passes through inhomogeneous sections of the medium. In other words, linear interaction appears as an exception to the geometric optics approximation of the propagation of waves in an inhomogeneous medium. This means that the ratio of the amplitudes and the phase difference of waves composing a radiation passing through an inhomogeneous medium change in a wave different to that which follows from the

---

[2] In the following presentation we shall follow review [19], in which the modern status of the topic of interacting electromagnetic waves in weakly anisotropic smoothly inhomogeneous media is presented.

WKB-approximation, that is the propagating waves cease to be independent.

The nature and scale of the medium's inhomogeneities in the interaction medium determine the transformation of the waves, and by studying this transformation it is possible to obtain information about the structure of the inhomogeneities. Moreover, by changing the inhomogeneity the effectiveness of the wave transformation may be controlled, and consequently the intensity and polarization of the forward and reflected waves may be governed.

Following [19] we formalized the linear interaction between waves in a more or less general physical situation for waves of arbitrary nature in an arbitrary anisotropic inhomogeneous medium.

Let a monochromatic wave propagate in a stationary medium without sources. We shall limit ourselves to a consideration of the one-dimensional case and omit the factor $\exp(i\omega t)$. The wave equations for N-components X_α of the field may then rewritten (see [20]) in the form of a system of first-order differential equations

$$d\mathbf{e}/d\xi = -iT\mathbf{e} \qquad (12.105)$$

Here we have introduced the following notation: $\mathbf{e}$ is an N-component column vector of the complex field variables X_α ($\alpha = 1, \ldots, N$); $T = T(x)$ is a square matrix which determines the local properties of the medium ($T(x)$ does not contain differential operators and has the same form for homogeneous and inhomogeneous media); $\xi = k_0 x$ is the dimensionless spatial coordinate in the x-direction, the wave propagation direction; $k_0 = \omega/c$; and $c = \text{const}$ is the characteristic wave velocity (in electrodynamics the speed of light in a vacuum). The frequency dispersion is included in $T(\omega)$.

We represent the column vector of the complex field variables in the form

$$\mathbf{e} = \sum_{i=1}^{N} f_i \vec{\mathcal{E}}_i, \quad \vec{\mathcal{E}}_i = \phi_i \mathbf{e}_i, \qquad (12.106)$$

where the $\mathbf{e}_i$ are normal waves and make up a complete system of the eigenfunctions for the T matrix at each point in the media; and $\mathbf{e}_i \mathbf{e}_i^* = 1$ ($i = 1, \ldots, N$). The eigenvalues, the refractive indices n_i, are determined from the equation $T\mathbf{e}_i = n_i \mathbf{e}_i$. Substituting (12.106) into (12.105) and going over to equations for the complex amplitudes of the

ADIABATIC INVARIANTS

interacting waves, assuming that $n_i = n_i(\xi)$, $\mathbf{e}_i = \mathbf{e}_i(\xi)$, and $f_i = f_i(\xi)$, we arrive at

$$df_i/d\xi + in_i f_i = \sum_{j=1}^{N} a_{ij} f_j \, , \quad a_{ij} = (d\vec{\mathcal{E}}_j/d\xi) \cdot \vec{\mathcal{E}}^{i*} \quad (12.107)$$

In order to obtain (12.107) we used the inverse system of vectors to $\vec{\mathcal{E}}_j$, i.e., $\vec{\mathcal{E}}^{i*}$ such that $\vec{\mathcal{E}}_j \cdot \vec{\mathcal{E}}^{i*} = \delta_{ij}$, where δ_{ij} is the Kronecker delta. This system defines the "transport" waves, which are the eigenvectors of the transport matrix $T^T \vec{\mathcal{E}}^{i*} = n_i \vec{\mathcal{E}}^{i*}$, and hence we have $T_{\alpha\beta} = \sum_{i=1}^{N} n_i \mathcal{E}_{i\alpha} \mathcal{E}^{i*}_{\beta}$.

It can be seen from (12.107) that there is a linear relation (f_i is a function of f_j when $i \neq j$) between the waves in an inhomogeneous medium, where $a_{ij} \neq 0$. The following equation follows from $a_{ii} = -(d\vec{\mathcal{E}}_i/d\xi)\vec{\mathcal{E}}^{i*} = 0$:

$$(1/\phi_i)(d\phi_i/d\xi) + (d\mathbf{e}_i/d\xi)\mathbf{e}^{i*} = 0 \quad (12.108)$$

which defines the factor $\phi_i(\xi)$. The condition $a_{ii} = 0$ means that the local refractive index for interacting waves is independent of the inhomogeneity of the medium (see (12.107)). It can be shown that in the geometric optics approximation, (12.105) has independent solutions $\phi_i \mathbf{e}_i \exp(-i\int n_i d\xi)$. The ϕ_i in the solutions are found from (12.108). Consequently according to (12.106) interactions between geometrically optic waves is described by changes in the amplitude f_i. When $a_{ij} \to 0$, we obtain a solution from (12.107) in the geometric optics approximation (a solution by the WKB-method), i.e.

$$f_i = f_i(\xi_{ex})\exp\left[-i\int_{\xi_{ex}}^{\xi} n_i(\xi)d\xi\right] . \quad (12.109)$$

The solution to (12.107) in a inhomogeneous medium, when $a_{ij} \neq 0$, differs from the WKB-solution of (12.109). This difference, as we have already said, is the manifestation of linear wave interaction, in that wave polarization in the geometric optics approximation (the polarization is given by the components of the wave field X_α/X_β) is not conserved adiabatically, as it should be locally for the given geometric wave. Thus, from the point of view of geometric optics various components of the field change inconsistently when the waves interact and thus disturb the local structure of the given normal wave e_1, which leads to the appearance of other waves.

In most cases, the interaction considered is between two waves (the forward and return), the interaction being described by

$$df_1/d\xi + in_1 f_1 = a_{12} f_2 ,$$
$$df_2/d\xi + in_2 f_2 = a_{21} f_1 .$$
(12.110)

The two-wave approximation is valid 1) for waves in the same direction, when only those two waves whose dispersion curves $n_1(\xi)$ and $n_2(\xi)$ interact and 2) for return waves when the curves are close together. It should be underlined that the phenomenon of linear interaction is not only related to the behavior of the waves' dispersion curves, it is also determined to a lesser extent, by the character of the polarization.

A qualitative analysis of linear interaction between waves as described by (12.110) is presented in [19]. This analysis makes it clear when the phenomenon is possible and the degree of efficient interaction between the waves. Moreover, typical dependencies of the wave transformation effect on the properties of the inhomogeneous medium are clarified.

We shall not dwell further on the mathematical side of this topic, referring the reader to [19]. We shall only describe, concisely, the features of mode interaction in waveguide systems, when wave transformation takes place due to the inhomogeneity of the boundaries. By way of example, we sight inhomogeneous long lines, irregular waveguides, and planar and optical fibers. The equations describing the interactions between waves in these systems are the same as (12.110), but the linkage coefficients of the waves a_{ij} are determined by the local properties of the boundaries. The phenomenon of wave interaction arises in these cases due to the irregularity of the mode coupling along the waveguide, and not just their coupling (this is in contrast to interactions between waves with unchanged propagation constants and unchanged linkage coefficients along the interaction region (see Chapter 10)).

To take a concrete example, we consider modes in a twisted lightguide. Because the technology for preparing optic fibers that conserve the polarization of radiation of several hundred meters or more has been mastered, and because of the future use of optic fibers in communications etc., the investigation of the polarized properties of monomode optic fibers has become very active (see [23]). Two fundamental modes with different phase velocities, polarized practically linearly and orthogonally (called LP-modes) propagate in a regular birefringent monomode optic fiber, which is analogous to an anisotropic medium [19]. The degeneracy of modes in a real optic fiber with a round cross-section is reduced only due to kinks, inevitable elliptical cross-sections of the core etc. The equation for the propagation of coupled LP-modes in a weakly directed and weakly anisotropic waveguide, as given in [19], has the form

$$dE_z/d\xi + in_z E_z = \alpha E_y,$$
$$dE_y/d\xi + in_y E_y = \alpha E_z.$$
(12.111)

where E_z and E_y are the amplitudes of the fundamental modes, z and y are the local major axes of the fiber (in the cross-section), and α is the real coefficient of linkage. This coefficient is determined by the deformation of the fiber core, the effect of external fields, glass anisotropy, and in a twisted waveguide, by azimuthal rotation of the optic axes. In the latter case, the linkage coefficient α approximately equals the local spatial velocity v of the rotation of the optic axes and may rise to the order of magnitude of $|n_z - n_y|$ (note that the spatial period of the mode beats is $2\pi/k_0|n_z - n_y|$). The properties of the normal waves of the waveguide are then significantly changed. It is important that the polarization of the modes becomes elliptical, which can be seen from the expression for the refractive indices $\tilde{n}^1$ and $\tilde{n}^2$ and the polarization coefficients $\tilde{K}_1$, $\tilde{K}_2$ ($\tilde{K}_i = -iE_{iy}/E_{iz}$, where $i = 1, \ldots, 2$) of the normal modes in an evenly twisted fiber. The corresponding relationships, which can be easily found from (12.111) when n_z = const, n_y = const and $\alpha = v$ = const, have the form

$$\tilde{n}_{1,2} = \frac{1}{2}\left(n_z + n_y\right) \pm v\sqrt{\tilde{q}^2 + 1},$$

$$\tilde{K}_{1,2} = \tilde{q} \mp \sqrt{\tilde{q}^2 + 1}, \quad \tilde{q} = \frac{n_z - n_y}{1}. \quad (12.112)$$

These are usually called spiral modes. When the fiber is not evenly twisted, when n_z, n_y and v are not constants, the spiral modes may interact, the interaction being described by (12.110). In this case, a qualitative analysis [19] of the system leads to the conclusion that effective interaction between spiral modes is only possible in sections of the waveguide where the fiber goes from being very twisted (in the scale of a mode beat period) to being weakly twisted, or vice versa[3].

In many applications (e.g., optic communications) wave transformation is not desirable and irregular sections of interaction must be eliminated. In other cases, for example, when measuring the local optic parameters of a wave, effective mode transformation is on the other hand needed.

To conclude we note that the applications of linear transformation of waves is expanding continuously.

---

[3] It is interesting that in holesteric nonhomogeneous liquid crystals, spiral waves are effectively transformed where there is a change from strongly twisted to weakly twisted spirals and vice versa [19]. What is meant here is cholosteric spiral, that is a spiral line described by a director, which is a unit vector characterizing the preferred direction of the long axis of the liquid crystal's molecule.

PART TWO

OSCILLATIONS AND WAVES IN NONLINEAR SYSTEMS

CHAPTER 13

THE NONLINEAR OSCILLATOR

13.1 Initial Remarks

We shall analyze the oscillation and wave phenomena, and their corresponding models, in nonlinear systems and media (i.e., nonlinear oscillations and nonlinear waves) in parallel, as we did in the first part of the book. We have a few short remarks to make, mainly historical.

Although the first nonlinear problems in the theory of waves appeared a long time ago (e.g., the Korteveg--De Vries equation describing isolated waves on a liquid surface was obtained in 1895), when the nonlinear theory of oscillations was only being born[1], the theory of nonlinear oscillations and the theory of nonlinear waves were developed practically independently for many decades. Wave theory despite certain exceptions remained mainly a linear science until very recently. The significant increase in interest in nonlinear processes arose a little after the theory of shock waves in gases began to be a widely applied. However, a truly "nonlinear" theory of waves only came into existence comparatively recently (in the 1960's), mainly due to problems in radio physics, the physics of plasmas, and

---

[1] Nonlinear mechanics was almost exclusively the source of all problems, even though the models were usually purely conservative. However, Poincare created the theory of limiting cycles at that time, and exotic examples in which undamped oscillations in systems with friction appeared separately [1].

nonlinear optics and acoustics.[2]

The theory of nonlinear oscillations came into being much earlier, with the foundations of the theory being formulated in the mid-1930's on the basis of topics that were of interest at the start of the century in radio technology, regulation theory, and of course, classical mechanics. A decisive contribution to this theory was made by L. I. Mandel'shtam and his pupils. The nonlinear oscillator was completely investigated, the effect of energy exchange in systems of linked oscillators had been observed, and work by Andronov and Van der Pol, the theory of periodic self-excited oscillations had been constructed. The phenomena of synchronization and concurrence had been discovered and White had already made attempts to construct a theory of the self-excited oscillations in distributed systems.

However, the classical theory of oscillations is based, with rare exceptions, on the theory for systems with small numbers of degrees of freedom and systems demonstrating simple periodic or quasiperiodic behavior. The considerable interest in very nonlinear systems and the investigation of complex behavior (including stochastic) of simple dynamic systems, and the behavior of ensembles, is characteristic of the modern theory.

Concerning our presentation of oscillation and wave theory in parallel, we want to underline once again that there are phenomena in the theory of waves that have exact analogs in the theory of oscillations. One such example is the analogy between the spatial wave beats when their interactions are stationary in a nonlinear medium and temporal beats in coupled nonlinear oscillators. This is now the place to address the question of why and up to what point can wave (distributed) phenomena be directly compared with finitely dimensional effects (more exactly, low-dimensional effects), i.e., why can a wave system be described by a model whose phase space has a small dimensionality?

An answer can be found by comparing the nonlinear wave processes in two limiting cases, i.e., in a medium with strong dispersion and small nonlinearity and in a nonlinear medium without dispersion [18, 19]. When a wave propagates, for example, in a compressible gas or on the surface of shallow water (no dispersion) the peaks of the wave move faster than the feet, and the wave continuously distorts. At some point in time the wave must tilt over and the profile

---

[2] It is interesting that now things have come full circle and the interest in the nonlinear theory of waves is once again moving towards gas dynamics and fluid mechanics, which is caused by the rapid expansion of investigations into atmospheric and oceanic processes.

THE NONLINEAR OSCILLATOR

must become ambiguous. Clearly this process cannot be described by a finitely dimensional model. The reason for this can be conveniently understood using a simple spectral approach. The phase velocities of small disturbances of any frequency in a medium without dispersion are the same. Hence any harmonic, however weak, appearing due to the nonlinearity will be in resonance with the fundamental wave and will effectively be disturbed by it. Thus, if we want to describe a process using a set of harmonics, we must allow for an infinite number of them.

If, however, the dispersion is large and the nonlinearity is weak (for example, for media in which nonlinear light waves propagate) then only a few waves turn out to be synchronous and hence it is possible to use a direct analogy with the processes in oscillating systems that have small numbers of degrees of freedom. Hence, these direct analogs are possible when the structure of the interacting waves is fixed and there are a few waves. We emphasize here that these waves do not necessarily have to be sinusoidal in space, as was the case in the example. The waves may themselves be the results of interactions between a large number of harmonic waves (e.g., nonlinear stationary waves in media with weak dispersion). It is only important that when they interact with one another over time they behave as well defined objects with known characteristics.

Having made these short comments we go directly to a discussion of the phenomena, effects, and models of the nonlinear theory.

13.2 Qualitative and Analytical Description. Examples of Nonlinear Systems

The "linear" part of the book began with a discussion of a linear oscillator, and we begin this part with an investigation of a nonlinear oscillator, whose equation has the form

$$\ddot{x} + f(x) = 0 . \quad (13.1)$$

If $f(x)$ is a linear function, then it is a linear oscillator (Chapter 1). We might ask what physical problems lead to equation (13.1) if $f(x)$ is a nonlinear function.

Equation (13.1) describes, for example, an oscillating circuit whose coil contains a ferrite core, which leads to a nonlinear relation between the magnetic flux Φ and the current I (Fig. 13.1a). The nonlinearity in the circuit may also be associated with the capacitance, if the charge Q depends nonlinearly on the potential U. In Fig. 13.1b $C(U)$ is the capacitor of a $p-n$-junction or a capacitor with a ferroelectric. In mechanics, the equation might describe a ball moving along a trough (Fig. 13.1c). That the ball is an oscillator is not in doubt. The question is under what conditions is it a linear oscillator and under what conditions is it nonlinear one.

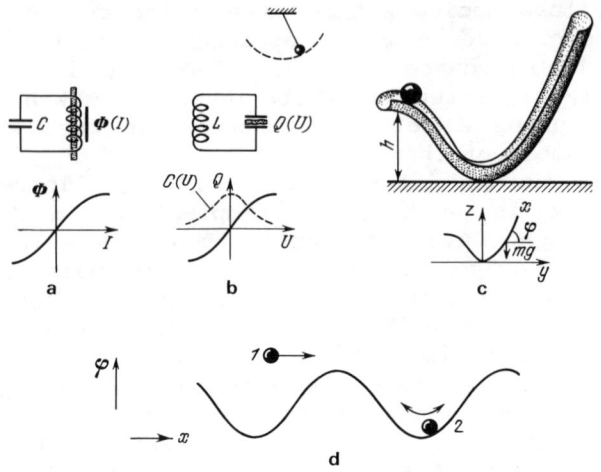

Fig. 13.1. Examples of nonlinear oscillators: a) an oscillating circuit whose coil contains a ferrite core ($\ddot{\Phi} + f(\Phi) = 0$, $\Phi = L(I) \cdot I$); b) an oscillating circuit whose capacitance contains a ferroelectric ($\ddot{Q} + f(Q) = 0$, $Q = C(Q)U_c$); c) a ball in a trough ($\ddot{x} + dW(x)/dx = 0$, $W(x) = gz(x)$); d) charged particle in the periodic electric field of a longitudinal wave (1 is the flying particle, 2 is the entrained particle, ϕ is the field potential, and x is the longitudinal coordinate).

The equation of motion for a ball with mass m has the form
$$m\ddot{x} = -F, \quad F = mg \sin\phi = mg\, dz/dx,$$
or
$$\ddot{x} + g\, dz/dx = 0. \tag{13.2}$$

where g is the acceleration of free fall. In the general case, (13.2) can be rewritten as

$$\ddot{x} + dW(x)/dx = 0, \quad W(x) = gz(x). \tag{13.3}$$

When $W(x) \sim x^2$ (the potential well is parabolic), our oscillator is linear. One important point is that the shape of the potential curve is not that of the profile of the trough in the (z, y)-plane. If, for example, the trough equation is $z = y^2$, then $dy = dz/2\sqrt{z}$, and it follows from the relation $(dx)^2 = (dz)^2 + (dy)^2$ that

THE NONLINEAR OSCILLATOR

$$x = \int [1 + 1/4z]^{1/2} dz .$$

This is not a parabolic relation for the potential energy, as is necessary for a linear oscillator.

Equation (13.3) can easily be integrated. If we make the substitution $\dot{x} = v$, then

$$\frac{dv}{dx}\frac{dx}{dt} + \frac{dW(x)}{dx} = 0 \quad \text{or} \quad v\frac{dv}{dx} + \frac{dW(x)}{dx} = 0 .$$

This yields $v^2/2 + W(x) = \mathcal{E}$, where $\mathcal{E}$ is the total energy of the nonlinear oscillator, and $W(x)$ is its potential energy.

A qualitative picture of the motion of the nonlinear oscillator (a conservative nonlinear system with one degree of freedom) can be presented without solving the specific problem simply in terms of its phase plane diagram. Using the conservation of energy as expressed above, we can write the velocity as

$$v = \dot{x} = \pm \{2[\mathcal{E} - W(x)]\}^{1/2} . \tag{13.4}$$

This is in fact the equation for the trajectory in phase space for our model. Because energy is conserved, we can give $\mathcal{E}$ at $t = 0$ and if $W(x)$ is known, it is easy to find $\dot{x}$ and draw the phase trajectory (Fig. 13.2). Motion with a small initial energy $\mathcal{E}_0 < W(x)$ simply does not exist because the quantity $\dot{x}$ would be imaginary.

Motion in the sections $0 < x < x_{02}$ and $x_{12} < x < x_{22}$, where x_{02}, x_{12} and x_{22} are determined from the condition $\dot{x} = 0$, i.e., $W(x) = \mathcal{E}_2$, correspond to the initial level of energy $\mathcal{E}_2$. The phase trajectory corresponding to this motion is labeled in Fig. 13.2 as 2. The points at which $dW(x)/dx = 0$ correspond to the equilibrium condition. By changing the initial values of the energy it is possible to construct all the trajectories on the phase plane.

Thus, the motion of a nonlinear oscillator is completely determined by the initial energy. Small amplitude oscillations will be harmonic. As the energy of the oscillations grows they diverge more and more from harmonic, that is most of the period is spent in the sections corresponding to the ball running up to the peaks and when it starts to move down from them (Fig. 13.2). Finally, when the initial energy is $\mathcal{E}_3 = mgh$, the motion of the ball ceases to be periodic (on the phase plane of Fig. 13.2, this is represented by the separatrix). This analysis lead to the very important conclusion that the motion of a nonlinear oscillator is asynchronous, that is, the frequency of the oscillation is dependent of the energy. It is possible to obtain the period of an oscillation from (13.4) for a finite motion in the form

$$T = \oint \{2[\mathcal{E}_2 - W(x)]\}^{-1/2} dx = 2 \int_{x_{12}}^{x_{22}} \{2[\mathcal{E}_2 - W(x)]\}^{-1/2} dx .$$

(13.5)

It can be said of motion, to which a closed trajectory not far from the separatrix corresponds, that $\omega = \omega(A^2)$, where A is the amplitude of the periodic oscillation. Let us consider in more detail the region of the phase plane containing the two equilibria, namely the saddle and the focus. The separatrix leaving the saddle returns to it (in the linear approximation it is easy to determine the slope of the separatrix close to the equilibrium). The separatrix loop shown in Fig. 13.3a is called a biasymptotic trajectory because it approaches the same equilibrium as $t \to +\infty$ and as $t \to -\infty$. The phase-plane diagram in Fig. 13.3a was drawn for the equation $\ddot{x} - x(1 - x/2) = 0$, which has the energy integral $\dot{x}^2 - x^2 + x^3/3 = $ const. The dependence of the variable x on time corresponding to the separatrix loop (Fig. 13.3b) is a single impulse or if (13.1) was obtained to describe a stationary wave (Chapter 19), it is a soliton. This impulse has an infinite tail if the time taken for motion close to the equilibrium $x = 0$ is infinitely long. Indeed, in the neighborhood of the saddle, the equation of motion has the form $\ddot{x} - x = 0$ or $\dot{x} = y$, $\dot{y} = ax$, i.e., as $x \to 0$, $y \to 0$, the system approaches the equilibrium state and leaves it at an infinitesimally small velocity. Consequently it takes infinitely long.

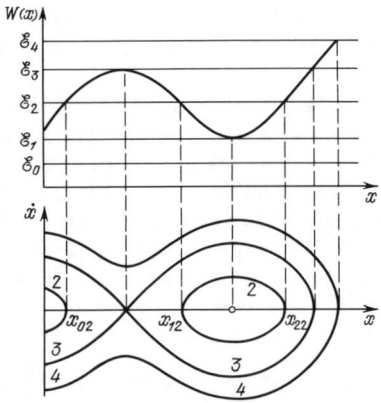

Fig. 13.2. The potential energy curve $W(x)$ and the phase-plane diagram corresponding to it for a nonlinear oscillator.

THE NONLINEAR OSCILLATOR

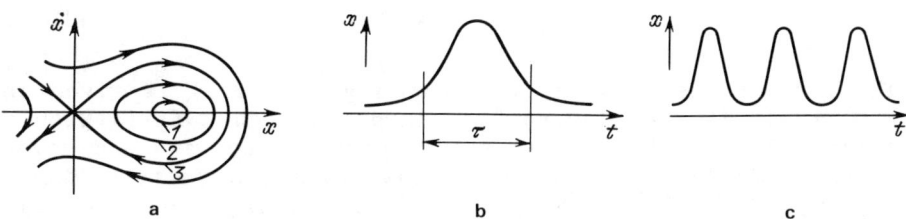

Fig. 13.3. a) The phase plane: *1*) the ellipse corresponds to a periodic, nearly harmonic oscillation; *3*) the separatrix loop corresponding to a soliton.
b) A soliton with infinitely long tails (the oscillogram of a variable x).
c) The oscillograph of a conoidal motion, which corresponds to trajectory 2.

As we shall see later, a special solution is the one to which the separatrix loop on the phase-plane diagram (Fig. 13.3a) corresponds. It is of great interest in the theory of nonlinear waves. For example, waves on the surface of shallow water can be approximated by the Korteveg--de Vries equation

$$u_t + v_0 u_x + u u_x + \beta u_{xxx} = 0 . \tag{13.6}$$

If we are only interested in waves travelling at a constant velocity without changing their shape, $u = u(x - Vt)$ (stationary waves), then after substituting this into (13.6) we obtain the equation for a nonlinear oscillator, whose phase-plane diagram is given in Fig. 13.3a.

Concerning the other trajectories, for example, trajectories like 1 in Fig. 13.3a, this motion is at the bottom of the potential well and consequently is almost linear; the frequency can be determined from the linearized equations. For trajectories like 2 in Fig. 13.3a, which are close to the separatrix, the function $x(t)$ is given in Fig. 13.3c. In this case if we have $dW(x)/dx = \sin x$ in (13.3), i.e., our oscillator is a simple pendulum, then the well-known exact equation expressed in terms of an elliptical integral is obtained [3].

One more example. Let us consider the behavior of an electron in the periodic electric field of a transverse wave (Fig. 13.1d). Let the potential field vary as $\phi = \phi_0 \cos(\omega t - kx)$. It is simplest to describe the motion of the electron by writing its equation of motion in the coordinate system associated with the wave. Then the potential is $\phi = \phi_0 \cos kx$ and the equation we want has the form

$$m_e \ddot{x} + e\phi_0 k \sin kx = 0 .\tag{13.7}$$

The phase-plane diagram corresponding to (13.7) is given in Fig. 13.4. If the electron is placed in such a field with a large velocity, then it is not trapped by the wave and moves along it following the oscillations. Thus a trajectory like 1 in Fig. 13.4 corresponds to such a "flying" particle. If however the initial velocity is less than a critical value determined from $m_e v_{cr}^2/2 < e\phi_0$, then the electron falls into the potential well and it will oscillate there. The trajectories like 2 in Fig. 13.4 corresponds to these oscillating motions. It is interesting that (13.7) is the asymptotic equation for the motion of an electron in the elementary theory of free-electron lasers with large Q resonators, such as the Fabry--Perot laser under conditions in which the wave amplitude can be considered constant [4] (Chapter 11).

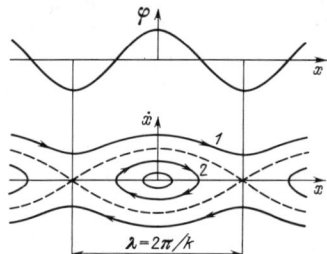

Fig. 13.4. Phase-plane diagram of a nonlinear oscillator describing the motion of a trapped particle (trajectory 2) and passing particle (trajectory 1) in a wave field (see Fig. 13.1d).

As we have seen, one of the main properties of a nonlinear oscillator is its asynchronicity. The asynchronicity will affect the motion's stability. Two neighboring points on close trajectories will, with time, separate in phase, i.e., there cannot be Lyapunov stability, however orbital stability for trajectories like 1 and 2 will be maintained (this can be seen from the phase-plane diagram). Close to the separatrices, as can easily be seen, there cannot even be orbital stability.

The system we have been looking at is conservative. In general, it is far from easy to establish whether a dynamic system whose phase-plane diagram is a planar, belongs to the class of conservative systems. Conservativity is the conservation of energy. However, in systems describing chemical reactions, or the coexistence of two biological species, for example, it is frequently impossible to introduce energy. Indeed, the system of equations (1.11) in

THE NONLINEAR OSCILLATOR

the notation of Chapter 1:

$$\dot{N}_1 = N_1(\varepsilon_1 - \gamma_1 N_2) , \quad \dot{N}_2 = -N_2(\varepsilon_2 - \gamma_2 N_1) ,$$

which describes the relationship between vegetarians and predators is at first glance nonconservative. At the same time, the phase-plane diagram of this nonlinear oscillator, given in Fig. 13.5a, seems to be like that of a mechanical conservative system. The point is that there are systems which have an integral of motion

$$\gamma_2 N_1 + \gamma_1 N_2 - \varepsilon_2 \ln N_1 - \varepsilon_1 \ln N_2 = \text{const} .$$

Thus, from the point of view of oscillation theory, we must regard the existence of a unique integral of motion like $\mathcal{F}(x, \dot{x}) = C$ as a necessary attribute for the conservativity of a two-dimensional system.

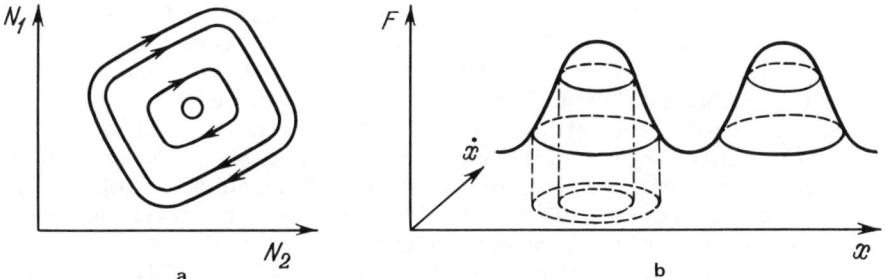

Fig. 13.5. a) Phase plane diagram of a nonlinear oscillator corresponding to the ecological situation of a vegetarian and a predator. b) An explanation of the phase plane diagram if the integral of motion [F] $(x, \dot{x}) \cdot = C$ is known.

It is possible to construct the surface $z = \mathcal{F}(x, \dot{x})$ and where it intersects the planes $z = C = \text{const}$, obtain the phase-plane diagrams by projecting the sections onto the $(x, \dot{x})$-plane (Fig. 13.5b). It is easy to show that only equilibria of the saddle and focus type can exist in two-dimensional conservative systems.

We shall now consider the behavior of an ensemble of a large number of noninteracting nonlinear oscillators. These oscillators may, for example, be electrons moving in the field of a longitudinal electric wave, (the behavior of an ensemble of linear oscillators was considered in Chapter 3). The first problem of this nature appeared at the end of the 1960's in microwave electronics when studying systems of excited nonlinear oscillators, such as a classical active medium for masers at cyclotron resonance [5], and in plasma physics, for instance, as related to the acceleration and heating of charged particles. We shall consider that the electrons' velocity distribution function in the beam is known and its curve is given in Fig. 13.6. In the system of coordinates associated with the wave $\phi = \phi_0 \cos(\omega t - kx)$, all the particles are divided into entrained and passing ones. Electrons whose velocities are in the interval $\omega/k \pm (2e\phi_0/m)^{1/2}$ do not have enough energy to overcome the potential barrier $e\phi_0$ and they oscillate in the potential well of the wave. Those electrons whose velocities lie outside this interval do not "notice" the wave (Fig. 13.4).

Every i-th electron in the sinusoidal wave's field behaves like a pendulum:

$$\ddot{x}_i + \omega_0^2 \sin x_i = 0 \quad (i = 1, 2, \ldots, N), \quad \omega_0^2 = k^2 e\phi_0/m.$$

(13.8)

The pendulum oscillations correspond to the trapped electrons and the rotations to the passing electrons (Fig. 13.4). Thus, particles in the field of a wave are an ensemble of identical nonlinear oscillators that only differ in terms of their initial energies. The way the ensemble behaves over time depends on the distribution function of the electrons with respect to their energies and velocities, the function $f(v)$. Because no interaction between the oscillators is taken into consideration, we can solve this problem quite simply by looking at the motion of the oscillators on the phase plane. We isolated the initial phase volume in the form of a region bounded by the separatrices on the $(x, \dot{x})$-plane (Fig. 13.7). If $(\partial f/\partial v)_{v=\omega/k} < 0$, then when $t = 0$ most of the trapped particles will be in the lower part of the region. Over time this region changes into a tight spiral due to the asynchronicity of the oscillators. The number of turns of the spiral increases continuously. Consequently, the number of particles with different velocities will change continuously and the distribution function $f(v)$ in the interval Δv will begin to fluctuate, becoming more and more divided (Fig. 13.6a). After a sufficiently long period of time, all the oscillators must once again gather in the initial phase volume because the motion of a conservative system (13.8) of N oscillators is reversible. Physically, however, it is clear that however long we wait a miracle does not happen: the particles will mix with one another and with the wave due to the smallest interaction, i.e., the whole region inside a separatrix will be evenly filled (this region is called a cat's eye). The number of particles moving faster than the wave ($\dot{x} > 0$) will become equal to the number of particles moving slower than the wave ($\dot{x} < 0$), and a plateau will form on the distribution function (Fig. 13.6b). Because the average kinetic energy of the particles with this sort of mixing grows, the sinusoidal wave in which the particles are oscillating looses some of its energy to accelerate the particles. This loss of energy from the monochromatic wave is called nonlinear Landau damping [6].

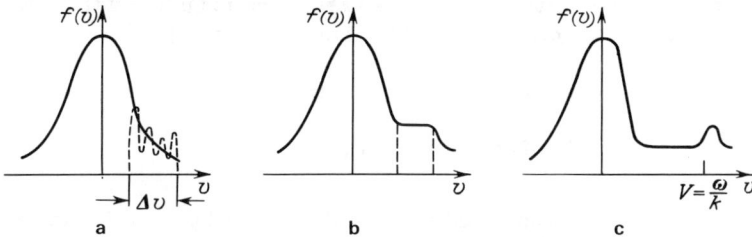

Fig. 13.6. The electron distribution function with respect to velocity: a) behavior of oscillations in the field of a periodic longitudinal wave; b) the formation of a plateau; c) the electron's velocity distribution in a system comprising of a plasma and a beam.

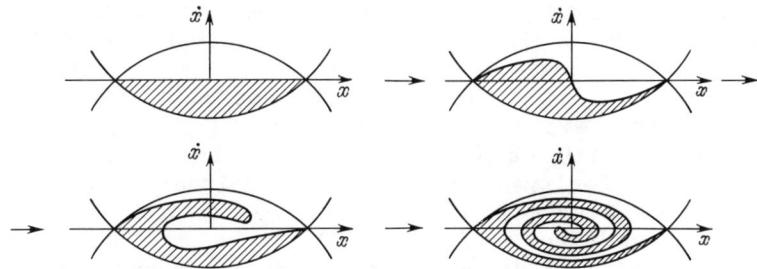

Fig. 13.7. Evolution of the phase volume in an ensemble of noninteracting electron-oscillators.

If the velocity distribution function of the particles is not uniform, as is the case in a system composed of an electron beam and a plasma, then the inverse process is possible, that is a wave of finite amplitude is amplified. When the wave's phase velocity falls in the interval of velocities corresponding to the left-hand slope of the nonuniform distribution function (Fig. 13.6c), then the waves amplitude grows as a result of linear Landau amplification (fewer slow particles take energy from the wave than fast particles give energy) and it traps passing particles. This amplification process clearly continues until the number of fast and slow particles corresponding to the left-hand slope of $f(v)$ becomes equal and the wave turns into nonlinear stationary wave (quasilinear relaxation). Thus the oscillator phases mix over time and in place of oscillations on the distribution function, a plateau becomes established. The time required for this plateau to form is of the same order of magnitude as the characteristic time of the particles moving along closed trajectories.

A more accurate relaxation time for the distribution function of the oscillators with respect to energy may only be determined by solving the self-consistent problem taking into account the changes in the wave amplitude over time. The appropriate equation is written in the form [14]

$$\partial f/\partial t + v\partial f/\partial \xi - (e/m)Ek \sin k\xi \, \partial \phi/\partial v = 0 ,$$

$$\partial E/\partial \xi = - 4\pi e \left(\int f(v)dv - n_0 \right) ,$$

where E is the amplitude of the longitudinal wave, and $\sin(\omega t - kx) = \sin k\xi$.

Many of the problems in the nonlinear theory of fluid mechanics stability reduce to a consideration of the interaction between an ensemble of oscillators and the wave, e.g., the nonlinear analysis of the disturbance to a surface wave by wind [14] and the theory of boundary layers [16]. The role of the nonequilibrial particles is taken by the particles of the medium moving with different velocities. In contrast to the example given above of an electron beam interacting with a plasma, the problems in fluid mechanics concerning the evolution of particle distributions in a liquid by velocity cannot in principle be one-dimensional, that is, the velocity at each point is uniquely determined in classical fluid mechanics. Consequently, if the particles of a medium move with different velocities in the field of one-dimensional wave (propagating along the x-axis), then the particles must be spaced out along the transverse coordinate y. Therefore the simplest consideration of the evolution of the velocity distribution function for the liquid's particles in the field of a fluid mechanics wave requires a consideration of the evolution of the profile of a two-dimensional flow with a transverse shift velocity [15].

It should be noted that the amplification of a wave with finite amplitude in a system consisting of a plasma and beam was only studied in detail relatively recently [17]. At the same time, the amplification of finite signals in travelling-wave tubes was known in microwave electronics some time ago, and theoretical and experimental investigations were carried out in the 1950's [7-9]. The phase-plane diagrams, both predicted and the experimentally measured ones, for a working travelling-wave tube are given in Figs. 13.8 and 13.9. The theoretical diagrams are interesting because it is possible to determine not only the phase state of the "machine" electrons relative to the wave, but also to determine their kinetic energy. This is important when selecting a method for increasing the efficiency of a travelling-wave tube. Several "machine" electrons are labeled by numbers. The various dimensionless lengths $\zeta (\zeta \sim x)$

THE NONLINEAR OSCILLATOR

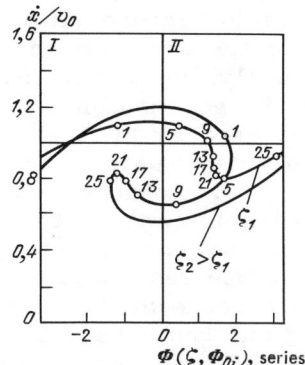

Fig. 13.8. Theoretical phase-plane diagram for a travelling-wave tube amplifying a finite signal.

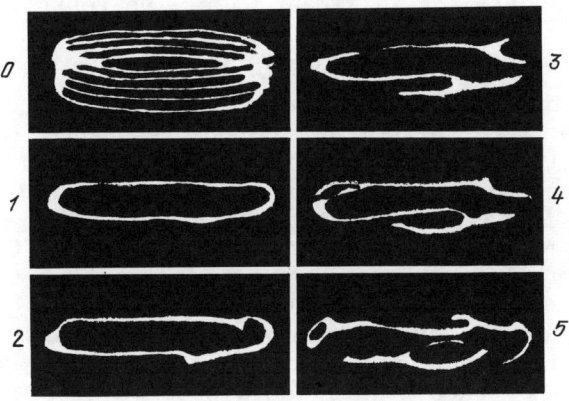

Fig. 13.9. Photographs of the velocity analyzer display for a working travelling-wave tube. Photograph 0 corresponds to the absence of a high-frequency signal. The amplitude of the high-frequency field increases with increasing photograph number [9].

correspond to the various intensities of the high-frequency electric field: when $\zeta_1 < \zeta_2$ the velocity modulation is closer to sinusoidal than at ζ_2, where eddies form. The equations used for the computer calculations correspond to "noninteracting" electrons ($\dot{x}/v_0$ is the relative velocity of the "machine" electrons; $\Phi(\zeta, \Phi_{0i})$ is the phase of a "machine" electron, that is its phase state with respect to the wave at a given coordinate ζ; Φ_{0i} is the initial phase of the i-th "machine" electron; and I, II are the areas of the accelerating and braking fields of the wave, respectively). Figure 13.9 is especially interesting. The model of the travelling-wave tube, which was studied by K. Katler (1956), contained a velocity analyzer: at the exit from each spiral the electron beam has to go through a crossed electrostatic and magnetic field and arrived at the fluorescent screen on which the electron velocity, which was proportional to the vertical deviation, and charge density, which was proportional to the radiation intensity, could be measured as the functions of the signal phase. Photographs taken for a working tube could be processed to yield diagrams like those in Fig. 13.8 [9]. The experiment is carried out on a unique tube that was 3 m long and had a beam 2.54 cm in diameter for a beam potential 400 V and frequency 100 MHz..

13.3 Nonlinear Resonance

If the oscillator is linear, i.e., we restrict ourselves to the first term in the expansion $\omega^2(x) = \omega_0^2 + \alpha x + \beta x^2 + \ldots$, then when the oscillator is acted upon by an external periodic force we observe, in essence, the basic effect of linear resonance (Chapter 1). The smaller the loss in the oscillator, the sharper and the higher the resonance curve appears to be (Fig. 1.9). We now look at how things change when the frequency depends on the amplitude. Suppose the frequency of the external force equals the frequency of rotation about one of the phase trajectories close to the center (Fig. 13.4). Then the system gains energy from the external source and oscillations first start and then grow. This means that the phase-plane corresponding to the system (the trace point) would seem to cross in sequence the phase trajectories corresponding to large energies. However, because the oscillator is not isosynchronous, other frequencies corresponds to larger energies. As the result the system leaves the resonance state and at some amplitude the oscillator ceases to feel the external force. Resonance ceases therefore due to a nonlinear change in frequency $\omega = \omega(A)$.

There is a new aspect to resonance for nonlinear oscillators. Whereas a linear oscillator only resonates at frequency close to the natural one, i.e. when $\Omega = \omega_0 + \varepsilon$, a nonlinear oscillator also resonates at harmonics. For example, a quadratic nonlinearity ($\omega^2 \sim \alpha x$) leads to the appearance in the nonlinear system of the spectral components 2Ω, 4Ω, etc. Consequently, if $2\Omega \approx \omega_0$, for example, then the system will resonate at a harmonic of the external force.

In general, even when the external disturbance is sinusoidal, some very nontrivial effects are possible in a nonlinear oscillator, that is the dynamics of the system may turn out to be exceedingly complex, even stochastic. These effects are only observed when there is nonlinearity. In order to investigate them, numerical studies of the phase space are necessary. Note that the phase space of a nonautonomous system with one degree of freedom is three-dimensional: x, $\dot{x}$, t (the third coordinate is time).

We shall, for the present, consider a system close to a linear autonomous oscillator, i.e., we shall consider weak nonlinearities, small dissipations, and a small amplitude for the external force. The solution method is then obvious, i.e., one of the asymptotic methods. The initial model can be described by

$$\ddot{x} + \omega_0^2 x = \mu[A_{ex}\cos \Omega t - \alpha x^2 - \beta x^3 - 2\lambda \dot{x}] , \quad \mu \ll 1 . \quad (13.9)$$

Because we wish to investigate resonance, we shall look for the solution with the frequency of the external force, i.e., in the form $x = A(t)\sin[\Omega t + \phi(t)]$. By assuming that $\omega_0 = \Omega - \varepsilon$ we transform (13.0) into

$$\ddot{x} + \Omega^2 x = \mu[A_{ex}\cos \Omega t - \alpha x^2 - \beta x^3 - 2\lambda \dot{x} + 2\varepsilon \Omega x] .$$

The Van der Pol method yields shortened equations for the amplitude and phase:

$$\dot{A} = \frac{A_{ex}}{2\Omega} \cos \phi - \lambda A , \quad A\dot{\phi} = -\frac{A_{ex}}{2\Omega} \sin \phi + \frac{3\beta}{8\Omega} A^3 - \varepsilon A . \quad (13.10)$$

Here ε is a linear disturbance. The resonance curve is the graph of the amplitude of the established oscillation versus the disturbance, i.e., the relation $A(\varepsilon)$, which is obtained from (13.10) under the conditions $\dot{A}$, $\dot{\phi} = 0$. These conditions determine equilibrium of (13.10)

$$\lambda A_0 = \frac{A_{ex}}{2\Omega} \cos \phi_0 , \quad -\varepsilon + \frac{3\beta A_0}{8\Omega} = \frac{A_{ex}}{2\Omega A_0} \sin \phi_0 .$$

It follows that $\lambda^2 + (\gamma A_0^2 - \varepsilon)^2 = A_{ex}^2/4\Omega^2 A_0^2$, where $\gamma = 3\beta/8\Omega$, and hence the link we are looking for has the form

$$\varepsilon = \gamma A_0^2 \pm \{(A_{ex}/2A_0\Omega)^2 - \lambda^2\}^{1/2} . \quad (13.11)$$

When constructing a resonance characteristic on the (A_0^2, ε)-plane (Fig. 13.10a), the amplitude of the external force A_{ex} is a parameter. When $A_{ex} < A_{ex}^*$, the resonance curves are graphs of unique functions, and reminiscent of the resonance curves for a linear oscillator with damping. The maximum of the curves are shifted towards larger frequencies if the fundamental frequency of the oscillator grows with increasing amplitude, and shifted towards smaller frequencies if the fundamental frequency decays. When $A_{ex} > A^*_{ex}$ the resonance curve is the graph of an non-single valued function

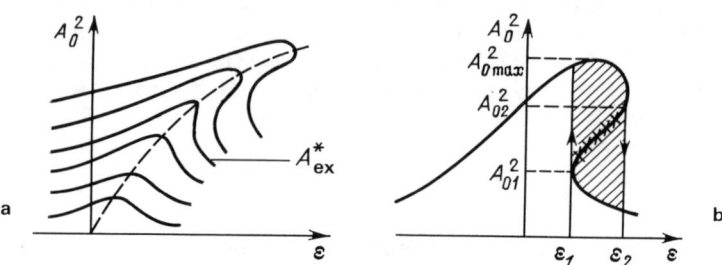

Fig. 13.10. Resonance curves for a nonlinear oscillator: a) when the amplitude is $A_{ex} < A^*_{ex}$ (A_{ex} is a parameter), the resonance curves are graphs of unique functions and look like the curves for a linear oscillator with damping but somewhat deformed (Fig. 1.9); b) the resonance curve when $A_{ex} > A^*_{ex}$, the crosses label the unstable section of the curve; the hysteresis region is shaded.

An investigation of the stability of the system's equilibria (13.10) corresponding to different sections of the resonance curve (we leave the reader to do this independently) indicates that the section labeled in Fig. 13.10b by crosses is unstable. As the disturbance is changed, we have the following regions: when $\varepsilon = 0$ there is exact resonance according to the linear approximation and the amplitude is far from being at a maximum; when ε is increased the amplitude is increased to a point A_{0max}; when $\varepsilon = \varepsilon_2$ there is a step-change in the oscillation's amplitude to a much lower value. Going in the opposite direction, the step-change occurs at $\varepsilon = \varepsilon_1$, the amplitude increasing (13.10b). The value A_{01} corresponds to the point of tangency between the curve $A_{02}^2 = \varepsilon_1$ and the resonance curve $A_0^2 = A_0^2(\varepsilon)$, while A_{02} corresponds to the point of tangency between the line $A_{02}^2 = \varepsilon_2$ and $A_0^2 = A_0^2(\varepsilon)$. By differentiating the equation for the resonance curve

$$(\lambda^2 + \gamma^2 A_0^4 - 2\gamma\varepsilon A_0 + \varepsilon^2)A_0^2 = \{(A_{ex})^2/4\Omega^2\}, \qquad (13.12)$$

for a vertical tangent, we find

$$dA_0/d\varepsilon = \left[-\varepsilon A_0 + \gamma A_0^3\right]\left[\varepsilon^2 + \lambda^2 - 4\gamma\varepsilon A_0^2 + 3\gamma^2 A_0^2\right]^{-1} = \infty \; .$$

The values A_{01} and A_{02} are then found by solving

$$\varepsilon^2 + \lambda^2 - 4\gamma\varepsilon A_0^2 + 3\gamma^2 A_0^2 = 0 \qquad (13.13)$$

and (13.12) simultaneously. Equating the expressions for the derivative $dA_0/d\varepsilon$ to zero, it is easy to find $A_{0max}^2 = \varepsilon/\gamma$. It follows from (13.12) that $A_{0max}^2 = A_{ex}/2\Omega\lambda$. The value of A_{ex}^* at which the hysteresis on the resonance curve can be determined by equating the roots of (13.13), i.e., equating its discriminant $4\gamma^2 A_0^4 - 3\gamma^2 A_0^4 - \lambda^2 = 0$ to zero, whence $A_0^4 = \lambda^2/\gamma^2$. At this point we have $\varepsilon_1 = \varepsilon_2 = 2\lambda$ and it follows from (13.12) that $A_{ex}^{*2} = 8\Omega^2\lambda^3/\gamma$.

We now look at what linearity does for parametric disturbance. In the Chapter 11 we investigated parametric resonance in an oscillator describable by, for example, Mathieu's equation (11.8), i.e.,

$$\ddot{x} + \omega_0^2(1 - \mu b \cos \omega_p t)x = 0 \; .$$

The development of parametric instability causes the oscillations in a system to grow. Linear damping, in this case, clearly does not exist: it only makes the strip of excitement narrower and does not lead to any bounding of the amplitude. At large oscillation amplitudes another type of nonlinearity may occur in an oscillator, appearing, for example, as the dependence of the frequency on the amplitude, viz., $\omega^2(x) = \omega_0^2 + \beta x^2$. Then the oscillation of a parametrically excited oscillator cannot grow without limit, despite the parametric increment. The frequency is incremented and due to this change in frequency, which we have already mentioned, the conditions for parametric resonance are violated. There is thus a bound on the amplitude (known as the maladjustment mechanism of bounding).

We rewrite the equation of such an oscillator in the form

$$\ddot{x} + \omega_0^2(1 - \mu b \cos \omega_p t)x + \mu\beta x^3 + 2\mu\lambda\dot{x} = 0 \; , \qquad (13.14)$$

where λ characterizes the linear losses. The fundamental parametric resonance where condition $\omega_p = 2\omega_0 + \mu\varepsilon$ must be fulfilled. By applying Van der Pol's method to (13.14) and searching for a solution in the form $x = 2A(t)\cos[\omega_p t/2 + \phi(t)]$, we attain the following for the amplitude and phase after averaging:

$$\dot{A} = -\mu\lambda A - \frac{\mu b\Omega}{8} A \sin 2\phi ,$$

(13.15)

$$\dot{\phi} = -\mu\varepsilon + \frac{3\mu\beta}{\Omega} A^2 - \frac{\mu b\Omega}{8} \cos 2\phi .$$

It can be seen from (13.15) that even in the absence of linear disturbances ($\varepsilon = 0$) there is a disturbance proportional to the square of the amplitude. This causes a change in phase such that the parametric increment falls to zero and the amplitude is thus bounded.

13.4 Overlap Between Nonlinear Resonances

Our investigation in the previous section of the properties of nonlinear resonance, in which we used Van der Pol's method is valid when there is only one unique (isolated) resonance bound in the nonlinear oscillator. All the events, if considered in a three-dimensional phase space ($\dot{x}$, x, t), must develop in a narrow circular layer. The projection of this layer onto the ($\dot{x}$, x)-plane is a closed band around the trajectory of the autonomous oscillator the period of motion around which $2\pi/\omega$ must be exactly equal or a multiple of the period of the external excitation $2\pi/\Omega$. In contrast to the case of a linear oscillator, a nonlinear oscillator may resonate at practically any frequency of a periodic disturbance, if of course the nonlinearity is large. This is explained by the asynchronicity and anharmonicity of the oscillations of a nonlinear oscillator (asynchronicity, which, as we know, is the dependence of the oscillation's frequency on its energy, whilst anharmonicity is the presence in the spectrum of periodic oscillations with higher harmonics).

If an oscillator, e.g., an simple pendulum, is affected by a small harmonic disturbance, then if second order effects and above are neglected, we get isolated resonance, namely resonance only at a certain amplitude level A, while the frequency $\omega(A)$ of the unperturbed motion will be about the same as the frequency Ω of the disturbance. The frequency and amplitude of the oscillation may vary about the resonance frequency within limits ΔA and $\Delta \omega$, these limits defining the width of the resonance. Higher order resonances, of the form

$$n\omega(A) \approx m\Omega ,$$

(13.16)

where n and m are integers, have small terms with orders of magnitude greater than $(m + n)$ in terms of the amplitude of the disturbance and nonlinearity; they can therefore be neglected.

The situation changes qualitatively when the small periodic disturbances cease to be harmonic, i.e., when

$$F_{ex}(t) = \sum_n A_n \cos n\Omega t . \qquad (13.17)$$

As you might guess, the same approximation gives rise to a situation in which the resonance condition (13.16) can be satisfied by various values of A (corresponding, naturally, to different values of n and m), i.e., several nonlinear resonances may exist simultaneously in a system, with each harmonic determining its own resonance in its part of the phase space. These resonances may be isolated and noninteracting, or they may overlap. It is not a trivial matter to determine what leads to overlapping [12,13,20], and here we shall only address this question qualitatively, leaving a more detailed discussion to Chapter 22. However, we shall first consider how resonance on the harmonics appears.

We shall do this for the model

$$\ddot{x} + \omega_0^2(1 + \alpha x + \beta x^2)x = A \cos \Omega t ,$$

utilizing perturbation theory for its analysis. The desired solution is represented in the form

$$x(t) = x^{(1)}(t) + x^{(2)}(t) + x^{(3)}(t) + \dots , \quad x^{(1)} = A \cos \omega t,$$

with the frequency ω being given in form of an expansion $\omega = \omega_0 + \omega^{(1)} + \omega^{(2)} + \dots$. We take the small disturbance $\xi = 1 - \omega_0^2/\omega^2$ and the nonlinear disturbance over to the right-hand side, and thus get an initial equation

$$(\omega_0/\omega)^2 \ddot{x} + \omega_0^2 x = -\alpha \omega_0^2 x^2 - \beta \omega_0^2 x^3 - \xi \ddot{x} + A \cos \Omega t . \qquad (13.18)$$

Let $\Omega = \omega_0/3 + \Delta\omega$, for example. Then at the first approximation there will be no resonance. The forced solution $x^{(1)}(t) \sim \cos[(\omega_0/3 + \Delta\omega)t]$ has a frequency far from the oscillator's fundamental frequency. However, in the second approximation the nonlinearity yields a term like $x^{(3)}$, i.e., there will be a resonance force on the right-hand side of the equation for $\{x^{(1)}\}^3 \sim \cos[3(\omega_0/3 + \Delta\omega)t]$ and a frequency

($\omega_0 + 3\Delta\omega$) (its amplitude is proportional to A^3), and consequently, a sort of parametric resonance arises; the respective harmonic appears due to the product $x^{(1)}x^{(2)}$.

In order to investigate the overlap of resonances, it is convenient to describe a nonlinear oscillator in force-angle variables. Let us look more closely at these new variables.

We introduced in Chapter 1 the dynamic variables p and q, which allowed us to write the equation of motion for a harmonic oscillator in canonical form, i.e.,

$$\frac{\partial q}{\partial t} = \frac{\partial \mathcal{H}}{\partial p} = p, \quad \frac{\partial p}{\partial t} = -\frac{\partial \mathcal{H}}{\partial q} = -\omega^2 q,$$

$$\mathcal{H} = \frac{1}{2}\left(p^2 + \omega^2 q^2\right).$$
(13.19)

Now let the Hamiltonian have the form

$$\mathcal{H}(q, p, t) = \mathcal{H}_0(q, P) + \mathcal{H}_1(q, p, t).$$

We change to new variables θ (in place of q) and I (in place of p) such that the equation of motion remains canonical

$$\partial\theta/\partial t = \partial\mathcal{H}'/\partial I, \quad \partial I/\partial t = -\partial\mathcal{H}'/\partial\theta \qquad (13.20)$$

with the new Hamiltonian $\mathcal{H}'(I, t)$. Now we find the variables I and θ following [10]. The principle of least action is valid for the equations of the Hamiltonian:

$$\delta \int (p\,dq - \mathcal{H}\,dt) = 0 \qquad (13.21)$$

(the coordinates and momentum vary independently). For the new variables we have

$$\delta \int (I\,d\theta - \mathcal{H}'dt) = 0. \qquad (13.22)$$

Equations (13.21) and (13.22) are equivalent when the integrands differ by the complete differential of some function F of coordinates, momentum, and time. Then after certain elementary transformations we get $d(F + I\theta) = p\,dq + \theta\,dI + (\mathcal{H}' - \mathcal{H})dt$. The function $\Phi(q, I, t) = F + I\theta$ is called a generating function [11]. Using it, we get

$$p = \partial\Phi/\partial q, \quad \theta = \partial\Phi/\partial I, \quad \mathcal{H}' = \mathcal{H} + \partial\Phi/\partial t. \qquad (13.23)$$

Suppose that to begin with $\mathcal{H}'$ is independent of t and θ. It follows from (13.20) that $\dot{I} = 0$, and $\theta = \omega t + \phi$, where $\omega(I) = \partial\mathcal{H}'/\partial I$, and ϕ is a constant. Thus, the new variable I

THE NONLINEAR OSCILLATOR

is conserved during motion. The quantity $\theta = \partial \Phi/\partial I$ is called the angular variable (or simply the angle). Its increment $\Delta\theta$ over a period of motion is 2π, i.e., $\oint (\partial\theta/\partial q)_{I=\text{const}} dq = 2\pi$. It can be seen, however, from the definition of θ that $\partial\theta/\partial q = \partial^2\Phi/\partial I \partial q$. Then using the first equation from (13.23) we get

$$\oint \left(\frac{\partial\theta}{\partial q}\right)_{I=\text{const}} dq = \frac{\partial}{\partial I} \oint p \, dq = 2\pi .$$

The definition of the other variable follows from this relation, i.e.,

$$I = (2\pi)^{-1} \oint p \, dq , \qquad (13.24)$$

which is called the action variable (or simply the action). The condition $I = \text{const}$ on the phase (p, q)-plane for a closed trajectory means that the area $2\pi I$ enclosed by the trajectory is conserved. For instance, the phase trajectory for a harmonic oscillator described by (13.19) is the ellipse $p^2/2 + \omega^2 q^2/2 = \mathcal{H}_0$, with the semiaxes $a = (\mathcal{H}_0)^{1/2}$ and $b = (2\mathcal{H}_0/\omega^2)^{1/2}$, and area $S = \pi ab = 2\pi(\mathcal{H}_0/\omega)$, i.e., $I = \mathcal{H}_0/\omega$ is an adiabatic invariant, as discussed in the last chapter.

Let us now consider a disturbance to a system with Hamiltonian

$$\mathcal{H}(I, \theta, \lambda) = \mathcal{H}_0(I) + \mu V(I, \theta, \lambda) , \quad \mu \ll 1 . \qquad (13.25)$$

Let the disturbance to the Hamiltonian be periodic both with respect to θ and λ and with period 2π; λ characterizes the external force which has frequency $\Omega = \lambda$.

Using (13.20) and (13.25) the equation of motion of the oscillator with the "undisturbed" variables is written as

$$\dot{I} = -\partial\mathcal{H}'/\partial\theta = -\mu\partial V(I, \theta, \lambda)/\partial\theta , \qquad (13.26)$$

$$\dot{\theta} = \partial\mathcal{H}'/\partial I = \partial(\mathcal{H}_0 + \mu V)/\partial I = \omega(I) + \mu\partial V(I, \theta, \lambda)/\partial I , \qquad (13.27)$$

where $\omega(I) = \partial\mathcal{H}_0/\partial I$ is the frequency of the nonlinear oscillations (show this for yourselves using definition (13.24)). We shall characterize the asynchronicity of the oscillator by the parameter [10]

$$\alpha = \left| \frac{I}{\omega} \frac{d\omega}{dI} \right| = \left| I \left(\frac{\partial \mathcal{H}_0}{\partial I} \right)^{-1} \frac{d}{dI} \left(\frac{\partial \mathcal{H}_0}{\partial I} \right) \right| = \left| \frac{1}{2} \frac{\partial^2 \mathcal{H}_0}{\partial I^2} \left(\frac{\partial \mathcal{H}_0}{\partial (I)^2} \right)^{-1} \right| . \quad (13.28)$$

Since the disturbances in θ and λ are periodic, we can expand μV in a double Fourier series

$$\mu V(I, \theta, \lambda) = \sum_{m,n} \left[V_{m,n}(I)/2 \right] \exp\left[i(m\lambda + n\theta) \right] + \text{c.c.} \quad (13.29)$$

The condition for resonance between the m-th harmonic of the oscillator and the n-th harmonic of the external force is analogous to (13.16), viz.,

$$m\Omega + n\omega \approx 0 \quad (\omega, \Omega > 0) . \quad (13.30)$$

When there is only one resonance ($m, n = \pm 1$ and $\omega \approx \Omega$), two terms remain in (13.29), one the resonance with argument $(\lambda - \theta)$ and the other the high-frequency disturbance with argument $(\lambda + \theta)$ We leave only the resonance term in (13.20) and rewrite (13.26) and (13.27) in the form

$$\dot{I} = -\mu V_{mn} \frac{\partial}{\partial \theta} \cos(m\lambda + n\theta) = \mu n V_{mn} \sin \psi_{mn} , \quad (13.31)$$

$$\dot{\psi}_{mn} = m\dot{\lambda} + n\dot{\theta} = m\Omega + n\omega(I) + \mu n (\partial V_{mn}/\partial I) \cos \psi_{mn} ,$$

where $\psi_{mn} = m\lambda + n\theta$ is the resonance phase. Suppose the value $I = I_p$ corresponds to exact resonance ($\psi_{mn} = 0$), with $|I - I_p| = |\Delta I| \ll I_p$. Given this assumption a universal Hamiltonian ($\omega(I_p) \neq 0$) is conserved independent of the concrete forms of the functions $\omega(I)$ and $V_{mn}(I)$, i.e.,

$$\mathcal{H}_y = n(d\omega/dI)_{I_p} (\Delta I)^2/2 + \mu n V_{mn}(I_p) \cos \psi_{mn} . \quad (13.32)$$

We shall show this. When $I = I_p$ we have

$$\dot{I}_p = 0 , \quad m\Omega + n\omega(I_p) + \mu n \, \partial V_{mn}/\partial I = 0 . \quad (13.33)$$

We subtract (13.33) from the corresponding equations in (13.31):

$$\Delta \dot{I} = \mu n V_{mn} \sin \psi_{mn} , \quad (13.34)$$

THE NONLINEAR OSCILLATOR

$$\dot{\psi}_{mn}^{p} = n[\omega(I) - \omega(I_p)] + \mu n \left\{ \frac{\partial V_{mn}}{\partial I} \cos \psi_{mn} - \frac{\partial V_{mn}(I_p)}{\partial I} \right\}.$$

If $|\Delta I| \ll I_p$, then we have

$$\omega(I) - \omega(I_p) \approx (\partial \omega / \partial I)_{I=I_p} \Delta I = \alpha \omega(I_p) \Delta I / I_p$$

and if the nonlinearity is moderate, i.e., $\mu \ll \alpha \ll 1/\mu$, then the terms proportional to μ in the second equation of (13.34) can be neglected. Therefore, finally we have

$$\Delta \dot{I} = \mu n V_{mn} \sin \psi_{mn} = -\frac{\partial \mathcal{H}_y}{\partial \psi_{mn}}, \quad \dot{\psi}_{mn} = n \left(\frac{\partial \omega}{\partial I} \right)_{I=I_p} \Delta I = \frac{\partial \mathcal{H}_y}{\partial (I)}.$$

(13.35)

It follows from (13.35) that the Hamiltonian in (13.32) is independent of time.

Hamiltonian (13.32) is similar to the Hamiltonian of a pendulum with mass $\left\{ n(d\omega/dI)_{I=I_p} \right\}^{-1}$ in a gravity field with acceleration $g = \mu n^2 (d\omega/dI)_{I=I_p} V_{mn}$. The phase-plane diagram corresponding to (12.32) can be constructed on the $(\Delta I, \psi_{mn})$-plane or given that $\omega(I) = \omega(I_p) + (d\omega/dI)_{I=I_p} \Delta I$ on the $(\omega(I), \psi_{mn})$-plane. We shall occasionally, omit the index from ψ_{mn} and simply write ψ. The phase trajectories for two nonlinear, noninteracting resonances are shown in Fig. 13.11. The region of the nonlinear resonance is bounded by the separatrices and inside this region the change in phase ψ_{mn} is bounded (phase oscillations). We find from (13.32) that $\mathcal{H}_y = \mu V_{mn}(I_p)$ when a separatrix goes through the point $\omega = \omega(I_p)$ (this means that $\Delta I = 0$) and $\psi_{mn} = 2k\pi$ (k is an integer). Then the maximum distance in terms of I for a region bounded by separatrices may again be found from (13.32) by setting $\psi_{mn} = \pi$, i.e.,

$$\mu n V_{mn} = [(\Delta I)_{\phi}^2 / 2] n (d\omega/dI)_{I=I_p} - \mu n V_{mn}$$

and consequently

$$(\Delta I_\phi) = 4 \left[\frac{\mu V_{mn}}{(d\omega/dI)_{I=I_p}} \right]^{1/2} . \qquad (13.36)$$

Using (13.36) we determine the maximum width of the nonlinear resonance:

$$(\Delta\omega)_\phi = (d\omega/dI)_{I=I_p} (\Delta I)_\phi = 4 \left[\frac{\mu V_{mn}}{(d\omega/dI)_{I=I_p}} \right]^{1/2} = \frac{4\Omega_{mn}}{n} , \qquad (13.37)$$

where Ω_{mn} is the frequency of the small oscillations. It follows from the condition $(\Delta i)/I_p \ll 1$ that

$$\frac{(\Delta I)_\phi}{I_p} = 4 \left[\frac{\mu V_{mn}}{(d\omega/dI)_{I=I_p}} I_p^{-2} \right]^{1/2} = 4 \left[\frac{\mu V_{mn}}{\alpha\omega I_p} \right]^{1/2} \ll 1 ;$$

the latter inequality only being valid for $\mu \ll \alpha$. When the asynchronicity ($\alpha \ll \mu^{-1}$) is moderate, we have $(\Delta\omega)_\phi/\omega = 4[\mu\alpha V_{mn}/I_p\omega]^{1/2} \ll 1$. Thus, for $\mu \ll \alpha \ll \mu^{-1}$ the use of a universal Hamiltonian (13.32) for investigating nonlinear resonance is rational because I and ω change very little.

Let us consider the physics of the phase oscillations. As we know, when there is nonlinear resonance the change in amplitude causes a shift in the frequency away from the resonance value, which stabilizes the amplitude of the oscillation. The shift in frequency naturally causes a change in the resonance phase, due to which the amplitude of the oscillation changes again, causing the frequency to return to the resonance value, because the amplitude changes in the opposite direction to the initial shift. All this leads to an isolated resonance. Interaction between resonances occurs through the relationship between the width of the resonance $(\Delta\omega)_\phi$ and the distance to the nearest-neighbor resonance $\Delta = |\omega_{i+1} - \omega_i|$ (Fig. 13.11); this relationship has been called the resonance linkage constant [10] and is introduced

THE NONLINEAR OSCILLATOR

thus[3]

$$s = (\Delta\omega)_\phi/\Delta. \qquad (13.38)$$

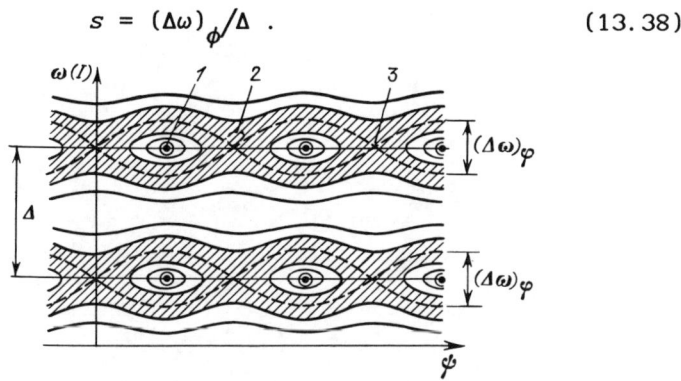

Fig. 13.11. Phase trajectories close to two resonances when $\mu \ll \alpha \ll 1/\mu$ (moderate nonlinearity): 1) neutral stable equilibrium, 2) ergodic layer; 3) unstable equilibrium; lower curves are for the i-th resonance, the upper ones for the $(i + 1)$-th resonance [10]; The dashed lines are the separatrices for the first approximation, which are violated in the subsequent approximations. This leads to the formation of stochastic layers (the shaded regions); Δ is the frequency distance between two resonances; $(\Delta\omega)_\phi$ is the width of the nonlinear resonance.

It should now be clear that an isolated resonance corresponds to $s \ll 1$ and resonance overlap corresponds to s 1. In the latter case, it follows from (13.32) that the averaged motion of the system in an isolated nonlinear resonance on the phase-plane $(\omega(I), \psi)$ is similar to the behavior of an electron in a potential well. Several potential wells (Fig. 13.11) correspond to several resonances. The overlap between resonances means that neighboring wells approach each other, which means the system may be able to leave one well to go into another. During these transitions a new form of instability arises for dynamic systems, that is stochastic instability (Chapters 22 and 23).

---

[3] In I, ψ variables the expression for the resonance linkage constant has the form $\bar{s} = (\Delta I)_\phi/(I_p^m - I_p^n)$, where I_p^m and I_p^n are the resonance values of the action. Strictly speaking, this expression is the one that should be used to find the resonance overlap condition.

CHAPTER 14

PERIODIC SELF-EXCITED OSCILLATIONS

14.1 Definitions

The majority of the dynamic nonlinear systems in technology and in nature are unconservative. Practically every system involves losses (friction, radiation, heating etc.), and usually the system is not energetically isolated: various external forces and fields act upon it, both static and variable. There are several new (with respect to conservative systems) phenomena which arise in dissipative systems in which the oscillation energy is both dissipated due to losses and replenished due to instabilities associated with the nonequilibrium of the system. The most important and remarkable of these is the generation of undamped oscillations whose properties are independent of the initial state of the system, i.e., undamped oscillations that are stable against both external disturbances and changes in the initial conditions. Systems which can generate such oscillations were called self-excited oscillating systems by A.A. Andronov [2] some 50 years ago. He was the first to define them rigorously mathematically by linking self-excited oscillations with limiting Poincare cycles [1].

A limiting cycle is a closed phase trajectory to which all the neighboring trajectories tend and is the image of the nonlinear self-excited oscillation we shall be discussing in this chapter. Self-excited oscillations in dynamic systems need not only be periodic, they may be quasiperiodic or even stochastic (Chapter 22). Therefore we shall give a very general definition.

A self-excited oscillation is undamped oscillation maintained by an external energy source in a nonlinear dissipative system, the form and properties of the self-excited oscillation being determined by this system and being independent of the initial conditions (at least, within finite limits).

Self-excited oscillations differ in principle from other oscillations in dissipative systems in that they are maintained, generally speaking, by an external force that does not need to be periodic. The oscillation of a violin string when the bow is moved evenly across it, the oscillation of current in a radio generator, the oscillation of the air in an organ pipe, a pendulum in a clock are all well-known examples of self-excited oscillations. The simplest self-excited oscillating systems, or self-excited generators, usually contain an oscillating system with damping, an amplifier, and a nonlinear limiter, the feed-back device. This can be seen, for example, in the classical Van der Pol generator (Figs. 14.1a and b). Self-excited oscillations in such a generator become established as follows: a small random oscillation occurring in the LC-circuit through the coil L' controls the anode current of the tube, which amplifies the oscillation in the circuit (given the appropriate positioning of the coils L and L'). When the losses in the circuit are less than the energy introduced into the circuit in this way, the oscillation's amplitude grows. The energy entering the circuit decreases with increasing amplitude due to the nonlinear relationship between the anode current and the voltage of the tube's grid. At some amplitude the oscillation becomes comparable with the losses. As the result, a periodic steady-state oscillation sets in such that the loss of energy is compensated by the anode battery.

Thus, the nonlinearity which governs the input of energy from the source and the losses of energy is essential for the establishment of self-excited oscillations. The frequency characteristics of the source are not important.

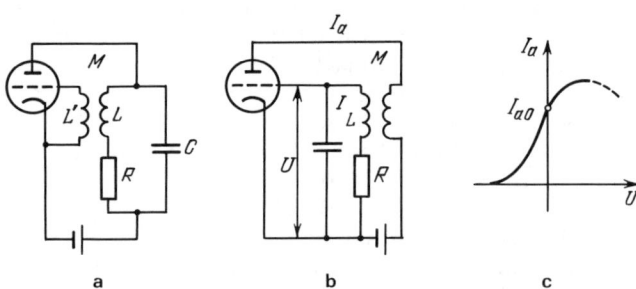

Fig. 14.1. Diagrams of Van der Pol generators: a) with a circuit in the anode chain; b) with the circuit in the grid chain; c) characteristics of the tube approximated by a cubic polynomial.

Self-excited oscillations also differ from fundamental oscillations, whose frequency is determined by the system's parameters and whose amplitude and phase are determined by the initial conditions, and forced oscillations, whose amplitude, phase, and frequency are determined by the external force, in that their amplitude and frequency are solely determined by the system's parameters and are independent of the initial conditions, while the phase is not important. (The characteristics of the source, naturally influence the system's parameters).

The regime we considered with reference to the Van der Pol generator yielded self-excited oscillations without requiring an initial impulse and is called a soft excitation regime. For generators with one degree of freedom, such regime corresponds to the phase-plane diagram in Fig. 14.2a. There are also systems that yield self-excited oscillations but with hard excitation. These are systems in which oscillations grow independently once they have reached a certain initial amplitude. In order to move a system with hard excitation into a regime with steady-state generation an initial disturbance with an amplitude greater than the critical one must be given. The phase-plane diagram of such a generator is given in Fig. 14.2b. It can be seen that in order to leave the trajectories in the stable limiting cycle, the initial point on the phase plane must lie outside the domain containing the stable equilibrium state. This demonstrates the physical meaning of unstable limiting cycles: they are the boundaries between different domains of initial conditions from which the system tends to different stable regimes of motion (on the phase plane such motions correspond to attracting trajectories or attractors, for example, stable equilibria or limiting cycles).[1]

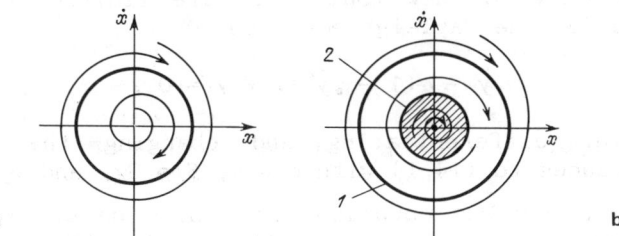

Fig. 14.2. Phase-plane diagrams of self-energized oscillatory systems: a) "soft" excitation; b) "hard" excitation. The initial point on the phase plane must lie outside the shaded area; 1) stable limiting cycle; 2) unstable limiting cycle.

The dimensions of a limiting cycle determine the amplitude of the self-excited oscillating generator, the time needed to circumscribe a cycle (its period), and the form of the limiting cycle (the form of the oscillations). Thus, when

---

[1] Other attractors in crude two-dimensional systems do not exist. In three-dimensional systems more complex attractors are possible and these will be discussed in Chapter 22 on stochastic oscillations in simple systems.

investigating periodic self-excited oscillations in a system we must find the limiting cycles in phase space and determine their parameters. No general method is known to find limiting cycles (for example, to determine the coordinates and type of equilibrium) even for second order systems. True, using the theory of Poincare indices (Chapter 15) we may formulate several criteria for the absence of limiting cycles on the phase plane. For example, if there is no equilibrium in a system, then it may not have limiting cycles, or if there is only one equilibrium and it is a saddle, then there can be no limiting cycles.

14.2 The Van der Pol Generator.
Self-Excited Oscillations as a Function of System Parameters

The Van der Pol generator (a diagram is given in Fig. 14.1b) and its description by the Van der Pol equation

$$\ddot{x} - \alpha(1 - \beta x^2)\dot{x} + \omega_0^2 x = 0 \qquad (14.1)$$

is now, half a century after its appearance, the main model of self-excited oscillation with one degree of freedom. It is relatively easy to show that there are limiting cycles for (14.1) and for the Rayleigh equation

$$\ddot{y} - \sigma(1 - \gamma \dot{y}^2)\dot{y} + y = 0 ,$$

which after differentiating and changing the variables $\dot{y} = x$, reduces to (14.1) with $\alpha = \sigma$, $\beta = 3\gamma$, and $\omega_0 = 1$.

The Van der Pol equation can easily be obtained for a tube generator, for example, with a circuit in the grid circuit, whose basic diagram is shown in Fig. 14.1b. We shall neglect the grid currents. Kirchhoff's laws for the oscillating circuit yield $I = -C\,dU/dt$, and $RI = U - L\,dI/dt - M\,dI_a/dt$. The quantity $-M\,dI_a/dt$ here is the emf introduced into the circuit due to the effect on the circuit of the anode current I_a, which flows through the coil in the anode chain (the term $-M\,dI_a/dt$ may be called the feedback emf). It follows from the above equations that

$$LC\,d^2U/dt^2 - [MS(U) - RC]dU/dt + U = 0 , \qquad (14.3)$$

where $S(U) = dI_a/dU$ is the slope of the tube's characteristic if the anode reaction is neglected. By assuming that the anode current I_a is dependent only on U, we have

$dI_a/dt = (dI_a/dU)dU/dt = S(U)dU/dt$) Equation (14.3) is a nonlinear equation for the tube generator. Assuming, further, that the anode grid characteristic of the lamp can be approximated by the polynomial $I_a = I_{a0} + S_0 U - S_2 U^3$ (Fig. 14.1c), we get $S(u) = S_0 - S_2 u^2$ and (14.3) takes on the form (14.1) where

$$\alpha = (MS_0 - RC)/LC, \quad \beta = 2MS_2/(RC - MS_0), \quad \omega_0^2 = 1/LC. \quad (14.4)$$

The parameter α shows how strongly the generator is disturbed (when $\alpha < 0$ the excitation condition is not fulfilled). The quantity β characterizes the amplitude of the self-excited oscillation, the smaller β the larger the amplitude. Introducing dimensionless variables and the parameters $\tau = \omega_0 t$, $x = \beta^{1/2} U$, $\mu = \alpha \omega_0$, we finally arrive at

$$\ddot{x} - \mu(1 - x^2)\dot{x} + x = 0. \quad (14.5)$$

We now look at the influence of the parameter μ on the form of the limiting cycle. When $\mu = 0$ the system becomes linear conservative. It is natural to expect that for small μ ($\mu \ll 1$) the self-excited oscillations would differ only a little from harmonic ones, the nonlinear friction only choosing the amplitude of the stable limiting cycle. At large μ, the oscillations may be very nonsinusoidal.

One method for finding limiting cycles is the graphical construction of the integral curves in the phase plane, the isocline method. An isocline is a locus of points at which the tangents to the integral curves have the same slope. We write (14.5) in a form $\dot{x} = y$, $\dot{y} = \mu(1 - x^2)y - x$. The equation of the integral curves will be

$$dx/dy = [\mu(1 - x^2)y - x]/y. \quad (14.6)$$

Let the slope of the integral curve at some point $M_0(x_0, y_0)$ be k, i.e., $(dy/dx)_{M_0} = k$. Then we obtain from (14.6)

$$k = [\mu(1 - x^2)y - x]/y. \quad (14.7)$$

If $x = 0$, we have $k = \mu$, i.e., the y-axis intersects the integral curves at some angle larger than μ. When $y = 0$, the tangents to the integral curves are vertical. By setting k to different values, we obtain the equations for different isoclines $y = x/[\mu(1 - x^2) - k]$ from (14.7). By constructing a family of isoclines, it is possible to construct the integral curves, and consequently the phase trajectories. The phase-plane diagrams obtained by this method for (14.5) are shown in Fig. 14.3 for different values of μ. Figure 14.4 contains oscillograms illustrating the establishment and form of self-excited oscillations in a system.

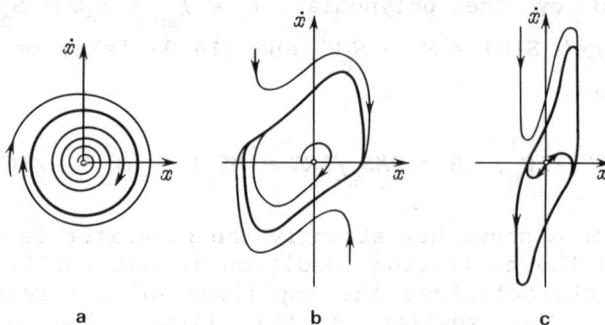

Fig. 14.3. Phase-plane diagrams corresponding to (14.5) for various values of the nonlinearity parameter μ: a) quasiharmonic oscillations ($\mu = 0.1$); b) strongly nonsinusoidal oscillations ($\mu = 1$); c) relaxation oscillations ($\mu = 10$).

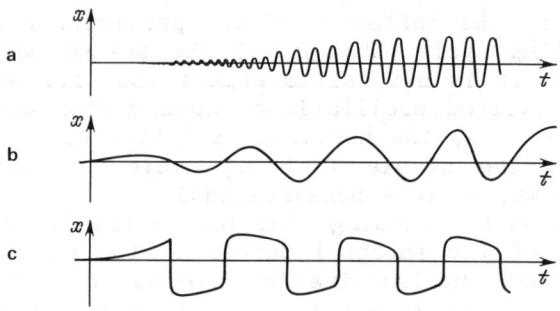

Fig. 14.4. Oscillograms illustrating the establishment and form of self-excited oscillations in a system described by (14.5). They correspond to the phase-plane diagrams in Fig. 14.3: a) $\mu = 0.1$; b) $\mu = 1$; c) $\mu = 10$.

The limiting cycles in Fig. 14.3 contain an internal critical point. For $\mu = 0.1$ and $\mu = 1$ this point is an unstable focus, and for $\mu = 10$ it is an unstable node. The form of the self-excited oscillations in this case changes from quasisinusoidal to relaxational. The quantity μ characterizes the amount of nonlinearity in a system: the larger the nonlinearity the greater the oscillation form differs from sinusoidal. In the physics literature, the quantity μ is sometimes called the strength of the limiting cycle; at small μ trajectories tend to the cycle only weakly, whereas at $\mu \gg 1$ this attraction is significant, i.e., the cycle is strong. When $\mu \ll 1$ and $\mu \gg 1$ it is simply to solve problems concerning self-excited oscillations using approximating analytical methods [3, 5].

14.3 Relaxational Self-Excited Oscillations. Fast and Slow Motions

When the nonlinearity is strong ($\mu \gg 1$) the oscillations become relaxational, which corresponds to sections of fast and slow motion. In order to find these discontinuous oscillations, Mandel'shtam and Papaleksi suggested using the step hypothesis. This means that at the step, the energy changes continuously. Let us consider Rayleigh's equation by way of example

$$d^2y/dt^2 - \varepsilon[1 - (dy/dt)^2]dy/dt + y = 0 ,$$

where ε is large. We introduce a new time $\tau = t/\varepsilon$ and variable $x = y/\varepsilon$, which therefore transform the parameter ε into a coefficient of a higher derivative:

$$\varepsilon^{-2}\dddot{x} - (1 - \dot{x}^2)\dot{x} + x = 0 . \tag{14.8}$$

Now the higher derivative will have a small parameter $\mu = 1/\varepsilon^2$. We shall attempt to define an asymptotic form for the solution of (14.8) as $\mu \to 0$. We write (14.8) in the form of a system of two first-order equations

$$\dot{x} = y , \quad \mu\dot{y} = (1 - y^2)y - x . \tag{14.9}$$

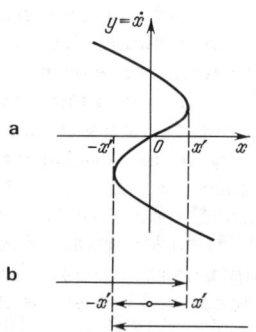

Fig. 14.5. The phase space for the system described by (14.9) when $\mu = 0$: a) the function $y = y(x)$ defines motion along x; b) the direction of motion in phase space for the interval $-x' < x < x'$.

Note that when $\mu = 0$ the phase space of the system will be a straight line x, movement along which being determined by the form of function $y = f(x)$, as shown in Fig. 14.5a. Because this function does not have unique values, the direction of motion in the interval $-x' < x < x'$ is not uniquely defined (see Fig. 14.5b). In other words the system has a dynamic contradiction in that the unique equilibrium state at $x = 0$ is unstable, but that it is uncertain to where the system will move from the points $x = \pm x'$.

We shall attempt to eliminate this contradiction by assuming the parameter μ has a finite value even when small. The equation for the integral curves of (14.9) have the form

$$dy/dx = [(2 - y^2)y - x]/\mu y .$$

As $\mu \to 0$ beyond the line $x = (1 - y^2)y$, the derivative $dy/dx \to \infty$ or $dx/dy \to 0$. The straight lines $x = $ const will be integral curves and the direction of motion along them will be defined by the second equation in system (14.9). It follows from the latter that the velocity of motion as $\mu \to 0$ will be very large. This is the "fast" motion. "Slow" motion occurs on the line $y(1 - y^2) = x$ itself; the motion being defined by the first equation in system (14.9). The phase-plane diagram is given in Fig. 14.6a. The upper and lower sections of the curve for the slow motions are stable with respect to fast motion, and the middle section is unstable.

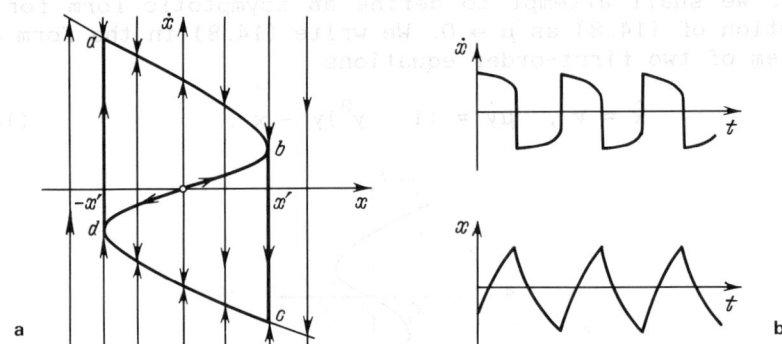

Fig. 14.6. a) Phase-plane diagram and b) shape of relaxation oscillations.

At the points $x = \pm x'$, there are "jumps" from one section of the curve $y(x)$ to another. The system leaves the limiting cycle abcd, consisting of the sections of the "fast" and "slow" motions, whatsoever the initial conditions. The system will, therefore, undergo relaxational oscillations whose form is graphed in Fig. 14.6b. The period of the oscillations T may be found by calculating the time taken to move along the limiting cycle [5]. The times for the fast motions may be neglected. We find the following from the equations for the slow motions $\dot{x} = y$, $(1 - y^2)y = x$

$$\frac{dx}{dt} = \frac{d[(1-y^2)y]}{y},$$

$$T = 2 \int_{y_a}^{y_b} \frac{d(u-y^3)}{y} = 2\left(\ln y - \frac{3}{2} y^2\right)\bigg|_{y_a}^{y_b}.$$

Thus, an account of a small parameter is significant when investigating the system's dynamics. As to whether this is always the case, it is clear that if the introduction of a small parameter does not increase the order of the equations then there is no point in taking it into account if the system is in some sense stable (Chapter 15). However, even if the order of the equation is increased, then it is insignificant when the whole of the curve of the slow motions is stable with respect to fast motions. The tracing point on the phase plane moves very quickly to the small neighborhood (order μ) of the curve of the slow motions, and the system's dynamics are determined solely by the slow motions [4]. An analytical condition is easily obtained. The general form of the system of two first-order equations with a small parameter μ besides the derivative is $\mu \dot{x} = P(x, y)$, $\dot{y} = Q(x, y)$. The equation for the fast motions has the form $y = $ const, $\mu \dot{x} = P(x, y)$; and the equation for the slow motions have the form $P(x, y) = 0$, $\dot{y} = Q(x, y)$. The curve for the slow motions $P(x, y) = 0$ is the locus of the equilibrium state for the fast motions. Clearly, all the sections of this curve will be stable with respect to the fast motions if $dP(x, y) < 0$ for all the points of the curve.

In conclusion, we note that because practically all the experience of the classical theory (at least for systems with small nonlinearities) has been associated with self-excited oscillation on the phase plane, the possibility of establishing periodic motions that correspond to the limiting cycle is associated exclusively with damped oscillations that are due entirely to a nonperiodic energy source. Even a few years ago no one could decide whether to call a nonlinear oscillator with friction affected by a periodic force

$$\ddot{x} + \gamma \dot{x} - \alpha x(1 - x^2) = f \sin \omega t \qquad (14.10)$$

a self-excited regenerator. However it is one, as it demonstrates undamped oscillations whose parameters (intensity, frequency, spectrum, etc.) are independent of finite changes in the initial conditions and weakly dependent on changes in the external force. For instance, in the $\ddot{x}, \dot{x}, t$ phase space of a nonautonomous system described by (14.10) there are stable periodic motions which, if considered stroboscopically with the period of the external force, correspond (in a Poincare mapping) to a stable stationary point.

Vigorous investigations of nonlinear dissipative systems with three-dimensional phase-plane diagrams have enabled us in recent years to find a completely new class of self-excited oscillatory systems. These are the self-excited generators of noise.

They are dissipative systems that undergo undamped chaotic oscillations, which have a continuous spectrum due to energy from non-noise sources. It is interesting that an oscillator as familiar to us as (14.10) is a self-excited generator of noise over a wide range of parameters. The discovery of stochastic self-excited oscillations is, perhaps, the most brilliant achievement of the modern theory. They have only appeared now because from Poincare's time until relatively recently the limiting cycle was the only example of a nontrivial attracting set, i.e., attractors in the phase space of a nonlinear dissipative system. True, complex multiloop limiting cycles had been observed quite a long time ago, and stable multiperiodic motions had been established when investigating the synchronization between self-excited generators.

The observation of complex limiting cycles and then bifurcation, which indicated that they would be further complications, should have lead to a widening of the concept of self-excited oscillation. In fact, this only occurred later, when the results of numerical experiments appeared proving the existence of "aperiodic phase flows" in dissipative nonequilibrial systems [6]. At practically the same time a new mathematical object, complex attractors, which Ruel and Takens called strange attractors, appeared in the abstract theory of dynamic systems.

An example of a strange attractor is a set in which there are no stable trajectories but all the trajectories are complex and confused. An attracting structure may be composed of saddle cycles (when all the trajectories uncoil around them tending to the cycles with these structures).

It is fascinating that now the new viewpoint on stochastic self-excited oscillations has been found, they are observed in very simple classical systems, for example coupled self-excited generators or relaxational generators with one and a half degrees of freedom. They are found because we now know what to look for.

CHAPTER 15

GENERAL PROPERTIES OF NONLINEAR
DYNAMIC SYSTEMS IN PHASE SPACE

15.1 Basic Types of Trajectory.
Fundamentals of Dynamic Systems (Structural Stability)

We begin with a system with one degree of freedom. Such a system is described by a second-order equation and may be completely investigated qualitatively by analyzing the behavior of the trajectory on the phase plane [1-6].

We have become familiar with the phase-plane diagrams of linear oscillators without friction (the equilibrium is either a focus or a saddle), linear oscillators with small damping (the equilibrium is a focus), and linear oscillators with large damping (the equilibrium is a node). Linear oscillators were discussed in detail in Chapter 1, but to make the presentation easier we list all the possible phase-plane diagrams of linear autonomous systems again; they are shown in Figs. 15.1a-d. We begin our investigation of nonlinear systems in the next two chapters with a discussion of the dynamics of a nonlinear oscillator and simple models of self-excited oscillation. Their fairly complicated phase-plane diagrams are also given in Fig. 15.1, which collates all the information we need at present.

The equation of a nonlinear oscillator is $\ddot{x} + f(x) = 0$ and as we mentioned in Chapter 13, it can be integrated to find an analytical expression for $x(t)$.

If there is damping in the system, i.e., the equation has the form $\ddot{x} + h\dot{x} + f(x) = 0$, then it is fairly difficult to find an analytical solution. However, it is clear from physical considerations that when the damping is small, a center-type equilibrium must be a focus. The corresponding phase-plane diagram is given in Fig. 15.2.

Now let us consider what will happen if the circuit is linear and the damping is nonlinear. Let, for example, the circuit have a nonlinear conductance (Fig. 15.3a). If the damping is sign constant, then oscillation in this nonlinear circuit will be little different from that of a linear oscillator with friction, and only the velocity at which the trace point approaches the equilibrium state will be changed.

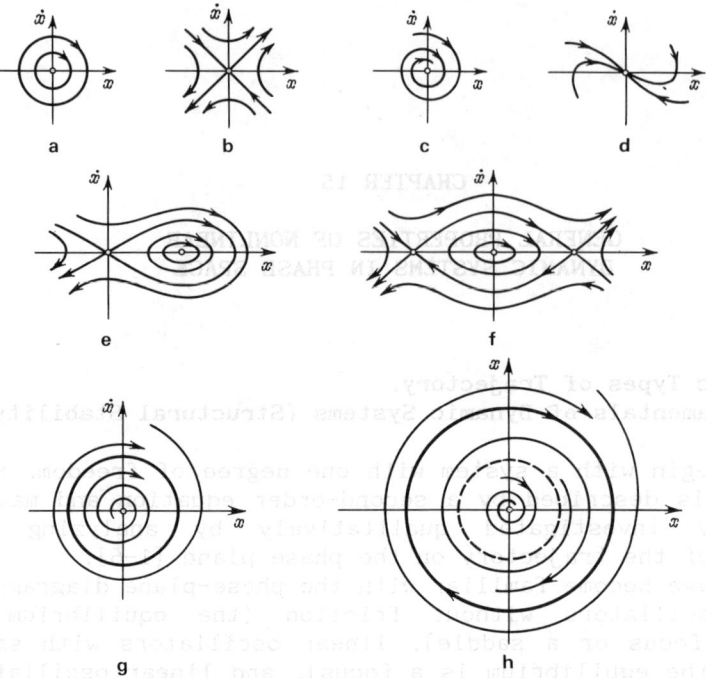

Fig. 15.1. Phase-plane diagrams of linear and nonlinear oscillators. The linear oscillators are: a) $\ddot{x} + \omega_0^2 x = 0$, the equilibrium is center; b) $\ddot{x} - a^2 x = 0$, the equilibrium is a saddle; c) $\ddot{x} + 2\gamma\dot{x} + \omega_0^2 x = 0$, $\gamma^2 < \omega_0^2$, the equilibrium is a focus; d) $\ddot{x} + 2\gamma\dot{x} + \omega_0^2 x = 0$, $\gamma^2 > \omega_0^2$ the equilibrium is a node (all the equilibria are at the origin). The nonlinear oscillators are: e) $\ddot{x} - x(1\ x/2) = 0$, the equilibria are saddle and a center; f) $\ddot{x} + \sin x = 0$, the equilibria are a saddle, a center, and saddle; g, h) self-excited oscillating systems.

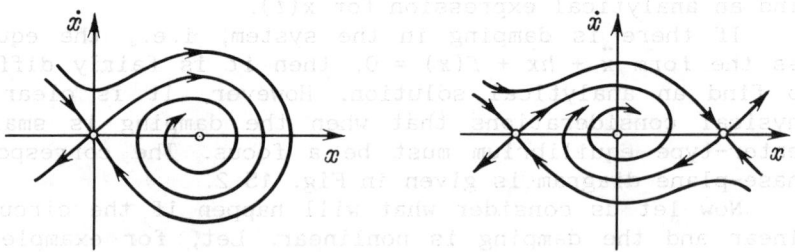

Fig. 15.2. Phase-plane diagrams with unconservative nonlinear oscillators with small dissipation ($\ddot{x} + h\dot{x} + f(x) = 0$, the equilibria are a saddle and a focus).

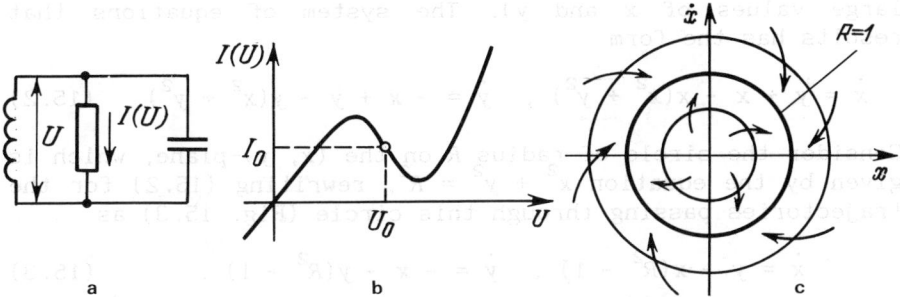

Fig. 15.3. a) Diagram of a linear circuit with nonlinear conductivity.
b) The current-voltage characteristic of a tunneling diode.
c) The circle $x^2 + y^2 = R^2$ and the trajectories passing through it ($R = 1$ is the trajectory of the limiting cycle; $R < 1$ is the trajectory leaving the circle R; and $R > 1$ is the trajectory entering the circle of radius R).

To see whether anything new arises when the damping is not sign constant, consider the example of a circuit with a tunnel diode, with which we are already familiar, and whose characteristics are given in Fig. 15.3b. If the working point is chosen in the falling section, then the characteristic may be approximated by the polynomial

$$I(U) = I_0 - g(U - U_0) + \alpha(U - U_0)^3.$$

Oscillation in a circuit with this conductivity will be described by the equation

$$\ddot{x} - \varepsilon\dot{x}(1 - x^2) + x = 0 \qquad (15.1)$$

or

$$\dot{x} = y, \quad \dot{y} = -x + \varepsilon(1 - x^2)y.$$

When x is large, this is the equation of an oscillator with nonlinear damping and the equilibrium in this system is unstable. In general, it is not possible to find an analytical solution to (15.1), but it is possible to investigate it qualitatively. We have seen that in such a system there is an isolated closed trajectory. This is a limiting cycle and corresponds to periodic self-excited oscillations, which we discussed in the last chapter.

We shall try to construct a model like (15.1), but one that is more convenient for analysis. To do this, we introduce "instability" into the linear system of equations $\dot{x} = y$, $\dot{y} = -x$ by adding terms x and y to the right-hand sides (the instability will obviously appear at small values of x and y), and introduce "damping" by adding the terms $-x(x^2 + y^2)$ and $-y(x^2 + y^2)$ (the damping will appear at large values of x and y). The system of equations that results has the form

$$\dot{x} = y + x - x(x^2 + y^2), \quad \dot{y} = -x + y - y(x^2 + y^2) \quad (15.2)$$

Consider the circle of radius R on the (x, y)-plane, which is given by the equation $x^2 + y^2 = R^2$, rewriting (15.2) for the trajectories passing through this circle (Fig. 15.3) as

$$\dot{x} = y - x(R^2 - 1), \quad \dot{y} = -x - y(R^2 - 1). \quad (15.3)$$

We find the integrals of motion for this system. We multiply the first equation by x and the second by $-y$ and multiply:

$$dR^2/dt = -2R^2(R^2 - 1). \quad (15.4)$$

It follows directly from (15.4) that there is a periodic solution corresponding to $R = 1$, i.e.,

$$x^2 + y^2 = 1, \quad x = \text{const}(t - t_0), \quad y = \sin(t - t_0) \quad (15.5)$$

The quantity $R^2 = x^2 + y^2$ characterizes the oscillations amplitude. If $R < 1$ then $dR/dt > 0$, and the R on the trajectory grows; if, however, $R > 1$, then $dR/dt < 0$ and all the trajectories leave the interior of the circle with radius R. When $R = 1$ we have $dR/dt = 0$, and (15.5) is an exact solution of (15.4). Thus, the circle $x^2 + y^2 = 1$ on the phase plane is a closed phase trajectory to which all neighboring trajectories tend, i.e., it is a limiting cycle (Fig. 15.3c). Let us see why this limiting cycle is stable. For $R > 1$ all the trajectories move into the region bounded by the radius R, but once inside the region the equilibrium (at the origin) is unstable and hence the trajectories enter the region but have nowhere to go except to coil around the limiting cycle (Fig. 15.3c).

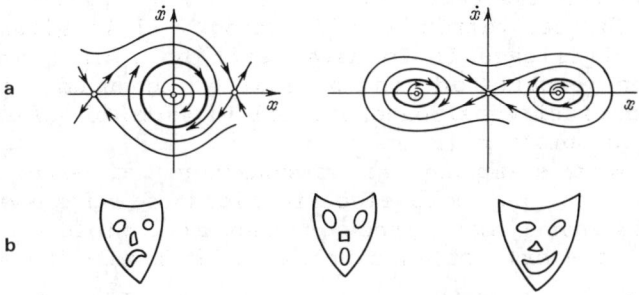

Fig. 15.4. a) Phase-plane diagrams of systems in which there are nonlinear elements other than nonlinear conductivity. b) Examples of topologically similar charts on a plane.

If an self-excited oscillating system has nonlinear elements other than nonlinear conductivity, e.g., nonlinear capacitance or inductance, then the phase-plane diagram will look something like those in Fig. 15.4a.

It is not necessary to know all the phase trajectories in order to give a qualitative description of the behavior of a system with one degree of freedom. It is sufficient to know the critical information, viz., a) the equilibria states, b) the separatrices of a saddle, c) the limiting cycles. When the positions of these are known, we may sketch the phase-plane diagram of any dynamic system, if only crudely.

The concept of a crude dynamic system was first introduced by Andronov and Pontryagin. The phase-plane diagram of a crude system is topologically unchanged by small changes in the system parameters. Informally, topological equivalence means that a diagram on the phase plane is qualitatively unchanged, i.e., it retains all its basic elements and all the basic relations. If there is a limiting cycle or the equilibrium is unstable on the phase plane, then a crude system will retain one cycle and one unstable equilibrium for all changes in its parameters. Examples of topologically equivalent charts are given in Fig. 15.4b. Mathematically, the concept of crudity can be defined (see [16]) for systems of two first-order equations of the form $\dot{x} = P(x, y)$ and $\dot{y} = Q(x, y)$ thus: a dynamic system is crude if there exists a small $\delta > 0$ such that all dynamic systems described by the equations

$$\dot{x} = P(x, y) + p(x, y) , \quad \dot{y} = Q(x, y) + q(x, y) ,$$

where $p(x, y)$ and $q(x, y)$ are analytical functions satisfying

$$|p(x, y)| + |q(x, y)| + |\partial p/\partial x| + |\partial p/\partial y| +$$
$$+ |\partial q/\partial x| + |\partial q/\partial y| < \delta$$

whose partitions of the phase plane by the trajectories have topologically identical structures. Thus, the concept of crudity is introduced as the mathematical form for the qualitative invariance of the moving system given small changes in its parameters. For certain changes the system will, however, cease to be crude, and such parameter changes are called bifurcations. A bifurcation is when a system acquires qualitatively new dynamics due to a small change in the system's parameters [17].

15.2 Basic Bifurcations on a Plane. Poincare Indices

A bifurcation is the mathematical expression for a restructuring of the way a system's (physical, chemical, etc.) movements are changed. The mathematical definition of of a bifurcation is based on the concept of the topological equivalence of dynamic systems. According to [17], for example, two systems are topologically equivalent if a motion in one can be transformed into a motion in the other via a continuous substitution of coordinates and time. Take for example the phase-plane diagram, which at first sight seem very different. By introducing a new system of coordinates, however, we can transform one into the other, i.e., a transition from the phase-plane diagram in Fig. 1.3 to the phase plane diagram in Fig. 1.4 is not a bifurcation since a bifurcation requires a transition from one system into a topologically nonequivalent one.

The phase plane of crude dynamic systems may only contain simple equilibria, such as foci, node and saddles, and attracting or repelling closed phased trajectories, i.e., stable or unstable limiting cycles.

Let us consider the simplest bifurcations of an autonomous system on the phase plane which arise when the system parameters are changed. The simplest bifurcations correspond to transitions through "noncrude systems with the first degree of noncrudity", when only one of the trajectories forbidden to crude systems appeared. These transitions are: a) a saddle-node equilibrium; b) a complex focus; c) a separatrix going from a saddle to the same saddle (loop separatrix) or to a different saddle; d) a double limiting cycle.

We shall look at these bifurcations in more detail. Suppose the change in a systems equilibrium occurs as the result of a change in some parameter α. The bifurcation value of the parameter we shall denote α_0. A type-one bifurcation is shown in Fig. 15.5a. When the parameter value is $\alpha < \alpha_0$ two equilibria exist in the system: a saddle and a node. When $\alpha = \alpha_0$ they merge, and a complex saddle-node critical point is formed. When the parameter α is increased, the equilibrium state disappears.

A type-two bifurcation is shown in Fig. 15.5b. The equilibrium (focus) looses its stability and a stable limiting cycle is born.

The third type of bifurcation is illustrated in Figs. 15.5c and d. In Fig. 15.5d a limiting cycle ($\alpha > \alpha_0$) is born from a separatrix loop ($\alpha = \alpha_0$). The formation of a double loop from what is called a merging of phase trajectories is shown in Fig. 15.5e. This cycle ($\alpha = \alpha_0$) is "semi-stable" in that all the phase trajectories inside the loop move away from it and all those outside the loop move towards it.

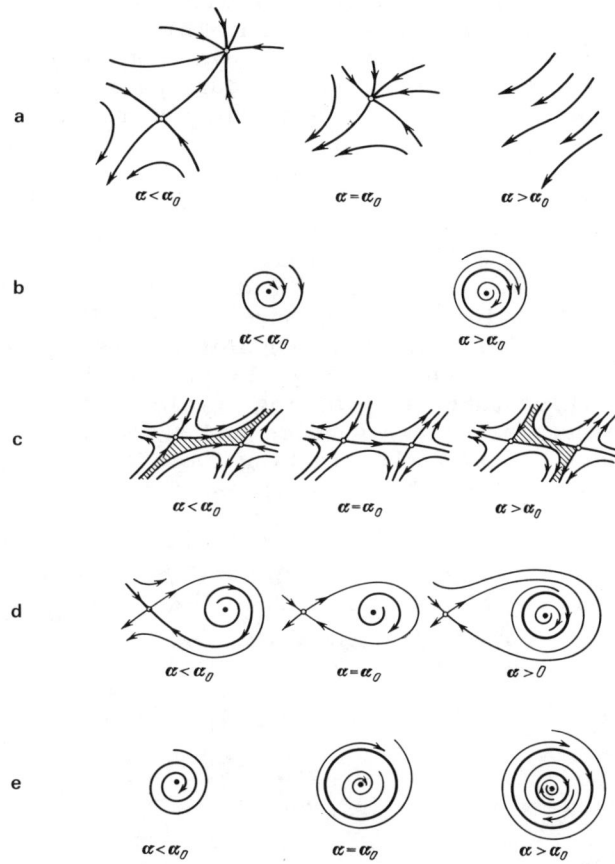

Fig. 15.5. a-d) Simple bifurcation of an autonomous system on the phase plane.
e) Birth of a stable and unstable cycle from a merging of phase trajectories.

Thus, we need to know the equilibrium states, saddle separatrices, and limiting cycles in order to construct the diagram of a dynamic system on the phase plane. If the parameters are varied, then we may always know how the phase-plane diagram will change. By knowing which bifurcations are possible, we can determine the qualitative changes that are possible in the phase-plane diagram. By drawing the phase plane, we can see which motions are possible, e.g., finite ones moving to infinity, those moving to a stable equilibrium etc.

Let us conclude this section with a consideration of the laws for the combined existence of various types of equilibrium and closed trajectories. Let there be a vector field in a plane (Fig. 15.6a). We draw a closed contour that does not pass through the equilibrium state. If we take a point S on this contour and move it along the contour, then the vector field passing through this point will be continuously rotated. When the point has completed one cycle, the vector field will have turned through an angle $2\pi j$, where j is an integer. The vector's direction of rotation will be considered positive if it coincides with the direction of motion of S. The integer j is called the Poincare index of the contour. For the contour in Fig. 15.6a, $j = 0$. If the equilibrium state is encircled by a closed contour, then it can be seen that the Poincare index for a center, node, or focus is +1, while that for a saddle is -1 (Figs. 15.6b and c). For a limiting cycle we have $j = +1$. The index of a closed curve containing several singular points is the sum of the indices for these points (Fig. 15.6d). It is immediately clear from this that a limiting cycle containing, for example, two saddles or two saddles and a focus cannot exist because we have $j = +1$ for a limiting cycle and yet for three equilibria the sum of the indices is $j = -1$ (Fig. 15.6e).

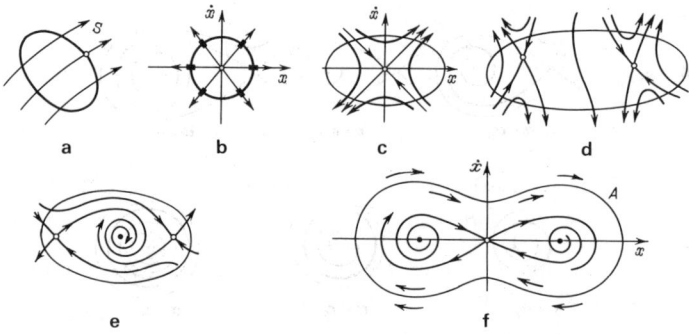

Fig. 15.6. An explanation of Poincare indices for a closed curve encircling one or more equilibrium points: a) $j = 0$ (no equilibrium points inside the curve); b) $j = +1$, a center (or a node of focus, which is the same thing); c) $j = -1$, a saddle; d) $j = -2$ ($j = -1 - 1 = -2$); e) $j = -1$ ($j = -1 + 1 - 1 = -1$); f) $j = +1$ ($j = -1 + 1 + 1 = +1$); A) limiting cycle.

Thus, a limiting cycle containing a saddle and two foci may exist (Fig. 15.6f) because in this case the sum of j is $-1 + 1 + 1 = +1$. Two equilibria (e.g., a saddle and a node) as shown in Fig. 15.5a, may merge and disappear because their combined index is $j = 0$, while the three equilibria in Fig. 15.6e may not disappear. Thus, we may use the theory of Poincare indices to state the following:

1. At least one critical point must exist inside a closed phase trajectory because the index of this trajectory is +1, whilst the index of a closed curve inside which no singular points exist is zero.

2. If only one singular point exists inside a closed phase trajectory, then it may not be a saddle, it must be a point with index +1.

3. If several simple singular points exist inside a closed phase trajectory, then there must always be an odd number of them, and the number of saddles must be one less than the number of the remaining singular points.

15.3 Point Transformations

A very convenient method for analyzing nonlinear dynamic systems is that of point transformations or the method of Poincare mapping [6]. This method can be used to reduce the effective dimension of a phase space and is especially productive for numerical experiments.

We shall be interested in the behavior of a trajectory in some region of phase space and reduce some surface Σ in which all, or nearly all, the trajectories of interest intersect. This surface will be called the secant. If a trajectory does not leave the region being investigated, then it will pass through the secant surface an even number of times (Fig. 15.7). The function S which links the coordinates x_{k-1} of the $(k-1)$-th intersection between the trajectory and the surface Σ with the coordinates x_k of the next intersection is called the sequence function:

$$x_k = S(x_{k-1}) . \qquad (15.6)$$

Because the x variable is only fixed at discrete times, this is a difference, and not a differential, equation. There is a completely determined mapping for each phase stream (i.e., a dynamic system describable by a differential equation) (15.6). If the stream is three-dimensional, then the mapping will be two-dimensional and the vectors x_k and x_{k+1} will only have two coordinates. If the stream is two-dimensional, then the secant will simply be a line and the mapping will be one-dimensional. The reduction of the space dimensionality by one considerably simplifies the investigation for three-dimensional systems and we arrive at a phase plane with which we are familiar. For example, Fig. 15.8b shows what a saddle limiting cycle and the trajectory close to it appear like in phase space. This corresponds to an unstable stationary point on the Σ surface.

A stable stationary point on the secant surface corresponds to a stable limiting cycle (Fig. 15.8a).

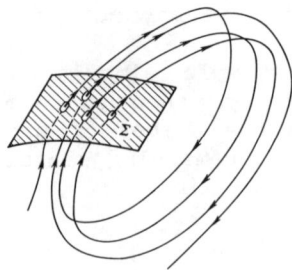

Fig. 15.7. Cross-section of a phase stream in a three-dimensional space: Σ is the secant surface, which the phase trajectories do not touch.

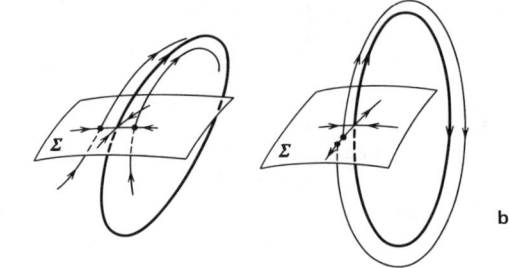

Fig. 15.8. a) A stable and b) a saddle stationary point on a secant corresponding to a stable and saddle cycle.

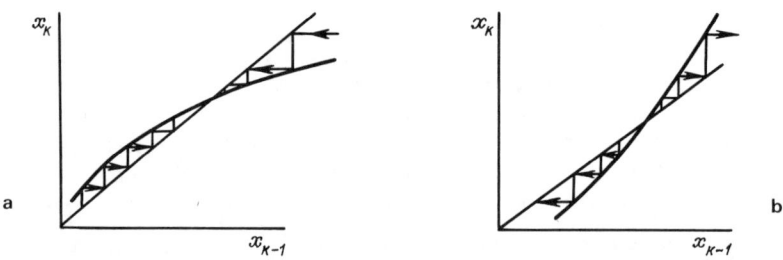

Fig. 15.9. Lamere diagrams for a) a stable and b) an unstable periodic motion.

The stability of the stationary point for a one-dimensional mapping can easily be illustrated with a diagram (a Lamere diagram) which illustrates the sequence (15.6). We plot the curve of $x_k = S(x_{k-1})$ on the (x_k, x_{k-1}) plane, and the stable point is where this curve intersects the straight line $x_k = x_{k-1}$ (Fig. 15.9). The Lamere ladder allows us to determine the stability of the stationary point. In Fig. 15.9a the ladder leads to a stable stationary point, for which $|dx_k/dk_{k-1}| < 1$; in Fig. 15.9b the ladder leads away from the stationary point, where $|dx_k/dk_{k-1}| > 1$, which means it is unstable.

Let us consider a general n-dimensional case. We choose the coordinate system $\xi = (\xi_1, \xi_2, \ldots, \xi_{n-1})$ on Σ. Then the Poincare mapping, which is determined by the system's trajectories, will link the coordinates of the k-th point of the intersection between the trajectories and Σ to the coordinates of the $(k + 1)$-th intersection point, i.e.,

$$\xi_{k+1} = F(\xi_k, \mu) , \qquad (15.7)$$

where μ_* is a parameter of the dynamic system. A stationary point $\xi^* = F(\xi^*, \mu)$ corresponds to a periodic motion in this mapping, while a torus corresponds to motion about an unclosed coil of a torus with a dimension one less. The stability of a periodic motion, i.e., a stationary point, is determined in the linear approximation by the eigenvalues of the matrix $B(\mu) = \partial F(\xi^*, \mu)/\partial \xi$, i.e., the roots of the characteristic equation $\det(B(\mu) - \lambda E) = 0$, where E is the unit matrix. These eigenvalues ρ_i are called the multiplicators (the name indicates the meaning, for a multiplicator is a transfer coefficient for a small disturbance on the Σ of one period). One of the multiplicators in an autonomous system corresponds to the evolution of the perturbation along the periodic trajectory and is always unity, i.e., the number of multiplicators that are of value from the point of view of an analysis of bifurcation in a system with an n-dimensional phase space will be $(n - 1)$.

If all $(n - 1)$ multiplicators have absolute values less than unity, i.e., they lie inside a unit circle on the complex plane, then all the perturbations will decrease in each step (orbit around the perturbation trajectory) and the periodic motion will be stable. If, however, even one of the multiplicators lies outside a unit circle, then the motion will be unstable. Thus, the bifurcation of a periodic motion occurs when one of the multiplicators passes through the unit circle.

We emphasize that because we have been discussing small perturbations against the background of periodic motion, the perturbations can be described by linear equations with periodic coefficients. Floquet's theorem (Chapter 11) $u(t) = \Phi(t)\exp(\Lambda t)$, where $\Phi(t)$ is a periodic matrix with period T, is valid for the fundamental matrix of the solutions $u(t)$ or this equation. The eigenvalues (λ_i) of matrix Λ are called characteristic indicators. The multiplicators are the eigenvalues of the matrix $\exp(\Lambda T)$, i.e., they are related to the characteristic indicators by the formula $\lambda_i = (\ln \rho_i)/T$.

15.4 Bifurcation of Periodic Motions

We have so far considered only one bifurcation of a periodic motion, the one corresponding to the birth (during a parameter change) of a limiting cycle from an equilibrium state (or when the parameter is changed in the reverse direction, the limiting cycle merges into an equilibrium state and thus disappears). This is the way a periodic regime arises or disappears in a Van der Pole generator when the feedback coefficient is increased.

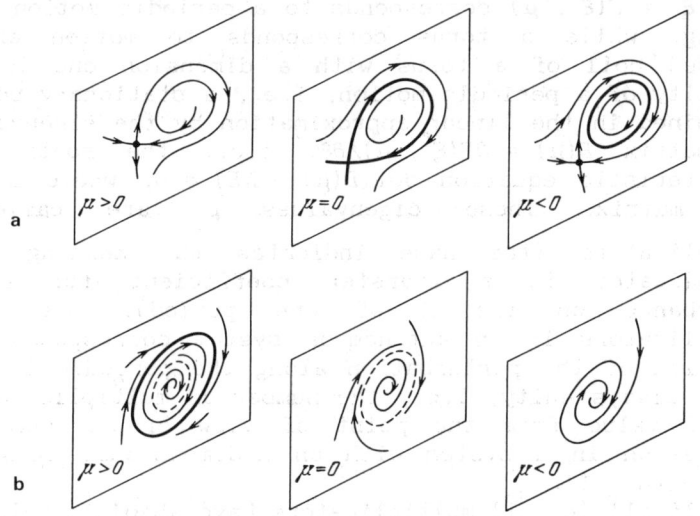

Fig. 15.10. Birth and death of a limiting cycle: a) from a saddle separatrix; b) from a merging of phase trajectories.

Two more complicated bifurcations of periodic regimes are encountered in systems with one degree of freedom: a) The birth of a limiting cycle from a saddle separatrix [7] (Fig. 15.10a). This bifurcation is typical for nonlinear oscillators with small perturbations by unconservative additives [12]. b) The birth (or corresponding death) of a pair of cycles (a stable and an unstable ones) from a merging of phase trajectories. This bifurcation is typical for self-excited generators with rigid oscillation regimes (Fig. 15.10b) and in the simplest case these generators are described by an equation of the form

$$\ddot{x} + \alpha(1 - \mu x^2 + \beta x^4)\dot{x} + \omega_0^2 x = 0$$

These bifurcations are also possible on the phase planes of systems with more than two dimensions. Systems with phase spaces whose dimensions are $n \geq 3$ have their own special bifurcations. The main one of these is the birth of an invariant torus from a limiting cycle (Fig. 15.11) and the bifurcations of period doubling. We shall look at these in some detail. In order to do this we use Poincare mapping.

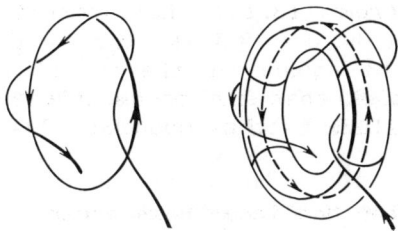

Fig. 15.11. Birth of a stable invariant torus.

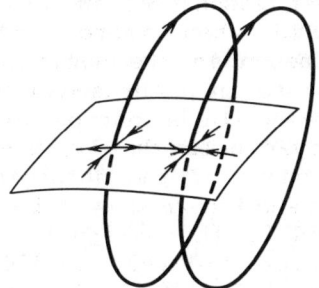

Fig. 15.12. A saddle and stationary point on the Σ plane merge.

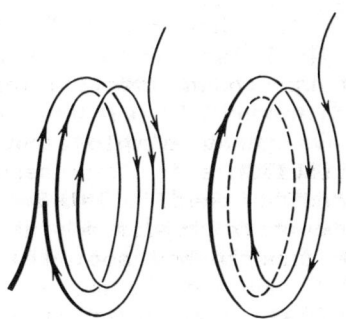

Fig. 15.13. Bifurcation of period doubling.

As we have seen, the bifurcation of a periodic motion is related to the transition of a multiplicator through the unit circle. We shall consider the following bifurcations: *a)* one of the multiplicators becomes equal to (+1); b) one of the multiplicators becomes equal to (-1); c) a pair of multiplicators take the values exp($\pm i\alpha$), where $\alpha \neq 0$, π, $\pi/2$, $2\pi/3$.

The first sort of bifurcation of a periodic motion is very much like the bifurcation of equilibrium states (Fig. 15.5a) in that the disappearance of two equilibria is similar to the merging and disappearance of two cycles. Indeed, they look the same on the secant Σ, the stationary points of the Poincare mapping acting as the equilibrium states (Fig. 15.12).

If one of the multiplicators of a stable periodic motion simply changes sign when the parameter passes through a (-1)-small perturbation over one rotation around the trajectory, then on the next orbit around the disturbed trajectory, it will clearly be closed (Fig. 15.13), i.e., a stable periodic motion with twice the period will have been born from the periodic motion and the original motion will be unstable. The resulting periodic motion may again loose the stability through a doubling period bifurcation when the parameter μ is changed again, etc. We shall say more about sequential bifurcations in Chapter 22 when dealing with the appearance of chaotic behavior in dynamic systems.

When a multiplicator of a periodic motion goes beyond the boundary of the unit circle at the points exp($\pm i\alpha$) when $\alpha \neq 0$, π, $\pi/2$, $2\pi/3$, a two-dimensional invariant torus either appears or disappears from the periodic solution (Fig. 15.11). The motion tends from being periodic into being quasiperiodic. This sort of bifurcation is observed in a system of two coupled self-excited generators when the system leaves a synchronization regime to a beat regime (Chapter 16).

Resonance with the loss of stability corresponds to values $\alpha = 0$, π, $\pi/2$, $2\pi/3$. This resonance is doubly degenerate (with respect to modulus and argument), and it must be investigated in a space with two parameters [10], namely, damping close to the periodic motion and frequency detuning from the resonance (in this case the detuning is the difference between the multiplicator argument and the resonance value of the argument).

Bifurcation that results in the disappearance of a static or periodic regime may cause the system to enter a "chaotic" or "stochastic" regime. Its mathematical form in phase space, called a strange attractor, may be constructed in topologically different ways, which is due, for example, to the manifold of the paths leading to its appearance. These bifurcations will be discussed in Chapter 22.

15.5 Homoclinic Structures

We shall now consider how a system of two nonlinear coupled oscillators behaves

$$\ddot{x}_1 + x_1 = -2\mu x_1 x_2, \quad \ddot{x}_2 + x_2(1 - \mu x_2) = -\mu x_1^2, \qquad (15.8)$$

This was investigated only relatively recently [13] using a detailed numerical simulation. The system is interesting for astrophysics, in particular, because it models the behavior of stars in a galactic field with a potential

$$U(x_1, x_2) = x_1^2/2 + x_1^2 x_2 - x_2^3/3 + x_2^2/2 .$$

when $\mu \ll 1$, the oscillators have a simple, quasiperiodic behavior. The same is true for moderate μ ($\mu \sim 1$), but small initial perturbation energies (Fig. 15.14). Figure 15.14 shows a cross section of the surface $x_1 = 0$ of a trajectory in phase space $(x_1, x_2, \dot{x}_2)$ of (15.8); and this space may be three-dimensional if we take into account the energy integral (for $\mu = 1$), i.e.,

$$\mathcal{E} = (1/2 + x_2)x_1^2 + (1/2 - x_2/3)x_2^2 + (\dot{x}_1^2 + \dot{x}_2^2)/2 .$$

All the trajectories appear to lie on a smooth surface, a torus, i.e., the system for any initial condition is tentatively periodic. If we increase the oscillator's energy, then the motion of the second oscillator first becomes very nonlinear and a motion close to the separatrix of a single nonlinear oscillator appears (cf. Fig. 15.1e). Since there is a forcing force proportional to $x_1^2(t)$, we may not say whether these motions will remain quasiperiodic or whether the type of motion will change, i.e., the trace point will move alternately from inside the separatrix region to outside it.

Numerical experiments with two linked nonlinear oscillators (15.8) and initial energies $\mathcal{E}_0 > 1/12$ are shown in Fig. 15.14. It can be seen from the figure that when the initial energy is raised to $\mathcal{E}_0 = 1/12$, which still corresponds to a simple motion, all but 0.04 of the phase trajectory does not coil on any surface, and seems to wander erratically in the bounded region of the phase space. When $\mathcal{E}_0$ is increased further, the region occupied by the random motions increases and that occupied by the periodic or quasiperiodic motions is compressed (Fig. 15.14b). Thus, motion in the three-dimensional phase space of the coupled nonlinear oscillators may be very complicated.

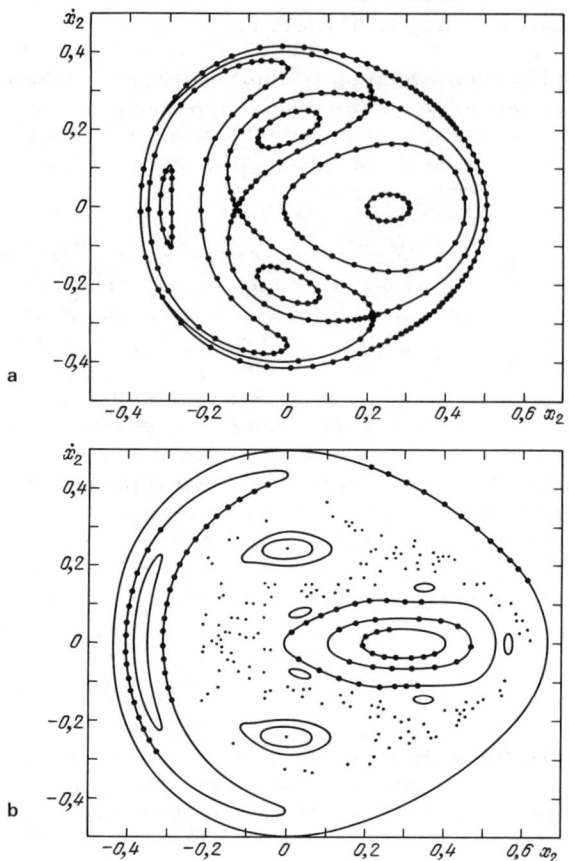

Fig. 15.14. a) Locus of the trajectories on the secant plane $x_1 = 0$ of the system's phase space (15.8) when $\mu = 1$ and $\mathcal{E}_0 < 1/12$. b) Complex motions of a system of two nonlinear oscillators (15.8) with $\mathcal{E}_0 = 0.125$.

In order to see where this complexity arises from, we return to the model of a nonlinear oscillator in a periodic field.

We shall assume that the motion of one of the oscillators is given and harmonic in the model of coupled nonlinear oscillators, i.e.,

$$\ddot{x}_1 - x_1 + x_1^3 = \mu \sin t \,. \qquad (15.9)$$

For $\mu = 0$ we know everything about this oscillator (Fig. 15.1). We shall consider its behavior for $\mu \ll 1$ in a three-dimensional phase space where the third coordinate is time t. Physically, it would seem clear that the qualitative difference between a nonautonomous motion and an autonomous motion will appear when the external force acting on the oscillator at various times moves it into a region with a qualitatively different type of behavior (on the phase plane this different motion will correspond to the regions inside or outside the separatrix). More simply, it can be seen if the sinusoid in (15.9) is substituted by a rectangular periodic momentum. Twice a period the phase-plane diagram (Fig. 15.1e) will be moved first to the left and then to the right by a distance of order μ. These pulsations will be almost indistinct for small-amplitude oscillations and the motions will remain simple. Motions closed to the separatrix, however, will become complex (Chapter 13). This complexity will be due to the presence in the space of system (15.8) of a homoclinic structure [5,6], which was discovered by Poincare during his investigations of the three-body problem in 1889. This structure only arises for spaces with $n \geq 3$ in the neighborhood of a homoclinic trajectory. The appropriate situation is shown in Fig. 15.5 for a three-dimensional case. The trajectory inside this structure was only fully described quite recently [8,14]. It was discovered, for instance, that the structure contains an even set of unstable (saddle) periodic trajectories, between which (for a wide range of initial condition) the oscillator wanders.

This descriptive example illustrates that three-dimensional dynamic systems may differ qualitatively from two-dimensional ones. These differences mainly arise because a finite or infinite number of nonwandering trajectories (which is typical of homoclinic structures) are possible in the phase space of three-dimensional systems. If the number of equilibria and periodic motions in a multi-dimensional system is finite, then its dynamics is very similar to the dynamics of a two-dimensional system, and it will only have simple motions. If there is an infinite set of different periodic motions in the phase space of the system (this includes the example we have just discussed), then the system behavior strongly differs from that of a two-dimensional system.

We shall consider the mechanism by which several properties of homoclinic structures appear. We return to the nonautonomous oscillator (15.9). When $\mu \neq 0$, saddle periodic motion appears (Fig. 15.15). In the x, $\dot{x}$, t phase space it corresponds to a trajectory passing at times $t = 2\pi n$ ($n = \ldots, -1, 0, 1, 2, \ldots$) through the origin. The stable and unstable separatrices will now become surfaces. The surface on which the trajectory tends to a periodic motion is called a stable manifold (W_s). The surface on which it tends away from a periodic motion (or tends to it at $t = -\infty$) is called an unstable manifold (W_u).

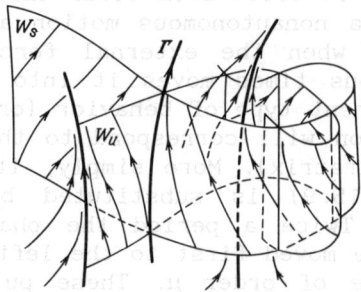

Fig. 15.15. Crude intersection of a stable (W_s) and unstable (W_u) manifold of a saddle periodic motion Γ in $(x, \dot{x}, t)$-space.

We shall describe the behavior of a trajectory using Poincare's mapping. We shall consider the coordinates $(x, \dot{x})$ on the secant plane $t = 2\pi(n + 1)$ to be a function of the coordinates $(x, \dot{x})$ on the plane $t = 2\pi n$. Let us map the plane $t = 2\pi n$ onto $t = 2\pi(n+1)$ and discuss point transformations of the plane $t = \text{const}$ onto itself. This is done by a formula $\xi_{n+1} = F(\xi_n)$, where we have a vector $\xi_n = (x_n, \dot{x}_n)$. If the motion is periodic then $\xi^*_{n+1} = \xi^*_n$ and $\xi^* = F(\xi^*)$, and a stationary point of the mapping ξ^* corresponds to the motion on the secant plane. A saddle stationary point corresponds to a saddle periodic motion on the secant plane, and stable and unstable separatrices correspond to stable and unstable manifolds. (These separatrices consist of the intersection points between the trajectories and the secants $t = \text{const}$). A point tends along a stable separatrix to ξ^* as a result of the infinite sequence of intersections with the secant surface as $n \to +\infty$ but it tends to ξ^* along an unstable separatrix as $n \to -\infty$. The behavior of a separatrix on the secant may be quite different to that on a phase plane. The most important difference is that a separatrix on a secant plane may be intersected (this is explained in Fig. 15.15), the intersection with the separatrix at one point entailing its intersection at an infinite number of points.

We shall look at this feature in more detail. A separatrix on the secant plane consists of the points obtained by a sequential application of the mapping. If the point belongs to two separatrices at once (a stable and unstable one), then all of its images (as $n \to \infty$) and anti-images (as $n \to -\infty$) must also belong to two separatrices at the same time. Consequently, unstable and stable separatrices must have an even set of common points, i.e., intersection points. It is clear that the intersection points of an invariant manifold must become more dense close to a stationary point, where the motion is exponentially slowed. As a result we get a situation such as that in Fig. 15.16 on the secant plane. The separatrix intersection points on the secant belong to a coaxial trajectory, which Poincare called homoclinic. As $t \to \pm \infty$ this trajectory coils and uncoils around the initial periodic motion. The neighborhood of a homoclinic trajectory in phase space is called a homoclinic structure. In such a structure there are an infinite number of different trajectories including periodic and random ones. Trajectories belonging to homoclinic structures were only recently fully described in terms of symbolic dynamics [14].

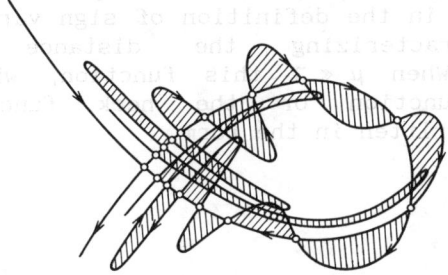

Fig. 15.16. A homoclinic trajectory on the secant plane t = const. The shaded sections of the homoclinic structure map onto each other.

We emphasize once again that separatrices in $(x, \dot{x}, t)$-space are surfaces that intersect along lines. These intersections do not disappear when the parameters of the physical system change, i.e., they are crude. A homoclinic structure is crude.

If we trace the evolution of a small phase volume in the neighborhood of a homoclinic curve, we note that it is deformed over time in a complicated way and that as $t \to \infty$ it is spread overthe whole structure (Fig. 15.16). The local instability of practically all trajectories inside a structure follows from this, i.e., points that are infinitesimally close at $t = 0$ diverge as t grows. This local instability of trajectories contained within a bounded phase volume is the reason for the complexity and confusedness of motions inside a homoclinic structure. These motions in a dynamic system will be considered again in Chapters 22 and 23.

Although the complex behavior of the dynamic system related to the existence of a homoclinic structure was discovered by Poincare, the corresponding image of the structure (Fig. 15.16) appeared much latter [9].

If a homoclinic structure exists in the phase space of a system, then this is almost a guarantee that the system's dynamics will be complex (Chapter 22). We shall present a useful criterion due to V.K. Mel'nikov [9] for the existence of a homoclinic structure for systems close to conservative (15.9). The initial conditions of a nonautonomous oscillator will be considered in the form

$$\dot{x} = p_0(x, y) + \mu p_1(x, y, \omega t, \mu),$$
$$\dot{y} = q_0(x, y) + \mu q_1(x, y, \omega t, \mu). \qquad (15.10)$$

Suppose at $\mu = 0$ in this system a separatrix loop exists on the phase plane (Fig. 15.10). The criterion for a homoclinic structure to appear for $\mu > 0$ in the phase space of (15.10) is to be found in the definition of sign variability in the function characterizing the distance between the separatrices. When $\mu \ll 1$ this function, which is called Mel'nikov's function or the neck function, may be approximately written in the form

$$\Delta_\mu(t_0) =$$

$$= \int_{-\infty}^{\infty} \Big\{ p_1[x_0(t - t_0), y_0(t - t_0)] q_0[x_0(t - t_0), y_0(t - t_0)]$$

$$- q_1[x_0(t - t_0), y_0(t - t_0)] p_0[x_0(t - t_0), y_0(t - t_0)] \Big\}$$

$$\times \exp\Big\{ -\int_0^{t-t_0} \frac{\partial}{\partial x} p_0[x_0(\zeta), y_0(\zeta)]$$

$$+ \frac{\partial}{\partial y} q_0[x_0(\zeta), y_0(\zeta)] \, d\zeta \Big\} \, dt. \qquad (15.11)$$

Here $x_0(t)$ and $y_0(t)$ are the solutions of the undisturbed system corresponding to a separatrix loop, and t_0 is a parameter characterizing the position of a point on this separatrix. We shall encounter actual applications of this criterion in Chapter 23 [11].

CHAPTER 16

SELF-EXCITED OSCILLATIONS IN MULTIFREQUENCY SYSTEMS

16.1 Forced Synchronization

The diagram of a two-circuit generator used by Van der Pol, Andronov and Vitt (see [5,11]) is shown in Fig. 16.1. At the time, the most important effects typical of the interactions between "elementary generators" had been observed, e.g., effects we considered in the previous chapter for the Van der Pol generator [6] such as mode competition, synchronization, and frequency pulling [3,4]. It was interesting that because of the nonlinearity in the Van der Pol generator even the fairly trivial effect of the simultaneous generation of two modes, that is possible when there is weak linkage (the case is typical, for example, for a gas laser with nonhomogeneous broadening of the lines in the spectra of the active material) went unnoticed in the papers by Andronov and Van der Pol [11].

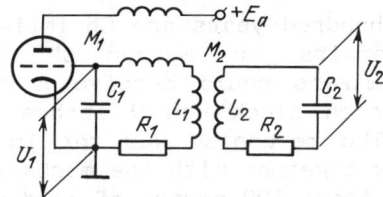

Fig. 16.1. Diagram of two circuit self-excited generator.

Competition, which is observed when there is stronger linkage between several self-excited oscillational modes, is explained by the relationship between the nonlinear damping of one of the modes and the energy of another. If the modes are equivalent and the link goes both ways, then a generation regime becomes established for the mode which dominates first. The dependence on the initial conditions means that in order for the system to go from one regime to another the frequency of one of the modes must be significantly changed, i.e., the detuning must be changed, the value of detuning being different in different directions (hysteresis). The detuning interval in which frequency generation depends on the prehistory is called the pulling interval.

Over the last two decades, interest in the classical effects has returned mainly due to the appearance of active distributed systems (e.g., molecular and optical lasers, and masers with cyclotron resonance), and the creation of systems with large number of active elements. When active devices are united in order to increase throughput or efficiency into ordered structures, the resultant systems become analogous to distributed ones.

Only the character of the dispersion in the resultant "medium" depends on the method of combining the active elements (Gann diodes, avalanche diodes, etc).

We begin a discussion of multifrequency systems with an analysis of the classical effect in nonlinear oscillations, namely the synchronization ("capture") of generator frequencies by an external sinusoidal signal, the frequency of which is close (but does not coincide with) the fundamental frequency of the generator. We shall assume that if completely determinate frequency relationships ("a single rhythm of a united form" [1]) become established when objects of arbitrary nature, considered to be equivalent, interact, then mutual or internal synchronization between the objects will take place.

If, however, one of the objects is so powerful that it forces its own frequency (given and unchanging) on the other self-excited oscillating systems, then we shall say there is external (forced) synchronization or frequency pulling[1].

---

[1] More than three hundred years ago Christian Huygens in his famous "Three Memoirs on Mechanics" [2] showed how unscrupulous watchmakers could deceive a credulous burger: "The pendulum of each clock was 9 inches long and half a pound in weight. The mechanism was set in motion by gears included in the box together with the mechanism. The case was 4 feet long and at least 100 pounds of lead was placed at the bottom to make the whole mechanism retain its vertical position. The following interesting observation could be made with these clocks. Two such clocks were hung on the same beam at rest on two supports. Both pendulum always moved in opposite directions and the oscillation would so exactly coincide that they would never in any way diverge. The tick from both clocks could be heard at the same instant. If this coincidence was artificially disturbed, then the situation would be re-established within a short time on its own. Initially I was amazed by this strange phenomenon, however I finally discovered, after diligent investigation, that the reason lay in very insignificant motions of the beam itself. The pendulum oscillations communicated a small amount of motion to the clocks, no matter how heavy they were. This

We shall give a quantitative theory for the synchronization of self-excited oscillation systems using the example of tube generators, whose main diagram is given in Fig. 16.2. We have already considered one example of how the investigation of a particular nonlinear dynamic system be reduced to numbers, namely self-excited oscillation in a system where we were able to distinguish between fast and slow motions. Formally, this distinction may be done if the equation has a small parameter next to the highest derivative. A small parameter in most cases (not only, of course, for the analysis of self-excited oscillation) allows us to reduce the order of the initial system and integrate in parts for the fast and slow motions. It should be noted that most methods for completely solving nonlinear problems without the numerical application of a computer are associated with the presence in the system of a small parameter, i.e., with the closeness of this system being studied to another, simpler, or rather integrable (if only approximately), system. The other case when it is possible to solve a problem analytically, and one more commonly encountered in physics and other applied areas, is when the initial nonlinear system is close to a linear oscillator, or several oscillators. Here, the solution will be close to a set of sinusoids, although their parameters will obviously not be numbers, they will instead be slowly changing functions of time.

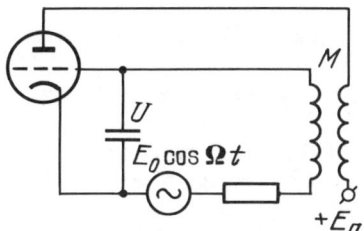

Fig. 16.2. Diagram of a valve generator synchronized with an external signal $E_0 \cos \Omega t$ whose frequency is close to the fundamental frequency of the generator.

---

motion was transferred by the beam and if the pendula were not moving in opposite directions, then they would be forced to do so and only then would the motion of the beam cease. However, this reason was not efficient if both clocks were not particularly homogeneous and consistent from the very beginning." Qualitatively, antiphase stable motion in clocks has rarely been so accurately described. Huygens could not have given a quantitative description because in his time there was no exact formulation of the laws of mechanics.

Let as consider one variant of the averaging method, Van der Pol's method, as applied to the diagram in Fig. 16.2, which is described by the equation

$$\ddot{U} - \alpha(1 - \beta U^2)\dot{U} + \omega_0^2 U = \varepsilon \cos \Omega t, \qquad (16.1)$$

where α, β and ω_0^2 are determined by (14.4), and $\varepsilon = E_0/LC$.

We shall consider that α and ε in (16.1) are small, i.e., the generator is weakly disturbed, and the amplitude of the external signal (or magnitude of the linkage with the external generator) is also small. We introduce a new time $t_N = \Omega t$, a dimensionless coordinate $x = \beta^{1/2} U$, and the parameters $\mu = \alpha/\Omega$, $\mu\xi = 2(\omega_0 - \Omega)/\Omega$, and $\mu E = \beta^{1/2} E_0/\Omega^2$. Then (16.1) is transformed into

$$\ddot{x} - \mu(1 - x^2)\dot{x} + \mu\xi x + x = \mu E \cos t, \qquad (16.2)$$

where ξ characterizes the relative detuning between the fundamental frequency of the generator and the frequency of the external signal, the subscript n being omitted. We rewrite (16.2) in the form of a system

$$\dot{x} = y, \quad \dot{y} = -x + \mu[(1 - x^2)y - \xi x + E \cos t]$$

or

$$\dot{x} = y,$$

$$\dot{y} = -x + \mu f(x, y, t), \qquad (16.4)$$

$$f(x, y, t) = (1 - x^2)y - \xi x + E \cos t.$$

Van der Pol departed from the method of varying arbitrary constants. The solution to (16.3) for $\mu = 0$ is known, i.e.,

$$x = A \sin t + B \cos t, \quad y = A \cos t - B \sin t$$

We shall look for a solution for a nonzero μ in the same form, but we shall consider that the amplitudes A and B are functions of time, i.e., $A(t)$ and $B(t)$. So far, this is a simple change of variables: from x and y we go to A and B. We differentiate $x(A, B)$ and $y(A, B)$ with respect to t to get

$$\dot{x} = \dot{A} \sin t - A \cos t + \dot{B} \cos t - B \sin t,$$

$$\dot{y} = \dot{A} \cos t - A \sin t + \dot{B} \sin t - B \cos t$$

and substitute these into (16.3). By solving for the derivatives $\dot{A}$ and $\dot{B}$ we obtain

$$\dot{A} = \mu f(A, B, t) \cos t, \quad \dot{B} = \mu f(A, B, t) \sin t.$$

These are called equations in Van der Pol variables. They are exact because no approximation has yet been made. Now we assume that μ is small. If $\mu \ll 1$, and $|f|$ is on average of the order of unity, then A and B in the first approximation will be slowly changing functions of time, remaining unchanged over a period $T = 2\pi$ of the functions on the right-hand sides of the equations for $\dot{A}$ and $\dot{B}$. Now we expand the periodic functions $f(A, B, t)\begin{pmatrix}\cos t\\ \sin t\end{pmatrix}$ in Fourier series and truncate to the zeroth harmonic, because it corresponds to slow changes in the derivatives $\dot{A}$ and $\dot{B}$. The more rapidly changing terms may be discarded relying on this slowness of A and B (they contribute in the next approximation). Thus, we have obtained approximate, averaged, or as they are known, the shortened, equations

$$dA/d\tau = \langle f(A, B, t)\cos t\rangle,$$
$$dB/d\tau = -\langle f(A, B, t)\sin t\rangle, \quad (16.4)$$

where $\tau = \mu t$ and

$$\left\langle f(A, B, t)\begin{pmatrix}\cos t\\ \sin t\end{pmatrix}\right\rangle = T^{-1}\int_t^{t+T} f(A, B, t)\begin{pmatrix}\cos t\\ \sin t\end{pmatrix} dt.$$

Now we apply these general results to our concrete systems (16.3). For our system we have

$$f(A, B, t) = (1 - A^2\sin^2 t - 2AB\sin t\cos t - B^2\cos^2 t)$$
$$\times (A\cos t - B\sin t) - \xi(A\sin t + B\cos t) + E\cos t.$$

whence we obtain

$$\langle f\cos t\rangle = (1/2)A - (1/8)A^3 - (3/8)AB^2$$
$$- (1/2)\xi B + (1/4)AB^2 + (1/2)E.$$
$$\langle f\sin t\rangle = -(1/2)B - (1/4)A^2 B + (3/8)A^2 B$$
$$+ (1/8)B^3 - (1/2)\xi A.$$

Finally, the shortened equations take the form

$$\frac{dA}{d\tau} = \frac{A}{2}\left[1 - \frac{1}{4}(A^2 + B^2)\right] - \frac{B}{2}\xi + \frac{E}{2},$$
$$\frac{dB}{d\tau} = \frac{B}{2}\left[1 - \frac{1}{4}(A^2 + B^2)\right] - \frac{A}{2}\xi. \quad (16.5)$$

We now consider various cases.

<u>An Autonomous Generator</u> ($E = 0$). a) If we search for a solution with a fundamental frequency ω_0, then $\xi = 0$. The parameters of the self-excited oscillation are determined by the steady-state solution to (16.5). This has an unstable equilibrium at the origin $A = B = 0$ and a continuous set of equilibria lying on the circle $\rho^2 = A^2 + B^2 = 4$. The phase-plane diagram is given in Fig. 16.3a. In considering this figure one might ask whether we have obtained an uncrude system from a crude one, and what values must the amplitudes A and B have for that to be self-excited oscillation. These questions are best answered by looking at the phase plane of the initial variables x and y. To do this, we must have a system of coordinates (x, y) rotating clockwise at an angle of velocity ω_0. A circle with radius $\rho = 2$ will become a limiting cycle, and phase trajectories which are straight lines on the A, B-plane will become spirals curving around the limiting cycle (Fig. 16.3b). In order to investigate this, remember that motion along a phase trajectory on the A, B-plane is at a velocity of the order μ and consequently a point may not be able to go round the radius in one orbit. Thus, the generator will have self-excited oscillations with amplitude $\rho = 2$ and arbitrary phase $\phi = \arctan(A/B)$.

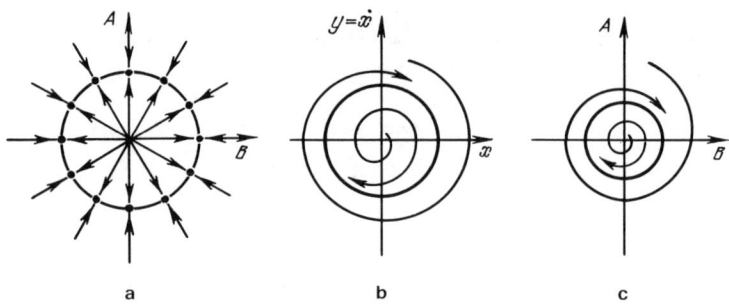

Fig. 16.3. Phase-plane diagrams of an autonomous generator ($E = 0$): a) on the A, B-plane the solution is sought with a fundamental frequency ω_0 (the origin is an unstable equilibrium; the circle $\rho^2 = A^2 + B^2 = 4$ is a set of equilibria); b) on the x, y-plane corresponding to a); c) on the A, B-plane the solution is sought for a frequency Ω.

b) If, however, we search for a solution with frequency Ω close to the fundamental frequency ω_0, then $\xi \neq 0$. It can be shown that for system (16.5), where $E = 0$, there is one stable limiting cycle, symmetric about the origin (Fig. 16.3c). As before $A^2 + B^2 = \text{const}$, but now A and B change with frequency ξ, and the change in phase are the same frequency, i.e., $d\phi/d\tau = \xi$. This means that the frequency shift is exactly ξ. If the limiting cycle had been asymmetric about the origin, then $\rho^2 = A^2 + B^2$ would not have been constant and the oscillation amplitude would have been periodically modulated, i.e., a beat would have arisen in the system. This is exactly what occurs in a nonautonomous situation.

Nonautonomous Generator. We shall attempt to find a synchronization regime, i.e., one in which the generator produces oscillations that are not at its fundamental frequency, but at the frequency of an external field. The presence of such a field means that a powerful generator's frequency can, for example, be controlled by a weak signal. We shall determine the parameters and boundaries of a synchronization regime and ascertain what will happen beyond these boundaries. In a pulling regime, the amplitude A and B must remain constant. For convenience, we introduce the amplitudes $a = A/2$, $b = B/2$ and $E_{ex} = E/2$. System (16.5) will then have the form

$$\dot{a} = a[1 - (a^2 + b^2)] - b\xi + E_{ex}, \quad \dot{b} = b[1 - (a^2 + b^2)] + a\xi. \tag{16.6}$$

The equilibrium is determined from the equation

$$a_0[1 - (a_0^2 + b_0^2)] - b_0\xi = -E_{ex},$$

$$b_0[1 - (a_0^2 + b_0^2)] + a_0\xi = 0,$$

whence it is easy to obtain the equation for the resonance curve for $\rho = a_0^2 + b_0^2$, i.e.,

$$\rho(1 - \rho)^2 + \xi^2\rho = E_{ex}^2. \tag{16.7}$$

This yields the amplitude of the oscillation at the external signal's frequency as a function of the signal's amplitude and the detuning. By solving (16.7) for the detuning, we obtain

$$\xi = \pm [[E_{ex}^2 - \rho(1 - \rho)^2]/\rho]^{1/2},$$

whence it follows that real values of ξ only exist for $E_{ex}^2 \geq \mathcal{F}(\rho) = \rho(1 - \rho)^2$ (Fig. 16.4). Resonance curves 1, 2, 3 correspond to the amplitude of the external force of $E_{ex_1}^2$, $E_{ex_2}^2$, $E_{ex_3}^2$. When $E_{ex}^2 < 4/27$, the resonance curves consist of two arms, this being a weak external signal. The force

$E^2_{ex} > 8/27$ corresponds to a strong signal and the resonance curve has the form of curve 3 in Fig. 16.4.

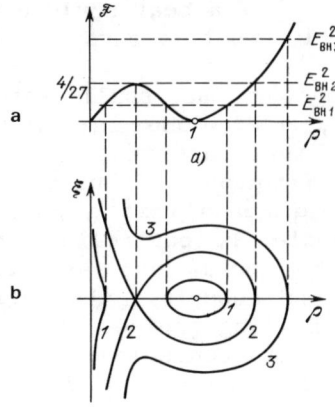

Fig. 16.4. Synchronization with an external signal: a) the amplitude of the external signal versus the amplitude of the oscillation at the external signal's frequency; b) resonance curves for a nonautonomous generator; the case of weak external signals corresponds to $E^2_{ex} < 4/27$ (curves 1 and 2); the case of strong signals corresponds to $E^2_{ex} > 8/27$ (curve 3).

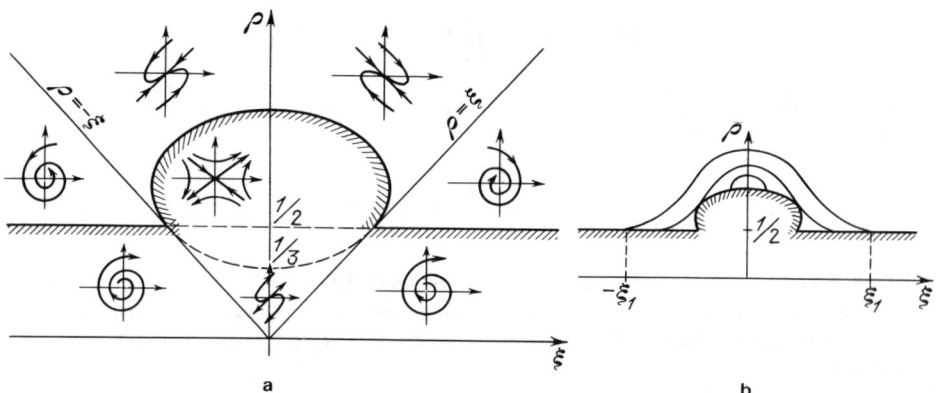

Fig. 16.5. Stability of the branches of the resonance curves and types of equilibria. a) Partition of the (ρ, ξ)-plane into regions with different equilibria; the stability boundary is shaded; b) combination of Fig. 16.4b and Fig. 16.5a (the arms of the resonance curve falling in the stability region are left).

All that remains is to ascertain which arms of the resonance curves are stable because only they will correspond to a real synchronization regime. To do this it is necessary to linearize (16.6) close to the equilibrium, and to find the boundaries of the stability and plot them on the (ρ, ξ) surface. By writing (16.6) in the form

$$\dot{a} = \phi(a, b), \quad \dot{b} = \psi(a, b),$$

we obtain linearized equations for $a_1 = a - a_0$ and $b_1 = b - b_0$, i.e.,

$$\dot{a}_1 = \phi'_a(a_0, b_0)a_1 + \phi'_b(a_0, b_0)b_1,$$

$$\dot{b}_1 = \psi'_a(a_0, b_0)a_1 + \psi'_b(a_0, b_0)b_1.$$

The characteristic equation has the form

$$\lambda^2 + p\lambda + q = 0,$$

where
$$p = -(\phi'_a + \phi'_b) = 4\rho - 2,$$

$$q = \begin{vmatrix} \phi'_a & \phi'_b \\ \psi'_a & \psi'_b \end{vmatrix} = (1 - 3\rho)(1 - \rho) + \xi^2.$$

The stability region is determined, consequently, by the inequalities

$$\rho > 1/2, \quad (1 - 3\rho)(1 - \rho) + \xi^2 > 0.$$

We determine the type of the equilibrium at the same time: when $q > 0$ we have a saddle region, and by equating the discriminant of the characteristic equation $D = p^2/4 - q > \rho^2 - \xi^2$ to zero, we obtain the boundary between the node and the focus. The partition of the (ρ, ξ) plane into regions with different types of equilibrium is shown in Fig. 16.5a. The stability boundary is shaded in the figure. Combining Figs. 16.4b and 16.5a, and leaving only those arms of the resonance curves that fall into the stability region (Fig. 16.5b), the synchronization region's boundaries for strong signals are determined by the intersection between the resonance curve and the straight line $\rho = 1/2$. By substituting this value of ρ into (16.7), we obtain

$$\xi_1^2 = 2E_{ex}^2 - 1/4. \qquad (16.8)$$

For very weak signals it is possible to consider that the boundary of synchronization region is determined by the intersection between the resonance curves and the straight line $\rho = 1$ (the coordinates of the locus of an ellipse does not differ much from this value), and consequently $\xi_1^2 = E_{ex}^2$. When leaving the synchronization regime, the generator behaves differently for strong and weak signals. A weak signal, as was shown above, corresponds to amplitudes of the external signal of $E_{ex}^2 < 4/27$. When $E_{ex}^2 = 8/27$ the resonance curve is tangent to the ellipse bounding the saddle region. If $E_{ex}^2 > 8/27$, then for any ξ in the system there is one equilibrium state and the external signal will be considered strong.

Let us consider each case.

<u>A Strong Signal</u> ($E_{ex}^2 > 8/27$). The amplitude-frequency diagram for this case is given in Fig. 16.6. The system contains a single equilibrium, which is stable in the detuning region $|\xi| < \xi_1$. The parameter value $\xi = \xi_1$ is bifurcational. On the phase-plane the system (16.6) for $\xi = \xi_1$ has a complex focus. When $\xi > \xi_1$ a limiting cycle arises and the focus becomes unstable. All that remains in order to prove the stability of the limiting cycle is to discover how the phase trajectories behave at large amplitudes a and b. From (16.6) the oscillation's amplitude at large a and b is reduced, consequently "infinity" is unstable, and all the phase trajectories converge to some region, i.e., they tend to a limiting cycle.

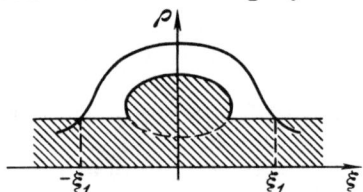

Fig. 16.6. Amplitude-frequency diagram for a nonautonomous generator with synchronization with a strong signal ($E_{ex}^2 > 8/27$).

Motion along a limiting cycle corresponds to a periodic change in amplitude a and b, which means that there is a biharmonic regime in the initial system (a beat regime). Beats arise furled with respect to amplitude (Fig. 16.7a) because the limiting cycle arises with zero radius. The frequency of the beats is finite because the limiting cycle arises from a focus and when it occurs, it has the frequency of the disappearing equilibrium. In order to determine the frequency we must find the roots of the characteristic equation $\lambda^2 + p\lambda + q = 0$ as $\xi \to \xi_1$. The value of the imaginary part of the root yields the desired frequency. It is fairly easy to prove that as the detuning increases the frequency of the beat increases (Fig. 16.7b). For large values of ξ we may assume that the amplitudes a and b change at some frequency ω^*. Moreover, they undergo even slower changes (small with period $1/\omega^*$). Then, by averaging repeatedly it is possible to find the cycle's amplitude on the plane for the Van der Pol variables and the frequency of rotation around them.

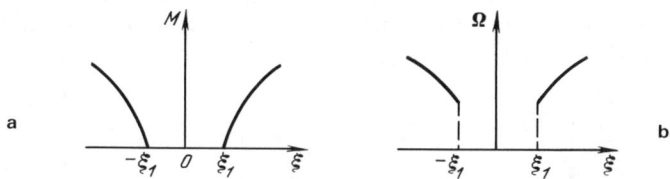

Fig. 16.7. The beat regime in a nonautonomous generator (biharmonic regime with strong external signal): a) the crude modulation M of the output signal versus detuning, which shows that the beat is induced furled around the amplitude; b) "rigid" perturbation of the beat frequency Ω and its relation to detuning.

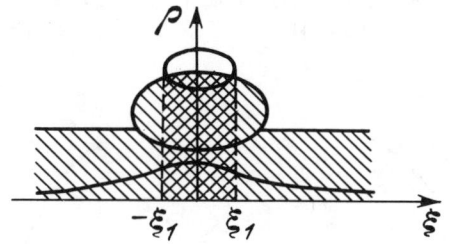

Fig. 16.8. Amplitude--frequency diagram of a nonautonomous generator with synchronization with a weak signal ($E_{ex}^2 < 4/27$). Synchronization occurs in the interval $|\xi| < \xi_1$ (double shading).

__Weak Signal__ ($E_{ex}^2 < 4/27$). The amplitude-frequency diagram for this case is given in Fig. 16.8. Synchronization arises in the detuning interval $|\xi| < \xi_1$, i.e., where there is a stable resonance curve arm. The presence in this region of unstable arms does not effect the pulling regime, but it does change the way the system leaves the regime. When $|\xi| < \xi_1$, the phase-plane diagram in Van der Pol variables is given in Fig. 16.9a. There are three equilibria: an unstable focus, a saddle, a stable node, with a total Poincare index of $j = +1$. "Infinity" is unstable, as we have shown. All the phase trajectories converge to a node. When $\xi = \xi_1$, there are two equilibria (a saddle and a node) and they merge to form a saddle-node singular point (Fig. 16.9b) with $j = 0$, which as the detuning increases disappears. A limiting cycle arises from the saddle's separatrix and all the phase trajectories tend asymptotically to the cycle. Because the radius of the limiting cycle when it arises is finite, and because the frequency with which the tracing point rotates around it when it arises is zero (the duration of motion in a loop of the separatrix is infinite), the changes in the depth of modulation and beat frequency when the detuning is increased (Fig. 16.10) will be different to those which occur for a strong signal.

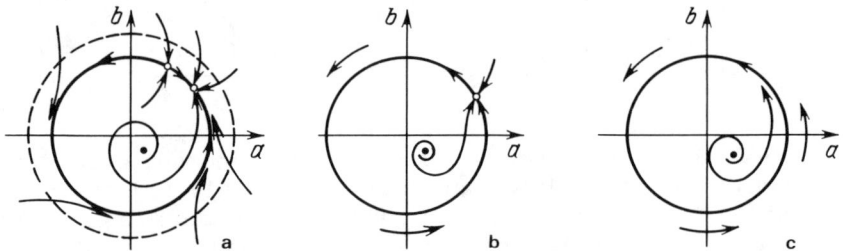

Fig. 16.9. Phase-plane diagrams of a nonautonomous generator in Van der Pol variables with a weak external signal, which is illustrated by the evolution of the equilibrium as the detuning is changed: a) $|\xi| < \xi_1$; b) $|\xi| = |\xi_1|$; c) $|\xi| < \xi_1$.

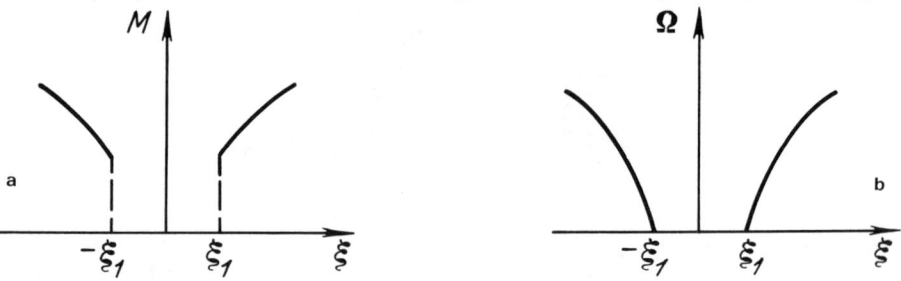

Fig. 16.10. Illustration of the basic beat regimes for weak and strong external signals (Fig. 16.7): a) the induced beats are rigid in amplitude (the limiting cycle dimension is immediately finite); b) a "soft" disturbance to the beat frequency (the frequency of rotation around a limiting cycle when it is born is zero).

We must note that the synchronization phenomenon does not have a lower level in terms of amplitude, and any signal, no matter how small, may synchronize a generator; the bands of synchronization become ever smaller.

We emphasise that if the nonlinearity of the generator is moderate, then the periodic force will not only affect a generator's synchronization or its operation in a beat regime (outside the pulling or synchronization bands), it will induce very complicated oscillation regimes or even oscillations with continuous spectra. These types of oscillation have recently been observed by Molchanov [13] in a nonautonomous generator described by an equation of the form $\ddot{x} - \mu(1 - x^2)\dot{x} + x^3 = B \cos \Omega t$. For instance, when $\mu = 0,2$, $\Omega = 4,0$ and $B = 17.0$, oscillations with continuous spectra in the interval $\omega \in [0; 4,5]$ were observed. The rise of stochastic oscillation in similar relatively simple dynamic systems will be discussed in detail in Chapter 22.

Synchronization is widespread in mechanics (for example, rotation synchronization of mechanical vibrators is similar to the effect Huygens observed in clocks), in radio electronics, in electronics and radio physics (synchronizing different self-excited generators in a vacuum or solid active elements, synchronizing lasers etc.), in chemistry, in biology, and in medicine [1]. There is even the hypothesis that the Solar System's dimensions can be attributed to synchronization. Molchanov's hypothesis that the major planets are orbitally synchronized and the Solar system is an example [14]. In celestial mechanics, a link between the average angular velocities ω_i of objects which can be described mathematically by "resonance" relations $\sum_i n_i \omega_i = 0$, where n_i are positive or negative integers and i is the number of rotations, is called synchronization or resonance. To see what can be extracted from such a synchronization, we hypothesize that the dissipative forces (attraction forces, braking forces by interplanetary dust etc.) can have over a period of thousands of millions of years during which the Solar system has evolved, in spite of the forces being small, have caused the planets to nearly steady-state (practically unchanged over millions of years) resonance orbits. Molchanov has compiled a table of the "Resonance relationships in the Solar System" for the nine major planets and their large satellites [15] (Table 16.1). The agreement between the theoretical frequencies, which satisfy the relationship $\sum_{i=1}^{9} \omega_i n_i = 0$, and the observed frequencies is indeed astounding.

TABLE 16.1.

Resonance Relationships in the Solar System [15].

No. Planet	ω_i^o observed	ω_i^t theoret.	$\dfrac{\Delta\omega}{\omega} = \dfrac{\omega_i^o - \omega_i^t}{\omega_i^o}$	n_1	n_2	n_3	n_4	n_5	n_6	n_7	n_8	n_9
1 Mercury	49.22	49.20	0.0004	1	-1	-2	-1	0	0	0	0	0
2 Venus	19.29	19.26	0.0015	0	1	0	-3	0	-1	0	0	0
3 Earth	11.29	11.828	0.0031	0	0	1	-2	1	-1	1	0	0
4 Mars	6.306	6.287	0.0031	0	0	0	1	-6	0	-2	0	0
5 Jupiter	1.0000	1.0000	0.0000	0	0	0	0	2	-5	0	0	0
6 Saturn	0.4027	0.400	0.0068	0	0	0	0	1	0	-7	0	0
7 Uranus	0.14119	0.14286	-0.0118	0	0	0	0	0	0	1	-2	0
8 Neptune	0.07197	0.07143	0.0075	0	0	0	0	0	0	1	0	-3
9 Pluto	0.04750	0.04762	-0.0025	0	0	0	0	0	0	0	-5	1

However, it remains only a hypothesis and raises many questions. We know that for there to be synchronization in a system of oscillators three conditions are essential: nonlinearity, coupling, and dissipation. The nonlinearity in the celestial oscillators has been known for several centuries ago and according to Kepler's third law the oscillation (orbital) frequency of a planet (in the well-known two-body problem) depends on the oscillators energy as $\omega \sim |\mathcal{E}|^{2/3}$. Coupling is by gravitation between the rotating bodies, and dissipation, as Molchanov noted, is caused by tidal forces. However, the dissipation is very small, and the disturbance (interaction) weak. Therefore, the synchronization band must be very narrow. Although the planets are not orbiting in synchronization (the system is not evolutionally mature), they have frequencies close to resonance. Why this should be so, Molchanov's hypothesis does not explain. It is possible that this frequency relation was induced by the geometry of the Solar System at its inception [12].

16.2 Competition

In the simplest formulation, competition is a purely energetic effect. It is not associated with the oscillation's phase and occurs even when there are no oscillations at all; recall competition between biological species. We begin with the example discussed at the beginning of this chapter of a two-circuit self-excited generator (Fig. 16.1), which can work in one or two generating modes depending on its parameters [11-13]. Its equation is written in the form

$$L_1 C_1 \ddot{U}_1 + [R_1 C_1 - MS(U_1)]\dot{U}_1 + U_1 + NC_2 \ddot{U}_2 = 0,$$

$$L_2 C_2 \ddot{U}_2 + R_2 C_2 \dot{U}_2 + U_2 + NC_1 \ddot{U}_1 = 0,$$
(16.9)

where $S(U_1) = S_0 - S_2 U_1^2$ is the slope of the valve's anode-net characteristic (we shall approximate the relation $i_a = i_a(U_1)$, as before, by a cubic parabola).

We introduce the dimensionless variables

$$t_N = \frac{1}{\sqrt{L_1 C_1}} t,$$

$$x_1 = \sqrt{\frac{MS_2}{MS_0 - R_1 C_1}} U_1, \quad x_2 = \frac{L_2 C_2}{NC} \sqrt{\frac{MS_2}{MS_0 - R_1 C_1}}$$

and the parameters as the partial frequencies of the circuit, $\xi = n_2^2/n_1^2$ and $\alpha = N^2/L_1 L_2 < 1$ as the coupling coefficient between the circuits $\mu = n_1(MS_0 - R_1 C_1)$, as the parameter characterizing the degree to which the generator is disturbed, and $\delta = R_2 L_1 C_1/L_2(MS_0 - R_1 C_1)$, as the ratio of the decrement in the decay in the second circuit to the increment in growth in the first circuit. Thus (16.9) takes the form

$$\ddot{x}_1 + x_1 + \alpha \ddot{x}_2 = \mu(1 - x_1^2)\dot{x}_1, \quad \ddot{x}_2 + \xi x_2 + \ddot{x}_1 = \mu \delta \dot{x}_2.$$
(16.10)

When $\mu = 0$ (a linear conservative system) in either of the circuits, we will observe oscillations with normal frequencies ω_1 and ω_2, i.e.,

$$x_1 = a_1 e^{i\omega_1 t_N} + a_2 e^{i\omega_2 t_N} + \text{c.c.},$$

$$x_2 = a_1 \Psi_1 e^{i\omega_1 t_N} + a_2 \Psi_2 e^{i\omega_2 t_N} + \text{cc} \tag{16.11}$$

The complex amplitudes a_1 and a_2 are determined by the initial conditions, while the distribution coefficients are given by the expressions $\Psi = (1 - \omega'^2)/\alpha\omega'^2 = \omega'^2/(\xi - \omega'^2)$ and $\omega' = \omega/n_1$ (the reader is left to prove this independently). The normal frequencies ω_1 and ω_2 satisfy the equation (this is easy to obtain by substituting the solution $x_{1,2} \sim \exp(i\omega t_N)$ into the system (16.10) for $\mu = 0$)

$$(1 - \alpha)\omega'^4 - (1 + \xi)\omega'^2 + \xi = 0.$$

The relation between the normal frequencies and the detuning, as determined by this equation, is given in a Venn diagram (Fig. 16.11). When $0 < \mu \ll 1$ (this means that the generator is weakly disturbed) and $\delta \approx 1$ (the second circuit's Q-factor is large) the system (16.10) may be solved, as in a previous section, using Van der Pol's method.

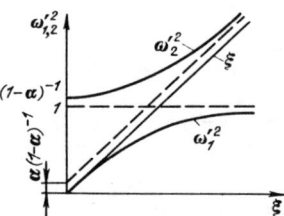

Fig. 16.11. Venn diagram. The asymptotes as $\xi \to \infty$ are $\omega_1'^2 = 1$ and $\omega_2'^2 = (\xi + \alpha)(1 - \alpha)^{-1}$.

Bearing in mind the smallness of μ, the solution to (16.10) will be sought in form (16.11), although the amplitudes a_i and a_i^* ($i = 1, 2, \ldots$) will be considered functions of time. By substituting this solution into (16.10) and solving it for the derivatives da_j/dt_N ($j = 1, 2, \ldots$), and by averaging the right-hand sides of the resultant equations with respect to time, we obtain short equations like (16.4), but for the two complex amplitudes a_1 and a_2. By using the substitution $a_j = A_j \exp(i\phi_j)$ ($j = 1, 2, \ldots$), we go to an equation for real amplitudes and phases, i.e.,

$$2\dot{A}_1 = \mu h_1 [1 - (A_1^2 + \rho_{12} A_2^2)] A_1 ,$$
$$2\dot{A}_2 = \mu h_2 [1 - (A_2^2 + \rho_{21} A_1^2)] A_2 ,$$
(16.12)

where
$$\dot{\phi}_1 = \dot{\phi}_2 = 0,$$
$$h_i = \sigma_i \lambda_i \quad (i = 1,2),$$
$$\sigma_1 = \omega_1^4 \frac{\omega_2^2 - 1}{4(\omega_2^2 - \omega_1^2)} , \quad \sigma_2 = \omega_2^4 \frac{1 - \omega_1^2}{4(\omega_2^2 - \omega_1^2)} ,$$
$$\lambda_1 = 4(1 - \delta\Psi_1/\xi\Psi_2) , \quad \lambda_2 = 4(1 - \delta\Psi_2/\xi\Psi_1) ,$$
$$\rho_{12} = 2\lambda_2/\lambda_1 , \quad \rho_{12} = 2\lambda_2/\lambda_1 .$$

It follows from (16.12) that oscillations in the generator will have the same frequencies as those in a linear system (the right-hand sides of the equations for the phase are zero, i.e., there are no corrections to the frequency).

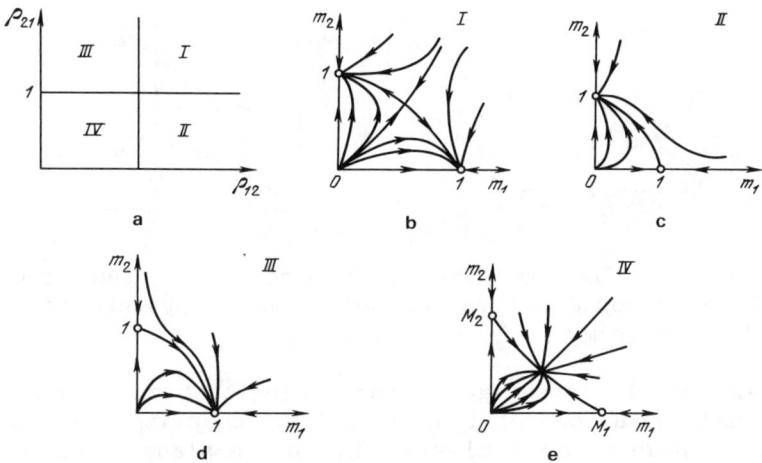

Fig. 16.12. The phase-plane diagrams of (16.13), illustrating the interaction between two modes.

Since the phase does not enter the equations for the amplitude, we can investigate the amplitude equation independently. We change variables to $\tau = \mu t_N$, $m_1 = A_1^2$ and $m_2 = A_2^2$. Thus, the equations for the squares of the amplitudes can be rewritten as

$$\dot{m}_1 = h_1[1-(m_1+\rho_{12}m_2)]m_1,$$
$$\dot{m}_2 = h_2[1-(m_2+\rho_{21}m_1)]m_2. \quad (16.13)$$

These equations describe interactions between two modes (normal oscillations) that take energy from one source. The coefficients ρ_{12} and ρ_{21} are the effects of the modes upon each other and are called the coefficients for nonlinear mode coupling. Physically, their meaning is obvious. At small ρ_{12} and ρ_{21} the modes do not affect each other and the self-excited oscillations in each mode behave independently. When ρ_{12} and ρ_{21} are large, by contrast, the level at which the oscillation amplitudes $|a_1|$ and $|a_2|$ stabilize is defined by the amplitudes of the other modes $|a_2|$ or $|a_1|$ (this is strong coupling). Finally, the link may not be mutual, when $\rho_{12} \neq \rho_{21}$. Here the effect of one mode on the other may be strong whilst the inverse effect may be weak.

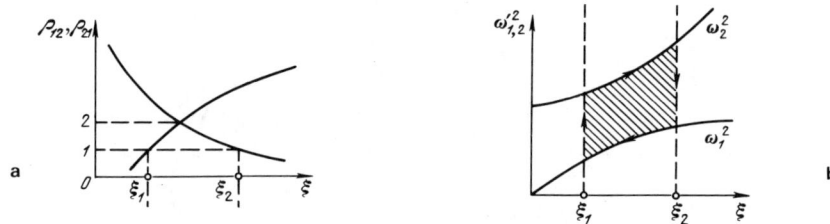

Fig. 16.13. The pulling phenomenon: a) the coupling coefficients for $\delta < 1$ versus detuning; b) hysteresis region on a Venn diagram.

We shall investigate the relationship between the stationary solutions of (16.13) and the coupling coefficients using a phase-plane analysis. The phase-plane diagrams for various ρ_{12} and ρ_{21} (Fig. 16.12a) are shown in Figs. 16.12b-e. It is easy to see that the system has four equilibria:

1) $m_1 = m_2 = 0$,

2) $m_1 = 1$, $\qquad m_2 = 0$,

3) $m_1 = 0$, $\qquad m_2 = 1$,

4) $m_1 = (1-\rho_{12})/(1-\rho_{12}\rho_{21})$, $m_2 = (1-\rho_{21})/(1-\rho_{12}\rho_{21})$.

The latter only exists when the parameters on the (ρ_{12}, ρ_{21})-plane are in regions *I* or *IV* and it is stable only for small coupling coefficients (region *IV*). It can be shown that for the generator we are considering the coupling coefficients may not all be simultaneously less than unity. Thus, a two-frequency regime (characteristic for the equivalent linear system) may not exist in a nonlinear self-excited generator. Whatever the initial conditions, oscillations with the second (first) normal frequency are established in region *II* (*III*). In region *I*, the "survival" of a mode depends on the initial conditions, and the size of the pulling region of a mode depends on the size of the coupling coefficients. This type of the interaction between the modes us usually called mode competition. We shall investigate the relationship between the system's operation and detuning between the circuits. It follows from the nature of the change in the coupling coefficients for $\delta < 1$ (Fig. 16.13a) that when $\xi < \xi_1$ only oscillations with a frequency ω_2 may exist, and that when $\xi > \xi_2$ only oscillations with the frequency ω_1 may exist (16.13b). In the region $\xi_1 < \xi < \xi_2$, any of the regimes (this region is shaded in the figure) may become established depending on the initial conditions. Thus, if the detuning is smoothly changed from small ξ, then the oscillations in the generator will first have large normal frequencies, at $\xi = \xi_2$ there will be a step change in the frequency to a value ω_1 (the amplitude will also undergo a step change), and as the detuning is increased further, the oscillations will be at lower normal frequencies. If ξ is changed in the opposite direction, there will be hysteresis (16.13b). This is called pulling, and it is well-known to experimentalists. In many cases it is damaging because the frequency can be changed when modifying the generator to alter some parameter. We shall not investigate the way the pulling interval width depends on the system's parameters because of the cumbersome calculations. We only note that in order to avoid pulling, weak feedback is needed in the generator or the *Q*-factor of the second circuit must be small.

The study of the interaction between two species of Volterra (Chapter 1) is central to theoretical ecology [7, 8].

Ecology usually studies three types of interaction:
a) the predator–prey interaction (Chapter 1);
b) the independent effect of one species on another;
c) competition between species (the interaction in which one of the species suppresses the population growth of the other whilst its own increases).

We applied the somewhat modified Volterra equations

which are used when considering these interactions. They are derived from the logistics equations $\dot{N} = \varepsilon N(1 - N/K)$, to which the terms $-\gamma_1 N_1 N_2$ and $-\gamma_2 N_2 N_1$ are added to describe the "suppression" of one species by another. The equations take the form

$$\dot{N}_1 = N_1 [\varepsilon_1 - (\varepsilon_1/K_1)N_1 - \gamma_1 N_2] ,$$
$$\dot{N} = N [\varepsilon - (\varepsilon/K)N - \gamma N] ,$$
(16.14)

where N_1 and N_2 are the species populations.

If we are considering the relationship between a prey population N_1 and a predator population N_2, which devours the prey, then (16.14) must be rewritten [8] as

$$\dot{N}_1 = \varepsilon_1 N_1 - (\varepsilon_1/K_1)N_1^2 - \gamma_1 N_1 N_2 ,$$
$$\dot{N}_2 = - \varepsilon_2 N_2 + \gamma_1' N_1 N_2$$
(16.15)

The solution to (16.15) is well-known [8]: the prey and predator populations oscillate, damping over time (the phase of the prey population leads the predator population oscillation) (Fig. 16.14).

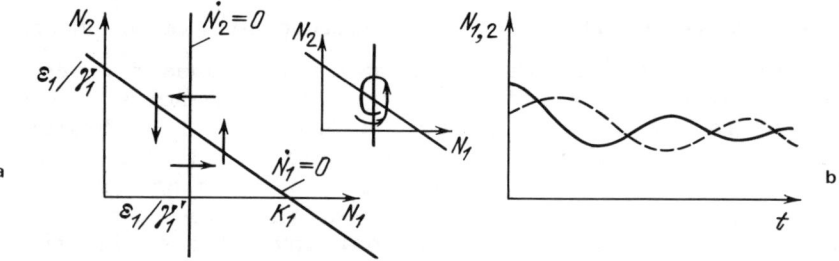

Fig. 16.14. a) Solution of (16.15) allowing for the self-restriction of the prey, the term $(\varepsilon_1/K_1)N_1^2$ and the first equation of (16.15); the arrows on the N_1, N_2-plane indicate the direction in which the system moves, i.e., its dynamics. b) Change in the predator population (dotted line) and prey population (solid line).

The dynamics of competition between two species is illustrated in Fig. 16.15, from which it follows that the equilibrium of (16.14) may be either stable or unstable (see Fig. 16.12).

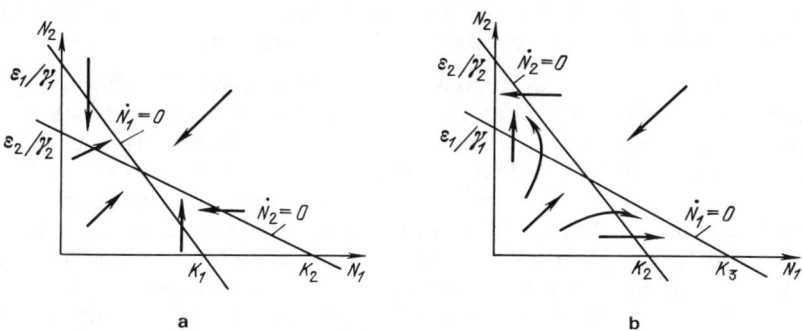

Fig. 16.15. Equilibrium of (16.14) when there is competition between two species [8]: a) stable ($K_1 < K_2$; $\varepsilon_2/\gamma_2 < \varepsilon_1/\gamma_1$); b) unstable ($K_1 > K_2$; $\varepsilon_2/\gamma_2 > \varepsilon_1/\gamma_1$). In b) the initial conditions determine which species survives.

When $\varepsilon_1 = \varepsilon_2$ for stable equilibrium we have $\gamma_2 > \gamma_1$, but $\varepsilon/K_1 > \varepsilon/K_2$. These inequalities between the coefficients of system (16.14) mean that when one of the competitors increases its population, its growth is more strongly suppressed than that of its competitor. If both species have the same requirements, then one quickly eliminates the other competitor.

16.3 Mutual Mode Synchronization

A very important question is which physical mechanisms interfere with the existence of multiperiodic motions in self-excited oscillatory systems with many degrees of freedom. To answer this question, we consider the behavior of an ensemble of quasiharmonic self-excited generators with weak coupling, i.e.,

$$\dot{A}_i = \gamma_i A_i (A_{i0}^2 - A_i^2) + \alpha_i (A_{j,i}, \phi_{j,i}),$$
$$\dot{\phi}_i = \omega_i + \beta_i (A_{j,i}, \phi_{j,i})$$
(16.16)

($i = 1, 2, \ldots, N$). Here the corrections $\alpha_i(A, \phi)$ and $\beta_i(A, \phi)$ reflect mode interaction. When $\alpha_i = \beta_i = 0$ ($i = 1, 2, \ldots, N$), all the oscillations are independent and the phase-space of (16.16) divides into N phase planes, on each of which there is a single stable limiting cycle with period $2\pi/\omega_i$ and amplitude A_{i0}. In the initial $2N$-dimensional space, these independent oscillations correspond to the attraction of the tracing point to an N-dimensional torus, the product of the independent cycles. If none of the ω_i are commensurable, then the phase trajectories on the torus will be dense and none of the loops will be closed, it will be quasiperiodic motion. When, however, there is coupling between the self-excited generators (or self-excited oscillating modes), such a simple quasiperiodic motion must, in general, be violated. The simplest, one-mode, self-excited oscillation is established in a multi-mode system as the result of competition, which is associated with the appearance in each or some of the modes of a nonlinear absorption that progressively increases as the energy of "strange" modes rises. This situation often occurs when all the modes take energy from one source. The coupling function is then dependent only on the mode energy, i.e.,

$$\alpha_i = -A \sum_{j \neq 1} \rho_{ij} A_j , \qquad (16.17)$$

where the ρ_{ij} are the coupling coefficients. When $\rho_{ij} < \gamma_i$, the link is weak and multi-frequency generation is possible. For example, in a gas laser with nonhomogeneously widened lines of the active substance, different resonator modes take energy from different active molecules. If, however $\rho_{ij} > \gamma_i$ (strong coupling), then a one-mode generation is established independent of the number of initially excited modes. As a rule, the mode with the greatest linear increment will be the victor. Thus, we have the fascinating situation in which a simple dynamic regime becomes established in a nonequilibrial system (medium) starting from noise generation, this being mainly due to competition.

Synchronization also leads to rarefaction and ordering of the oscillational spectra. During synchronization, the modes do not suppress each other, but both shift in frequency so that the nonlinear corrections either become equal or commensurable. Limiting cycles appear on the torus instead of the quasiperiodic coils. Mutual mode synchronization is possible both in terms of frequency and in terms of wave number. In the latter case, synchronization appears particularly nontrivial and it is spatial mode synchronization that explains the occurrence of complex ordered structures in self-excited oscillatory systems with greater than one dimension (for instance, the hexahedral prismatic Benar cell during thermal convection, about which we shall talk later).

A graphic example of synchronization in an ensemble of many self-excited generators is given in Fig. 16.16. Here the results of a numerical experiment with (16.16) are shown for

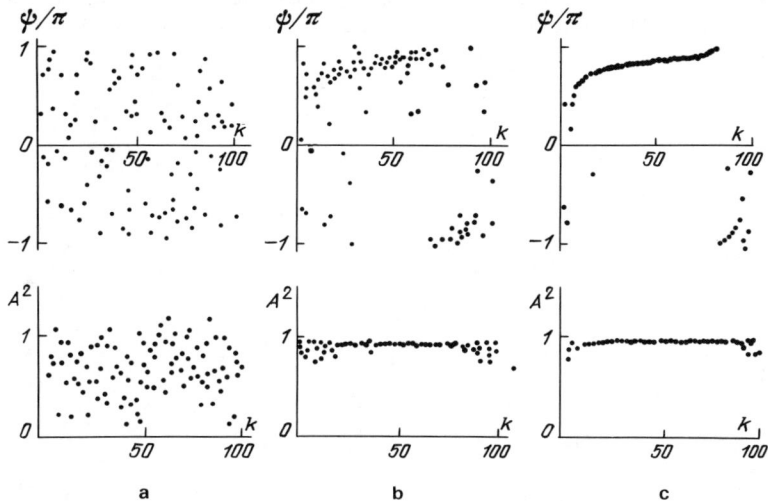

Fig. 16.16. Synchronization of an ensemble of 100 coupled self-excited generators with Lorentzian frequency distribution (ψ is the total phase of the oscillator, A^2 is the square of the amplitude, and $T = 2\pi/\omega_0$). Time: a) 0 T; b) 40 T; c) 80 T. k is the generator number.

a linear coupling between oscillators with close frequencies at $\gamma_i = A_{i0} = 1$ [9, 10]:

$$\alpha_n = N^{-1} \sum_{k=1}^{N} [A_k \cos(\phi_k - \phi_n) - A_n],$$

$$\beta_n = 0{,}2 N^{-1} \sum_{k=1}^{N} A_k \sin(\phi_k - \phi_n).$$

It was assumed that in the absence of interaction the self-excited generators are distributed with respect to frequency according to the Lorentz function

$$f(\omega) = (g/\pi)[g^2 + (\omega - \omega_0)^2]^{-1}.$$

It is clear that if the amplitudes A_i and phases $\Psi_i = \phi_i - \omega_0 t$ of the modes were initially random, then after 80 periods a regime with practically a single frequency is established due to synchronization. If the coupling between the generators is large, the inverse to synchronization will take place, i.e., chaos. Two linked self-excited generators demonstrate this behavior, i.e., when the coupling in the system is very strong stochastic self-excited oscillations are possible (in this case the system becomes a self-excited generator with stochastic behavior, see Chapter 22).

CHAPTER 17

RESONANCE INTERACTIONS BETWEEN OSCILLATORS

17.1 Interaction Between Three Coupled Oscillators in a System with Quadratic Nonlinearity

We shall consider one of the basic, but at the same time one of the most elementary, problems in the theory of nonlinear oscillations and waves, namely the interaction between three coupled oscillators with quadratic nonlinearity. In the absence of nonlinearity, as we know, a system of three coupled oscillators will undergo motions that are the simple superpositions of oscillations at the three normal frequencies (ω_1, ω_2, ω_3). The equation of the system written in normal coordinates takes the form $\ddot{x}_j + \omega_j^2 x_j = 0$ ($j = 1, 2, 3$). The presence of weak nonlinearity leads to a small right-hand side in the equation, i.e.,

$$\ddot{x}_j + \omega_j^2 x_j = \mu f_j(x_1, x_2, x_3), \quad \text{where } \mu \ll 1. \tag{17.1}$$

Two questions naturally arise: 1) why have we chosen three oscillators for analysis and 2) why have we limited ourselves to quadratic nonlinearity? These questions are related. In fact, if any of the quantities are nonlinear functions, e.g., of the potential field or potential (the nonlinearity, whatever it is, being weak), then this function may be expanded in the form of a power series of the potential. In our case (Fig. 17.1a), the charge Q depends on the potential U, and hence

$$Q(U) = C_1 U + C_2 U^2 + C_3 U^3 + \ldots$$

Thus, if the nonlinearity is weak then the quadratic term is the first one to give a nontrivial effect. At the same time, the nonlinearity causes new combination frequencies in the system. When the nonlinearity is quadratic, the simplest process is the formation of the sum $\omega = \omega_n + \omega_m$ or difference $\omega = \omega_n - \omega_m$ of the frequencies. Combination components that appear (again due to nonlinearity) may of course take part in the interaction, but only if their amplitudes are not small. In order for the "new" components generated by the weak nonlinearity to have a moderate amplitude, the frequencies must be resonant, i.e., they must be close to the system's normal frequency. It follows that the simplest interaction with quadratic nonlinearity may only proceed if the system's normal frequencies satisfy the resonance condition

$$\omega_1 \pm \omega_2 = \omega_3 . \qquad (17.2)$$

It is also true that there may be degeneration when $\omega_1 = \omega_2$ and that we may consider the system to have two normal frequencies ω and 2ω, but this is a special case (and we shall consider it). Without loss of generality we may write in the resonance condition (17.2) only the "+" sign, i.e., $\omega_1 + \omega_2 = \omega_3$.

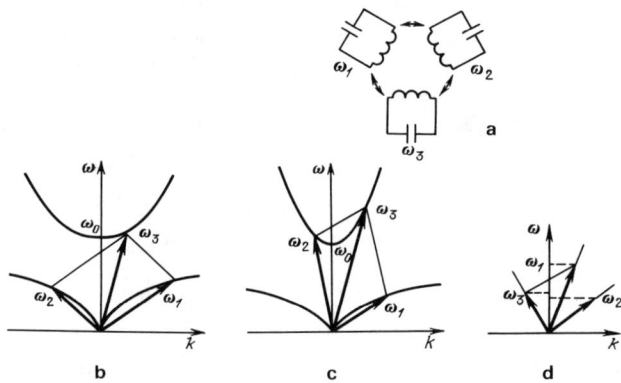

Fig. 17.1. a) A possible model for the interaction between three coupled oscillators. b,c) Dispersion diagrams illustrating resonance interactions between three couple waves/oscillators (e.g., interaction between the high and low frequency electromagnetic waves in a medium consisting of oscillators with a natural frequency of ω_0). d) Relationship between frequency and wave vectors for forced Mandel'shtam--Brillouin scattering.

Thus, when the nonlinearity is weak, the three oscillators in the system react with lumped parameters or three normal resonator modes may be effective only when condition (17.2) is fulfilled. Moreover, if we consider a medium with the dispersion characteristic shown in Figs. 17.1b or c, then the resonance condition must be fulfilled both by the frequencies and by the wave numbers,

RESONANCE INTERACTIONS   355

i.e., $\omega_1 + \omega_2 = \omega_3$, $k_1 + k_2 = k_3$ [1]. Thus, the initial equations for a weakly nonlinear conservative system with three degrees of freedom may be written as in (17.1). We shall use the asymptotic method (Chapter 16) to solve it, looking for a solution in the form

$$x_j(t) = a_j(\mu t)\exp(i\omega_j t) + cc + \mu w_j(t) . \qquad (17.3)$$

After substituting the solution (17.3) into (17.1) and separating terms with different orders, we obtain an equation for the correction w_j, which characterizes the degree to which the approximate solution departs from the exact one:

$$\ddot{w}_j + \omega_j^2 w_j = -2i\omega_j \dot{a}_j e^{i\omega_j t} + 2i\omega_j \dot{a}_j^* e^{-i\omega_j t}$$
$$+ f_j\left(a_{1,2,3} e^{i\omega_{1,2,3} t} + cc\right) . \qquad (17.4)$$

As we have seen, in order for the error not to grow, it is necessary and sufficient for the right-hand side of (17.4) not to be resonant with frequency ω_j, or for the right-hand side of the equation to be orthogonal to the eigenfunctions of (17.4) at $\mu = 0$. We obtain an equation for the amplitudes from this condition, i.e.,

$$2i\omega_j \dot{a}_j = T^{-1} \int^{t+T} f_j(ae^{i\omega t} + cc) e^{-i\omega_j t} dt ,$$

or

$$t\dot{a}_j = (2\omega_j)^{-1} \left\langle f_j(ae^{i\omega t} + cc) e^{-i\omega_j t} \right\rangle , \qquad (17.5)$$

where the symbols <> mean averaging over the period T. In our case of quadratic nonlinearity, f_j may be represented by the relation

$$f_j = \sum_{k,l} \alpha_{jkl} x_k x_l .$$

Since we are searching for a solution for $x(t)$ in the form (17.3), then an oscillating factor $\exp\{i(\omega_k \pm \omega_l)t\}$ will enter the nonlinear function f_j. It is clear that a contribution to the right-hand side of (17.5) will give terms with $\omega_k \pm \omega_l \approx \pm \omega_j$ because all the other combinations will contain factors of $\exp\{i(\omega_k \pm \omega_l \mp \omega_j)t\}$, which cause the appropriate terms in f_j to be averaged to zero. Finally after averaging we obtain three equations for the complex amplitudes, i.e.,

$$i\dot{a}_1 = \sigma_1 a_3 a_2^*, \quad i\dot{a}_2 = \sigma_2 a_3 a_1^*, \quad i\dot{a}_3 = \sigma_3 a_1 a_2.$$

This system can be exactly integrated in Jacobian elliptic functions, however we shall try to understand the system's behavior qualitatively without solving it. We make the substitution

$$a_{jN} = a_{jst}/\sigma_{jst}^{1/2}, \quad \sigma_{jN} = (\sigma_{1st} \sigma_{2st} \sigma_{3st})^{1/2}.$$

Then we obtain the following in the new variables (the subscript N will be omitted):

$$i\dot{a}_1 = \sigma a_2 a_3^*, \quad i\dot{a}_2 = \sigma a_3 a_1^*, \quad i\dot{a}_3 = \sigma a_1 a_2. \tag{17.6}$$

Without loss of generality, the quantity σ may be considered positive. We multiply each equation in (17.6) by a_j^* and add the complex conjugate using a fact that $a_j \dot{a}_j^* + a_j^* \dot{a}_j = d|a_j|^2/dt$. As the result we obtain

$$\dot{N}_1 = -i\sigma(a_3 a_2 a_1^* - a_3^* a_2^* a_1),$$

$$\dot{N}_2 = -i\sigma(a_3 a_2 a_1^* - a_3^* a_2^* a_1), \tag{17.7}$$

$$\dot{N}_3 = i\sigma(a_3 a_2 a_1^* - a_3^* a_2^* a_1),$$

where $N_j(t) = a_j a_j^* = |a_j|^2$ characterizes the intensity of the oscillations in the j-th mode or with the j-th normal frequency. By analogy with quantum mechanics N is often called a quantum number. It is easy to obtain two independent integrals of motion from (17.7) and the third is a consequence of the first two, i.e.,

$$\frac{d}{dt}(N_1 - N_2) = 0, \quad N_1(t) - N_2(t) = \text{const} = C_1,$$

$$\frac{d}{dt}(N_2 + N_3) = 0, \quad N_2(t) + N_3(t) = \text{const} = C_2; \tag{17.8}$$

$$N_4(t) + N_3(t) = \text{const} = C_3.$$

These are called the Manley-Rowe relations. The following important conclusions follow from them.

1. If the energy is mainly stored in only one of the first (or second) modes when $t = 0$, i.e., $N_1(t) \gg N_2(0)$ and $N_3(0)$ (or $N_2(0) \gg N_1(0)$ and $N_3(0)$), then at any t the oscillation intensity at the summed frequency (N_3) will be insignificant. In fact, if $(N_3) = 0$, then it might grow due to a decrease in $N_1(t)$ because $N_1(t) + N_3(t) = \text{const} = N(0)$. However, $N_2(t)$ would then have to decrease because $N_1(t) - N_2(t) = N_1(0) - N_2(0)$. However, $N_2(t) + N_3(t) = N_2(0)$ is a small quantity and consequently $N_3(t)$ may not grow by more than $N_2(0)$. At time $t = t'$, we would then have $N_2(t') = 0$ and $N_3(t') = N_2(0)$. Thus, the energy of a high-frequency oscillation may not grow at the expense of another, even if it is a low-frequency oscillation, even though this does not in principle contradict the law of the conservation of energy. The energy conservation law may be obtained by multiplying (17.7) by respectively ω_1, ω_2, ω_3, and then summing them to get

$$\frac{d}{dt}(\omega_1 \dot{N}_1 + \omega_2 \dot{N}_2 + \omega_3 \dot{N}_3) =$$

$$= i\sigma(a_3 a_2^* a_1^* - a_3^* a_2 a_1)(-\omega_1 - \omega_2 + \omega_3).$$

However $\omega_1 + \omega_2 = \omega_3$ and consequently

$\omega_1 N_1 + \omega_2 N_2 + \omega_3 N_3 = \text{const}$ or $\omega_1 A_1^2 + \omega_2 A_2^2 + \omega_3 A_3^2 = \text{const}$,

$a_j = A_j e^{i\phi_j}$.

2. If at $t = 0$ the energy is stored mainly in the high-frequency mode, i.e., $N_3(0) \gg N_1(0)$ and $N_2(0)$, then the situation is different. It follows from the integrals in (17.8) that N_1 and N_2 may simultaneously grow at the expense of N_3, i.e., it is possible for energy to go from a high-frequency mode to a low-frequency mode. This process is called decay or decay instability. We have found that a low-frequency mode may not transfer energy to a high-frequency one, but a high-frequency mode may decay, i.e., its energy may be transferred to a low-frequency mode. This is easily explained in terms of quasiparticles. Although the energy-conservation law does not prohibit the decay of low-frequency modes, it is still important that the law is

fulfilled in each elementary interaction between quasiparticles: $\hbar\omega_1 + \hbar\omega_2 = \hbar\omega_3$ ($\hbar\omega_j$ is the energy of a quantum with frequency ω_j). When the number of quanta with frequency ω_1 is large at the initial moment in time, and the number of quanta with frequency ω_2 is small, the numbers of quanta with frequency ω_3 during the interaction remains small (quanta with frequency ω_1 cannot combine with them). The decay of quanta with frequency ω_3 to quanta frequency ω_2 and ω_1 is not forbidden. The oscillograms N_j for the case when the low-frequency mode is maximum ($N_1(0) \gg N_2(0)$ and $N_3(0)$) is shown in Fig. 17.2a. Figure 17.2b illustrates the case when $N_3(0) \gg N_1(0)$ and $N_2(0)$.

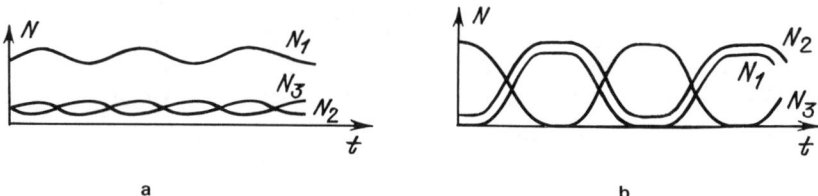

a   b

Fig. 17.2. Oscillograms of oscillation intensities with the j-th normal frequency (j = 1, 2, 3): a) $N_1(0) \gg N_2(0)$ and $N_3(0)$, b) $N_3(0) \gg N_1(0)$ and $N_2(0)$.

We can make another important conclusion on the basis of a quantum analogy, namely, if the system parameters change slowly, then the quantity defined by (17.8) is an adiabatic invariant. Thus, quantum oscillators do not change the quantum number when the parameters change slowly, i.e., the number of quanta in the absence of merging or decay is an adiabatic invariant. However, where there is merging or decay then the difference $N_1(t) - N_2(t)$, which is not used during merging, is conserved when the parameters change slowly and obviously the sum of the quanta N_3 arising at time t and the quanta N_3 of those that have not disappeared at time t, i.e., $N_2(t) + N_3(t)$, is also an adiabatic invariant.

In the special case of a constant phase difference, it is easy to construct the phase-plane portrait for (17.6), which describes interactions between weakly nonlinear oscillators. Assuming that $a_j = A_j \exp(i\phi_j)$ in (17.6), we obtain

$$\dot{A}_1 = \sigma A_2 A_3 \sin \Phi, \quad \dot{A}_2 = \sigma A_1 A_3 \sin \Phi, \quad \dot{A}_3 = -\sigma A_1 A_2 \sin \Phi,$$

$$\dot{\Phi} = -\sigma\{A_1 A_2/A_3 - A_1 A_3/A_2 - A_2 A_3/A_1\}\cos \Phi, \qquad (17.9)$$

where $\Phi = \phi_3 - \phi_2 - \phi_1$. We shall assume that $\Phi = \pi/2 = $ const. Then (17.9) may be written as

$$\dot{A}_1 = \sigma A_2 A_3, \quad \dot{A}_2 = \sigma A_3 A_1, \quad \dot{A}_3 = -\sigma A_1 A_2. \qquad (17.10)$$

In this special case, we can explain the phase trajectories in a three-dimensional phase space (A_1, A_2, A_3). They will lie on a surface of constant energy, viz., the ellipsoid

$$\omega_1 A_1^2 + \omega_2 A_2^2 + \omega_3 A_3^2 = \text{const}$$

with semiaxes $(\omega_2 \omega_3)^{1/2}$, $(\omega_1 \omega_3)^{1/2}$ and $(\omega_1 \omega_2)^{-1/2}$. The phase trajectories are obtained when this ellipsoid intersects the surfaces $A_1^2 + A_3^2 = $ const, $A_1^2 - A_2^2 = $ const, and $A_1^2 - A_2^2 = $ const. Close to the A_1 and A_2 axes (Fig. 17.3), the phase trajectories (type 1) are ellipses, i.e., each of the modes A_1 and A_2, when lightly disturbed, indeed undergoes small oscillations about the initial value. The mode with the highest frequency may decay, i.e., completely transfer its energy to modes A_1 and A_2 (type 2 trajectory). Our equations (17.10) coincide with Euler's equations, which describe the free motion of a solid with a fixed point, and a moment of inertia relative to the major axes satisfying $I_3 > I_2 > I_1$. These equations have the form [2]

$$\frac{d\Omega_1}{dt} = -\frac{I_3 - I_2}{I_2} \Omega_2 \Omega_3, \quad \frac{d\Omega_2}{dt} = \frac{I_3 - I_1}{I_1} \Omega_1 \Omega_3,$$

$$\frac{d\Omega_3}{dt} = -\frac{I_2 - I_1}{I_3} \Omega_1 \Omega_2.$$

The rotation of a body around an axis with an average moment of inertia I_2 is unstable, i.e., it is the analog of the decay of mode ω_3.[1]

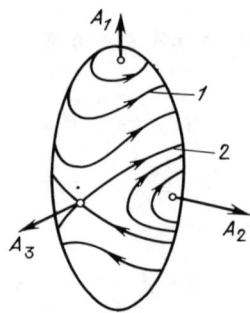

Fig. 17.3. Phase-plane portrait of a system describing the interactions between three weakly nonlinear oscillators in three-dimensional phase space (A_1, A_2, A_3).

When $N_1(0) \gg N_2*0)$ and $N_3(0)$, system (17.6) may be integrated by assuming that $a_1(0) = a_1^0$ = const. This is called the given field approximation. Then we may write (17.6) as $\dot{a}_2 = -i\sigma a_3 (a_1^0)^*$ and $\dot{a}_3 = -i\sigma a_1^0 a_2$. It follows from here that $\ddot{a}_2 + \sigma^2 |a_1^0|^2 a_2 = 0$ and $\ddot{a}_3 + \sigma^2 |a_1^0|^2 a_3 = 0$, i.e., a_2 and a_3 will change periodically thus

$$a_2 = a_2(0)\sin[\sigma|a_1^0|t + \phi_2(0)]$$

$$a_3 = a_3(0)\sin[\sigma|a_1^0|t + \phi_3(0)] \ .$$

If, however, $N_3(0) \gg N_2(0)$ and $N_1(0)$, then the change in the amplitudes a_2 and a_1 can be understood for small t. Thus, we have

$$\dot{a}_1 = -i\sigma a_3^0 a_2^* \ , \quad \dot{a}_2 = -i\sigma a_3^0 a_1^* \ .$$

It follows that $\ddot{a}_1 - \sigma^2|a_3^0|^2 a_1 = 0$ and $\ddot{a}_2 - \sigma^2|a_3^0|^2 a_2 = 0$, i.e., a_1 and a_2 grow exponentially. However, it follows from the Manley-Rowe relations (17.8) that this growth must be limited by $a_3(0) = a_3^0$.

We shall give the solution to (17.9) in the general form

---

[1] Many interesting problems like the one presented arise when analyzing the oscillations of Earth satellites around their relative equilibria in orbit [3].

for an arbitrary initial phase difference $\Phi = \phi_3 - \phi_2 - \phi_1$ [9,12]. The last of the equations in (17.9) is written in the form

$$\frac{d\Phi}{dt} = \sigma \left[\frac{A_3 A_2}{A_1} + \frac{A_3 A_1}{A_2} - \frac{A_1 A_2}{A_3} \right] \cos \Phi = \cot \Phi \frac{d}{dt} \ln(A_1 A_2 A_3) \ .$$

By integrating this equation we find $A_1 A_2 A_3 \cos \Phi = G = \text{const}$. This integral and the Manley-Rowe relations (17.8) can be used to obtain an equation for $N_3(t)$ from (17.9), i.e.,

$$dN_3(t)/dt - 2\sigma [N_3(C_3 - N_3)(C_2 - N_3) - G^2]^{1/2} \ .$$

If the three roots of the equation $N_3(C_3 - N_3)(C_2 - N_3) = G^2$ are arranged in descending order, then the equation for $N_3(t)$ may be transformed to

$$\sigma(t - t_0) = $$

$$= - (1/2) \int_{N_3(t_0)}^{N_3(t)} [(N_3 - N_c)(N_3 - N_b)(N_3 - N_a)]^{-1/2} dN_3$$

$(N_c \geq N_b \geq N_a \geq 0)$. The integral on the right-hand side can be transformed by a change of variables $y(t) = [(N_3(t) - N_c)/(N_b - N_a)]$ to an elliptic form, $(y(t_0) = 0)$, i.e.,

$$\sigma(t - t_0)(N_c - N_a) = - \int_0^{y(t)} [(1 - y^2)(1 - \alpha^2 y^2)]^{-1/2} dy \ ,$$

whence $y(t) = \sinh[\sigma(N_c - N_a)^{1/2}(t_0 - t); \alpha]$ and $\alpha = [(N_b - N_a)/(N_c - N_a)]^{1/2}$.

Finally, we obtain a general solution for $N_3(t)$,

$$N_3(t) = N_a + (N_b - N_a)\sinh^2[\sigma(N_c - N_a)^{1/2}(t_0 - t); \alpha] \ .$$

We leave the reader to obtain the general solution for the special cases we considered above.

To complete this section, we shall dwell briefly on degenerate resonance interactions between oscillators with frequencies ω and 2ω. We shall consider the resonance interaction between nonlinearly coupled oscillations using a simple model of a spring pendulum (Fig. 17.4a), whose equations neglecting friction are

$$\ddot{u}_1 + \frac{k}{m} u_1 = l\left[\dot{u}_2^2 - \frac{g}{2l} \dot{u}_2^2\right],$$

$$\ddot{u}_2 + \frac{g}{l} u_2 = -\frac{1}{l}\left(\frac{g}{l} u_1 u_2 - 2u_1 \ddot{u}_2\right).$$

When this was solved by the averaging method it was noted that when the parameter ratio was $k/m \approx 4g/l$, i.e., when $\omega_{vert} \approx 2\omega_{ang}$, there was a periodic jump in energy from the angular oscillations to the vertical oscillations and vice versa (Fig. 17.4b), which was immediately confirmed experimentally.

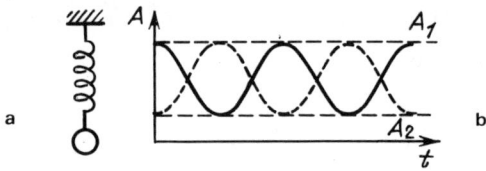

Fig. 17.4. a) A spring pendulum. b) Periodic exchange of energy between the angular oscillations and the vertical oscillations.

R.V. Khokhlov discovered when studying the stationary nonlinear operation of a parametric, travelling wave amplifier that when propagating along the amplifier the pumping wave $2\omega_0$ parametrically amplified the initial wave ω_0, transferring nearly all of its energy to it [1,4,5,8].

As it propagates further, the reverse occurred, i.e., the intense wave ω_0 generates a secondary harmonic and then the whole sequence is repeated, i.e., the periodic transfer of energy between harmonics was observed (Fig. 17.4b).

In the simplest case corresponding to (17.6), when $\omega_1 = \omega_2 = \omega$ and $\omega_3 = 2\omega$ (the synchronicity condition $\omega + \omega = 2\omega$ is fulfilled exactly and the interaction is called degenerate) we have

$$i\dot{a}_\omega = \sigma a_{2\omega} a_\omega^*, \quad i\dot{a}_{2\omega} = \sigma a_\omega^2.$$

The character of the interaction described by this system is quite different to the degenerate case considered above. The differences are:

1) If only an oscillation with the fundamental frequency ω is excited at the initial moment in the system, i.e., $a_\omega(0) = a_\omega^0$ and $a_{2\omega}(0) = 0$, then as time goes by an oscillation with the harmonic frequency 2ω appears, and the energy from the first mode is transferred to energy in the second harmonic. The process of merging between the quasiparticles $\omega + \omega = 2\omega$ always occurs.

2) The rate of growth of each mode in (17.6) only depends on "strange" amplitudes, and in the degenerate case the change in a_ω depends on the fundamental amplitude.

3) If at $t = 0$ we have $a_\omega = 0$, then the oscillation with this frequency does not appear.

Assuming weak nonlinearity, the short (averaged) equations for the amplitudes and phases of the oscillators ω and 2ω interacting over time or in space, are written as

$$\dot{A}_1 = -\sigma_1 A_1 A_2 \sin \Phi, \quad \dot{A}_2 = \sigma_2 A_1^2 \sin \Phi,$$

$$\dot{\Phi} = -(2\sigma_1 A_2 - \sigma_2 A_1^2/A_2)\cos \Phi - \Delta\omega,$$

where $\Phi = 2\phi_1 - \phi_2 - \Delta\omega t$. These equations can easily be transformed to those of nonlinear oscillator if the integral of energy $\sigma_2 A_1^2(t) + \sigma_1 A_2^2(t) = \text{const} = \sigma_1 A_0^2$ is utilized and the new variables $X = A_2 \sin \Phi$ and $Y = A_2 \cos \Phi$ are introduced. The phase-plane diagrams of the resultant oscillator for various detunings are given in Fig. 17.5.

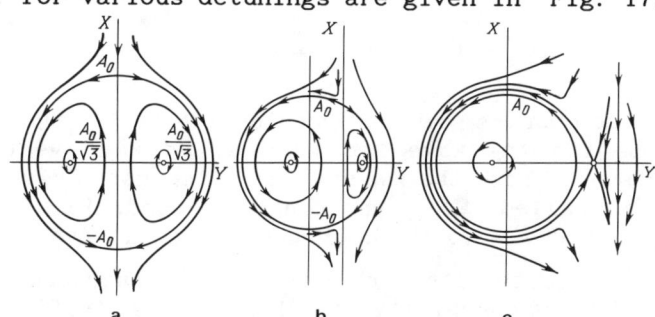

a b c

Fig. 17.5. Phase-plane diagrams of a nonlinear oscillator describing the exchange of energy between harmonics in a system with quadratic nonlinearity: a) $\Delta\omega t = 0$; b) $|\Delta\omega|/2\sigma_1 A_0 < 1$; c) $|\Delta\omega|/2\sigma_1 A_0 > 1$.

We can see that when the assumption of weak nonlinearity (or small initial excitation energy, which is the same thing) is made a system of two nonlinearly coupled oscillators only demonstrates simple, quasiperiodic motions. From the physical point of view the difference between these motions (Fig. 17.5) is that there are different depths of the energetic beats between the oscillators and these beats have different periods.

17.2 Resonance Interactions Between Waves in Weakly Nonlinear Media with Dispersion

In our analysis of the interactions in a system of three coupled oscillators we noted that in a medium with dispersion and for weak nonlinearity, three waves with fixed spatial structures will interact in the same way. True, the resonance condition must be fulfilled both for the frequencies and for the wave numbers. However, the method for investigating multiwave interactions in a medium with dispersion has its own features, which we shall now discuss.

Suppose waves with $\omega = \omega_j$ and $k_j = k(\omega_j)$ are propagating in a medium with dispersion and small nonlinearity, where the relation $k(\omega)$ is determined by the dispersion equation. As the result of interactions due to the nonlinearity in the medium a forced combination wave will arise with frequency $\omega_N = \sum_{j=1} n_j \omega_j$ and wave number $k_N = \sum_{j=1} n_j k_j$. This wave remains small (of the order of the nonlinearity) if ω_N and k_N do not satisfy the dispersion equation $D(\omega_N, k_N) \neq 0$, and will grow if the equation $D(\omega_N, k_N) = 0$ is satisfied, or in other words when

$$\omega_N = \omega_i \, , \quad k_N = k_i(\omega_N) \, . \qquad (17.11)$$

These relationships can be considered to be the resonance conditions for the wave frequencies and wave vectors which are necessary for the waves to interact. They are also frequently called the synchronicity conditions since the phase velocity $\mathbf{v} = (\omega_N/k^2)\mathbf{k}$ of the combined wave is close to the phase velocity of one of the medium's fundamental waves.

If the synchronicity conditions are fulfilled for a large number of waves, then interactions will cause the waves to have forms very different from sinusoidal. The quasiharmonic approximation will then not work. However, frequently the number of interactive waves is quite small. Problems such as these are very important for nonlinear optics, the physics of solids and plasma physics. For example, the classical problem of nonlinear optics concerns forced Mandel'shtam–Brillouin scattering [4,5]. A light wave with frequency ω_1 incident on a crystal causes a modulation in the density of the medium (the electrostriction effect), and an acoustic wave with frequency ω_2 arises. The light is reflected from the resultant inhomogeneity, leading to a wave with frequency $\omega_3 = \omega_1 - \omega_2$ propagating in the opposite direction (Fig. 17.1d). The interaction between the waves in the one-dimensional case (a light wave with electromagnetic field $E = E_y$ and $H = H_z$ is propagating in the x-direction) is described by the system of equations

$$\frac{\partial E}{\partial x} + \frac{1}{c}\frac{\partial H}{\partial t} = 0,$$

$$\frac{\partial H}{\partial x} + \frac{\varepsilon}{c}\frac{\partial E}{\partial t} - \frac{1}{c}\left(\frac{\partial \varepsilon}{\partial \rho}\right)\frac{\partial}{\partial t}(\rho E) = 0,$$

$$\frac{\partial \rho}{\partial t} + \rho_0 \frac{\partial u}{\partial x} = -\frac{\partial}{\partial x}(\rho u),$$

$$\frac{\partial u}{\partial t} + \frac{c_{snd}^2}{\rho}\frac{\partial \rho}{\partial x} = -u\frac{\partial u}{\partial x} - \frac{c_{snd}^2}{\rho_0^2}\frac{\partial \rho}{\partial x} + \eta\frac{\partial^2 u}{\partial x^2} + \frac{1}{8\pi}\left(\frac{\partial \varepsilon}{\partial \rho}\right)\frac{\partial E}{\partial x}.$$

(17.12)

The first two equations describe a change in the light waves' electromagnetic field taking into account the changes in the dielectric permittivity of the medium ε due to the presence in the medium of density perturbations. The last two determine the change in the density ρ and the particle velocity u in the sound wave taking into account the ponderomotive force that arises due to electrostriction effect. The first of these last two equations is the continuity equation and the second is the equation of motion. Let us look at how (17.2) should be solved if the right-hand sides of the equations, which characterize the nonlinear coupling, are small? Since the changes in the amplitudes and phases of quasiharmonic waves that interact effectively must change slowly due to the smallness of the nonlinearity, it is natural to apply a method that is in some way associated with averaging over a time and the space variable (we recommend the reader going through this section recalls section 17.1).

We shall consider a general procedure for constructing such a method assuming that the waves are one-dimensional [12]. Let the field in a weakly nonlinear medium be described by equations of the form

$$AU_t + BU_x + CU = \mu f(U, U_t, U_x), \quad \mu \ll 1, \qquad (17.13)$$

where A, B, and C are square matrices, U and f are n-dimensional vector functions with f being a polynomial in U, U_t, U_x. The variables E, H, ρ, and u are components of the vector U for (17.12). When $\mu = 0$, the field in the medium is a superposition of harmonic waves

$$U = a\psi \exp(i\omega t - ik(\omega)x) + c.c. , \qquad (17.14)$$

where a is a complex amplitude that depends on the initial and boundary conditions, and ψ is a polarization vector defined by the system

$$(i\omega A - ikB + C)\psi = 0 , \qquad (17.15)$$

and ω and k are linked via the dispersion equation

$$D(\omega, k) = \text{Det}| |A\omega - Bk - iC| | = 0 . \qquad (17.16)$$

One of the components of the ψ-vector may always be set equal to unity and the rest found from (17.16). We shall considered the interaction of a finite number of waves in the form (17.14), for which the synchronicity conditions (17.11) are fulfilled. (That the synchronicity conditions are only fulfilled for a finite number of waves means that the system has dispersion.) When $\mu = 0$, the solution will be sought in the form

$$u(x, t) = \sum_j a_j(\mu x, \mu t)\psi(\omega_j, k_j)\exp(i\omega_j t - ik_j t) + cc$$
$$+ \mu w(x, t) , \qquad 17.17)$$

assuming beforehand that the amplitude changes slowly in space and overtime. In order for solution (17.18) to be valid the correction $w(x, t)$ may not grow over time. By substituting the solution into (17.13) we obtain an equation for w in the form

$$Aw_t + Bw_x + Cw = h(x, t) , \qquad (17.18)$$

$h(x, t) =$

$$-\sum_j \exp(i\omega_j t - ik_j x)\left[A\psi(\omega_j, k_j)\frac{\partial a_j}{\partial t} + B\psi(\omega_j, k_j)\frac{\partial a_j}{\partial x}\right]$$

$$+ cc + f\left(\sum_j a_j\psi_j \exp(i\omega_j t - ik_j x) + cc\right) .$$

In order for the function $w(x, t)$ to be bounded given any x and t, it is necessary and sufficient that resonance times be

RESONANCE INTERACTIONS   367

absent from the right-hand sides of (17.18), i.e., the right-hand side must be orthogonal to the eigenfunctions of the linear problem. Since the nonlinearity is polynomial, the right-hand sides of (17.18) are periodic functions in x and t and they may be represented as Fourier series

$$h(x, t) = \sum_{r=1}^{N} F_r \exp(i\omega_r t - ik_r x) + \text{c.c.} , \quad (17.19)$$

$$F_r = (\lambda T)^{-1} \int_{x}^{x+\lambda} \int_{t}^{t+T} h(x, t)\exp(-i\omega_r t + ik_r x)dx\, dt .$$

The functions w also can be represented as

$$w(x, t) = \sum_{r=1}^{N} W_r \exp(i\omega_r t - ik_r x) + \text{cc} \quad (17.20)$$

After substituting (17.19) and (17.20) into (17.18), we obtain a nonhomogeneous system of algebraic equations for W after equating the coefficients of exponentials with the same exponents, i.e.,

$$(i\omega_r A - ik_r B + C)W_r = F_r ,$$

whence the l-component of the vector W_r is written in the form

$$W_1(\omega_r, k_r) = - iD^{-1}(\omega_r, k_r) \sum_{j=1}^{N} D_{j1} F_j(\omega_r, k_r) ,$$

where the D_{j1} are the cofactors of the matrix elements $(A\omega - Bk - iC)$ defined in (17.6).

If the terminus of the vector (ω_r, k_r) does not lie on the dispersion curve, i.e., it is not a fundamental wave of the system, then $D(\omega_r, k_r) \neq 0$ and the additive corrections are bounded. In the opposite case, W will grow secularly. In order for this not to be so, it is necessary to eliminate the resonance term from the equation for the correction. Mathematically, this reduces to a fulfillment of the equality

$$\sum_{j=1}^{N} D_{j1} F_j(\omega_r, k_r) = 0 . \quad (17.21)$$

Since $D_{j1} = \gamma \zeta_j^* \psi_1$, where ζ_j^* is an eigenfunction of the system and conjugate with (17.15), condition (17.21) may be written in the form $\sum_{j=1}^{N} \zeta_j^* F_j(\omega_r, k_r) = 0$. This is the orthogonality condition. Using (17.18) and (17.19) for the complex amplitudes a_j we obtain from here the equation

$$(\zeta, A\psi)\partial a_j/\partial t + (\xi, B\psi)\partial a_j/\partial x = (\xi, <f(\omega_j, k_j)>) , \quad (17.22)$$

where
$$<f(\omega_j, k_j)> =$$

$$= (\lambda T)^{-1} \int_x^{x+\lambda} \int_t^{t+T} f(\omega_j, k_j) \exp(-i\omega_j t + ik_j x) \, dx \, dt .$$

Using the relations

$$\frac{(\zeta, B\psi)}{(\zeta, A\psi)} = -\frac{D'_k}{D'_\omega} = \frac{d\omega}{dk} = v_{gr} , \quad \frac{\zeta_1}{(\zeta, A\psi)} = \frac{D_{1k}}{\psi_k D'_\omega} ,$$

we write (17.22) as

$$\frac{\partial a_j}{\partial t} + v_{gr}(\omega_r, k_r) \frac{\partial a_j}{\partial x} = (\psi_j D'_\omega)^{-1} \sum_l D_{lj} <f_l> . \quad (17.23)$$

This is the equation we want for the complex amplitudes of interacting quasiharmonic waves.

By way of example, let us consider the interaction between high-frequency and low-frequency electromagnetic waves in a medium whose dispersion characteristic is shown in Fig. 17.1c.[2] This medium consists of oscillators with fundamental frequency ω_0 and each element of the volume has a polarizability χ. When the nonlinearity is quadratic, it is natural to consider the interaction between three waves as an elementary process. The synchronicity condition has the form

$$\omega_1 + \omega_2 = \omega_3 , \quad k_1 - k_2 = k_3 . \quad (17.24)$$

The equations for the electromagnetic field components and

---

[2] Resonance interactions when applied to waves on water are well presented in [6], and for plasmas there is [7].

the medium's polarization are written as

$$\frac{\partial E}{\partial t} + 4\pi \frac{\partial P}{\partial t} + c \frac{\partial H}{\partial x} = -\mu j(E), \quad \frac{\partial H}{\partial t} + c \frac{\partial E}{\partial x} = -\mu \frac{\partial B(H)}{\partial t},$$

$$\frac{\partial P}{\partial t} - R = 0, \quad \frac{\partial R}{\partial t} + \omega_0^2 P - \omega_0^2 \chi E = 0. \quad (17.25)$$

where $(\chi = (\varepsilon - 1)/4\pi)$. The dispersion equation of (17.25) is easily found and is

$$D(\omega, k) = \begin{vmatrix} \omega & -ck & 4\pi\omega & 0 \\ -ck & \omega & 0 & 0 \\ 0 & 0 & \omega & i \\ i\omega_0^2 \chi & 0 & -i\omega_0^2 & \omega \end{vmatrix}$$

$$= \omega^2(\omega^2 - \varepsilon\omega_0^2) - c^2 k^2 (\omega^2 - \omega_0^2) = 0$$

We set $\psi_E = 1$ and thus we have

$$\psi_N = \frac{1}{D_{11}} \begin{vmatrix} ck & 0 & \omega \\ 0 & \omega & t \\ i\omega_0^2 \chi & -i\omega_0^2 & \omega \end{vmatrix},$$

where $D_{11} = \omega(\omega^2 - \omega_0^2)$, i.e., $\psi_N = ck/\omega$. By analogy we have $\psi_P = \omega_0^2 \chi/(\omega_0^2 - \omega^2)$, $\psi_R = -i\omega\omega_0^2\chi/(\omega_0^2 - \omega^2)$. We shall look for the solution in the form of a sum of waves, i.e.,

$$E, H, P, R =$$

$$= \sum_{j=1}^{3} \psi_{E,N,P,R}(\omega_j, k_j) a_j(\mu x, \mu t) \exp(i\omega_j t - k_j x) + \text{c.c.}$$

$$(17.26)$$

The equation for a complex amplitude has the form

$$\frac{\partial a_j}{\partial t} + v_j \frac{\partial a_j}{\partial x} = (\psi_E D'_\omega)^{-1} \sum_{l=1}^{4} D_{lj} \langle f_l \rangle = f_j, \quad (17.27)$$

where

$$f_j = -\left.\frac{D_{11}}{D'_\omega}\right|_{\omega_j, k_j} \left\langle j\left(\sum_{j=1}^{3} a_j \exp(i\omega_j t - ik_j x) + cc\right)\right.$$

$$\left. \cdot \exp(-i\omega_j t + ik_j x)\right\rangle$$

$$-\left.\frac{D_{21}}{D'_\omega}\right|_{\omega_j, k_j} \left\langle \frac{dB(H)}{dH} \frac{\partial H}{\partial t} \left(\sum_{k=1}^{3} \psi_N(\omega_k, \omega_k) a_k \exp(i\omega_k t - ik_k x)\right.\right.$$

$$\left.\left. + cc\right) \exp(-i\omega_j t + k_j x)\right\rangle ,$$

and

$$D_{21} = -ck(\omega^2 - \omega_0^2), \quad D'_\omega = 2\omega(\omega^2 - \omega_0^2)^2 + \omega_0^4 4\pi\chi/(\omega^2 - \omega_2^2) .$$

We now give the concrete form of the nonlinear relationships, i.e., $j(E)$ and $B(H)$. Let

$$J(E) = gE + g_N E^2, \quad B(H) = \mu_N H^2 . \tag{17.28}$$

If the synchronicity conditions are not fulfilled exactly then we have

$$\omega_1 + \omega_2 = \omega_3 + \Delta\omega, \quad k_1 + k_2 = k_3 + \Delta k , \tag{17.29}$$

where $\Delta\omega$ is the frequency detuning, and Δk is the detuning from resonance in terms of a wave number. Using (17.26) and (17.28), we find from (17.27) the following short equations:

$$\frac{\partial a_1}{\partial t} + v_1 \frac{\partial a_1}{\partial x} = \mu\left[g_1 a_1 + \sigma_1 a_3 a_2^* e^{i\Delta\omega t} + i\zeta_1 a_3 a_2^* e^{i\Delta\omega t}\right] ,$$

$$\frac{\partial a_2}{\partial t} + v_2 \frac{\partial a_2}{\partial x} = \mu\left[g_2 a_2 + \sigma_2 a_3 a_1^* e^{i\Delta\omega t} + i\zeta_2 a_3 a_1^* e^{i\Delta\omega t}\right] ,$$

$$\tag{17.30a}$$

$$\frac{\partial a_3}{\partial t} + v_3 \frac{\partial a_3}{\partial x} = \mu \left[g_3 a_3 + \sigma_3 a_1 a_2 e^{-i\Delta\omega t} + i\zeta_3 a_1 a_2 e^{-i\Delta\omega t} \right],$$

$$g_j = \frac{D_{11}}{D'_\omega} g \bigg|_{\omega_j, k_j}, \quad \sigma_j = \frac{D_{11}}{D'_\omega} g_N \bigg|_{\omega_j, k_j}, \quad \zeta_j = \frac{D_{21}}{D'_\omega} \mu_N \bigg|_{\omega_j, k_j}.$$

(17.30b)

Note that D_{11}/D'_ω and D_{21}/D'_ω, and consequently g_j, σ_j and ζ_j depend only on the frequency squared and the wave numbers, i.e., they always have the same sign and are real. The function $j(E)$ determines the dissipative nonlinearity, and it is related to σ_j in (17.30). The nonlinear relation $B(H)$ determines the conservative nonlinearity and enters equations with coefficients ζ_j.

Let us consider several cases.

1. If we assume that the field is spatially homogeneous, i.e., $\partial a_j/\partial x = 0$, then the wave interaction can be described by the same equations as the oscillations in a system of three coupled oscillators. This description is called the given field structure approximation. We know that when $\Delta\omega = 0$ in a conservative system (i.e., when $j(E) = 0$) energy will be exchanged between the modes if the high-frequency mode has a large initial energy. If, however, the synchronicity is not exact, i.e., $\Delta\omega \neq 0$, then for small $\Delta\omega$ it is natural to assume that the nature of the interactions will be analogous, although complete energy exchange is improbable (Fig. 17.6a).

2. Let the system be conservative, as before, i.e., $j(E) = 0$. We now assume that the process is steady-state, i.e., $\partial a_j/\partial t = 0$. This is already a wave problem because we are considering interactions between waves in space. In this case, as before, system (1730) reduces to an ordinary differential equation.

a) If all the v_j in (17.30) have the same sign, i.e., all three waves are propagating in the same direction, then the problem reduces to the previous one. Thus, if the medium is nonlinear and conservative (e.g., a crystal), and we place a wave with frequency ω_3 at the boundary of the medium, then when there are fluctuations two different waves with frequencies ω_1 and ω_2 will be excited, and the amplitudes will change in space as shown in Fig. 17.6b. Thus, it is possible to choose a crystal length l such that we obtain a low-frequency wave with maximum amplitude at the output.

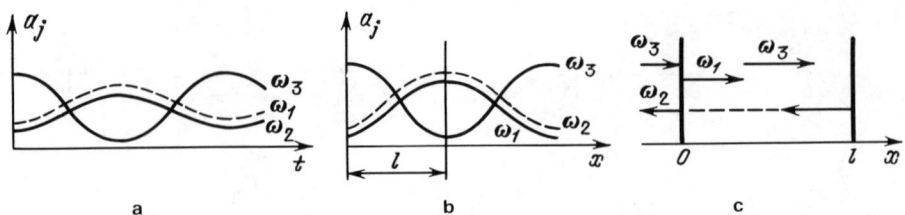

Fig. 17.6. An investigation of (17.30) in a conservative case when $j(E) = 0$; $\omega_1 + \omega_2 = \omega_3 + \Delta\omega$; $k_1 + k_2 = k_3 + \Delta k$: a) the approximation of a given field structure $\partial a_j/\partial x = 0$, $\Delta\omega \neq 0$, incomplete energy exchange between the modes; b) amplification of waves with frequencies ω_1 and ω_2 when a signal with frequency $\omega_3 (v_j > 0)$ is placed at the input of a nonlinear system; c) diagram of scattering backwards for v_1, $v_3 > 0$, but $v_2 < 0$.

b) If, however, v_1, $v_3 > 0$, and $v_2 < 0$, then a wave with frequency ω_2 will be scattered backwards. When considering this case the presence of a boundary is crucial (Fig. 17.6c).

17.3 Explosive Instability

We consider the interaction between spatially homogeneous fields in a nonconservative medium, i.e., $\partial a_j/\partial x = 0$, $B(H) = 0$, $j(E) \neq 0$. An example of such a medium is an active transmission line, one of whose possible realizations is given in Fig. 17.7a. A tunnel diode is an active element in the circuit. The equations for the complex amplitudes of the interacting waves in the system have the form (the nonlinearity brought by a tunnel diode is considered to be quadratic)

$$\dot{a}_1 = a_3 a_2^* \exp(i\Delta\omega t), \quad \dot{a}_2 = a_3 a_1^* \exp(i\Delta\omega t),$$

$$\dot{a}_3 = a_1 a_2 \exp(-i\Delta\omega t).$$

We change to real amplitudes and phases by the substitution $a_j = A_j \exp(i\phi_j)$ and thus get

RESONANCE INTERACTIONS 373

$$\dot{A}_1 = A_2 A_3 \cos \Phi, \quad \dot{A}_2 = A_3 A_1 \cos \Phi, \quad \dot{A}_3 = A_1 A_2 \cos \Phi,$$

(17.31)

$$\dot{\Phi} = -(A_1 A_3 / A_2 + A_1 A_2 / A_3 + A_3 A_2 / A_1) \sin \Phi + \Delta \omega,$$

where $\Phi = \phi_3 - \phi_1 - \phi_2 - \Delta \omega t$.

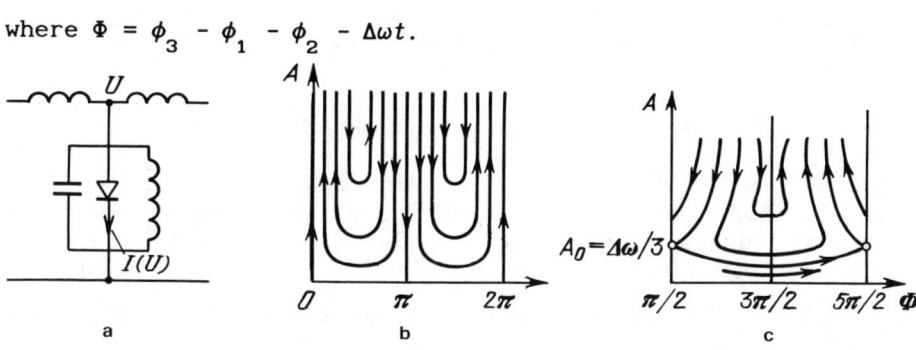

Fig. 17.7. Interaction between spatially homogeneous fields in a nonconservative medium: a) an active transition line as an example of the medium; b) the integral curves $A^3 \sin \Phi = $ const on a phase cylinder ($\Delta \omega = 0$); c) the integral curves $\sin \Phi = (\Delta \omega A + \text{const}) A^{-2}$ on a phase cylinder ($\Delta \omega \neq 0$).

1. Let $\Delta \omega = 0$, i.e., the waves are synchronized. Then (17.31) has two independent integrals of motion $A_1^2 - A_2^2 = $ const and $A_3^2 - A_2^2 = $ const, and a third follows from these, i.e., $A_3^2 - A_1^2 = $ const. The existence of these relationships means that if $A_i(0) = A(0)$ ($i = 1, 2, 3$) and $\phi_3 - \phi_1 - \phi_2 = 0$ when $t = 0$, then for $t \neq 0$ the amplitudes of all the waves will be the same. On this basis, we need only consider two equations in (17.31):

$$\dot{A} = A^2 \cos \Phi, \quad \dot{\Phi} = -3A \sin \Phi. \quad (17.32)$$

By multiplying the first equation by $A\dot{\Phi}$, and the second by $A\dot{A}$, and taking one equation from the other, we obtain

$$A^3 \cos \Phi \cdot \dot{\Phi} + 3A^2 \sin \Phi \cdot \dot{A} = 0 \quad \text{or} \quad d(A^3 \sin \Phi)/dt = 0$$

We have found, in this way, an equation for the integral curves $A^3 \sin \Phi = $ const on a phase cylinder (Fig. 17.7b). It follows from (17.32) that there is a whole straight line of equilibria ($A = 0$) in the system. If $\cos \Phi > 0$ then A grows, whilst when $\cos \Phi < 0$ we see that A dies away. As $\Phi(t) \to 0$, we have $2\pi \sin \Phi(t) \to 0$, i.e., the phase difference of the interacting waves decays to zero and the phases synchronize. Here $\dot{A} = A^2$ and $A(t) = 1/(1/A(0) - t)$. When $t' = 1/A(0)$, the amplitudes of the interacting waves rise to infinity. Note that the amplitudes grow more quickly than exponential because they reach infinity in finite time. This is called explosive instability [7]. The phenomenon of explosive instability can be seen, for instance, in media where the dissipative nonlinearity is quadratic ($\sim E^2$).

2. Let us consider the case of odd synchronicity, i.e., $\Delta\omega \neq 0$. Using the same assumptions, we obtain the following instead of (17.32)

$$\dot{A} = A^2 \cos \Phi, \quad \dot{\Phi} = -2A \sin \Phi + \Delta\omega.$$

Whence we get $d(A^2 \sin \Phi)/dt = \Delta\omega dA/dt$ or $\sin \Phi = (\Delta\omega A + \text{const})/A^2$, which are the equations for integral curves on a phase cylinder (Fig. 17.7c). There is only one equilibrium here, $\Phi_0 = \pi/2$ (i.e., $\cos \Phi_0 = 0$) and $A_0 = \Delta\omega/3$. The phases also synchronize and the amplitudes tend to infinity, i.e., as before we have explosive instability. Thus, linear detuning cannot stabilize explosive instability.

Explosive instability, which appears as a one-time growth in amplitude of all the resonance-linked waves, is also possible in media without dissipation if the media are nonequilibrium [7,10]. An example is the interaction between waves with different signed energies (Chapter 10) in a plasma—electron beam system. If a wave which decays (ω_3) or a pair of low-frequency waves ($\omega_{1,2}$) has negative energy, then the right-hand sides of the equations for $\dot{a}_i$ ($i = 1,2,3$) will have the same signs, and we shall again come to an equation of the form (17.31) instead of (17.9). Because the waves have negative energies, when the energy is given to other waves (increasing their amplitude), the negative-energy waves will grow in amplitude and there will be one-time growth in all the interacting waves, viz., there will be explosive instability [11].

CHAPTER 18

SIMPLE WAVES AND THE FORMATION OF DISCONTINUITIES

18.1 Kinematic Waves

For nonlinear systems with lumped parameters the nonlinear oscillator described by the equation $\ddot{x} + f(x) = 0$ is, as we have seen (Chapter 13), the basic model. The solution of this equation is the basis on which approximate solutions can be constructed to account for disturbances in various factors, such as the external force, positive or negative dissipation (Chapters 15-17), and nonstationary parameters. In the theory of nonlinear waves there are several basic models. Primarily, there is the one-wave approximation model, whose equation is

$$\partial u/\partial t + V(u)\partial u/\partial x = 0 , \qquad (18.1)$$

which describes a plane travelling wave in a nonlinear medium without dissipation or dispersion. Then there is Burgers equation for media with damping:

$$\partial u/\partial t + V(u)\partial u/\partial x - \alpha \partial^2 u/\partial x^2 = 0 ; \qquad (18.2)$$

Finally, there is a generalized Korteveg--de Vries equation for a travelling wave in a medium with dispersion in the high-frequency range:

$$\partial u/\partial t + V(u)\partial u/\partial x + \beta \partial^3 u/\partial x^3 = 0 . \qquad (18.3)$$

The most widespread model allowing for interactions between colliding waves is the model described by the Klein--Gordon equation (for a medium with dispersion in the low-frequency range), i.e,

$$\partial^2 u/\partial t^2 - V^2 \partial^2 u/\partial x^2 + \mathcal{F}(u) = 0 . \qquad (18.4)$$

In this chapter and the next we shall be considering problems leading to these models, and the phenomena and effects which they describe [1-6, 14-16].

Before actually going on to waves in continuous media,

we shall consider a simple model that is well-known in electronics. Suppose there is a beam of noninteracting particles moving along the x-axis such that the velocity in Euler variables satisfies the equation

$$dv/dt = \partial u/\partial t + v\, \partial v/\partial x = 0 \,. \tag{18.5}$$

In electronics, (18.5) describes the behavior of an electron beam in the drift tube of a klystron device (we qualitatively considered a very simple two-resonator klystron in Chapter 1) in terms of kinematic theory. The difference between the electron velocities causes electron clusters, grouping of the electron stream, in the drift tube. Superficially, (18.5) seems very similar to the equation for a simple wave, although, of course, a beam of noninteracting particles is a not nonlinear medium.

We shall first consider waves of small amplitude, when $v = v_0 + v'$, $v' \sim \exp\{i(\omega t - kx)\}(v_0 \gg v')$. In this approximation we find from (18.5) that $\partial v'/\partial t + v_0 \partial v'/\partial x = 0$, and consequently $\omega = v_0 k$ ($v_0 = $ const), i.e, in the linear case there is no dispersion in the system. Suppose at time $t = 0$ the beam's velocity is disturbed by $a \sin kx$. We change coordinates to a system moving at a velocity v_0 and consider the evolution of the initial disturbance. By substituting in $x = x_{st} - v_0 t$ and $v = v_0 + u$, and ignoring the subscript, we obtain $\partial u/\partial t + u\, \partial u/\partial x = 0$ in this system. The solution of this nonlinear equation has the form of a "simple" wave $u = U(t - x/u)$, where the expression for U is determined by the initial perturbation. When such a wave propagates in a nonlinear medium, its profile changes over time because various points on the profile propagate with different velocities. In the case of the beam, this occurs because the particles jostle each other due to their different velocities, and because some particles overtake others. As a result, the function $u(x,t)$ becomes multivalued [7].

We shall follow a beam on the phase plane (u,x), on which each particle is shifted by its own velocity. The upper half-plane $(u > 0)$ corresponds to movement to the right and the lower half-plane $(u < 0)$ corresponds to movement to the left, the velocity of each part being proportional to its distance from the x-axis. Figure 18.1 illustrates the evolution of a beam on the phase plane (u,x). The initial state of the beam is a sinusoid $a \sin kx$ on the (u,x)-plane. Here the dashed line shows the density of the beam's volume charge as a function of x (Fig. 18.1a). As time goes by, the wave profile is distorted, particles with $u > 0$ moving forward and those with $u < 0$ lagging behind the wave. At the same time, concentrations of particles form near points 1 and

2, where $u = 0$, beam grouping thus taking place (Fig. 18.1b). The wave gradually becomes steeper and finally the derivative $\partial u/\partial x$ tends to infinity at the leading edge (the density $\rho(x)$ of the beam's volume charge also tends to infinity at this point). At the next moment the wave topples and thus ceases to be single-valued (Figs. 18.1c and d). A turning point appears on the wave, i.e, a return beam is formed. After the wave topples the function $p(x)$ doubles the number of critical points (Figs. 18.1c and d). As time grows on, the flow structure becomes more complicated and many flows arise. However, we shall not dwell on this point [2, 4].

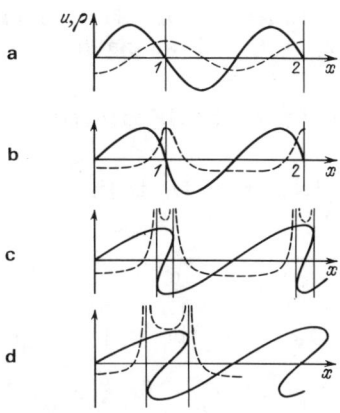

Fig. 18.1. The evolution over time of a sinusoidal disturbance in a beam of noninteracting particles (the velocity u is the continuous curve, the density ρ is the dashed curve): a) initial state of the beam corresponding to the initial velocity disturbance; b) formation of electron density, grouping of the particles close to the points 1 and 2; c,d) the velocity "wave" topples and twice the number of critical points form on the curve $\rho = \rho(x)$.

In microwave electronics, a description of electron grouping in drift space is not done with Euler variables (x, t), and instead Lagrange variables (t, t_0) or (x, t_0) are used, where t_0 is the point in time the electron flies into the drift tube. Using the kinematic approach, the residence time of an electron in the grouping space (the drift tube) is defined as $t - t_0 = \int_0^l dx/v(x, t)$, where l is the length of the drift, and x is the integration variable. The group current may be found from the charge conservation law $I(0, t_0)dt_0 = I(x, t)dt$, in which dt_0 is the time it takes the group of electrons to pass through the plane $x = 0$, and dt is the time it takes the group to pass through the x-plane. Let us assume that the device operation, which modulates the velocity before the drift tube (Chapter 1), is described by

$$mv^2/2 = mv_0^2/2 + eV_1 \sin \omega t_0,$$

where v is the velocity at the inlet to the drift tube, v_0 is the velocity in the absence of a governing high-frequency action, V_1 is the amplitude of the high-frequency controlling voltage, m and e are the mass and charge of an electron, and ω is the angular frequency of the harmonic controlling action. Thus when $V_1/V_0 = \xi \ll 1$ we have

$$v = v_0\{1 + (1/2)\xi \sin \omega t_0\},$$

where $V_0 = mv_0^2/2e$. It is clear that in this approximation

$$t = t_0 + l/v \approx t_0 + (l/v_0)\{1 - (1/2)\xi \sin \omega t_0\}$$

or for the angle of flight in the drift space

$$\omega t - \omega t_0 = \theta = \omega l/v_0 - (\omega l \xi/2v_0)\sin \omega t_0.$$

The quantity $\xi\theta_0/2$ where $\theta_0 = \omega l/v_0$ characterizes the difference between the residence times of different electrons in the drift tube. It is called the grouping parameter (the kinematic theory of grouping is elegantly presented in [8]).

A clear understanding of grouping in a drift tube comes from a space-time diagram on the $(x, \omega t_0)$-plane (Fig. 18.2). Because the stream is homogeneous in terms of velocity and density until it intersects the plane of the modulating device, the electron trajectories up to the device are divided by identical temporal (angular) intervals $\Delta \omega t_0$ (the stream is homogeneous in density) and have the same slope (a stream is homogeneous in velocity). The controlling voltage modulates the electron velocities and there is a periodic change in trajectory slope. The slope of the straight lines in Fig. 18.2 does not change for type-2 electrons because they intersect the modulator plane when the controlling voltage is zero. The slopes of the trajectories for type-1 electrons, which fall into the field's braking phase, diminish, and the slopes for the type-3 electrons increase (they fall into the accelerating phase of the field). During one period of the high-frequency force the trajectories either converge (forming a particle compression) or diverge (forming a particle rarefaction), and this illustrates the process of grouping. Sometimes the literature uses the term phase focusing by analogy with the focusing of beams of light in geometric optics instead of grouping.

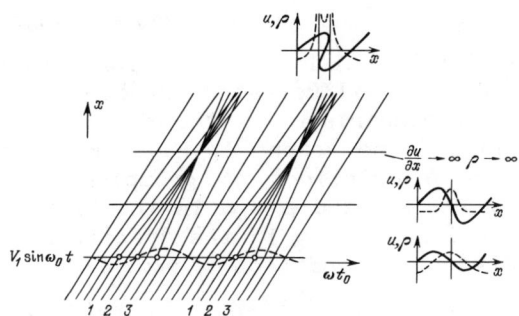

Fig. 18.2. Space-time diagram of electron grouping in a drift tube: 1) an electron, which is braked by the field; 2) an electron not affected by the field; 3) an electron accelerated by the field. The figure on the right indicates the correspondence between the "wave" (Fig. 18.1) and particle descriptions of grouping.

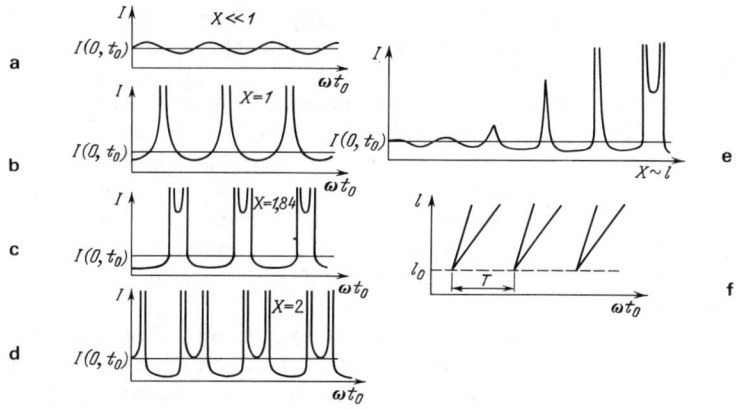

Fig. 18.3. a-e) Beam current grouped in drift space versus the initial phase of the electron entering, and versus the length of the drift. f) Trajectory of a phase focus formed in the plane $l_0 = 2v_0/\xi\omega$.

From the conservation of charge and the expression for the flight time $(t - t_0)$ we get $dt/dt_0 = 1 - (\theta_0\xi/2)\cos \omega t_0$ and the grouped current is

$$I = I(0, t_0)/[1 - (\theta_0 \xi/2)\cos \omega t_0] .$$

Note that although we assumed that $\xi \ll 1$, the grouping parameter $X = \omega_0 \xi/2$ may not be small because θ_0 may take any value. The behavior of the grouped current versus ωt_0 and versus the parameter X, which is proportional to the length l of the drift tube, is illustrated in Fig. 18.3, which is taken from [8]. A comparison of the density versus length graph (Fig. 18.1) and the graphs in Fig. 18.3 shows that the wave and particle descriptions of grouping are equivalent. The conditions $\partial u/\partial x \to \infty$ and $\rho \to \infty$ in the "wave" description correspond to $X = 1$ and $I \to \infty$ in the "particle" description. We find from the last condition $X = \omega l_0 \xi/2v_0 = 1$ that the phase focus, a compression with an infinitely large value, is formed at a distance $l_0 = 2v_0/\xi\omega$. The trajectories of such foci are given in Fig. 18.3f.

18.2 Travelling Waves in a Nonlinear Medium Without Dispersion

The absence of dispersion means that different physical variables occur instantaneously after changes in other variables as the waves move through the medium, i.e, there is no independent space-time scale (relaxation time, structural periodicity, etc.) in the medium. For instance, this is possible for electromagnetic waves only when the material equations express a functional relationship between the polarization and the field, i.e., the link between them is local in time and space. This locality of the linkage means that the phase velocity for small sinusoidal perturbations is independent of the frequency or wave number of the perturbation.

In linear media without dispersion (Chapter 4), we know that propagation without distortion and at a constant wave velocity is possible for an arbitrary wave shape and each component of the field in the wave satisfies the same equation $\partial u_j/\partial t + V_s \partial u_j/\partial x = 0$. Different physical variables (components u_j) change in proportion to each other, i.e, $u_k \sim u_j(x - V_s t)$. It is clear that such a wave may not generally exist in a nonlinear medium because even when there is small nonlinearity perturbations accumulate and cause a continuous deformation to the wave profile. However, in the absence of dispersion when one of the properties mentioned above of travelling waves would seem to be conserved in a nonlinear medium, namely different variables in the wave may be related algebraically, i.e, locally. It is easy to see that such a solution does in fact exist in nonlinear media without dispersion. They are called simple waves.

Because all the components of the field in a simple wave are expressed in terms of each other algebraically, a single first-order equation in one of the components can be used to describe the wave instead of the initial equation system. This equation must be written for a standing wave, whose velocity depends on the field amplitude, i.e,

$$u_j = u_j[x - V_s(u_j)t] \ . \qquad (18.6)$$

It is clear that the solution to (18.6) will satisfy (18.1) given above and this is the equation for a simple wave. Because the wave velocity depends on amplitude slow perturbations at various points on the profile will propagate with different speeds, as we saw in the example, and this will lead to a change in the wave shape. Naturally (18.1) describes a simple wave of any physical nature, i.e, in this sense it is universal.

In many cases, describing the evolution of nonlinear waves as the interaction between separate quasiharmonic waves is easiest both in terms of the mathematics and from a physical understanding of the mechanism of the nonlinear processes. We shall use this spectral approach to look at the basic phenomena of the nonlinear propagation of waves in media without dispersion, namely, the deformation of a simple wave and the appearance of discontinuities.

When there is strong nonlinearity there is no point in discussing interaction between separate harmonics because the lifetimes are of the order of magnitude of the interaction times and of the period. Therefore, we shall consider the nonlinearity to be small. Then the field in the medium may be described by a system of equations of the form

$$Au_t + Bu_x = \mu f(u, u_x, u_t) \ , \qquad (18.7)$$

where A and B are constant matrices, u is a vector of the field components, and f is a vector function containing nonlinear terms (in general both dispersive and dissipative).

We assume that initially (at $t = 0$) we create a periodic disturbance in the medium $U_0 \exp(-ikx) + cc = U(x, 0)$, and then for $t > 0$ and $\mu = 0$ the disturbance will take the form of a travelling wave

$$U(x, t) = U_0 \exp(i\omega t - ikx) + cc \qquad (18.8)$$

with a frequency $\omega = V_s(k)k$ which can be determined from the dispersion equation of the medium $D(\omega, k) = \text{Det}(A\omega - Bk) = 0$.

To make the discussion simple we shall consider that a given real k corresponds to only one real solution of the dispersion equation in ω (the remaining normal waves being strongly damped, viz., they correspond to complex roots $\omega(k)$).

When the nonlinearity is small, it is natural to search for a solution to (18.7), given these initial conditions, in a form close to that in a travelling sinusoidal wave, i.e,

$$u(x, t) = U_0 \exp(i\omega t - ikx) + \mu w(x, t) + c.c. , \qquad (18.9)$$

where $w(x,0) = 0$. Let $f(u, u_x, u_t)$ be a polynomial in u, u_x, u_t, and then by searching for a correction $w(x, t)$ we obtain from perturbation theory the following for the amplitude of the corresponding harmonics:

$$W_j^{(m)} = \sum_{l=1}^{n} \frac{\Delta_{lj}(m\omega, mk)}{D(m\omega, km)} F_l^{(m)} . \qquad (18.10)$$

Here $W_j^{(m)}$ is the j-th component of the m-th harmonic of the vector function $w(x,t)$, $F_l^{(m)}$ is the l-th component of the m-th harmonic of the vector function $f(U, U_x, U_t)$, D is the determinant of the matrix $|Am\omega - Bmk|$ and is the polynomial on the left-hand side of the dispersion equation, and Δ_{lj} is an algebraic addition to the a_{ij} element of the matrix.

If there is no dispersion in the medium, then the phase velocities of all the harmonics are the same and the m-th harmonic of the fundamental wave, which here may be considered to be an external field, is in resonance with the fundamental wave of the medium, i.e, the dispersion equation $D(m\omega, mk) = 0$ is fulfilled. The function $w(x,t)$, is therefore secular and close to resonance the solution has the form of a beat,

$$w(x, t) \sim \frac{1}{D'_\omega \Omega} \sin \Omega t \cdot \sin(\omega t - kx) ,$$

and as the frequency difference $\Omega = \omega(mk) - m\omega$ tends to zero (exact resonance) the solution grows linearly, i.e,

$$w(x, t) \sim \frac{\sin \Omega t}{\Omega t} t \sin(\omega t - kx) \to t \sin(\omega t - kx) .$$

Therefore, in the absence of dispersion the amplitudes of all the harmonics of the fundamental wave in a nonlinear medium grow continuously and the solution close to the sinusoidal wave (18.9) quickly becomes invalid. Moreover, the growing harmonics react together nonlinearly and new combination waves are born. Due to the absence of dispersion, these waves are resonant and their amplitudes increase, as the result of which new harmonics are formed. The number of interacting sinusoids, in this case, avalanche and the wave spectrum increases continuously. We emphasize that because of the infinite number of resonances, such a spectrum widening leads to a continuous decrease in the energy initially stored in a spectral interval arbitrarily bounded from above. The dissipation of this energy in the repeatedly arising harmonics with frequencies tending to infinity again corresponds to a continuously increasing steepness in the profile of the propagating wave and the formation of a domain where the field is infinitely quickly changing, viz., a discontinuity.

We obtain the equation of a simple wave for a transmission line with nonlinear capacitance (Fig. 18.4a). The initial equations have the form

$$\partial I/\partial x = - \partial Q(U)/\partial t , \quad \partial U/\partial x = - L\partial I/\partial t . \quad (18.11)$$

A typical graph of the charge in the capacitor versus voltage is shown in Fig. 18.4b. We shall seek a solution in the form of a simple wave, i.e, we shall assume that $I = I(U)$. Then, by introducing a nonlinear capacitance $C_N = dQ/dU$, we have

$$C_N(U) \frac{\partial U}{\partial t} + \frac{dI}{dU}\frac{\partial U}{\partial x} = 0 , \quad L \frac{dI}{dU}\frac{\partial U}{\partial t} + \frac{\partial U}{\partial x} = 0 .$$

These are two equations for one variable and so the coefficients of the derivatives must be the same, i.e, $LdI/dU = C_N(U)/(dI/dU)$ or $(dI/dU)^2 = C_N(U)/L$, which is the analog of wave conductivity. Whence $I(U) = \pm \int (C_N(U)/L)^{1/2} dU$, and the plus and minus signs refer to waves travelling to the right and to the left, respectively. Thus we obtain the following equation for waves propagating to the right

$$\partial U/\partial t - (LC_N(U))^{-1/2} \partial U/\partial x = 0 . \quad (18.12)$$

This is the desired equation for a simple wave, where $(LC_N(U))^{-1/2} = V(U)$ is its velocity. The solution $U = U[x - V(U)t]$ corresponds to this equation. If $C_N(U)$ is a monotonically decreasing function, then $V(U)$ is monotonically increasing (Figs. 18.4c and d). Thus, the points at the peaks of a simple wave will move faster then those at the feet (this is shown by the arrows in Fig. 18.4e). The trailing edge of a wave will be stretched and the leading edge will become more curved. At some point in time the peaks will topple and the U versus x,t function will become multivalued (Fig. 18.4e). This multivaluedness clearly has no physical significance for electric fields and a solution in the form of a simple wave simply becomes inapplicable. Note that the occurrence of a region of infinitely increasing changes in the physical values over time and space is the result of the neglection of dispersion and dissipation in the medium being investigated.

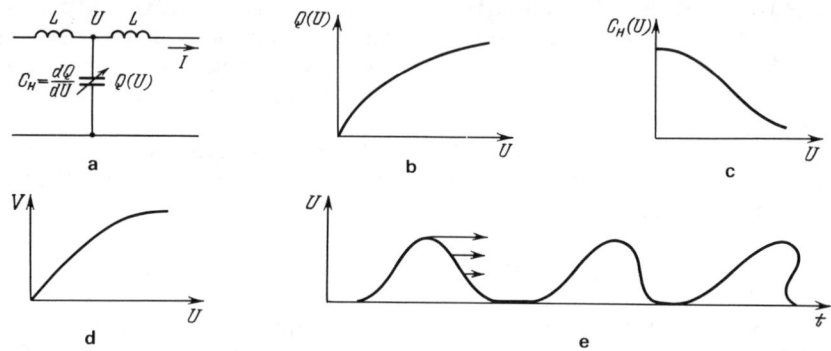

Fig. 18.4. a) Diagram of a transmission line with nonlinear capacitance. b-d) Characteristics of the "medium" or model. e) Change in the wave profile during propagation down such a line.

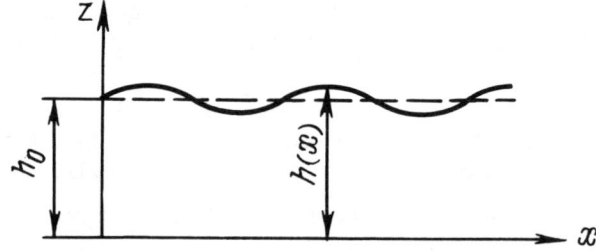

Fig. 18.5. A wave disturbance in a layer of liquid above a solid bottom.

We shall present several examples of simple waves propagating in complex media relying on the corresponding linear problems in Chapter 5. We begin with an analysis of wave propagation in a liquid layer above a solid bottom with an average depth h_0 (Fig. 18.5). Let us consider a gravitational wave with wavelength $\lambda \gg h_0$ (this is the shallowness condition) propagating along the x-axis. Because the waves are long, the horizontal velocity V for all depths may be considered the same and independent of z, and hence the Euler equation can be written for V in a form

$$\partial V/\partial t + V\partial V/\partial x + \rho^{-1}\partial p/\partial x = 0 . \quad (18.13)$$

The pressure p is here understood in a sense of being an average value over depth. It is higher the greater the liquid depth, being $(h - h_0)\rho g$ greater than the pressure in the undisturbed layer. Thus $\rho^{-1}\partial p/\partial x = g\partial h/\partial x$. Because the channel is shallow we need not consider the way the density ρ depends on the depth z, i.e, we can consider an incompressible liquid ρ = const. We need to add the continuity equation for the depth h, i.e,

$$\partial h/\partial t + \partial(Vh)/\partial x = 0 , \quad (18.14)$$

which expresses the fact that the rate of change in layer height $\partial h/\partial t$ is related to the flow differences between the infinitesimally close cross sections x and $x + \Delta x$.

For convenience, we rewrite (18.13) and (18.14) as

$$\partial V/\partial t + V\partial V/\partial x + g h\partial x = 0,$$
$$\partial h/\partial t + V\partial h/\partial x + h\partial V/\partial x = 0. \quad (18.15)$$

This is a system of nonlinear equations. By linearizing it in the vicinity of the equilibrium values of V_0 and h_0, and by obtaining the dispersion equation $\omega = k(V_0 \pm \sqrt{gh_0}$, it is easy to see that the system has no dispersion. We may therefore suggest that the variables in the wave will be functionally related, i.e, $h = h(V)$. Because of this we obtain the following instead of (18.15):

$$\frac{\partial V}{\partial t} + V\frac{\partial V}{\partial x} + g\frac{dh}{dV}\frac{\partial V}{\partial x} = 0 , \quad \frac{dh}{dV}\left(\frac{\partial V}{\partial t} + V\frac{\partial V}{\partial x}\right) + h\frac{\partial V}{\partial x} = 0 .$$

It follows from here that $V + gdh/dV = V + h/(dh/dV)$ or $dh/dV = \pm\sqrt{h/g}$. Thus, we have obtained one equation instead of the two in (18.15), viz.,

$$\partial V/\partial t + (V \pm \sqrt{gh})\partial V/\partial x = 0 . \quad (18.16)$$

We find from (18.16) that the velocity $u = V \pm \sqrt{gh}$ depends on the height of a point in the wave profile. A completely analogous equation is obtained for the propagation of a sound wave in a gas [9]. The initial equations in this case were the Euler equation

$$\partial \mathbf{v}/\partial t + (\mathbf{v}\ \mathrm{grad})\mathbf{v} + \rho^{-1}\mathrm{grad}\ \pi = 0$$

and the continuity equation

$$\partial \rho/\partial t + \mathrm{div}(\rho \mathbf{v}) = 0 ,$$

which in one-dimensional case and given that $dp/d\rho = c_{snd}^2$ the initial equations transform into the system

$$\frac{\partial u}{\partial t} + u \frac{\partial u}{\partial x} + \frac{c_{snd}^2}{\rho} \frac{\partial \rho}{\partial x} = 0 , \quad \frac{\partial \rho}{\partial t} + \frac{\partial}{\partial x}(\rho u) = 0 . \qquad (18.17)$$

By multiplying the first equation in (18.17) by $d\rho/du$ and subtracting the resultant equation from the second equation in the system, we discover that $c_{snd} d\rho/du = \pm \rho$. Using this result we obtain from (18.17)

$$\partial u/\partial t + (u \pm c_{snd}) \partial u/\partial x = 0 . \qquad (18.18)$$

It is easy to see that (18.18) coincides with (18.5) if we change coordinates to a system moving at the speed of sound.

An equation analogous to (18.18) is obtained for long-wave perturbations such as ion sound in plasmas with hot electrons, if it is assumed that the electron temperature of the plasma is constant due to the large thermal conductivity of the electrons. Thus we obtain the following equations from (5.90)-(5.92) for the propagation of a wave in such a plasma:

$$\partial v/\partial t + v \partial v/\partial x + c_{snd}^2 n^{-1} \partial n/\partial x = 0 , \quad \partial n/\partial x + \partial(nv)/\partial x = 0 ,$$

where $c_{snd}^2 = \sqrt{k_B T_e / m_i}$. By manipulating these equations as we did above, we arrive at (18.18) with a variable v instead of u.

Equations (18.16) and (18.18) are those of a simple wave and the solutions are simple, or Riemann, waves. These waves are said to be simple because they can be described by a single first-order equation instead of a system of equations.

We shall find the general equation for a simple wave. Let the vector function U characterizing a medium satisfy

$$A(U)U_t + B(U)U_x = 0 , \qquad (18.19)$$

where $A(U)$ and $B(U)$ are square matrices. We shall consider that (18.19) is a hyperbolic system. In scalar form, it can be written as

$$\sum_{j=1}^{n} \left[a_{ij}(U) \frac{\partial U_j}{\partial t} + b_{ij}(U) \frac{\partial U_j}{\partial x} \right] = 0 \quad (i = 1, 2, \ldots, n).$$

Assuming that $U_j = U_j(U_k)$, we obtain

$$\sum_{j=1}^{n} \frac{dU_j}{dU_k} \left[a_{ij}(U_k) \frac{\partial U_k}{\partial t} + b_{ij}(U_k) \frac{\partial U_k}{\partial x} \right] = 0.$$

This is a system of linear equations in dU_j/dU_k. In order for a nontrivial solution to exist, it is necessary and sufficient that

$$\text{Det} \| a_{ij} V(U_k) + b_{ij} \| = 0,$$

where $V(U_k) = (\partial U_k/\partial t)/(\partial U_k/\partial x)$ ($k = 1, 2, \ldots, n$). In the general case, the equations yield n different values $V_1(u_k)$, which correspond to n different simple waves.

18.3 Determining the Discontinuity Coordinates

We shall determine the coordinates of a discontinuity arising due to the evolution of a simple wave for an example of waves in a stream of cars [6].

We shall assume that it is one-way traffic and that there are no traffic lights. We use q to denote the flow rate of automobiles and equate it to the number of automobiles passing a given point of the highway per unit time. We use ρ to denote the density (concentration) of the automobiles equating it to the number of automobiles per unit length. If the total number of automobiles on the highway is conserved (there are no sources or sinks), then $\partial\rho/\partial t + \partial q/\partial x = 0$, and $q = q(\rho)$ or

$$\partial\rho/\partial t + u(\rho)\partial\rho/\partial x = 0, \quad u(\rho) = dq/d\rho. \qquad (18.20)$$

This is the equation of a simple wave, whose solution is often called a kinematic wave. A typical graph of the flow rate of automobiles versus the density is given in Fig. 18.6a. Initially the flow rate is proportional to the number of automobiles per unit length, then it reaches a maximum and begins to fall, dying away to zero for a very large concentration (the automobiles are bumper to bumper and stop).

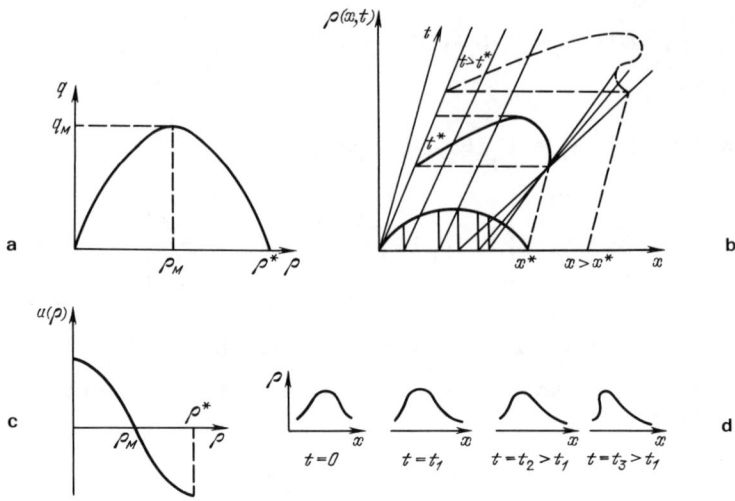

Fig. 18.6. Propagation of waves in a stream: a) graph of automobile flow rate versus density; b) the onset of a discontinuity in the wave profile for $du/d\rho > 0$; c) the rate of propagation of a disturbance versus automobile density; d) formation of a discontinuity on the trailing edge of an impulse from a group of automobiles.

Observation indicates that for one-way traffic without traffic lights we have $\rho^* = 140$ km^{-1}, $\rho_M = 50$ km^{-1}, and a maximum flow of $q_M \sim 1500$ h^{-1}. The largest flow rate is obtained for the very small speed of about 30 km/h. If a density disturbance arises in the automobile traffic (for example someone brakes, then the disturbance will propagate at a velocity $u(\rho) = dq/d\rho$ (the stream velocity is $q/\rho = V$). The solution to (18.20) is sought in the form

$$\rho(x, t) = \rho[x - u(\rho)t]$$

or if we use an inverse function as

$$x - u(\rho)t = \Psi(\rho) . \qquad (18.21)$$

It is easy to find a solution using this notation as each point in the wave profile will move along a straight line in the (x, t)-plane (Fig. 18.6b) with this own velocity $u(\rho)$. These straight lines are called characteristics (Chapter 7). A discontinuity in the wave profile arises where characteristics intersect, that is where $\partial\rho/\partial x$, $\partial\rho/\partial t$, $\partial^2\rho/\partial x^2$, and $\partial^2\rho/\partial t^2 \to \infty$ (because this is an inflection). The coordinates of a discontinuity (i.e, the time t^* and x^* when the discontinuity is formed and the value ρ^* at the

SIMPLE WAVES

overtaking point) are easily found using (18.21). Suppose we are given $\rho(x)$ and $t = 0$. By differentiating (18.21) with respect to the coordinate, we obtain

$$1 - \frac{du}{d\rho}\frac{\partial \rho}{\partial x} t = \frac{d\Psi}{d\rho}\frac{\partial \rho}{\partial x}. \qquad (18.22)$$

At $t = 0$ the value of $d\Psi/d\rho = (\partial\rho/\partial x)^{-1}_{t=0}$ is the initial density profile. Since Ψ is not a function of t, as time goes on it does not change. Because of this (18.22) takes the form

$$\frac{\partial \rho}{\partial x} = \left(\frac{\partial \rho}{\partial x}\right)_{t=0} \left[1 + t\frac{du}{d\rho}\left(\frac{\partial \rho}{\partial x}\right)_{t=0}\right]^{-1}.$$

Thus, a discontinuity is formed on the leading edge of the wave for $t > 0$ if $(\partial \rho / \partial x)_{t=0} < 0$, and on the trailing edge $(\partial \rho / \partial x)_{t=0} > 0$ if $du/d\rho < 0$. Because $u(\rho)$ is a monotonically decreasing function for a stream of automobiles (Fig. 18.6c), the discontinuity (large automobile concentration) tends to form on the trailing edge of the impulse of a group of automobiles (Fig. 18.6d). Note that when $u(\rho) > 0$, the wave moves in a direction of the stream of automobiles, and when $u(\rho) < 0$ it moves in the opposite direction. Automobiles (moving faster than the wave) catch up to the density step change and increase the density (in order not to make the traffic jam more dense, the driver must rapidly brake in the transition region and then gradually increase his velocity to leave the jam behind).

Returning to the determination of the discontinuity coordinates, we rewrite the equation system whose solutions define the coordinates. Since at a discontinuity $\partial x/\partial \rho$ and $\partial^2 x/\partial \rho^2 = 0$, then by differentiating (18.21) with respect to ρ at constant t, we obtain

$$-(du/d\rho)|_{\rho=\rho*}t^* = \Psi'(\rho^*), \quad -(d^2u/d\rho^2)|_{\rho=\rho*}t^* = \Psi''(\rho^*).$$

By including (18.21) at the discontinuity point, i.e,

$$x^* - u(\rho^*)t^* = \Psi(\rho^*),$$

we obtain a system of three equations from which the unknown variables x^*, t^*, ρ^* can be found.

In a case with boundary conditions (at $t = 0$ the form of the wave at the boundary is given) the discontinuity coordinates can be found from the conditions $\partial t/\partial \rho = 0$ and $\partial^2 t/\partial^2 \rho = 0$ by analogy with the above.

Thus, in a linear medium without dispersion, any travelling wave is stationary, i.e, its shape does not change as it propagates. Moreover, all the physical variables in such a wave are algebraically related. At the same time, even when there is weak nonlinearity in a medium, in the absence of dispersion all the harmonics arising out of the nonlinearity are in resonance with the fundamental wave and they all propagate at the same velocities. Therefore, after a long period of time, even when there is very weak nonlinearity, their amplitudes will increase, and so lead to a considerable change in the wave shape, i.e, stationary waves are impossible in nonlinear media without dispersion. In terms of spectra, this means that the spectrum of the initial disturbance will continuously widen to the right. As the result, infinitely higher frequencies appear in the spectrum of wave, which corresponds to an infinitely rapid fall at the wave fronts.

18.4 Weak Shock Waves.
Boundary Conditions at a Discontinuity

After the formation of a discontinuity or shock wave (Chapter 19) it is impermissible, in general, to apply (18.1) or (18.20) to describe the propagation of a wave in a nonlinear medium without dispersion. However, if the discontinuity occurs in a very narrow region in space, then because the solutions outside the discontinuity region are smooth, it is natural to try to conserve (18.1) as a description of the evolution of the wave and exclude the discontinuity region by substituting it with appropriate boundary conditions. This approach is analogous, in idea, to the introduction of fast and slow motions for the analysis of relaxation oscillations (Chapter 14).

In order to obtain the boundary conditions, initial equations such as (18.19) must be written as conservation laws

$$\frac{\partial u_i}{\partial t} + \frac{\partial}{\partial x} F_i(u_1, \ldots, u_n) = 0 . \qquad (18.23)$$

If, now, we assume the discontinuity to be infinitely thin, then its propagation must be characterized by a single velocity $v_p(t)$. By moving at the discontinuity velocity, we integrate (18.23) in a small neighborhood of the discontinuity with respect to x and t:

$$\int\int_Q \left(\frac{\partial u_i}{\partial t} + \frac{\partial}{\partial x} F_i(u_1, \ldots, u_n) \right) dx\, dt = 0 .$$

SIMPLE WAVES

Using Green's formula, we go to an integral around a contour

$$\int_\Gamma (u_i dx - F_i dt) = 0 . \qquad (18.24)$$

By choosing the contour as in Fig. 18.7, and bearing in mind that the discontinuity is infinitesimally thin, we find from (18.24) that

$$\int_{t_1}^{t_2} \{[u_2(t) - u_1(t)]v_p(t) - [F_2(t) - F_1(t)]\}dt = 0 . \qquad (18.25)$$

We assumed that the discontinuity moved along a trajectory described by the equation $dx/dt = v_p(t)$, the indices 1 and 2 labelling the physical values before and after the discontinuity, respectively. Since the integration interval in (18.25) is arbitrary, it is necessary to require that the integrand is zero, i.e, to require that

$$\Delta u_i(t) = \Delta F_i(t)/v_p(t) . \qquad (18.26)$$

This is the boundary condition for the discontinuity that we are looking for.

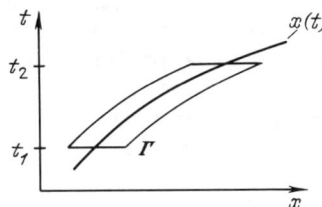

Fig. 18.7. Finding the velocity of a discontinuity ($x(t)$ is the discontinuity's trajectory).

If the physical variables are known at the discontinuity, then it is possible to determine the discontinuity velocity from (18.26). We give an example of the propagation of a wave in a transmission line with a nonlinear capacitance (Fig. 18.4a) [11]. The corresponding equations (18.11) already have the form of conservation laws. All that remains is to integrate then along the discontinuity trajectory, i.e,

$$I_2 - I_1 = v_p[Q(U_2) - Q(U_1)] , \quad U_2 - U_1 = Lv_p(I_2 - I_1) .$$

Assuming that the "amplitude" of the discontinuity is known (for example, equal to the amplitude of the initial wave input onto the line), we can find the discontinuity's propagation velocity:

$$v_p^2 = (U_2 - U_1)L^{-1}[Q(U_2) - Q(U_1)]^{-1} . \qquad (18.27)$$

The boundary conditions are obtained in an analogous way for a discontinuity arising when a planar electromagnetic wave propagates in a half-space filled with a ferrite whose magnetic induction is a function of the field $B(H)$, as in Fig. 18.8a, i.e.,

$$E_2 - E_1 = (v_p/c)(B_2 - B_1) ,$$
$$H_2 - H_1 = (v_p \varepsilon/c)(E_2 - E_1) \qquad (18.28)$$

(ε is the dielectric permeability of the medium, and c is the velocity of light) [10]. We recommend that the reader obtains these expressions independently.

Since small perturbations in front of the discontinuity move more slowly than the discontinuity (to take an actual example, we shall discuss the situation when the discontinuity is formed on the leading edge of the wave), i.e, the discontinuity catches up and absorbs the small perturbations, while small perturbations moving behind the discontinuity catch up with it (and also disappear into it), the total energy of the wave with a discontinuity must decrease over time. In other words, a discontinuity may only be stable if its energy is dissipated. We shall show this for the example we touched on, namely, a plane electromagnetic wave in a nonlinear medium filled with a ferrite.

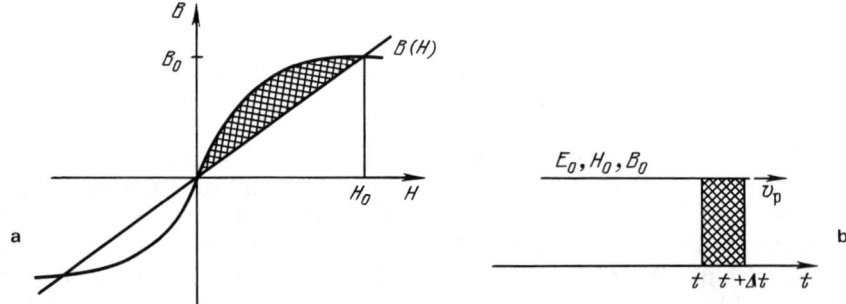

Fig. 18.8. Calculating the energy being dissipated at the front of a shock wave: a) nonlinear characteristic of the medium corresponding to the evolution of a discontinuity; b) energy loss over time Δt.

The conservation of energy in a dielectric volume is

SIMPLE WAVES

$\partial W/\partial t + \text{div } S = 0$, where $S = (c/4\pi)[EH]$ and $dW = (4\pi)^{-1}(EdD + HdB)$. Without loss of generality, we can assume that the field in front of the discontinuity is zero (Fig. 18.8b). Let us consider the change in energy in the discontinuity region of a time Δt. In order to do so, we write the energy balance in the shaded volume in Fig. 18.8b. The stored energy is

$$W_s = (4\pi)^{-1}\left[\varepsilon E_0^2/2 + \int_0^{B_0} HdB\right] v_p \Delta t,$$

and the arriving energy is

$$W_a = (c/4\pi) E_0 H_0 \Delta t = (4\pi)^{-1} H_0 B_0 v_p \Delta t.$$

We have taken into account the boundary conditions at the discontinuity $E_0 = (v_p/c) B_0$. Using the equations

$$\int_0^{B_0} HdB = H_0 B_0 - \int_0^{H_0} BdH, \quad \frac{\varepsilon E_0^2}{2} = \frac{1}{2}\frac{cH_0}{v_p}\frac{v_p}{c}B_0 = \frac{1}{2}H_0 B_0,$$

we find that

$$W_a - W_s = (v_p \Delta t/4\pi)\left(\int_0^{H_0} BdH - (1/2)H_0 B_0\right). \quad (18.29)$$

One can easily imagine that if the function $B(H)$ (Fig. 18.8a) is convex, then this difference will always be positive. The power dissipating in the discontinuity will be

$$P = (v_p/4\pi)\left(\int_0^{H_0} BdH - (1/2)H_0 B_0\right),$$

and in the general case it will be

$$P = (v_p/4\pi)\left[\int_0^{H_0} BdH - (1/2)(H_2 - H_1)(B_2 - B_1)\right]$$

$$= (v_p/4\pi)\left[(1/2)(H_2 - H_1)(B_2 - B_1) - \int_{B_1}^{B_2} HdB\right].$$

Although in principle we can construct a discontinuity solution for the initial nonlinear equations in which there is no dissipation, it is very easy to show that in such a case the discontinuity will be unstable [12, 13]. All the discontinuities arising as the result of a simple wave toppling (mathematically these are called evolutionary) are unstable and the energy dissipation in them is positive.

CHAPTER 19

STATIONARY SHOCK WAVES AND SOLITONS

19.1 Structure of a Discontinuity

Depending on the physical situation, different things happen after an infinite gradient arises in the profile of a simple wave. For example, if the wave is on a liquid surface, then it simply caves in, forming foam; if it is a stream of noninteracting particles, then a multivalued situation arises in the wave profile and after the formation of "discontinuity" in the main flow, several different flows moving at very different velocities form. For sound or electromagnetic fields, however, where multivaluedness is impermissible, the way the nonlinear wave develops depends on which effect predominates in the region of the rapidly changing field, i.e., dissipative or dispersion effects. We shall now study travelling waves in nonlinear media with dissipation and dispersion.

When dissipative, nonlinear and dispersion corrections are added to the initial equations describing wave propagation, and they are of the same order of magnitude and small in comparison to the linear terms, it is not difficult to use the perturbation method to find an equation in the one-wave approximation, viz.,

$$\partial u/\partial t + V(u)\partial u/\partial x + \beta \partial^3 u/\partial x^3 - \nu \partial^2 u/\partial x^2 = 0 . \qquad (19.1)$$

The Korteveg–de Vries equation (for $\nu = 0$) and the Burgers equation (for $\beta = 0$), which are the canonical equations in

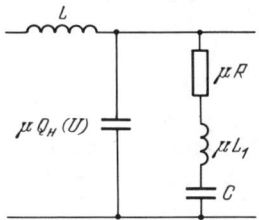

Fig. 19.1. Equivalent diagram for a transmission line, a model of a nonlinear medium with dissipation and dispersion.

the theory of nonlinear waves (Chapter 18), are special equations of (19.1). Many of the results in this chapter are obtained for (19.1).

We begin with a consideration of a "medium", a transmission line model as shown in Fig. 18.4a, but with corrections to the components allowing us to include dispersion (μL_1) and dissipation (μR). The equivalent circuit is shown in Fig. 19.1. The telegraph equations will be our starting point

$$\partial I/\partial x = - \partial Q/\partial t - \mu C_N \partial U/\partial t , \quad C_N = dQ_N/dU ,$$

$$\partial U/\partial x = - L \, \partial I/\partial t , \quad Q/C + \mu R \, \partial Q/\partial t + \mu L_1 \partial^2 Q/\partial t^2 = U .$$

We rewrite the last equation in the form

$$\partial Q/\partial t = C \, \partial u/\partial t - \mu [RC \, \partial^2 Q/\partial t^2 + L_1 C \partial^3 Q/\partial t^3]$$

and assume that $\mu \ll 1$. When $\mu \to 0$ we have $\partial Q/\partial t = C \, \partial U/\partial t$, i.e., when $\mu \ll 1$ it is possible to substitute $\partial Q/\partial t$ in the square brackets by the zeroth approximation; thus yielding

$$\frac{\partial I}{\partial x} = - C \frac{\partial U}{\partial t} - \mu C \left[\frac{C_N(U)}{C} \frac{\partial U}{\partial t} - RC \frac{\partial^2 U}{\partial t^2} - L_1 C \frac{\partial^3 U}{\partial t^3} \right] ,$$

$$\frac{\partial U}{\partial x} = - L \frac{\partial I}{\partial t} .$$

When $\mu \ll 1$ we search for solution in the form of a wave in which U and I are related as if in linear medium: $I = \sqrt{C/L} \, U$. By substituting in the relation $I = \sqrt{C/L} \, U$, we obtain the following equation for a travelling wave moving to the right

$$\frac{\partial U}{\partial t} + \frac{1}{\sqrt{LC}} \frac{\partial U}{\partial x} = - \mu \left\{ \frac{C_N(U)}{C} \frac{\partial U}{\partial t} - RC \frac{\partial^2 U}{\partial t^2} - L_1 C \frac{\partial^3 U}{\partial t^3} \right\} .$$

This equation is usually written in another form, as the equation for the zeroth approximation, i.e., the single $\partial/\partial t$ in the equation for a simple wave is substituted by $-(1/\sqrt{L/C})\partial/\partial x$, yielding

$$\partial U/\partial t + V_0 \partial U/\partial x + v(U)\partial U/\partial x - \nu \partial^2 U/\partial x^2 + \beta \partial^3 U/\partial x^3 = 0 .$$

Finally, by going over to a system of coordinates moving at the velocity $V_0 = (LC)^{-1/2}$ ($t_N = t$, $x_N = x - V_0 t$), we obtain

$$\partial U/\partial t_N + v(U)\partial U/\partial x_N - \nu \partial^2 U/\partial x_N^2 + \beta \partial^3 U/\partial x_N^3 = 0 .$$

In our case we have $v(U) = -\mu V_0 [C_N(U)/C]$, $\nu = \mu V_0 RC$, $\beta = \mu V_0 L_1 C$. The resultant equation is the same as the reference equation (19.1) of the one-wave approximation.

In the last chapter we established that in a nonlinear medium without dissipation or dispersion, the steepness of the profile of a propagating wave increases continuously and a discontinuity forms, that is, a region of infinitely rapid change in the physical properties over time and space [1-3]. In order for the discontinuity to be conserved as the wave propagates, we have seen that energy dissipation is necessary in the discontinuity, thus ensuring the irreversibility of nonlinear evolution. In terms of spectra this means that the energy flow is directed toward higher frequencies where the energy losses are significant.

Thus, the discontinuity may be stable due to high-frequency dissipation alone. We shall assume, initially only qualitatively, how dispersion affects the discontinuity.

The phase velocity of the harmonics generated by the nonlinearity, even when there is weak dispersion, differs somewhat from the velocity of the fundamental wave. When the harmonic number is high, this difference becomes so strong that the harmonic is no longer in resonance with the fundamental wave of the medium and its amplitude remains small (proportional to the nonlinearity). Such a wave has a negligible effect in a process and the spectrum of the nonlinear wave is thus bounded. In terms of space—time diagrams, this means that the width of the domain of rapid change in the field will be finite. Thus, dispersion also bounds the width of the discontinuity.

Naturally, when the number of harmonics forming a nonlinear wave is bounded, irreversible deformation in the wave profile is impossible in a medium without dispersion. The energy stored initially (at $t = 0$ or at $x = 0$) in the first harmonic becomes the energy of a finite number of harmonics. Thus, in view of the conservativeness of the system, the energy is collected back, after which it is transferred back to the harmonics and so on (it is assumed that the harmonic phases do not become chaotic and there is no irreversible mixing; we shall discuss this later). Thus, a sinusoidal wave which is deformed due to the nonlinearity must, as it propagates, be restored and then the profile is again distorted, after which it returns to normal.

However, when there are certain relationships between

the amplitudes and phases of the interacting harmonics, energy may not be exchangeable between the harmonics (in the corresponding phase space this will be an equilibrium). We have met similar solutions, e.g., when analyzing the interactions between a sinusoidal wave and its second harmonic in a weakly nonlinear medium (Chapter 17). A wave whose profile does not change as it propagates corresponds to such a spectral equilibrium in a real (x,t)-space. This is a stationary wave. The velocity at which a stationary wave propagates V is constant and so solutions in the form of stationary waves are described by ordinary differential equations whose arguments are the running coordinate $\xi = x - Vt$.

Although stationary waves are a special class of solution, their role in the theory of nonlinear waves is extraordinarily large. This is, of course, because of the simplicity of finding them (integrating ordinary, and not partial, differential equations), but perhaps more importantly it is because waves close to stationary ones arise during the evolution of a wide class of nonstationary perturbations. Moreover, the stability of stationary waves is characteristic both of systems with dissipation and for conservative systems. A remarkable example of this is the stability of solitons. We add that once a solution in the form of a stationary wave is known, it is possible to investigate nonstationary solutions, so they must be locally (in time and space) close to the stationary solutions [4-6].

We now return to the basic model (19.1) and qualitatively investigate the possible solutions when there is a loss in the medium ($\nu \neq 0$). Dissipation, as we have said, makes the process irreversible. It is clear that when the losses are very small ($\nu \ll \beta$), the solutions to (19.1) will differ little from the conservative case. Dissipation (like dispersion) leads to a certain breakdown in the wave shape and finally may equalize the nonlinear increase in the profile's steepness. The discontinuity may be approximated to being stationary, i.e., it propagates with constant velocity and almost without changing form.

An account of high-frequency dissipation and dispersion allows us to investigate the nature of the field changes at the front of a shock wave, i.e., the structure of a discontinuity, using the stationary wave approximation. Because all the variables in the medium outside the shock front change very slowly, we can assume that in general they remain constant, i.e., their values correspond to equilibrium states on the phase plane of the system of ordinary differential equations describing the stationary wave. Thus, the investigation of the front structure for a shock wave reduces to the location of the unique phase trajectory that unites these equilibria.

If we change to a moving coordinate system in (19.1)

$\xi = x - Vt$, then because $\partial u/\partial t = -V du/d\xi$ and $\partial u/\partial x = du/d\xi$, i.e., $\delta u/\partial t = -V\partial u/\partial x = -V du/d\xi$, when $v(u) = u$ we obtain

$$\beta d^2 u/d\xi^2 - \nu du/d\xi = Vu - u^2/2 . \qquad (19.2)$$

The model (19.2) may be thought of as a nonlinear oscillator with damping, where ξ is the analog of time and u is the coordinate of a point mass. The equation of the potential well has the form $W(u) = -Vu^2/2 + u^3/6$. The equilibria positions are found at the points $u_{01} = 0$ and $u_{02} = 2V$. In order to determine the type of equilibrium we compile the characteristic equation $\beta p^2 - \nu p + (u_0 - V) = 0$ (assuming that $u = u_0 + u'$, $u' \sim \exp(p\xi)$). Hence the equilibrium u_{01} is a saddle, while u_{02} is a node when $(\nu^2 - 4\beta V) > 0$ and a focus when $(\nu^2 - 4\beta V) < 0$. The phase-plane diagrams for various ν and the corresponding field changes at the shock wave front are illustrated in Fig. 19.2. The dependence u on ξ of the whole of the ξ-axis is obtained from an analogy between the model (19.2) and a nonlinear oscillator [6]. The solution begins when $\xi \to +\infty$, and then the point mass falls into the potential well and oscillates in it with damping until it reaches the value $u = 2V$ as $\xi \to -\infty$.

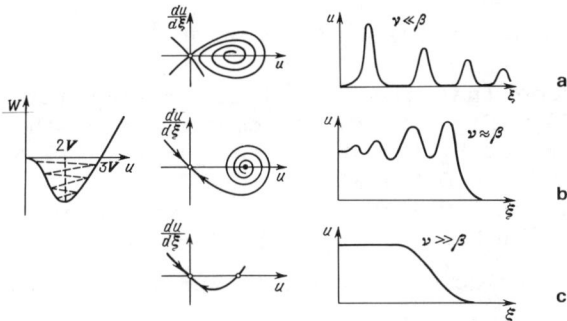

Fig. 19.2. The form of the potential well, the phase-plane diagrams, and the way a shock wave propagates for various values of ν: a) $\nu \ll \beta$ in a nonlinear oscillator's wave; b) $\nu \approx \beta$; c) $\nu \gg \beta$ is a shock wave without oscillations.

We shall not analyze the structure of a discontinuity using (19.1), and instead we shall do the analysis directly for a medium whose equivalent circuit is given in Fig. 19.3. According to the definition of a shock wave, the length of the front is small in comparison to the characteristic temporal and spatial scales in which the voltage and current change (which depend on the medium) outside the sharp drop in the shock wave's profile. This allows us to distinguish between "fast" and "slow" motions, and by defining a region in which the quantities change quickly, investigate the structure of the region (the structure of the shock wave front) by assuming that the wave is stationary [6-8].

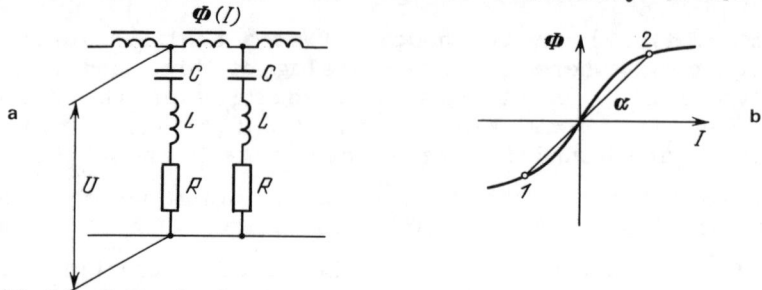

Fig. 19.3. a) Equivalent circuit for a transmission line with temporal dispersion. b) The function $\Phi(I)$.

The telegraph equations will be our starting point:

$$\partial U/\partial x = - \partial \Phi/\partial t , \qquad \partial I/\partial x = - \partial Q/\partial t , \qquad (19.3)$$

where I, U, Φ and Q are the current, voltage, line induction flow, and line charge in the line. In general, these quantities are related in terms of an integral equation of the form

$$Q = \hat{Q}\{U, U\} , \qquad \Phi = \hat{\Phi}\{I, U\} . \qquad (19.4)$$

We shall assume that the flow Φ is related to the current quasistatically (Fig. 19.3b) and consequently, the dispersion will be determined by the induction L in Fig. 19.3a. For this circuit we have $U = Q/C + R\partial Q/\partial t + L\partial^2 Q/\partial t^2$.

We change to a coordinate system moving at the velocity of the discontinuity v_p. Then all the quantities will only depend on one variable $\xi = x - v_p t$, and we have $\partial I/\partial x = dI/d\xi$ and $\partial Q/\partial t = - v_p dQ/d\xi$, i.e., it follows from (19.3) that

$$dI/d\xi = v_p dQ/d\xi , \qquad (19.5)$$

$$dU/d\xi = v_p d\Phi/d\xi , \qquad (19.6)$$

$$U/Q/C + Rv_p dQ/d\xi + Lv_p^2 d^2Q/d\xi^2 . \qquad (19.7)$$

by differentiating (19.7) with respect to ξ and using (19.5) and (19.6), we find

$$LCv_p^3 d^3I/d\xi^3 + RCv_p^2 d^2I/d\xi^2 + dI/d\xi - Cv_p^2 d\Phi/d\xi = 0 . \qquad (19.8)$$

We integrate (19.8) with respect to ξ from $-\infty$ to ξ, which is the running coordinate inside the discontinuity region. Finally, we find

$$LCv_p^2 d^2I/d\xi^2 - RCv_p dI/d\xi + \{(I - I_1) - Cv_p^2[\Phi(I) - \Phi(I_1)]\} = 0, \quad (19.9)$$

where $I(-\infty) = I_1$ and $I(\xi) = I$. The coordinates of the equilibrium state and the velocity v_p at which the discontinuity moves are related by

$$v_p^2 = (I_2 - I_1)/C[\Phi(I_2) - \Phi(I_1)] . \quad (19.10)$$

This is the boundary condition for a discontinuity, which we have already met (see (18.27)). It admits of a simple graphical interpretation (Fig. 19.3b): $\tan \alpha = 1/(Cv_p^2)$, where α is the slope of the straight line connecting points 1 and 2 on both sides of the drop in the curve $\Phi = \Phi(I)$. The characteristic equation corresponding to (19.9) under the condition $I_1 = 0$ has the form

$$h^2 p^2 + 2\delta p + (1 - Cv_p^2 \Phi_I') = 0 , \quad (19.11)$$

where $h^2 = LCv_p^2$, $2\delta = RCv_p$, $\Phi_I' = \partial\Phi/\partial I$. The critical point $I = I_1 = 0$ is always a saddle $\Phi_I' > 1/Cv_p^2$, and it corresponds to the "foot" of the wave. The critical point $I = I_2$ corresponds to the "peak" of the wave and is either a focus, when $\Phi_I' < 1/Cv_p^2$, $\delta^2 < h^2(1 - Cv_p^2\Phi_I')$, or a node, when $\Phi_I' < 1/Cv_p^2$, $\delta^2 > h^2(1 - Cv_p^2\Phi_I')$. When there is no dissipation, i.e., $R = 0$, we have from (19.8)

$$LCv_p^2 d^3I/d\xi^3 + dI/d\xi - Cv_p^2 d\Phi/d\xi = 0 ,$$

$$h^2 d^2I/d\xi^2 + I - Cv_p^2\Phi(I) = \text{const} .$$

If we represent the function $\Phi(I)$ by two parabolas, then we find that $\Phi(I) \to I^2$ and

$$h^2 d^2I/d\xi^2 = -I + Cv_p^2 I^2 ,$$

whence $I = A \sinh^2\sqrt{A/12h}$, where $A = $ const. In this case, the critical point is a center. All these situations are gathered together in Fig. 19.4 [6]. The results for a concrete model fully correspond to the qualitative investigation of (19.1) in the one-wave approximation.

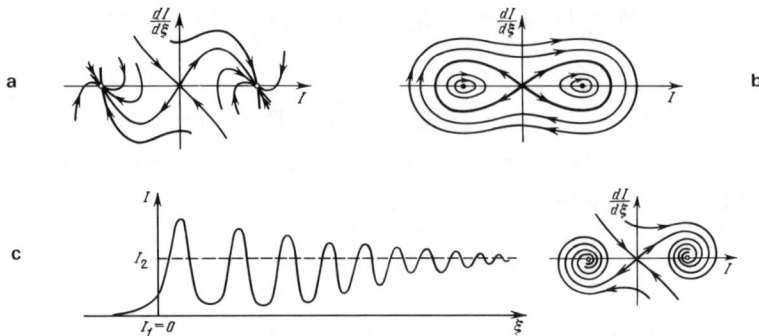

Fig. 19.4. Phase-plane diagrams of stationary wave in lines with temporal dissipation: a) strong damping; b) no dissipation in the line; c) weak damping, the structure of the wave front and the phase-plane diagram.

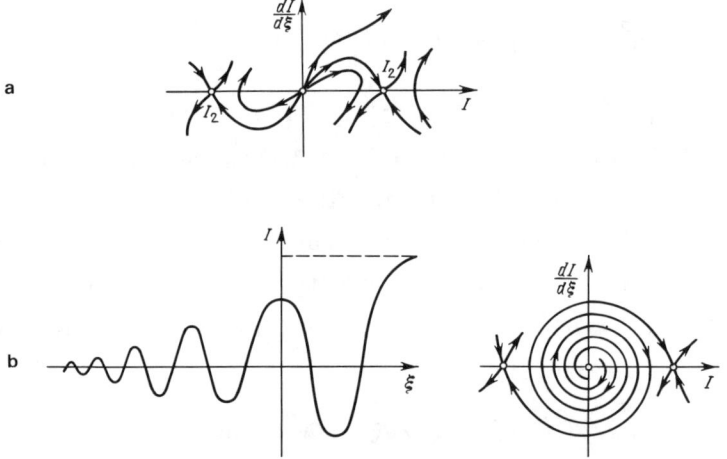

Fig. 19.5. The phase-plane diagram and front structure of an electromagnetic shock wave in a line with spatial dispersion: a) strong damping; b) weak damping.

In a line with spatial dispersion, i.e., the line flow Φ and charge Q not being locally related with the current and voltage, the velocity v_p of the shock wave may be lower than the group velocity of the disturbances arising in front of the shock wave. As a result, oscillations overtake the front and waves are observed in the stationary wave near the crest, in the leading section of the front. If, for example, an inductive coupling is introduced between the cells in the transmission line given in Fig. 19.3a, then we may assume for systems with small cells (with respect to the spatial scale of the disturbance) that

$$\Phi = \Phi(I) - M\partial^2 I/\partial x^2 , \quad Q = CU - RC\partial Q/\partial t , \tag{19.12}$$

where M is the coefficient that allows for the inductive connection between cells (for simplicity we shall assume that $L = 0$ in Fig. 19.3a). In this case (19.3) using (19.12) turns into the equation in [6], i.e.,

$$MCd^2 I/d\xi^2 - RCdI/d\xi - \{(I - I_1) - Cv_p^2[\Phi(I) - \Phi(I_1)]\} = 0 . \tag{19.13}$$

when $M < 0$, the phase-plane diagram for (19.13) is similar to the phase-plane diagrams in Fig. 19.4. The influence of spatial dispersion on the structure of an electromagnetic shock wave is qualitatively the same as that for temporal dissipation. However, when $M > 0$ and R is small, the critical point $I_1 = 0$ becomes an unstable focus (Fig. 19.5b). Oscillations arise in front of the shock wave front (Fig. 19.5b) and the group velocity of the oscillations is faster than v_p. Stationary electromagnetic shock waves have been experimentally investigated for coaxial wave guides filled with ferrite and multi-link artificial lines [7, 8].

19.2 Solitary Waves - Solitons[1]

We consider a medium without dissipation ($\nu = 0$). Suppose, for the moment, that the nonlinearity in the medium is quadratic, i.e., $V(u) = u$, then instead of (19.1) we will have the equation derived by Korteveg and de Vries for waves on the surface of a liquid:

$$u_t + uu_x + \beta u_{xxx} = 0 . \tag{19.14}$$

The solutions (both stationary and nonstationary) to this equation have now been very well studied, but we shall only discuss the simplest, supplementing our discussion with

---

[1] There is a large literature on solitons. The main sources are [9-12]. Good informal presentations of the physics of solitons are given in [13] and [15].

qualitative considerations. First, let us think what effect adding terms which describe dispersive spreading to the equation for a simple wave will have. We already know that dispersive atomization may compensate for wave overtaking, and thus stabilize its profile, i.e., stationary travelling waves may exist whose profiles do not change over time. These waves are defined over all space and travel with the constant velocity V, i.e., all the variables in the wave are functions of the running coordinates $\xi = x - Vt$. We have for these waves $\partial u/\partial x = du/d\xi$ and $\partial u/\partial t = - Vdu/d\xi$, i.e., the stationary waves of (19.14) are described by an ordinary differential equation $\beta d^3u/d\xi^3 + (u - V)du/d\xi = 0$, or after integration,

$$\beta d^2u/d\xi^2 - (Vu - u^2/2) = C_1 , \qquad (19.15)$$

Thus, the equation of a conservative nonlinear oscillator corresponds to stationary waves from the Korteveg–de Vries equation. We shall consider the constant to be zero (this can always be done by introducing a new variable), whence (19.15) becomes $\beta d^2u/d\xi^2 = - \partial W/\partial u$, where $W = - (Vu^2)/2 + u^3/6$. The potential energy W of a stationary wave and its phase-plane diagram is given in Fig. 19.6.

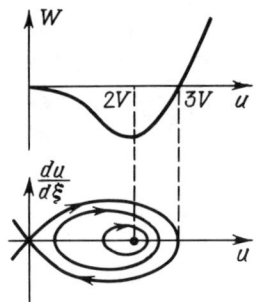

Fig. 19.6. Potential energy $W = - Vu^2/2 + u^3/6$ and phase-plane diagram of stationary waves. The equilibrium $u = 2V$ is a center. The soliton corresponds to the separatrix.

There are various classes of solutions to the Korteveg–de Vries equation.
1. Quasisinusoidal oscillations with small amplitudes (their phase trajectories are close to the center), correspond to barely noticeable nonlinearity (Fig. 19.7a).

STATIONARY SHOCK WAVES AND SOLITONS

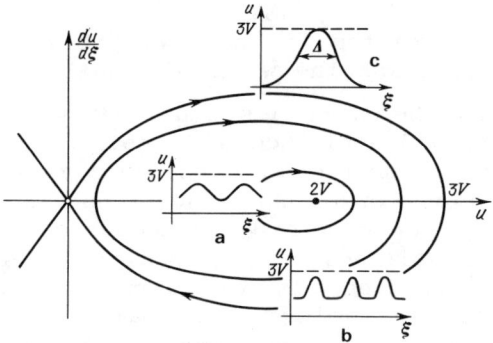

Fig. 19.7. Various classes of solution to the Korteveg–de Vries equation and the corresponding phase-plane diagrams of the stationary planes: a) quasi-sinusoidal oscillations with small amplitude are near the center; b) Conoidal waves (periodic soliton nets) are close to the separatrix; c) a soliton (solitary wave) is near the separatrix.

2. Motion near a separatrix and along the separatrix itself. These are strongly nonlinear waves and are what interest us. Periodic motions close to the separatrix (Fig. 19.7b) are called conoidal waves. The separatrix corresponds to a solution localized in space in the form of a single ridge or solitary wave, viz., a soliton (Fig. 19.7c) with amplitude $u_{max} = 3V$. This solution is analytically written in the form

$$u(x, t) = u_{max} \cosh^{-2}[(x - Vt)/\Delta] ,$$

where Δ is the characteristic width of the soliton. The validity of the solution is easily verified by directly substituting it into (19.2) with $C_1 = 0$. After substituting, we use the identity $\cosh^2(\xi/\Delta) - \sinh^2(\xi/\Delta) = 1$ and obtain

$$\frac{\beta^2}{\Delta^2}\left\{\frac{4u_{max}}{\cosh^2(\xi/\Delta)} - \frac{6u_{max}}{\cosh^4(\xi/\Delta)}\right\} \equiv V\frac{u_{max}}{\cosh^2(\xi/\Delta)} - \frac{u_{max}}{2\cosh^4(\xi/\Delta)} .$$

(19.16)

It is possible to find Δ and u_{max} from here. The identity (19.16) is fulfilled for any ξ, and consequently the coefficients for identical orders of $\cosh(\xi/\Delta)$ must be equal, i.e.,

$$4\beta/\Delta^2 = V , \quad 6\beta/\Delta^2 = u_{max}/2 .$$

Thus, we have found: 1) $u_{max}\Delta^2 = 12\beta$ = const, which means that the higher soliton the narrower it is; 2) $\Delta^2 = 4\beta/V$, $u_{max} = 3V$, which means the wider the soliton, the slower it will travel and the lower its amplitude will be. Thus, the width, velocity and amplitude of a soliton described by the Korteveg–de Vries equation are uniquely related, i.e., the family of solutions in a form of solitons are one-parametric, and by changing, for example V, we obtain different solitons.

The reason solitons, i.e., special forms of stationary waves, are interesting, is the same as the reason other stationary waves are interesting, namely, a wide class of nonstationary perturbations asymptotically approach the soliton as they propagate. This fact was observed experimentally a long time ago when more than 100 years ago Scott-Russell[2] observed the soliton and poetically described it [10].

The soliton, one of the most fascinating objects in modern physics, obtained new life mainly due to the derivation of exact solutions for many of the equations in the nonlinear theory of waves. The key to these derivations is the method of inverse scattering [11], which started life in a paper by Gardner, Green, Kruskal and Miura [21], who in 1967 established a link between the Korteveg–de Vries and the Schrodinger equations. We shall now look briefly at this link. It is known [14] that Schrodinger's equation $\partial^2\Psi/\partial x^2 + [U(x) + \varepsilon]\Psi = 0$ when the potential $U(x)$ is positive-definite and decays to zero as $x \to \pm\infty$, and has finite solutions tending to zero with the derivatives at infinity, moreover the spectrum of its eigenvalues ε is discrete. Let us consider Schrodinger's equation

$$\partial^2\Psi/\partial x^2 + [u(t, x)/6\beta + \varepsilon] = 0 , \quad \beta > 0 , \quad (19.17)$$

where $u(t,x)$ is dependent on time and a parameter. The eigenvalues are generally dependent on t. We shall show that the eigenvalues ε do not depend on t if the function $u(x,t)$ satisfies the Korteveg–de Vries equation (more exactly, if $u(x,t)$ is some positive-definite solution of the Korteveg–de Vries equation that decays to zero at $\pm\infty$, then the corresponding spectrum of eigenvalues remains unchanged). We find from (19.17) that

$$u(x, t) = -6\beta(\Psi^{-1}\partial^2\Psi/\partial x^2 + \varepsilon) .$$

We substitute this expression into (19.14). After some

---

[2] A biography of John Scott-Russell, a great engineer and boat-builder of the Victorian age, may be found in [12].

calculations we obtain

$$\Psi^2 \partial \varepsilon / \partial t = (\Psi' A - \Psi A')' , \qquad (19.18)$$

where $A(t,x) = 6\beta[\beta^{-1} \partial \Psi / \partial t - 3\Psi'\Psi''/\Psi + \Psi''' - \varepsilon\Psi'/6]$, the primes being the corresponding derivatives with respect to x.

We integrate the left and right-hand sides of (19.18) with respect to x from $-\infty$ to $+\infty$. The right-hand side of the resultant equation will equate to zero because the eigenvalues (and the derivatives) of the discrete spectrum of the Schrodinger's equation disappear at infinity. Thus,

$$(d\varepsilon/dt) \int_{-\infty}^{\infty} \Psi^2 dx = 0 .$$

Because of normalization $\int_{-\infty}^{\infty} \Psi^2 dx \neq 0$, we have $d\varepsilon/dt = 0$, i.e., $\varepsilon =$ const. Since the solution $u(t,x)$ is arbitrary, the spectrum ε is what interests us. We now show that if $u(t,x)$ is a soliton, then Schrodinger's equation has a single eigenvalue. When $u(t,x) = u_{max} \cosh^{-2}[(x - Vt)/\Delta]$ is a soliton, (19.17) takes the form

$$\Psi'' + (U_0 \cosh^{-2}\alpha\xi + \varepsilon)J = 0 .$$

Here $U_0 = u_{max}/6\beta$, $\alpha = 1/\Delta = [u_{max}/12\beta]^{1/2}$. The discrete eigenvalues of Schrodinger's equations are given by the formula ([14], section 23, problem 4):

$$\varepsilon_n = -\alpha^2(s - n)^2 , \quad n = 0, 1, 2, \ldots,$$

where $s = \frac{1}{2}(-1 + \sqrt{1 + 4U_0/\alpha^2}$, and $n < s$ must be fulfilled. By substituting into the expression for s the values of U_0 and α given above, we obtain $s = 1$, i.e., only one eigenvalue exists, viz., $\varepsilon_0 = u_{max}/12\beta$. Thus, we have found that a) the spectrum of eigenvalues is independent of t, even though $u(t,x)$ changes over time; b) each eigenvalue corresponds to a soliton. We therefore conclude that any local positive disturbance is a set of solitons and if we wait long enough these solitons form and the disturbance will turn into a sequence of solitons with rapidly rising amplitudes (Fig. 19.8c). Because the "soliton composition" or set of solitons comprising the disturbance is independent of time, solitons may only change places in space. The number of solitons depends on the form of the initial disturbance, and their peaks lie along a single straight line because the distance between the solitons is proportional to their velocity, whilst the velocity, as we have seen, is proportional to their amplitude.

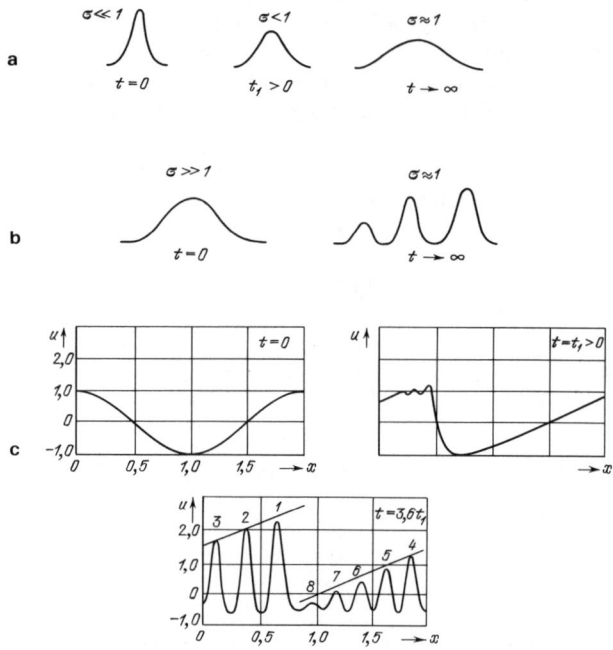

Fig. 19.8. Evolution of an initial disturbance for various values of the parameter $\sigma = \Delta^2 u_{max}/12\beta$, which characterizes the ratio of the nonlinearity and dispersion in a system: a) dispersive spreading dominates; b) initially there is a tendency to topple but due to dispersion, disturbances with different wavelengths scatter and the disturbance scatters into short impulses. c) The results of numerical simulation [15] (for one period).

This method of solving the Korteveg–de Vries equation is called the inverse scattering method because we determine the eigenvalues for Schrodinger's equation with potential $u(t,x)$, where t is a parameter. In a quantum mechanical equation, ε is the energy level, and Ψ is the wave function. The direct quantum mechanical problem of scattering is to solve (19.17) with a given potential u. This allows as to calculate, for example, the reflection coefficient of a wave (the wave is defined as being dependent on the coordinates of the wave function Ψ), which comes from infinity to be incident on a charge image $u(x)$. If the wave arriving from infinity is planar with unit amplitude, then the amplitude of the reflected wave is called the reflection coefficient. However, we determined the potential. This is the converse problem to the quantum scattering problem; the charge image $u(x)$ can be reestablished from a known reflection coefficient. The method of the inverse scattering is presented in detail in [10-12].

We shall now see why the soliton is a stable disturbance. We introduce the dimensionless parameter $\sigma = \Delta^2 u_{max}/12\beta$. This parameter describes the ratio between the nonlinearity and the dispersion in a system because the larger the amplitude u_{max}, the more strongly the nonlinearity is manifest, whilst β characterizes the high-frequency dispersion. For a soliton $\sigma = 1$, i.e., the effects of nonlinear evolution and dispersive atomization exactly cancel each other out. When $\sigma \ll 1$ (Fig. 19.8a) a disturbance with a steep front behaves as if in a linear dispersing medium. The main effect is the appearance of a relatively long-wave oscillation, which causes Δ to increase, and consequently σ to increase, i.e., to the re-establishment of a wave with $\sigma = 1$. When $\sigma \gg 1$, the dispersion effects are insignificant and the main influence is nonlinearity, which causes the formation of short impulses. Only then does dispersion come into play to equalize the process (Fig. 19.8b). Thus an initial large-amplitude perturbation is broken down into a sequence of solitons whose crests lie along a single straight line (in Fig. 19.8c the results of the numerical calculation in [15] are graphed).

19.3 Solitons as Particles

Given that they are quite complex formations, solitons and soliton periodic solutions (conoidal waves) should behave in a very complicated manner. However, judging from many physical and numerical experiments, this is not always true. Often, solitons when they interact, behave surprisingly simply, repelling each other, attracting each other, or oscillating with respect to each other (Fig. 19.9), as if they were classical particles. Quite recently it was established that this superficial analogy is, in fact, very profound for weakly interacting solitons (or conoidal waves). If the velocity difference (or energy, which is the same thing) of the solitons is small and the distance between the solitons' maxima is large with respect to the effects of width throughout the process, then their interaction is analogous to the interaction between particles and can be described by Newton's equations. The soliton in the field of the "tail" of another soliton behaves like a ball in a trough. For example, the following equation is obtained for pairs of solitons [16]

$$d^2u/dt^2 - v(\mathcal{E})f(v, u) = 0 ,\qquad(19.9)$$

where u is the distance between the soliton maxima, $f(u)$ is the force field in the tail of one soliton, where another is positioned, and $v(\mathcal{E})$ is the soliton's velocity as the function of energy. Equations like (19.19) can be derived for small interactions from the initial equations for waves by representing the field in the vicinity of each soliton (its parameters being considered to change slowly) in the form of an asymptotic series whose terms have been truncated.

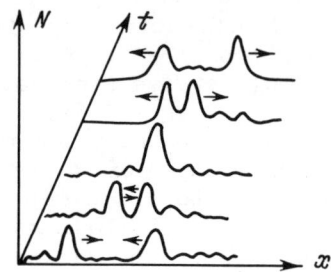

Fig. 19.9. Collision between ion-acoustic solitons (N is the particle concentration).

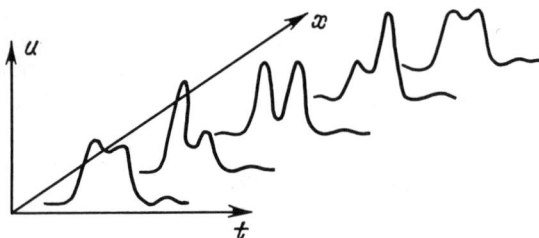

Fig. 19.10. Oscillating pair of solitons.

Once the analogy between solitons and particles had been established (i.e., equation (19.19) had been obtained), all that needed to be known in order to describe the interaction between solitons was the force function $f(u)$, i.e., the nature of the soliton "tails". If $f(u)$ is monotonic, then the solitons either repel or attract each other. The majority of accurate solutions illustrate the repulsion of solitons. If, however, the solitons have oscillating "tails", as for example the solitons in a capillary-gravity waves in shallow water or in a nonlinear artificial transmission line with inductive connections between the chains, then the function $f(u)$ is sign variable and the solitons sometimes attract and sometimes repel, thus forming an oscillating pair (the bound state, Fig. 19.10).

The interaction between large numbers of identical solitons can be considered in the same way because the nature of the "tails" does not depend on the number of solitons in them.

We add that this analogy between nonlinear waves and oscillations is not as trivial as the mode analogies we considered earlier.

19.4 Higher-Dimensional Solitons

We have only considered the simplest examples of solitons, i.e., one-dimensional stationary solitary waves either in a one-dimensional distributed system (transmission line), or plane waves whose profiles only change in the direction of propagation (for example, solitons in shallow water and those described by the Korteveg–de Vries equation). At the same time, it is obvious that even in shallow water and on falling films of liquid (Chapter 24), and when ion-sound solitons are propagating in a plasma, solitons and soliton-like solutions must in general also depend on the transverse coordinate, i.e., at the minimum they must be two-dimensional. The simplest model within which such solitons can be described is the generalization of the Korteveg–de Vries equation suggested by Kadomtsev and Petviashvili, i.e.,

$$\frac{\partial}{\partial x}(u_t - uu_x - \delta u_{xxx}) = \alpha u_{yy} . \qquad (19.20)$$

The Kadomtsev–Petviashvili equation may be obtained from the wave equation[3] for potential acoustic waves assuming weak dispersion:

$$\partial^2 \phi/\partial t^2 - c^3 \Delta\phi = -\beta\Delta\Delta\phi - \partial(\Delta\phi)^2/\partial t . \qquad (19.21)$$

Here ϕ is the potential of the velocity, c is the speed of sound in the medium and $\beta \sim D^2 c^2$ characterizes the dispersion (D has the sense of the spreading length of the wave packet). The dispersion sign may be either positive ($\beta > 0$) or negative ($\beta < 0$). We shall be interested in waves whose profiles get more curved due to nonlinearity. This change in profile only takes place in the direction of propagation and so it will be considered a slowly changing function in the other coordinates, i.e., we search for a solution in the form

$$\phi \approx \phi(x - ct, \mathbf{r}_\perp, t) . \qquad (19.22)$$

---

[3] For a medium with dissipation the equation for wave packets, namely, the Khokhlov–Zabolotskii equation [24], is as universal.

Substituting (19.22) into (19.21) and retaining only the first order terms (the order of the nonlinearity and dispersion), we obtain

$$2c \frac{\partial^2 \phi}{\partial x \, \partial t} + c^2 \Delta_\perp \phi = \beta \frac{\partial^4 \phi}{\partial x^4} - c \frac{\partial}{\partial x} \left(\frac{\partial \phi}{\partial x} \right)^2 . \qquad (19.23)$$

This equation has the same form as (19.20) if we set $u = \partial\phi/\partial x$, i.e.,

$$\frac{\partial}{\partial x} \left(\frac{\partial u}{\partial t} + u \frac{\partial u}{\partial x} - \frac{\beta}{2c} \frac{\partial^3 u}{\partial x^3} \right) = -\frac{c}{2} \Delta_\perp u . \qquad (19.24)$$

Equation (19.24) has a solution in the form of a one-dimensional soliton, i.e.,

$$u = u_0 \equiv -\frac{3\beta}{2c\Delta^2} \cosh^{-2}\left[\frac{1}{2\Delta} \left(x + \frac{\beta t}{2c\Delta^2} \right) \right] , \qquad (19.25)$$

which has a width $\sim \Delta$ ($|\beta|/\Delta^2 \ll c^2$ is the condition for (19.21) and (19.24) to be applicable), and a velocity $\beta/2c\Delta^2$. The solution (19.25) describes a stationary wave in a system of coordinates moving along the x-axis at the speed of sound c. Therefore, when the dispersion is positive ($\beta > 0$) the soliton moves slower than the speed of sound and has a negative amplitude. If, however, $\beta < 0$ then the amplitude of the soliton is positive and its speed is faster than that of sound.

The sign of the dispersion in this case also determines the stability of a one-dimensional soliton against higher-dimensional disturbances [17], thus, when $\beta > 0$ a higher-dimensional disturbance grows, and when $\beta < 0$ a one-dimensional soliton is stable. A more rigorous result on the stability of a one-dimensional soliton in the Kadomtsev–Petviashvili model is obtained by an inverse problem [22]. Here, we shall only look at the result using some simple considerations. By linearizing (19.24) about the trivial solutions, we find that the phase velocity of quasiharmonic higher-dimensional disturbances with wave vector $\mathbf{k}(k_\perp, k_x)$ is

$$v_{ph}(\mathbf{k}) = c(k_\perp k_x^2/2 + \beta k_x^2/2c .$$

It can be seen that in a medium with positive dispersion, the velocity of disturbances is always greater than the velocity of a soliton, i.e., it must give energy to the two-dimensional small disturbances in the medium that overtake it. This explains the instability of a soliton in a medium with $\beta > 0$. When $\beta < 0$, the soliton oscillation decays due to the radiation of sound lagging behind it; in a medium with negative dispersion, a soliton is more stable than higher-order disturbances.

We give here the exact solution to (19.24) in the form of a two-dimensional soliton. It was first obtained numerically and then analytically [17]. Equation (19.24) for stationary solutions is first rewritten in dimensionless form ($\zeta \sim x$, $\eta \sim y$), i.e.,

$$\frac{\partial^2 u}{\partial \zeta^2} + \frac{\partial^2 u}{\partial \eta^2} - \frac{1}{12}\frac{\partial^4 u}{\partial \zeta^4} = \frac{1}{12}\frac{\partial^2}{\partial \zeta^2} u^2 \;. \tag{19.26}$$

Then the two-dimensional soliton is given by the expression

$$u(\zeta, \eta) = 8(1 + 4\eta^2 - 4\zeta^2)/(1 + 4\eta^2 + 4\zeta^2)^2 \;.$$

To conclude this section on higher-dimensional solitons, we note that a very convincing argument has been made that the remarkable feature of Jupiter's atmosphere, namely the Great Red Spot, is a two-dimensional Rossby soliton. In Chapter 5, we only looked at linear Rossby waves in a rotating atmosphere. If for a simple idealization (the atmosphere is an incompressible fluid whose depth is much smaller than the characteristic scale of the disturbance, and the angular velocity of the planet ω_0 is large) we also take into account nonlinearity, then we obtain the following two-dimensional nonlinear equation for departures of the atmospheric depth from its equilibrium value $h = h(t, \phi, \alpha)$

$$\frac{\partial}{\partial t}\left(I^2 h - r_0^2 \Delta h\right) - \frac{v_0}{R}\frac{\partial}{\partial \phi}\left(h + \frac{h^2}{2}\right) = 2\omega_0 \frac{r_0^4}{I}\left[\zeta \nabla h\right] \nabla \Delta h \;, \tag{19.27}$$

$H_0 \ll L \ll R$ where L is the characteristic scale of the disturbances, R is planet's radius, $I = \sin \alpha$, α is the latitude, ϕ is the meridional angle, r_0 is the Rossby–Obukhov scale, v_0 is the Rossby drift velocity caused by inhomogeneities in the Coriolis force with latitude ($\sim \omega_0 \sin \alpha$), and ζ is a unit vector along the vertical. Equation (19.27) has a solution in the form of a two-dimensional soliton [18]:

$$u(r, \alpha) \approx 1.6 \left(\frac{r_0}{LI_0}\right)^2 \cosh^{-4/3}\left\{\frac{3}{4}\frac{r_0}{L}\left[1 + \xi(\alpha - \alpha_0)\right]\right\},$$
(19.28)

where $L \sim r_0 h^{-1/2}$. It can be seen that characteristic dimension (radius) of a soliton L exceeds r_0 and diminishes as its dimension's amplitude h increases. The velocity of the soliton is approximately $(1 + h)$ times greater than the Rossby velocity γ_0. If we neglect the dependence of I on latitude, then the soliton becomes one-dimensional, that is, variables only change with r (distance on the horizontal plane to the center of the soliton).

It can be seen from (19.28) that the smaller I_0 is, the larger the velocity of the soliton, the larger its amplitude, and the smaller the dimension of the soliton, i.e., the vortex. It can be shown [18] that the rotation of a vortex and the direction of its motion are opposite to the rotation of the planet, i.e., it is an anticyclone and in an anticyclone the pressure in the middle of the vortex is larger than that at the periphery.

A similar soliton has recently been modeled in the laboratory [19]. "Shallow water" was investigated in a cylinder with a parabolic floor and rotating around the vertical axis of symmetry. Inhomogeneities in the floor were necessary to imitate the effects caused by gradients in the Coriolis force[4].

The parabolic profile of the vessel was necessary because it is only in a paraboloid (at a constant rotation speed) that a rotating layer of liquid can be achieved whose depth is independent of the coordinates, and the drift of solitons with $H_0 = $ const can be studied. In Fig. 19.11 several results are given from the interesting paper [19]. Its authors observed, for instance, that long-lived vortex solitons (19.28) can exist in a thin layer of an evenly rotating liquid. They survive without decaying for a life-time determined by the viscosity and at a distance of an order greater than their diameter. It is remarkable that if the disturbance initially created is too large, then it

---

[4] It is known [20] that if the depth of the atmosphere is independent of the geographic coordinates, then Rossby waves (in this case the soliton we are interested in) arise and exist due only to a latitudinal gradient in the Coriolis force. It is this gradient that creates vortices in the medium that drift in the opposite direction to the planetary rotation.

brakes down into several smaller flows (zonal) (Fig. 19.11b).

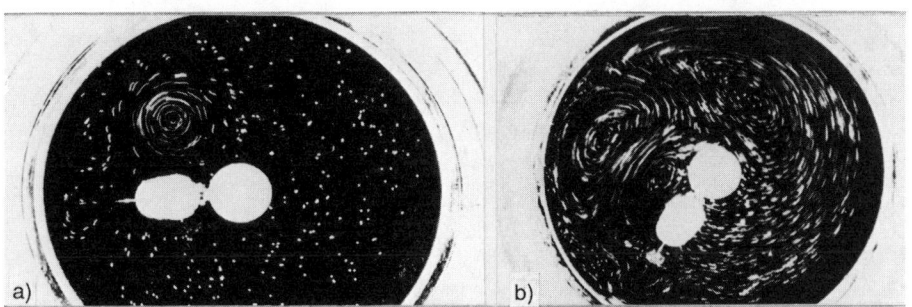

Fig. 19.11. Rossby solitons: a) steady-state soliton; b) disintegration of initial disturbance into two solitons [19].

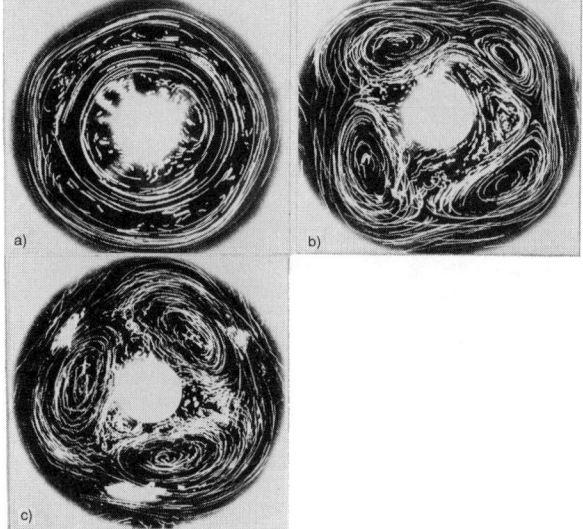

Fig. 19.12. Flow pattern in a thin film of a rotating liquid with a velocity shift: a) given a cyclonic shift; b), c) given an anticyclonic shift. The Rossby solitons can be distinctly seen against the background of the black floor [23].

We emphasized that such solitons must arise independently and due to the development of inherent instability in the system. Thus, recent experiments [23] have shown that Kelvin–Helmholtz instability, which is associated with the presence of a velocity shift in the flow profile, may lead to the generation of anticyclonic Rossby solitons (Fig. 19.12). These solitons drift in the opposite direction to planetary rotation. This type of soliton is similar in its properties and existence conditions to the Great Red Spot of Jupiter [23].

CHAPTER 20

MODULATED WAVES IN NONLINEAR MEDIA

20.1 General Remarks

Practically all oscillations and waves are modulated. Modulation is by definition a slow change in the parameters of the carrier wave, its amplitude, phase, frequency, and even form. It may be due to the action of an external force or field (forced modulation) or it may arise independently due to some sort of instability (self-modulation or auto-modulation). We already know of examples of both forced modulation and self-modulation. Changes in the wavelength and amplitude of quasiharmonic waves in smoothly nonhomogeneous media are forced modulation and determined by the "modulation" relationship linking the medium's parameters in space. The appearance of beats outside the synchronization band in an self-excited generator to which a periodic signal has been fed is an example of modulation which appears due to the interaction between unmodulated oscillations. On a plane of slowly changing amplitudes, such a modulation corresponds to a stable limiting cycle, as we have seen. Modulation obviously arises due to the interaction between oscillators even in conservative systems and media (Chapter 17). For example, when the resonance condition $2\omega_0 = \omega_1 + \omega_2$ is fulfilled, this process is naturally called mutual modulation; if, however, $|\omega_0 - \omega_{1,2}| \ll \omega_{0,1,2}$ and $N_0(0) \gg N_{1,2}(0)$, then the decay of pairs of quasiparticles ω_0 into satellites ω_1 and ω_2 is called self-modulation.

Because only modulated oscillations and waves can carry information, the process of "creating a modulation" and transferring the modulation on a carrier wave is very important in various applications. In this chapter we shall only consider processes leading to modulation. Mainly, we shall discuss the modulation of waves that arises when they propagate and interact in nonlinear media. There are a great variety of nonlinear phenomena and effects associated with the modulation of waves, e.g., the self-focusing of wave beams [1, 25], the self-compression of wave packets [2, 15], the inversion of a wave front [3,4], and other examples are given in [4].

Although we shall only be considering quasiharmonic modulated waves, it should be said that a much wider class of modulated waves exists, namely the nonsinusoidal, and even not necessarily periodic, waves with slowly changing parameters. As we already know, the behavior of waves in a nonlinear medium depends on the relationship between the dispersion (D) and nonlinearity (N) parameters. When $N \ll D$, the wave will be quasiharmonic and its harmonics will move at considerably different velocities (there is no synchronicity) and therefore the fundamental wave will effectively not excite them, i.e., its shape will not be significantly affected. Here the wave may be written in the form $A(\mathbf{r}, t)\exp(i\psi) + $ cc, where A is the slowly changing amplitude and ψ is the total phase (characteristic function). Using this description it is possible to construct "a nonlinear geometric optics" (like the linear geometric optics in [5] and Chapter 12), in which the equation for the wave amplitude and total phase, as opposed to the linear case, are related. The nature of a wave's modulation as the wave propagates depends on its amplitude (this is self-excitation and it is this class of phenomena which includes self-focusing of wave beams and self-modulation leading to the formation of wave packets).

If, however, the dispersion and nonlinearity are of the same order, then the wave will be significantly nonsinusoidal (the harmonics growing due to the energy of the fundamental component will change the shape of the wave). In media with $N \sim D$, as we have seen, it is possible for stationary nonlinear waves to exist (Chapter 19) and propagate without distortion in the profile at a constant velocity. These waves obviously belong to a special but important class of waves in nonlinear media. However, if these waves are considered the basis for constructing a wider class of solutions, assuming that their parameters are smoothly modulated over time and space, then it is possible to describe a much wider circle of nonlinear phenomena, such as the onset of modulation against the background of periodic soliton nets, and the deformation in the profile of a nonlinear wave as it propagates in a nonhomogeneous medium [6]. A similar approach seems to be fruitful for $N \gg D$, when shock waves can form. If, when the relation $N \gg D$ is retained, the nonlinearity itself is small, then the evolution of the wave may be considered to be a slow modulation because it takes place over intervals much longer than the characteristic length [6,7]. Now, we return to the quasiharmonic waves. The first question that arises when discussing the behavior of modulated waves in a nonhomogeneous medium is how the modulation propagates.

In equilibrium transparent (without dissipation) media, the evolution of a one-dimensional modulated wave

$A(x,t)\exp(i\psi(x,t))$ is described by

$$\partial k/\partial t + \partial \omega/\partial x = 0 ,\qquad (20.1)$$

$$\partial W/\partial t + \partial S/\partial x = 0 ,\qquad (20.2)$$

where $k(x,t) = -\psi_x$, $\omega(x,t) = \psi_t$ are respectively the wave number and frequency of the modulated wave, and W and S are the density and energy flow of the wave averaged over the period [6]. Equation (20.1) is clearly obtained from the definition of k and ω, while (20.2) is simply an expression of the conservation of energy in the medium over a period. In order to close (20.1) and (20.2) we must add the dispersion equation of the medium. If the medium is nonlinear, then the frequency (or wave number) must depend on the wave energy (remember the anisochronous oscillator), i.e., we must write

$$\omega = \omega(k, A^2) , \quad \text{or} \quad k = k(\omega, A^2) . \qquad (20.3)$$

Thus, in a nonlinear medium the equations describing the propagation of phase and energy are not independent [5, 6]. We now take into account that our wave is quasiharmonic, and the relationship between ω or k on A^2 is weak, therefore (20.3) can be expanded into series

$$k(\omega, A^2) \approx k(\omega, 0) + \alpha A^2 + \ldots \qquad (20.4)$$

After substituting this expression into (20.1), we obtain

$$\partial \omega/\partial x + v^{-1}(\omega)\partial \omega/\partial t + \alpha \partial A^2/\partial t = 0 \qquad (20.5)$$

which is the approximation of nonlinear geometric optics [5,6,10]. We now restrict ourselves to small frequency modulations and introduce a relative detuning ξ from the fundamental frequency ω_0: $\xi = (\omega - \omega_0)/\omega_0$. By changing to a moving system of coordinates $\tau = t - x/v$, $\chi = x$ we obtain the following equation from (20.5) (a simple substitution and expansion of $v(\omega)$ into series):

$$\partial \delta/\partial \chi + \delta \partial \delta/\partial \tau + \kappa \alpha \partial m/\partial \tau = 0 , \qquad (20.6)$$

where $(\delta = \xi \omega d^2 k/d\omega^2)$, $\kappa = v d^2 k/d\omega^2$ and $m = A^2$. Here we have discarded the term $\sim \alpha \delta \partial m/\partial \tau$ because it is smaller by a higher order than the remaining terms. In order to obtain an equation for m, we must use an explicit expression for the energy and energy flux of a wave in a nonlinear medium. Since we are discussing waves with small amplitudes, it is generally valid to expand W and S in series $W = \zeta_2(\omega)A^2 + \zeta_1(\omega)A^4 + \ldots$ (and the analogous one for S), and the equation for the transfer of energy (20.2) can be represented in the form

$$v\partial m/\partial \chi - \partial(\delta m)/\partial \tau + (1/2)\zeta \partial m^2/\partial \tau = 0 , \qquad (20.7)$$

where ζ may be considered a constant when the frequency modulation is small. Equations (20.6) and (20.7) describe the propagation of a modulation wave for the given assumptions.

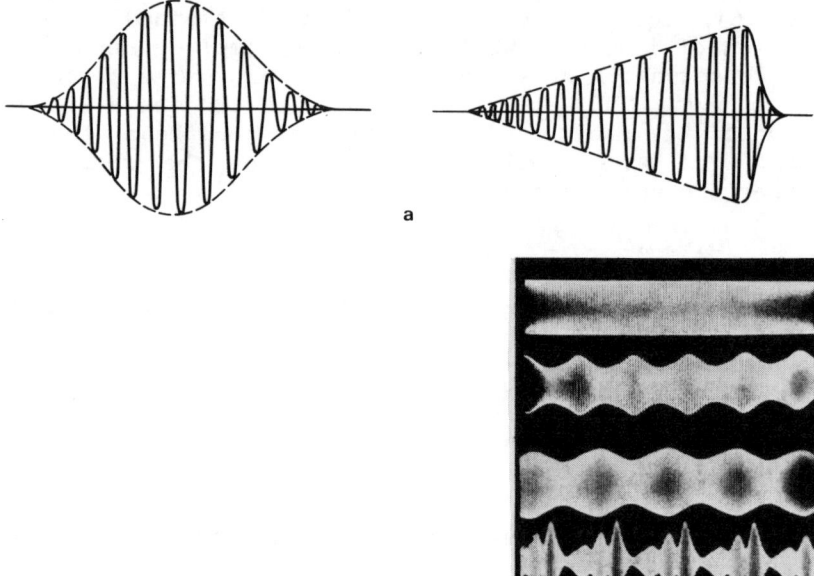

Fig. 20.1. a) The evolution of a simple wave envelope when propagating in a nonlinear medium ($\beta v'_k > 0$).

b) Self-modulation of a wave in a nonlinear transmission line with $\beta v'_k < 0$ [15].

Several features of this type of propagation can already be seen in (20.7). Suppose there is no dispersion in a narrow spectral interval close to ω_0[1], then $d^2k/d\omega^2 = 0$, i.e., $\delta \equiv 0$ and (20.7) is the equation of a simple wave, which we know so well (Chapter 18) and whose solution is $m = m(t - x/u)$, where $u = \zeta m + v$. Thus, a small perturbation of the envelope in this approximation evolves as a simple wave [15] and a region

---

[1] In general of course, dispersion exists in the medium and indeed it is this that allows us to ignore the harmonics (they are not synchronous with the fundamental wave).

where there is a fast change in the modulation may be formed (Fig. 20.1a) (toppling in the modulation wave is hindered by the dispersion ~ $d^2k/d\omega^2$, which we shall neglect).

It is not difficult to see from (20.6) that amplitude modulation in a weakly nonlinear medium will give rise to frequency modulation.

An observant reader will, no doubt, have observed that (20.6) and (20.7) are reminiscent of the one-dimensional equations of hydrodynamics (δ being the velocity in a sound wave, and m being density). The principal difference is that in our case the quantity $\kappa\alpha$ acts as the square of the speed of sound ($c^2 = dp/d\rho$, see Chapter 5) and it may be negative (if such a "medium" could be created, then the increase in pressure would decrease the wave's density). When $\alpha\kappa > 0$, as in fluid mechanics, equations (20.6) and (20.7) have solutions in the form of two families of simple waves, i.e., fast waves and slow waves. The leading edge of fast waves grows more curved, and the trailing edge of slow waves grows more curved (as we have seen, a modulation wave cannot topple and our equation simply becomes inapplicable). If, however, $\alpha\kappa < 0$, then the wave velocity becomes complex (convince yourselves of this for the example of a modulation wave with a small amplitude described by the linearized equations (20.6) and (20.7)). What this all means and what physical phenomenon it correspond to, we shall look at now.

20.2 Self-Modulation. Reversibility

Let us set up a simple experiment. We feed in a sinusoidal oscillation whose frequency is in the strong dispersion region $\omega(k)$ (e.g., the flat section in the dispersion curve in Fig. 4.18) to the boundary of a LC-transmission line (see Fig. 4.6, where one should assume that $Q(U) = C_0 + C_N U^3$), i.e., the harmonics arising due to the nonlinearity are not synchronized with the fundamental wave (it follows that they do not increase). We ask what oscillation will be observed at the output of the line. An oscillogram is shown in Fig. 20.1b [8] and shows that the oscillations are modulated. This is the phenomenon of self-modulation we discussed in the Introduction; the modulation arises due to the development along the transmission line of parametric instability, which in this case induces satellite waves with frequencies close to the carrier one. This is the instability that corresponds to complex propagation velocities of the modulation wave, about which we have spoken. A similar variety of parametric instability (in the terms used in Chapter 11, this is a second instability zone) arises in the theory of nonlinear waves and is called modulational instability.

In order to describe modulational instability and related phenomena, we return to the basic equation in the theory of modulated waves in nonlinear media, the nonlinear parabolic equation or nonlinear Schrodinger equation (it includes (20.6) and (20.7) as special cases), i.e.,

$$\left(\frac{\partial a}{\partial t} + v \frac{\partial a}{\partial x}\right) - \frac{i}{2} \frac{d^2\omega}{dk^2} \frac{\partial^2 a}{\partial x^2} - \frac{i}{2k_0} \frac{d\omega}{dk} \Delta_\perp a = i\alpha \varepsilon_N(|a|^2)a \qquad (20.8)$$

Here a is the complex amplitude $\exp[-i(\omega t - \mathbf{kr})]$ of the wave, $\mathbf{k}$ is its wave number, $\Delta_\perp$ is the Laplacian in the transverse y, z coordinates, and $\varepsilon_N(|a|^2)$ characterizes the form and size of the medium's nonlinearity. For example, for a light wave $\sqrt{\varepsilon_N}$ is a value proportional to the nonlinear correction to the refractive index. For the simple case of planar waves, we use the following equation instead of (20.8):

$$\left(\frac{\partial a}{\partial t} + v \frac{\partial a}{\partial x}\right) - \frac{i}{2} \frac{d^2\omega}{dk^2} \frac{\partial^2 a}{\partial x^2} + i\beta a |a|^2 = 0 , \qquad (20.9)$$

where $\alpha \varepsilon_N = \beta$. The terms in the parenthesis describe the modulation waves travelling in a linear medium without dispersion and with a group velocity v, while the term in the second derivative (parabolic term) is proportional to $d^2\omega/dk^2$ and corresponds to dispersion spreading, while the coefficient β is of the same magnitude and sign as the nonlinearity ($\beta \sim \alpha$, in (20.6) and (20.7)).

Equations (20.8) and (20.9) are the equations for the second approximation in the asymptotic method for quasiharmonic waves (Chapter 17). By being patient and doing all the arithmetic carefully, it is not difficult to obtain them in the form of (17.30) (see [9]). However, here we shall use a simpler and clearer derivation [6, 10] based on the equations we have already met, i.e., (20.1) and (20.2). We rewrite (20.2) in the following form (remember we are considering quasiharmonic waves)

$$\partial A^2 \partial t + \partial[v(\omega, A^2)A^2]/\partial x = 0 \qquad (20.10)$$

and introduce a complex amplitude (envelope)[2]

---

[2] We shall now, in fact, repeat the derivation of (20.6) and (20.7) only in a complex form and we shall not neglect any of the small terms with second order.

$$a(x, t) = A(x, t)\exp[i\phi(x, t)], \qquad (20.11)$$

where $\partial\phi/\partial t = \omega_0 - \omega$, $\partial\phi/\partial x = k - k_0 = k_1$ (k_0, ω_0 are the wave number and frequency of the harmonic waves against to background of which our modulation wave exists). If we now expand the right-hand side of the dispersion equation (20.3) into series about k_0 and $A^2 = 0$, then after substituting the expansion and (20.11) into (20.1) and (20.10), we obtain the desired equation (20.9). We shall do this for the example of weakly nonlinear gravity waves, whose nonlinear dispersion equation was obtained in the middle of the last century by Stokes:

$$\omega^2 = gk(1 + k^2 A^2). \qquad (20.12)$$

Assuming that $k = k_0 + k_1$, we expand the right-hand side of the expression into series

$$\omega(k) = \omega_0 + \frac{\omega_0}{2k_0} k_1 - \frac{\omega_0}{8k_0^2} k_1^2 - \frac{1}{2}\omega_0 k_0^2 A^2 + \ldots, \qquad (20.13)$$

where $\omega_0 = (gk_0)^{1/2}$ is the dispersion equation of gravity waves in the linear approximation (Chapter 5). After substituting (20.13) and (20.11) into (20.10) and (20.1), we find a nonlinear parabolic equation (20.9), i.e.,

$$\left(\frac{\partial a}{\partial t} + \frac{\omega_0}{2k_0}\frac{\partial a}{\partial x}\right) + \frac{i}{2}\frac{\omega_0}{4k_0^2}\frac{\partial^2 a}{\partial x^2} + i\frac{\omega_0 k_0^2}{2}|a|^2 a = 0,$$

in which we have $v = \omega_0/2k_0$, $d^2\omega/dk^2 = -\omega_0/4k_0^2$, and $\beta = \omega_0 k_0^2/2$ for modulated gravity waves on deep water.

Equation (20.8), which describes higher-dimensional modulation waves, was obtained in exactly the same way except that k was considered a vector and in the series expansion about k_0 the transverse components were accounted for (here we have $\partial\psi/\partial x = k_{1x}$, $\partial\psi/\partial y = k_{1y}$, and $\partial\psi/\partial z = k_{1z}$). We suggest the reader does this for the example of gravity waves which we have considered.

Equation (20.9) is one of the basic equations of nonlinear physics. It describes the evolution of an optic wave in nonlinear crystals, Langmuir waves in plasmas, heat waves in solids, and many other types of wave. In particular, it is related to the Ginsburg–Landau equation [12] in the physics of superconductors.

We shall now use these equations to describe three basic phenomena observed when one-dimensional quasiharmonic waves propagate in weakly nonlinear media, namely modulation instability, the existence of stationary envelope waves (including solitons), and the periodic (in time and space) return of weakly modulated waves (which approximate a periodic sequence of solitons as they evolve) to the initial, weakly modulated state.

Modulational instability, as we shall see, is only possible for certain relationships between the signs of the nonlinearity and the dispersion of the group velocity:

$$\beta d^2\omega/dk^2 < 0 . \qquad (20.14)$$

It is easiest to understand the physical mechanism of this restriction (usually called Lighthill's condition) if we consider the effect of self-modulation in terms of spectra rather than space-time, restricting ourselves to an analysis of the interaction between only three oscillator waves which form a wave with sinusoidal modulation.

Equations like the following are obtained for the complex amplitude of the carrier ω_0 and the spectral satellites $\omega_\pm$ that are symmetrically positioned around it from (20.9)) (the amplitude of the satellites is assumed to be small):

$$\dot{a}_0 = - i\beta |a_0|^2 a_0 ,$$
$$\dot{a}_\pm = - i\beta a_0^2 a_\mp^* - 2i(\beta |a_0|^2 + (d^2\omega/dk^2)k_1^2/4)a_\pm . \qquad (20.15)$$

It is assumed that because the satellites are spectrally close, the detuning is

$$\delta = 2\omega_0 - \omega(k_0 + k) - \omega(k_0 - k_1) \approx - (d^2\omega/dk^2)_1^2.$$

Thus, we have returned to a consideration of parametric instability. The parametric increment by which the amplitude of the satellites increases in the carrier field is

$$\gamma = \pm k_1 \left[- \beta |a_0|^2 d^2\omega/dk^2 - k_1^2/4(d^2\omega/dk^2)^2 \right]^{1/2} . \qquad (20.16)$$

Because the spatial scale of the modulation is arbitrary, a necessary (and as $k_1 \to 0$ a sufficient) condition for modulation instability is $\beta \cdot d\omega^2/dk^2 < 0$.

Now it should be obvious what the physical sense is: in order for modulation instability to appear, the nonlinear detuning from synchronicity, which is proportional to $\beta |a_0|^2$, must compensate the linear desynchronism, which is proportional to $(d^2\omega/dk^2)k_1^2$. Naturally, this is only possible when k_1 is not large: $|k_1| < k_0$. According to (20.16) the parametric increment almost linearly increases as $|k_1|$ increases from zero. Until it reaches a maximum and then it quickly dies away to zero for $|k_1| \to k_0$ where $k_0 = 4|\beta(d^2\omega/dk^2)^{-1}||a_0|^2$. For short-wave modulation ($\Lambda < 2\pi/k_0$), the nonlinear detuning cannot compensate for dispersive spreading and modulation compressions will not take place (the increment becomes imaginary).

The phenomenon of self-modulation can be observed by the reader on the sea, by looking at a chain of wind waves. The phenomenon helps explain the certainty of the "ninth wave" [13].

The nonlinear stage in a development of a modulational instability depends on asymptotes of the initial disturbance as $|x| \to \infty$. If this disturbance dies away rapidly at infinity, as was the case with wave impulses of the field itself (their evolution in the one-wave approximation being described by the Korteveg–de Vries equation), then whatever its form the initial impulse of the modulation wave as $t \to \infty$ decays into solitons (these are obviously "radio solitons" and have a high-frequency filling) plus an oscillating "tail". As in the similar situation described by the Korteveg–de Vries equation, this "tail" contains much less energy than that in the solitons, and is only crucial when considering the interaction between solitons (Chapter 19). The number of solitons depends on the form of the initial profile. The evolution of an initial disturbance localized in space is more regressively studied using the inverse radiation method [14] and here we shall only give the solution to (20.9) in a form of solitary stationary waves of modulation (envelope waves)

$$a(x, t) = A \operatorname{sech}\left\{\left|\frac{\beta}{2d^2\omega/dk^2}\right|^{1/2} A\left[(x - x_0) - (v + V)t\right]\right\} \quad (20.17)$$

$$\times \exp\left\{i\frac{\beta}{2}A^2 t - i\left(\frac{d^2\omega}{dk^2}\right)^{-1}\left[(x - x_0) - (v + V)t + \psi_0\right]\right\},$$

where A is the amplitude of the soliton, V is its velocity in the coordinate system moving with the group velocity v, and x_0 and ψ_0 are the initial coordinate and phase of the soliton. This solution was obtained by changing to the real variables A and ψ in (20.9), and then by assuming that the amplitude and phase move with constant velocities V and V_ψ, respectively (for the existence of a solution in the form of a soliton it is necessary that $V > 2V_\psi$ [16]), we can obtain ordinary differential equations for the amplitude and phase. These equations are easily integrated and reduce to the single equation of a nonlinear oscillator [15], whose solution is written in the form (20.13). These equations together with the soliton solution, i.e., an impulse of an envelope, have another solution in a form of a solitary trough, a shadow wave [15]. We suggest the reader does the integration, independently, using the information in [16].

We emphasize that in contrast to the usual soliton of the Korteveg–de Vries equation, the velocity and amplitude of a soliton envelope are independent parameters. These solitons were experimentally observed and the interactions investigated for waves over deep water in [11].

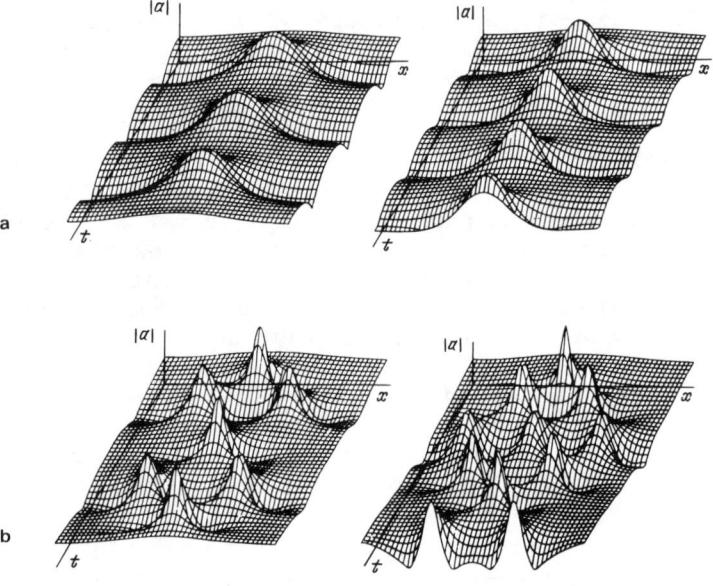

Fig. 20.2. Modulation waves on the surface of a deep liquid: a) stationary waves; b) the returnability phenomenon.

If, however, the initial disturbance is not localized in space, for example being periodic, then its evolution will be completely different. The sinusoidal modulation waves growing as the result of the modulational instability will be nonlinearly distorted. One or more solitons may form during the period of a wave and then the solitons will merge, and the wave will once again return to its initial condition, after which the sequence will be repeated. This phenomenon of returnability was experimentally observed on gravity waves on deep water, which we have discussed [17,11,45]. The appropriate numerical results are given in Fig. 20.2 [18, 11,45]. The results of physics experiments with nonlinear LC-circuits, which can be approximated by equations such as the Korteveg-de Vries equation with cubic nonlinearity are shown in Fig. 20.3. When the cells are sinusoidally excited at the boundary almost complete returnability along the chain was observed. The sinusoid was transformed into a periodic sequence of solitons, i.e., a large number of harmonic oscillators were excited, and then the solitons returned to the sinusoidal wave. All the harmonics returned the energy to the first harmonic.

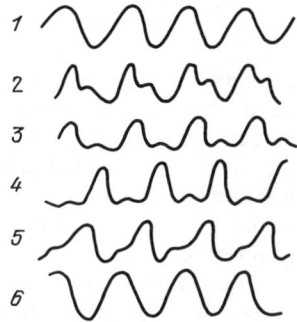

Fig. 20.3. The periodic evolution of nonlinear waves in LC-circuits: the sinusoidal wave input onto the line is turned into a sequence of impulses whose form is close to that of the soliton (1-4); after which the very nonlinear wave once again returns to a sinusoidal form (5. 6).

This effect was first observed in a numerical experiment by Fermi, Pasta, and Ulam [20]. They were trying to confirm the hypothesis that in systems with a very large number of degrees of freedom the presence of even weak nonlinearity is sufficient to distribute energy stored in separate degrees of freedom (modes) equally among all the modes (mixing) and thus establish thermodynamic equilibrium (thermalization). Fermi, Pasta, and Ulam experimented with models of nonlinear chains containing large numbers of particles and they did not observe thermalization, the system periodically returned to the condition with the initial energy distribution (the Fermi-Pasta-Ulam paradox).

Two questions arise when considering this paradox: the first and foremost is why is there no mixing, and the second is why does the system not go to some equilibrium state (for example a sequence of solitons) instead of periodically oscillating. Before considering these problems, let us recall that the systems being considered are conservative, i.e., in phase-space neither asymptotic stable equilibria nor any other sort of attractor (the limiting trajectories or set of trajectories possible in systems where there is a compression of the phase volume) are possible for the finite-dimensional models (from N interacting harmonics) corresponding to these conservative systems. However, we saw in Chapters 22 and 23 that a set possessing a very complex structure can exist in the phase space of such systems even when the number N is not large. The set need not be attracting but it will take up a large volume in the phase space and motion inside it corresponds to our intuitive understanding of mixing. The systems we are discussing belong to the class of completely integrable systems in which such complex (mixing) regions may not exist in the phase space[3]. The proximity of the model that Fermi, Pasta, and Ulam were experimenting on to a completely integrable system explains why they did not observe thermalization.

The complete integrability of a nonlinear Schrodinger equation with periodic boundary conditions was proved in [21]. A nonlinear modulation wave in this case has a discrete spectrum, and because the group velocity is dispersed, the spectrum may be considered bounded (satellites with large numbers are not resonant and therefore do not grow). In this situation, it is natural to change from space-time descriptions to a spectral one and to consider the interaction between several (in the simplest case three, ω_0 and $\omega_\pm$), spectral components. It is assumed that the synchronicity condition $2k_0 = k_- + k_+$ and $2\omega_0 = \omega_+ + \omega_- + \Delta\omega$, where $\Delta\omega$ is a small detuning from exact synchronicity, is fulfilled for a medium with cubic nonlinearity.

The equations for the amplitudes A_0 and $A_\pm$ can be found in exactly the same way as the equation for the amplitude of the fundamental wave and its second harmonic interacting in a medium with quadratic nonlinearity (Chapter 17). We shall give the resultant system for the special case when the satellites are equivalent, i.e., $A_+ \equiv A_- \equiv A_1$, $\phi_+ = \phi_- = \phi_1$ [22], i.e.,

$$\dot{A}_0 = 2A_0^2 A_1^2 \sin \Phi, \quad \dot{A}_1 = - A_0^2 A_1^2 \sin \Phi,$$
$$\dot{\Phi} = - s + A_1^2 - A_0^2 + (2A_1^2 - A_0^2)\cos \Phi, \quad (20.18)$$

---

[3] Complete integrability means that the $(N-1)$-th independent integral of motion exists for a finite-dimensional system with N degrees of freedom.

where $\Phi = [\Delta\omega t + 2(\phi_0 - \phi_1)]\sin\beta$, $s = \text{sign}(\beta d^2\omega/dk^2)$, and the differentiation is with respect to a dimensionless time $\tau = \Delta\omega t$ (the modulation instability corresponds to $s = -1$). By using the integrals

$$A_0^2 + 2A_1^2 = E, \quad A_1^2(2A_0^2 \cos\Phi + 2A_0^2 + A_1^2 - 2s) = F, \quad (20.19)$$

it is possible to reduce the investigation of (20.18) to an analysis of motion on the phase-plane.

The resultant phase-plane diagrams are given in Fig. 20.4. For convenience, the phase variables are taken to be $x = (2A_0)^{1/2}\cos(\Phi/2)$ and $y = (2A_0)^{1/2}\sin(\Phi/2)$. Only trajectories laying inside the circle $x^2 + y^2 = 2E$ have any physical meaning and an exit into this circle corresponds to a reduction of the energy of the fundamental mode and the generation of satellites. It can be seen that this is only possible when $E > 2$ (20.14) and when the energy in the fundamental mode is large; a full transfer of the energy to the satellites is possible and the separatrices move towards the circle in the equilibrium $x = y = 0$ (Fig. 20.4c), and when $1/2 < E < 2$, only a partial transfer of energy is possible (Fig. 20.4b). When the energy of the fundamental wave is small (or when $s = +1$), there is no modulation instability (Fig. 20.4).

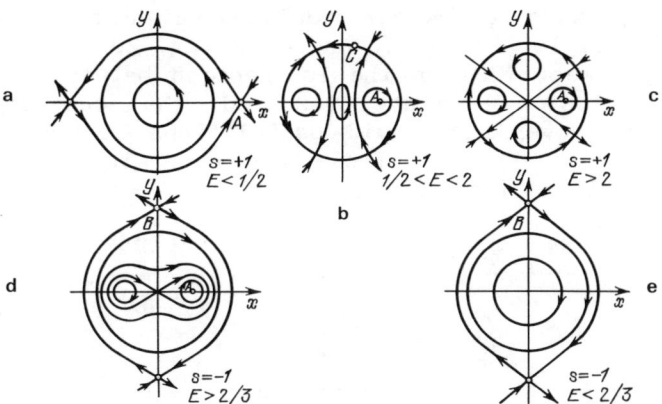

Fig. 20.4. Phase-plane diagrams of a conservative system described by a modulational decay in the fundamental harmonic into a pair of identical satellites. The coordinates of the equilibria are: A) $x = [(6E + 4s)/7]^{1/2}$, $y = 0$; B) $x = 0$, $y = (2E - 4s)^{1/2}$; C) $x = s^{1/2}$, $y = (2E - s)^{1/2}$.

The phenomenon of returnability is described by the trajectories in Fig. 20.3c, which are close to the separatrices. Initially, the energy is almost completely transferred to the satellites, and then it is returned to the fundamental mode, and so on. Solitons in this model correspond to equilibrium states whose coordinates are given in the caption to Fig. 20.4.

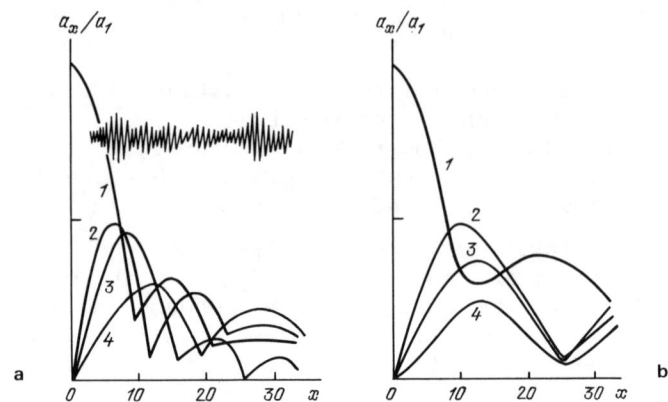

Fig. 20.5. a) Aperiodic and b) periodic energy exchange between the 1) fundamental mode and 2, 3, 4) the satellites.

We note in the conclusion of this section that the effect of returnability is also observed in more complex situations, such as when modulated high-frequency waves interact with low-frequency ones. In Fig. 20.5, the results of an experiment [23] on the interaction between such waves in a transmission line are presented. The equations, which are averaged over the rapid oscillations, describing this "medium" are similar to those for Langmuir and ion-sound waves in a plasma [24], i.e.,

$$\frac{\partial a}{\partial t} - \frac{i}{2} \alpha \frac{\partial^2 a}{\partial x^2} + \frac{i}{2} \omega_p na = -v_1 \frac{\partial^2 a}{\partial x^2},$$

$$\frac{\partial^2 n}{\partial t^2} - c^2 \frac{\partial^2 n}{\partial x^2} = \zeta \frac{\partial^2 |a|^2}{\partial t^2} + \delta \frac{\partial^4 n}{\partial t^4} + v_2 \frac{\partial^3 n}{\partial x^2 \partial t}.$$

(20.20)

Here a is the amplitude of the high-frequency (Langmuir) wave and is the field of the low-frequency (ion-sound) wave, $\omega_p = -1 + (C_1 + C_2/L_2C_1C_2)^{1/2}$ is the frequency about which the averaging is made $\alpha = 3\omega_p r_D/2$, where $r_D = 2C_2c^2/(3C_1\omega_p^2)$, $c = [L_1(C_1 + C_2)]^{-1/2}$ is the velocity of the low-frequency waves, $\zeta = c_2/c_1$, $\delta = r_D^2/l^2$, where l is the characteristic dimension of the high-frequency wave packet, and $v_1 = (R/2)\sqrt{C_1/L_2}$ and $v_2 = RC_1/(2L_1C_2)$ characterize the wave damping.

In the experiment, a monochromatic "Langmuir" (with frequency ω) and an ion-sound wave (with frequency Ω) were excited at the start of a "semi-infinite" line (the line consisted of 50 cells, and was balanced at the end). As the Langmuir wave propagated, it was modulated and several tens of satellites appeared. Then, depending on the ratios ω/ω_p and Ω/ω_p, either a stationary propagation of Langmuir solitons (Fig. 20.5a) or a regime corresponding to reversibility, with a periodic transfer of energy between the satellites and the carrier wave (Fig. 20.5b), became established. More complicated aperiodic exchanges of energy were observed in the system and we shall return to a discussion of them in Chapter 23.

20.3 Self-Focusing

We now turn to a discussion of the evolution of higher-dimensional modulated waves using equation (20.1). We obtain the following dispersion equation for waves with small-amplitude modulations $\sim \exp[i(\Omega t - kx - \mathbf{k}_\perp \mathbf{r})]$ by linearizing (20.1), where $\varepsilon_N(|a|^2) = -\beta |a|^2$:

$$\Omega(k, k_\perp) = vk \pm \left[\left(\frac{v}{2k_0} k_\perp^2 + \frac{1}{2} \frac{d^2\omega}{dk^2} k^2\right)\left(2\beta |a_0|^2 \right.\right.$$

$$\left.\left. + \frac{v}{2k_0} k_\perp^2 + \frac{1}{2} \frac{d^2\omega}{dk^2} k^2 \right)\right]^{1/2}$$

(20.21)

(Ω, k and $k_\perp$ are the frequency, and longitudinal and transverse wave numbers of the modulation wave). The increment in the modulational instability (20.12), which we have met already, can be obtained from this equation as a special case for a planar wave. Now, by contrast, we shall for simplicity assume that there are no one-dimensional perturbations, i.e., $k \equiv 0$. Then it follows from (20.21) when $\beta < 0$ that for all

$$k_\perp^2 < 4|\beta||a_0|^2 k_0/v \qquad (20.22)$$

the quantity $\Omega(k_\perp)$ is purely imaginary, that is, higher-dimensional disturbances with frequencies equal to the filling grow in the direction of the propagation. The physical manifestation of this can be described as follows: if we input a planar wave with frequency ω_0 at the boundary of a nonlinear medium whose dielectric permeability grows with increasing field strength (to be specific, we shall discuss an electromagnetic wave), then as the wave propagates it is transformed into a system of wave beams that is periodic in the transverse direction. This is the phenomenon of self-focusing [25, 26]. In other words, the instability that causes self-focusing is a stationary special variant of parametric instability or the decay of a pair of quasiparticles in one state into a pair of quasiparticles with the same energy but different pulse directions, i.e., $2\mathbf{k}_0 \to \mathbf{k}_1 + \mathbf{k}_2 + \Delta\mathbf{k}(|a|^2)$ (Fig. 20.6).

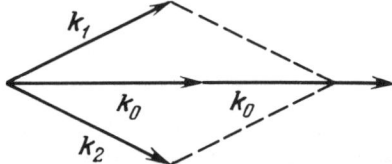

Fig. 20.6. Decay of a pair of quasiparticles with momentum $\mathbf{k}_0$ into a pair with the same energy and momentum $\mathbf{k}_1$ and $\mathbf{k}_2$ in a self-focusing medium.

It can be seen from (20.22) that self-focusing only begins when the amplitude (or power) of the wave entering the nonlinear medium via a finite aperture is greater than some critical value. For example, for cylindrical perturbation with radius R self-focusing begins when the power of the energy inside a circle of radius R is greater than $P_{cr} \sim a_0^2 R^2 = \pi^2 v/k_0 |\beta|$. Clearly, if we were to discuss a wave unbounded in the transverse direction, then (20.22) would simply mean that at small $|a_0|^2$ only perturbations with very large transverse scales would grow. However, real beams have a finite width and so the threshold is apparent at an amplitude defined by (20.22), i.e., the dimension of the perturbation does not have to be larger than the beam dimensions.

As was the case for space-time packets propagating in a one-dimensional weakly nonlinear medium, dispersion has a stabilizing effect and as a result stationary modulation waves may become established. When higher-dimensional perturbations arise, diffractional spreading may in principle hinder the nonlinear focusing of waves perpendicular to the propagation direction (described in (20.8) by the terms proportional to $\Delta_\perp a$). As the result of the combined effect of diffraction and nonlinearity, stationary focused wave beams may exist [27]. These beams, e.g., cylindrical waveguides, are extremely interesting for practical purposes since if they could be realized it would be possible to transfer the energy of, say an electromagnetic field, over long distances in a nonlinear medium without the danger of losses caused by diffraction. However, these waveguides are unstable.

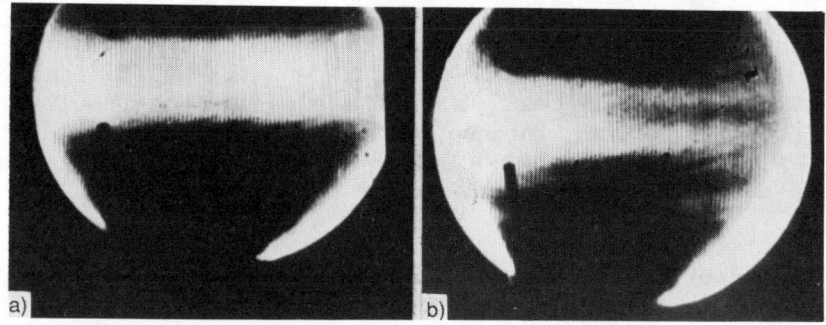

Fig. 20.7. Self-focusing in an acoustic packet in water with gas bubbles: a) self-focusing threshold not reached; b) self-focusing threshold exceeded.

It is possible to use the method of moments [27] to understand this instability. For simplicity, we shall assume that all the coefficients in (20.8) are unity: $ia_t + \Delta a - |a|^2 a = 0$[4]. This equation may be written in the form $ia_t = \delta\mathcal{H}/\delta\alpha$ [28], where the Hamiltonian is $\mathcal{H} = \int (|\nabla a|^2 - |a|^4/2)dr$. Let us consider the evolution over time of the effective width of the beam $\langle r^2 \rangle = N^{-1} \int_{-\infty}^{\infty} r^2 |a|^2 dr \left(N = \int_{-\infty}^{\infty} |a|^2 dr \right)$, i.e.,

$$d^2\langle r^2\rangle/dt^2 = 4d \int_{-\infty}^{\infty} |\nabla a|^2 dr - 2d \int_{-\infty}^{\infty} |a|^4 dr , \qquad (20.23)$$

where d is the dimension of the space, $d \geq 2$. This expression is obtained by averaging $\langle r^2 \rangle$ using the equation of motion for the beam in Hamiltonian form and integrating by parts. The first term in (20.23) describes diffractional spreading

---

[4] We shall investigate the stability of a beam homogeneous along the propagation direction with respect to homogeneous disturbances.

of the wave, and the second term its nonlinear compression. Thus, it follows from (20.23) [27,28] when $d \geq 2$, that $d^2<r^2>/dt^2 \leq 8\mathcal{H}$ (the case $d = 2$ in the second relationship corresponds to the equality sign). A stationary waveguide (nonlinear compression is compensated by diffractional divergence) corresponds to the value $\mathcal{H} = 0$). It can be seen from this identity that two-dimensional (axisymmetric) wave guides are unstable and any small perturbation causes either a compression of the wave (when the initial energy of the perturbation is negative and, hence $\mathcal{H} < 0$) or the spreading of the beam when $\mathcal{H} > 0$.

To conclude we are emphasize that self-focusing of beams is not only of interest to optics; it has applications in acoustics, plasma physics, etc. The experimental results on self-focusing of an intense ultrasound wave in distilled water with gas bubbles [29] (the bubbles appear in the water due to cavitation) are given in Fig. 20.7.

20.4 Interaction Between Wave Beams and Packets

The phenomena that occur during nonstationary interactions between quasiharmonic waves are very diverse. For example, the merging of impulses and beams in resonantly interacting waves in nonequilibrial media [30], the existence of coupled (three-wave) modulation solitons [31], and the inversion of a wavefront [32,33]. A quantitative description of these and similar effects is very complicated because a system of coupled nonlinear parabolic equations must be solved. However, many of these phenomena can be easily explained qualitatively, which is what we shall do in this section.

Inversion of a Wavefront[32,46]. In the very first experiments on the forced scattering of electromagnetic waves on a sound net created by the waves (the synchronicity condition is $\omega_0 = \omega_c + \Omega$, $\mathbf{k}_0 = \mathbf{k}_c + \mathbf{q}$, where ω_0, $\mathbf{k}_0$ and ω_0, $\mathbf{k}_c$ are respectively the frequency and wave number of the incident and scattered electromagnetic waves, and Ω and $\mathbf{q}$ are the frequency and wave number of the acoustic wave), it was observed that when the scattered wave left the nonlinear interaction region the evolution of the beams of the incident pumping wave was repeated by the wave beam scattered backwards. This was then discovered that in many experimental situations, viz., the scattered wave exactly reproduces the complex conjugate of the incident wave, which is strongly modulated in the transverse direction [3]. The repetition of the wave scattered backwards along the same optical path which the pumping wave travelled to the nonhomogeneous (in general random) medium, but in the opposite direction, means that the region of nonlinear interaction is an effective mirror. However, the mirror is quite unusual in that the wave reflected backwards repeats the optical path of the incident wave except that its phase front is the complex conjugate of the phase front of the pumping wave, i.e., $a_c^*(\mathbf{r}) \sim a_0(\mathbf{r})$. The complete phase of the quasiharmonic wave ($i\omega t - ikx + i\phi$) when propagating in the x-direction changes as if the incident wave were travelling backwards in time. The reproduction of the transverse modulation of the incident waves in the radiation leaving the nonlinear interaction zone gave this phenomenon its name of wavefront inversion.

The nonlinear interaction volume operates as an inverting mirror because of the selective amplification of the scattered (Stokes) wave, which grows out of noise, in the field of the pumping beam, which is strongly nonhomogeneous in the transverse direction. If the phase front of the pumping wave is not modulated, then the waves scattered backwards are amplified in the field of the pumping wave with an arbitrary transverse structure. If, however, the pumping wave has a sharp cut, then the scattered wave modulated in the transverse direction such that its amplitude maxima correspond to the amplitude minima of the pumping wave and vice versa is not amplified as much as a wave which duplicates the pumping wave's profile. Such an "inverting mirror" would be very useful, for instance, to implement self-correcting transport for powerful electromagnetic radiation of long distances. To do this the wave front of a weak signal arriving down the transport pathway from the desired receiver of the radiation needs to be inverted. The inverted wavefront of the signal could then be used to "yield" information about the pathway from the phase front.

Coupled Solitons [31]. We saw in Chapter 17 that when there is resonance interaction between three (or two) spatially homogeneous or stationary waves in a medium with quadratic nonlinearity, the energy exchange, and consequently wave amplitude change, occurs for certain phase relationships. Given certain phase differences a stationary state is possible (in Fig. 17.5 this corresponds to an equilibrium) in which the wave amplitude is not changed. It is natural to suggest that such a state should occur for the interactions between modulated waves, wave packets, if the phase change due to the effects of dispersive atomization are stabilized by nonlinear interactions. In terms of spectrum this would be the same, in essence, as a nonlinear frequency shift compensating linear asynchronicity, which we discussed in our treatment of the generation of satellites and the establishment of soliton envelopes during the propagation of wave packets in a medium with cubic nonlinearity. In the simplest case, when the fundamental wave ω interacts with the second harmonic 2ω, and the dispersion effects inside the narrow spectral interval only affects the fundamental frequency, we arrive at the standard equation describing solitons and two-dimensional waveguides in a medium with cubic nonlinearity, viz.,

$$d^2 a_\omega / dx^2 - \alpha a_\omega + \gamma a_\omega |a_\omega|^2 = 0.$$

It has now been shown [31] that coupled solitons are stable with respect to perturbations that do not change the energy of interacting packets.

The merging of wave impulses during explosive instability is an example of a purely energetic interaction between waves. The wave phases during explosive instability, as we know (Section 17.3) rapidly synchronize. Therefore it is possible to write an equation for the amplitudes of three waves satisfying the condition $\omega_1 + \omega_2 = \omega_3$ by adding terms to (17.31) that are proportional to the group velocities v_1, v_2, and v_3 of the waves, i.e.,

$$\frac{\partial A_1}{\partial t} + v_1 \frac{\partial A_1}{\partial x} = A_2 A_3, \quad \frac{\partial A_2}{\partial t} + v_2 \frac{\partial A_2}{\partial x} = A_1 A_3,$$

$$\frac{\partial A_3}{\partial t} + v_3 \frac{\partial A_3}{\partial x} = A_1 A_2 \qquad (20.24)$$

(Here the phase difference of the waves is $\Phi = \phi_1 + \phi_2 - \phi_3 = 0$.) Let us suppose that at $t = 0$ one of the waves A_3 considerably predominates over the others. Then linear equations are found for the weak waves A_1 and A_2, from whose solution it follows that A_1 and A_2 exponentially grow at the initial stage. When the amplitudes of all three waves become of the same order of magnitude, the instability changes over to the faster nonlinear stage and an "explosion" takes place, viz., the amplitudes of the wave packets tend to infinity within a finite time $t_\infty \sim \ln(A_{30}/A_{01})$, $t_\infty \sim (\ln A_{03}/A_{02})$ (A_{03}, A_{01}, and A_{02} are the initial amplitudes of the waves). This situation clearly must be similar to what happens to spatially homogeneous waves during an explosion. This is understandable because the wave packets do not have time to scatter due to the difference between group velocities ($v_1 \neq v_2 \neq v_3$) and wave packets that are overlapping at some moment in time continue to overlap as $t \to t_\infty$ because they grow too quickly in time. This has been confirmed by both analytical and numerical investigations [30,34] (Fig. 20.8). It is natural that if the initial amplitudes of the impulses are small and the group velocity difference is large, then the impulses will pass through each other in such a short period of time that the instability simply cannot enter the nonlinear (explosive) stage.

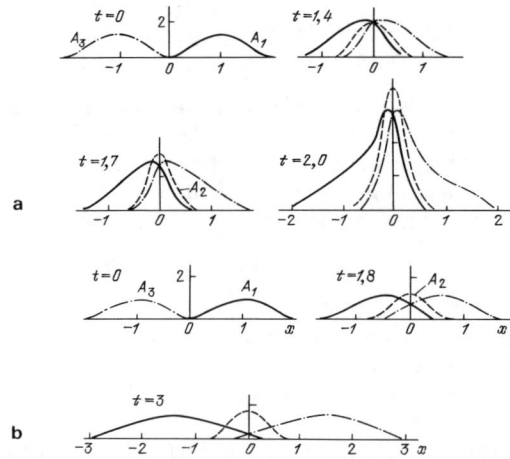

Fig. 20.8. Merging of impulses during explosive instability: a) amplitude of impulses exceed the critical value and the impulses merge; b) the amplitudes are smaller than the critical value and the impulses scatter.

20.5 Interactions Between Waves Having Randomly Modulated Phases. Wave Kinetics

In general, when quasiharmonic waves interact in a weakly nonlinear medium, the amplitude and phase changes may take place at considerably different characteristic time scales. For example, we have seen that during explosive instability, the wave phases quickly synchronize after which their difference may be considered practically constant and against this background the nonlinear evolution of the amplitudes is considered (Chapter 17). We have seen on several occasions how the division into fast and slow motions enables us to investigate many phenomena quite deeply without applying numerical methods (for example, the method of distinct oscillations and asymptotic methods based on the slowness with which the wave parameters change and the consequent averaging).

We shall now use this approach to consider the elementary process of resonance interaction between waves: the decay interaction of three waves in a medium with quadratic nonlinearity for which the synchronicity conditions must be fulfilled, i.e.,

$$\mathbf{k}_1 + \mathbf{k}_2 = \mathbf{k}_3 \; , \quad \omega(\mathbf{k}_3) = \omega(\mathbf{k}_1) + \omega(\mathbf{k}_2) \, \Delta\omega \; .$$

We saw in Chapter 17 that in order for the interaction to be effective, the detuning $\Delta\omega$ must be small. As the detuning increases the waves exchange less and less of their energy (Fig. 17.7). At the limit of detuning, large in comparison with the nonlinearity which is proportional to σa, the phase difference $\Phi = \phi_3 - \phi_1 - \phi_2$ of the interacting waves change rapidly in time. A large $\Delta\omega$ appears in the equation for Φ in comparison to which the nonlinear terms may be neglected, i.e., $\dot{\Phi} \approx \Delta\omega$ and the motions can be divided into fast and slow ones. Then by averaging the equation for the amplitudes $\dot{A}_j \sim A_i A_k \sin \Phi$ over the rapidly gyrating phase Φ, we find that $\dot{A}_j \approx 0$, i.e., no interaction takes place. This is correct if $\Delta\omega = $ const. However, if there is detuning that from to time to time rises to large values although remaining close to zero on average, then there must still be effective interaction even if it is small despite the rapid phase fluctuations.

Before convincing ourselves of this, note that a similar situation is quite often encountered in the physics of nonlinear waves [35-44]. Random inhomogeneities in a medium, fluctuations in its parameters over time, and the action of external irregular fields are all basic factors leading to "drift" of the natural frequencies in interacting waves over time or space. Such "drift" is possible even when the waves forming a resonance triplet participate in a large number of other interactions whose influence on the initial process may be crudely represented as the effect of an effective external field. I this case, the approximation of chaotic phases has a certain justification in that it is possible for individual anharmonic waves to be made chaotic due to the action of regular external forces (see [42] and Chapter 22). Of course, random fluctuations in a medium's parameters over time and space cause fluctuations in the amplitudes of waves (if only because the energy of the field at a given frequency is somewhat redistributed in space). However, because the energy of the waves on average does not change this energy redistribution cannot be large over the wave packet. Changes in the phase are not in any way limited. For example, due to small fluctuations in the group velocity, leading to a shift in the waves of only $\lambda/2$, the phase is changed by $\pi/2$.

Using these considerations, let us look at the interaction between three waves (17.30) assuming that their amplitudes change slowly and that their phases are rapidly changing functions of time. Equations (17.30) when $\partial a_j / \partial x = 0$, $g_j = g_N = 0$, $\zeta_1 = \zeta_2 = \zeta_3$, and $\Delta\omega = 0$ are

MODULATED WAVES IN NONLINEAR MEDIA   439

$$\dot{a}_1 = \sigma a_3 a_2^*, \quad \dot{a}_2 = \sigma a_3 a_1^*, \quad \dot{a}_3 = -\sigma a_1 a_2 \qquad (17.30a)$$

We rewrite the equations for the squares of the amplitudes $N_j = a_j a_j^*$ ($a_j = A_j \exp(i\phi_j)$), i.e.,

$$\dot{N}_1 = \sigma a_1^* a_2^* a_3 + cc, \quad \dot{N}_2 = \sigma a_1^* a_2^* a_3 + cc,$$

$$\dot{N}_3 = -\sigma a_1 a_2 a_3^* + cc \qquad (20.25)$$

If the phases were completely uncorrelated, the waves would not interact and their amplitudes would remain the same as the initial amplitudes $a_j = N_j^{1/2} \exp(i\phi_j)$. There is, however, some weak interaction because of partial correlation between the phases and therefore we must add to the undisturbed amplitude a_j (here N_j is a slow function of time than ϕ_j) a small correction $\mu a_j'$ which arises from other waves:

$$a_j(t, \mu t) = N_j^{1/2}(\mu t) \exp[i\phi_j(t)] + \mu a_j'(t, \mu t). \qquad (20.26)$$

By calculating $\mu a_j'$ from the initial equations (17.30a) in the given field approximation for two other waves, we obtain for the first order of magnitude in the interaction (i.e., first order in μ)

$$\mu a_{1,2}' = \sigma(N_{2,1} N_3)^{1/2} \int_{t_0}^{t} \exp[i(\phi_3 - \phi_{2,1})] dt,$$

$$\mu a_3' = -\sigma(N_1 N_2)^{1/2} \int_{t_0}^{t} \exp[i(\phi_3 + \phi_{2,1})] dt, \qquad (20.27)$$

where t_0 is a large negative time when the interaction was started (i.e., $\mu a t < t_0$). When integrated with respect to rapid changes in time, the slow variables $(N_i N_j)^{1/2}$ can be taken outside the integrand. By substituting (20.26) and (20.27) into (20.25) and retaining only the first terms that do not disappear when averaging over the phases, we obtain

$$\dot{N}_3 = \sigma^2 \Big[N_1 N_2 \langle \exp[i(\phi_1 + \phi_2)] \int_{t_0}^{t} \exp[-i(\phi_1 + \phi_2)] dt' \rangle$$

$$- N_3 N_1 \langle \exp[i(\phi_3 - \phi_1)] \int_{t_0}^{t} \exp[-i(\phi_3 - \phi_1)] dt' \rangle \quad (20.28)$$

$$- N_2 N_3 \langle \exp[i(\phi_3 - \phi_2)] \int_{t_0}^{t} \exp[-i(\phi_3 - \phi_2)] dt' \rangle \Big]$$

The angle brackets here mean averaging over the phase. By assuming that $t_0 \to -\infty$ and integrating by parts, it can be shown that $\langle \exp[i\theta(t)] \int_{-\infty}^{t} \exp[-i\theta(t')] dt' \rangle = 2\tau$, where τ is the correlation time of the function $\theta(t)$. Thus, the equations describing changes in the intensity of three waves interacting in resonance using the approximation of chaotic phases have the form

$$\dot{N}_3 = W(N_1 N_2 - N_2 N_3 - N_1 N_3) ,$$
$$\dot{N}_{1,2} = - W(N_1 N_2 - N_2 N_3 - N_1 N_3) , \quad (20.29)$$

where $W = 2\sigma^2 \tau$. The first thing to follow from these equations are the Manley–Rowe relations $N_1 - N_2 = $ const and $N_3 + N_{1,2} = $ const. Although the nature of the interactions is very different from the case of dynamic phases.

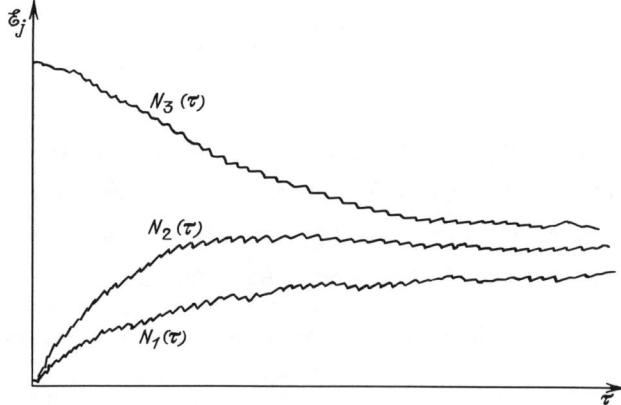

Fig. 20.9. Establishment of an equilibrium when three waves with randomly modified phases interact.

At a given level of the summed energy $\mathcal{E}$ of the interacting waves (the energy is clearly conserved), (20.29) has a single equilibrium $N_{1,2,3} = \text{const}/\omega_{1,2,3}$, which is stable (convince yourself of this independently). Consequently as $t \to \infty$ and independently of the initial conditions, a state is established in our system with an equal distribution of energy over all the degrees of freedom (Fig. 20.9)

$$\mathcal{E}_j = \omega_j N_j = T_0 = \text{const} . \qquad (20.30)$$

This equilibrium distribution corresponds to the well-known Rayleigh-Jeans law [40]. If the synchronicity condition is fulfilled straight away for many triplets of waves, then instead of (20.29) being summed over all possible resonances such as $\omega(\mathbf{k}) = \omega(\mathbf{k}') + \omega(\mathbf{k}'')$, we will have [36, 38]

$$\dot{N}_\mathbf{k} = \int [|\sigma_{\mathbf{k},\mathbf{k}',\mathbf{k}''}|^2 (N_{\mathbf{k}'} N_{\mathbf{k}''} - N_\mathbf{k} - N_{\mathbf{k}'} - N_\mathbf{k} N_{\mathbf{k}''})$$
$$\cdot \delta(\mathbf{k} - \mathbf{k}' - \mathbf{k}'')]d\mathbf{k}'\mathbf{k}'' . \qquad (20.31)$$

Here $\sigma_{\mathbf{k},\mathbf{k}',\mathbf{k}''}$ is a coefficient defining the level of nonlinear interaction between three waves with wave numbers $\mathbf{k}$, $\mathbf{k}'$ and $\mathbf{k}''$, and $\delta(\mathbf{k} - \mathbf{k}' - \mathbf{k}'')$ is a delta-function which selects from all the wave triplets that fulfill the frequency resonance conditions those that also fulfill the wave number resonance conditions. This equation was obtained for waves whose dispersion equation is obtained for both positive and negative frequencies. Moreover, it was assumed that the condition $\omega(-\mathbf{k}) = -\omega(\mathbf{k})$ was fulfilled.

It can be seen that the equilibrium spectrum will, for an arbitrary number of wave triplets, be characterized by a uniform distribution of energy over all the degrees of freedom. Equation (20.31) has a solution $N_\mathbf{k} = T_0/\omega_\mathbf{k}$ (this is easily confirmed by a direct substitution and the use of the relation $\omega_\mathbf{k} = \omega_{\mathbf{k}'} + \omega_{\mathbf{k}''}$).

Equations like (20.31) are called the kinetic equations for waves. The first term in the parenthesis describes the merging of quasiparticles with momentum $\mathbf{k}'$ and $\mathbf{k}''$, i.e., the inception of a quasiparticle with momentum $\mathbf{k}$, and the second two terms describe the extinction due to decay into quasiparticles with momenta $\mathbf{k}'$ and $\mathbf{k}''$. These equations were first derived by Payerls to describe "gas" phonons, namely acoustic waves in a solid (dielectric) [41].

In many cases, for example, waves on a fluid surface [36,37], the dispersion equation of the waves is such that the condition for three-frequency interactions is not fulfilled. In this case it is said that the spectrum is undecaying. The main process, therefore, determining the nature of the nonlinear wave phenomenon in a weakly nonlinear medium will be a four-quantum process such as $\omega_k = \omega_{k_1} + \omega_{k_2} + \omega_{k_3}$ or $\omega_k + \omega_{k_1} = \omega_{k_2} + \omega_{k_3}$. In the approximation of chaotic phases the following kinetic equations to describe such a process are obtained by repeating the operations used to derive (20.29) (equations of the type $a_j \sim a_i a_l a_k$ being the starting point):

$$\dot{N}_k =$$

$$\int |\sigma_{kk_1k_2k_3}|^2 \left(N_k N_{k_1} N_{k_2} N_{k_3} + N_k N_{k_1} N_{k_2} N_{k_3} - N_k N_{k_1} N_{k_2} N_{k_3} - N_k N_{k_1} N_{k_2} N_{k_3} \right)$$

(20.32)

$$\times \delta\left(\omega_k + \omega_{k_1} - \omega_{k_2} - \omega_{k_3}\right) \delta\left(\mathbf{k} + \mathbf{k}_1 - \mathbf{k}_2 - \mathbf{k}_3\right) d\mathbf{k}_1 d\mathbf{k}_2 d\mathbf{k}_3 .$$

Here the wave spectrum is assumed to be continuous (N_k is the number density of quanta in a spectral interval from $\mathbf{k}$ to $\mathbf{k} + \Delta\mathbf{k}$). Again, direct substitution indicates that the equilibrium state corresponds to a Rayleigh-Jeans spectrum, i.e., an equilibrium spectral distribution in an ensemble of a large number of quasiparticles is independent of the nature of the interactions (collisions) between them, as a result of which this equilibrium becomes established.

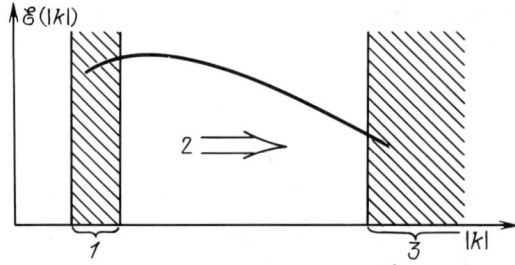

Fig. 20.10. Spectrum of weak turbulence: 1) region of source, 2) flow of energy in the inertial interval, 3) outflow region.

If the medium is dissipative, then undamped waves may only exist if the wave energy loss is compensated by an external source. In many cases (e.g., when gravity waves are excited on the surface of water by wind [36]) the energy is accumulated in a system of interacting waves and then taken away from the system due to dissipation into completely separate spectral regions (Fig. 20.10). The energy flows from the source region to the energy sink region via an inertial interval (a spectral region where the energy source and energy sink are absent) due to interactions between waves of different scales. If the wave phases become chaotic as the result of the interaction, then this ensemble of waves, which have random phases in a dissipative medium and are supported by an external energy source, is called weak wave turbulence [36-38].

Weak wave turbulence is described by a kinetic equation for the waves

$$\dot{N}_k = I_{sink}\{N_k\} + D(\mathbf{k}, N_k) - \Gamma(\mathbf{k}, N_k), \qquad (20.33)$$

which is the balance equation for the quasiparticles: $D(\mathbf{k}, \dot{N}_k)$ describes the inflow of energy into the system, and $\Gamma(\mathbf{k}, N_k)$ the energy loss. If the energy inflow is related to instability, then $D(\mathbf{k}) = \gamma(\mathbf{k})N_k$. The energy sink is usually losses due to friction or viscosity (e.g., for waves on a water surface $\Gamma(\mathbf{k}, N_k) \sim \nu k^2 N_k$). $I_{sink}\{N_k\}$ is the collision integral and takes into account interactions between waves in the chaotic phase approximation. For three-wave interactions this integral coincides with the right-hand side of (20.31), and for four-wave interactions the integral is the same as the right-hand side of (20.32).

In the inertial interval, which in a spectrum of $\mathbf{k}$ space is between the source and the sink regions, the stationary solution of the kinetic equation — the spectrum of weak-wave turbulence — is only determined by the collision integral. The influence of the energy source and sink regions may be considered as boundary conditions. Thus, finding the spectrum of turbulence reduces to finding the distribution $N(k)$ which makes the collision integral die away to zero. We already know one solution to $I_{sink}\{N_k\} = 0$, mainly the Rayleigh--Jeans distribution. However, this distribution corresponds to a situation in which there is no energy flow from the source to a sink, i.e., the system is at equilibrium. The universal power-low distribution $N(\omega)$ of the type $\mathcal{E}(\omega) \sim \omega^\beta$ [44] corresponds to nonzero energy flows through the inertial interval. At present universal methods for solving (20.33) are being developed and the reader may find out more in [36].

CHAPTER 21

SELF-EXCITED OSCILLATIONS IN DISTRIBUTED SYSTEMS

21.1 General Remarks

Distributed self-excited oscillational systems are extremely wide spread in nature and technology. Lasers, important functional systems in living organisms (blood circulation, breathing, and speech), wind and string musical instruments, variable stars (the cepheide variables), and autocatalytic chemical reactions are all examples. Some processes associated with the coexistence of different biological species also have an self-excited oscillational nature [1]. In microwave electronics, generators with inverse waves and several Cherenkov (including relativistic) generators are typical distributed self-excited oscillational systems.

We already know (Chapter 14) that any nonconservative system in which undamped wave or oscillational motion whose parameters are determined by the system and are independent of finite changes in the initial conditions is self-excited oscillational.

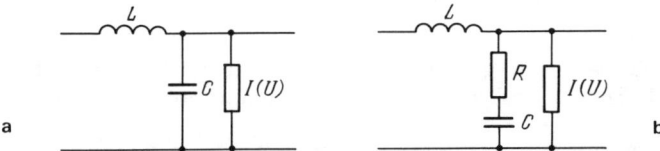

Fig. 21.1. Examples of long lines, models of active media: a) line without dispersion; b) line with high-frequency linear losses and linear current $I(U)$.

Let us consider a concrete example of the induction and instability restriction in a distributed self-excited oscillational system. The transmission line shown in Fig. 21.1a is a model of a medium with amplification. If the medium is active (i.e., $I(U) = -gU$ where the conductivity g is large), but linear, then any disturbance input into the system will grow. However, when the amplitudes are large, one of several restriction mechanisms must come into play, i.e., the nonlinearity in the conductivity or capacitance will appear or the amplification will be replaced by the distortion or restriction. For example, an active nonlinear medium for which $I(U) = -gU + \beta g U^3$ is an amplifier at small amplitudes (the conductivity is negative, i.e., $I(U) \approx -gU$), but at large amplitudes the medium is a nonlinear absorber ($U \approx [I(U)/\beta g]^{1/3}$). Here the restriction mechanism is nonlinear damping.

We may hope that we can construct a more or less complete theory of self-excited oscillation for a one-dimensional problem since for nonlinear wave processes in a nonequilibrium medium there are only a finite number of ways of combining medium characteristics such as the dispersion, nonlinearity, or dissipation. This allows us to use a single approach to the description of nonlinear waves in nonequilibrium media and to employ a small number of basic (model) problems to give a fairly general presentation of wave phenomena in such medium.

A nonlinear travelling wave in an active medium may be described using the one-wave approximation, as we have already discussed, when we can restrict our consideration to only one type of wave disturbance due to the smallness of the nonlinearity.

We shall consider several types of problem and in each case try to investigate the effect of nonlinearity, dissipation, and dispersion alternately. We first look only at the nonlinearity in a system equivalent to the diagram given in Fig. 21.1a. The equation for waves in such a line without dispersion but with a nonlinear active filling has the form

$$\partial U/\partial t + V_0 \partial U/\partial x = -I(U)/C . \qquad (21.1)$$

If there are high-frequency linear losses in the line (Fig. 21.1b), then the resultant equation will be

$$\partial U/\partial t + V_0 \partial U/\partial x - \nu \partial^2 U/\partial x^2 = -I(U)/C . \qquad (21.2)$$

If we also include a reactive nonlinearity (for simplicity quadratic), then the equation will take on a more general form, i.e.,

$$\partial U/\partial t + V_0 \partial U/\partial x + U \partial U/\partial x - \nu \partial^2 U/\partial x^2 = -I(U)/C . \qquad (21.3)$$

We shall consider each of these cases.

21.2 Medium Without Dispersion. Discontinuous Waves

Suppose we have $I(U) = -g(1 - \beta U^2)U$ in (21.1). We introduce $m = \beta U^2$ and change to a new coordinate $\xi = x - V_0 t$ and time $\tau = t$, which yields

$$\partial m/\partial \tau = (2g/C)m(1 - m) = \sigma m(1 - m).$$

This equation can be integrated to

$$\sigma \tau = \ln \frac{m(\tau, \xi)}{1 - m(\tau, \xi)} - \ln \frac{m_0(x - V_0\tau)}{1 - m_0(x - V_0\tau)},$$

$$m(x, \tau) = \frac{m_0(x - V_0\tau)}{m_0(x - V_0\tau) + [1 - m_0(x - V_0\tau)]\exp(-\sigma\tau)}.$$

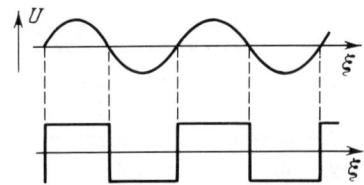

Fig. 21.2. Conversion of a sinusoidal wave input to the line (Fig. 21.1a) into a sequence of square pulses.

Let us see how a signal is changed as it propagates in such a medium. If $m_0 = 0$, then $m(x, t) = 0$, i.e., the initial value is conserved. If, however, $m_0 \neq 0$, then over time ($t \to \infty$) the amplitude tends to a constant value $m = 1$. When we send a pulse to the input, whatever its form, as it propagates it turns into a square wave with standard amplitude. If, for example, at the boundary of the medium a sinusoidal wave is given, then it is turned into a sequence of square pulses with amplitude $\beta^{-1/2}$ (Fig. 21.2). Thus, we have found that in this nonlinear medium, an arbitrary initial disturbance is transformed either into a specially homogeneous disturbance or into a discontinuous one, the discontinuities arising where $m_0(x) = 0$. The discontinuities clearly arise because of our neglection of the dispersion in a region of fast changes in the field.

21.3 Stationary Waves

The presence of dispersion in regions where the frequency is rapid (small scales) causes the higher harmonics of the initial disturbance to be unsynchronized with the fundamental wave, and the spectrum of the nonlinear wave will be restricted. It is, unfortunately, impossible to trace analytically the evolution of a wave in an active nonlinear medium with dispersion because it has not yet been possible to solve even the simplest equations describing wave propagation in such a medium. The investigation of stationary waves, waves propagating at a constant velocity and without change in form, is of particular interest. These waves are established as the result of competition between the active nonlinearity and active dispersion.

We now take into account high-frequency losses (imaginary dispersion), i.e., we return to (21.2). In this case, it is clear that the front is smoothed. Note that in self-excited oscillational systems (we are discussing circular or unrestricted systems) stationary waves clearly have a special role, somewhat like the role played by limiting cycles in lumped systems. It is best to do the analysis in terms of spectra because a stationary wave may be considered as the sum of harmonic waves whose amplitudes and phases are algebraically related, i.e., a stationary wave may be related to an equilibrium of the system of equations for complex harmonic amplitudes.

The period of a developed stationary travelling wave is determined either by the boundary or by the initial conditions. The velocity of a stationary wave depends on the nonlinear and dispersion properties of the medium and is a parameter different values of which correspond to different types of stationary wave. In certain cases however a periodic wave in a nonequilibrium medium may only propagate with one certain velocity.

We change the coordinates in (21.2) to a running coordinate $\xi = x - \bar{V}t$, where $\bar{V}$ is the propagation velocity of the stationary wave. The equation can then be rewritten as

$$(V_0 - \bar{V})dU/d\xi = \nu d^2U/d\xi^2 + (g/C)U(1 - \beta U^2) . \qquad (21.4)$$

This equation describes a stationary travelling wave. In form it is the same as the equation for a lumped nonlinear oscillator with damping $\delta = \bar{V} - V_0$. Clearly, the periodic solution that we are interested in only exists when $\bar{V} = V_0$. The phase-plane diagram of the system for this case is given in Fig. 21.3. A continuum of close trajectories corresponds to self-excited oscillations in the form of periodic stationary waves. The amplitude of such a wave is determined by its period. The reduction of a study of self-excited oscillations in a distributed system to the investigation of the equations of a nonlinear oscillator, a typical situation for conservative systems, seems paradoxical. However, there is a simple physical explanation. The dissipation processes and the removal of energy from the active medium in this case are only balanced for the continuous set of stationary waves propagating at a velocity V_0. This is only possible when there is no reactive dispersion in the medium, as this would cause the amplitude of the periodic wave to be dependent on its velocity.

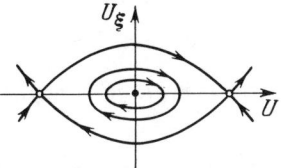

Fig. 21.3. Phase-plane diagram of a system describing stationary waves in a nonlinear active medium with imaginary dispersion.

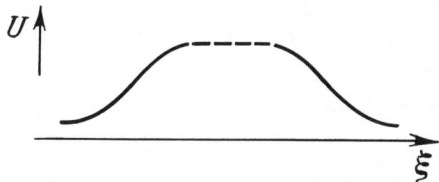

Fig. 21.4. Stationary pulse with a finite front width which corresponds to a separatrix going from a saddle to a saddle.

The presence of separatrices in a system (Fig. 21.3) going from a saddle to a saddle means that stationary pulses with a finite front width may propagate in the system (Fig. 21.4), these being dissipative solitons.

The amplitude and form of a periodic wave is described by the period (boundary conditions) and the form of nonlinearity. For example, in a transmission line with tunnel diodes, the operating point of which lies on the downwards section of the characteristics close to the maximum, the nonlinearity is quadratic (U^2 in (21.4) is replaced by U) and the stationary waves may have the form of a sequence of solitons or conoidal waves. An example of solitons in a nonequilibrium dissipative medium are waves on a thin film of water flowing down a sloped asphalt pavement. These waves develop due to the nonlinearity and are stabilized by surface tension. The curvature of the wavefront is increased due to the nonlinearity (Chapter 24).

Let us consider a general one-dimensional case when the medium has both nonlinear losses and nonlinear reactance (capacitance). The equation for a wave in such a medium has the form (21.3)

$$\partial U/\partial t + V_0 \partial U/\partial x + U\partial U/\partial x - \nu \partial^2/\partial x^2 = (g/C)U(1 - \beta U^2) .$$

We simplify the task somewhat by assuming that the unconservative and dissipative nonlinearities act in different directions U. Suppose for the moment that the dissipative nonlinearity does not exist, i.e., we neglect U^2 (β is small). Then the reactive nonlinearity becomes manifest at small amplitudes and the active nonlinearity at large amplitudes. We again restrict ourselves to a consideration of stationary waves:

$$\xi = x - \bar{V}t (V_0 - \bar{V})dU/d\xi + UdU/d\xi - \nu d^2U/d\xi^2 - (g/C)U = 0$$

(note that this equation, given several simplifying assumptions, describes periodic changes in the populations of a coexisting prey and predator species). If $g = 0$, then all disturbances in the medium are damped; if $\nu = 0$, then when $g > 0$ the opposite is true, namely the disturbances grow. In order for the motions to be finite, it is necessary that the effect of both factors is equalized. This is clearly only possible when $\bar{V} = V_0$. When $\bar{V} \neq V_0$, it is fairly easy to see that as $\xi \to \pm \infty$ we have $U \to \infty$. When $\bar{V} = V_0$ we get the equation

$$\nu d^2U/d\xi^2 - UdU/d\xi + \gamma U = 0 \quad (\gamma = g/C) . \tag{21.5}$$

This equation is easily integrated by assuming that $dU/d\xi = V$ and $\nu dV/d\xi = U(V - \gamma)$. The equation of the integral curves is

$$\nu dV/dU = U(V \quad \gamma)/V . \tag{21.6}$$

It can be seen from (21.5) that the integral curves, which are the phase trajectory $V = \gamma$, exist. We obtain from (21.6)

$$\nu dV/dU = U[1 + \gamma/(V - \gamma)]^{-1} ,$$

$$(U^2 - U_0^2)/2 = \nu[V + \gamma \ln(\gamma - V)/\gamma] ,$$

where U_0 is the value of U when $V_0 = 0$. The phase-plane diagram of this system is given in Fig. 21.5a. Waves with large amplitudes have a region of slow change, and on the phase-plane this corresponds to motion near the straight line $V = \gamma$, and a region of fast changes, which corresponds on the phase plane to motion along the loop that descends far below.

SELF-EXCITED OSCILLATIONS IN DISTRIBUTED SYSTEMS

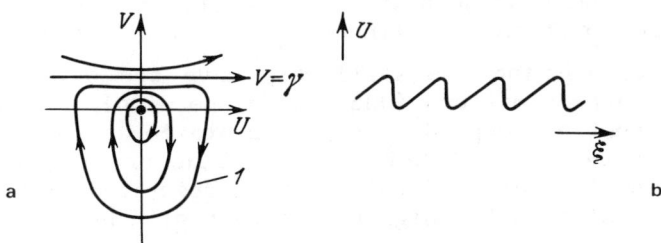

Fig. 21.5. a) Phase-plane diagram and b) form of the stationary waves described by (21.5).

Let us now consider an experiment corresponding to this situation. We put a sinusoidal wave at input of the medium described by (21.4). Far from the boundary, this wave becomes close to stationary and it can be described using (21.5) on the phase plane (Fig. 21.5a). When the trace point moves along a type-one trajectory close to $V = \gamma$, the function $U(\xi)$ changes as $\gamma\xi$, i.e., it grows linearly. Motion along a closed trajectory, which occurs very rapidly, corresponds to a step change leading the wavefront, and the wave becomes saw-toothed (Fig. 21.5b).

We can easily imagine that if we were to feed high-frequency oscillations into a transmission line on which a wave is described by (21.4), then when $\omega > \omega_{cr}$ the oscillations would in general be damped and then as ω was made smaller they would turn into undamped sinusoidal waves and only fairly low-frequency waves would be saw-toothed.

21.4 The Existence and Role of Limiting Cycles

If the nonlinearity is cubic, i.e., $V(U) = V_0 - \alpha U^2$, then we have the following four stationary waves

$$\nu \partial^2 U/\partial \xi^2 + [(V - V_0) + \alpha U^2]\partial U/\partial \xi + \gamma U = 0 ,$$

whence it can be seen that the periodic waves will only exist for $V \neq V_0$ ($V > V_0$ given $\alpha < 0$ and $V < V_0$ given $\alpha > 0$). The phase-plane diagrams corresponding to stationary waves will be traditionally self-excited oscillational but with a limiting cycle. The physical difference between the properties of stationary waves in media with reactive cubic nonlinearity and those with reactive quadratic nonlinearity is that in a medium with a cubic nonlinearity, the velocity of the harmonics forming the nonlinear periodic wave will depend on the harmonic's amplitude (the self-influencing effect), and consequently the velocity of the nonlinear wave must differ from the linear V_0. If the velocities of the harmonics forming the stationary wave are different in the linear approximation (due to dispersion), then limiting cycles must correspond to periodic stationary waves.

For example, an active wave guide or transmission line with tunneling diodes, which can be described with an account of dispersion in the high-frequency region by the equation

$$\hat{M}_2(U) - \beta \frac{\partial^4 U}{\partial x^2 \partial t^2} = \gamma \frac{\partial^2}{\partial x \, \partial t} \left[\frac{\partial U}{\partial x} - \left(\frac{\partial U}{\partial x}\right)^2 \right], \quad (21.7)$$

where $\hat{M}_2(U) = \partial^2 U/\partial t^2 - V^2(U)\partial^2 U/\partial x^2$, will contain stationary waves with the equation

$$\beta \frac{\partial^2 U}{\partial \xi^2} - \frac{\gamma}{V} \left[1 - \left(\frac{\partial U}{\partial \xi}\right)^2 \right] \frac{\partial U}{\partial \xi} + \frac{V_0^2 - V^2}{V^2} U = 0. \quad (21.8)$$

This equation already allows for the existence of codirected waves in a medium with dispersion. When $V < V_0$, it becomes the Van der Pole equation and has one limiting cycle, which also corresponds to self-excited oscillations in the form of periodic stationary waves. It can be seen that when $\beta(V_0^2 - V^2)/\gamma^2 \ll 1$, these waves will be relaxational (on the phase plane the cycle will be broken). When the dispersion is weak $\beta \to 0$, the condition is fulfilled for all $V^2 < V_0^2$, i.e., the slow (short) and fast (long) waves $(V_0^2 - V^2)/\beta V^2 = (2\pi/\Lambda)^2$ will be relaxational, where Λ is the length of the stationary waves). If, however, the dispersion is strong, then an extraordinarily interesting situation occurs: both a sinusoidal $(V^2 \ll V_0^2)$ and a relaxational $(V^2 \sim\!\!\leq V_0^2)$ stationary wave may exist at the same point. The physics of this feature is quite simple in that the dispersion in this case is only manifest for small scales (i.e., for slow waves), as a result of which the fast waves behave as if they were in a nonlinear medium without dispersion.

Thus, a limiting cycle only corresponds to self-excited oscillations in the form of stationary waves in the phase plane of the system describing stationary motions when the active medium has dispersion (linear $V = V(\omega)$ or nonlinear $V = V(U^2)$).

In the general case, the stability or instability of such a cycle does not mean the stability or instability of the corresponding periodic stationary waves. Real disturbances that evolve over time cannot be described by the equation for stationary waves. Aperiodic stationary waves corresponding to trajectories arriving and departing from the cycle are defined over all space from $-\infty$ to $+\infty$ and cannot be realized in a restricted system. However, in some cases, a link between the stability of a limiting cycle and a periodic wave can nevertheless be traced. For example, if the limiting cycle is unstable in the phase plane for stationary waves $U(t - x/v)$ as $v \to \infty$, then a periodic stationary wave (for $v = \infty$ this is already not a wave but an oscillation) is unstable.

21.5 Competition Between Stationary Waves in an Active Medium

A solution to a travelling wave problem yields the possibility of investigating interactions between waves. Naturally, it is only sensible to speak of interaction when we can trace the evolutions of the separate waves in the process, i.e., when the transformation of separate waves takes place more slowly than the space-time scale characterizing the waves. This is only possible when the medium has small nonlinearity and the local field can be represented as the superposition of the individual waves.

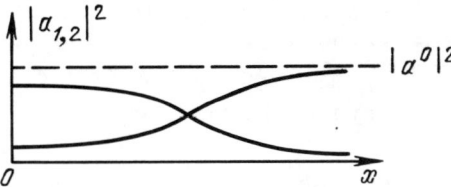

Fig. 21.6. Spatial competition between waves in an active nonlinear medium with low-frequency or high-frequency viscosity.

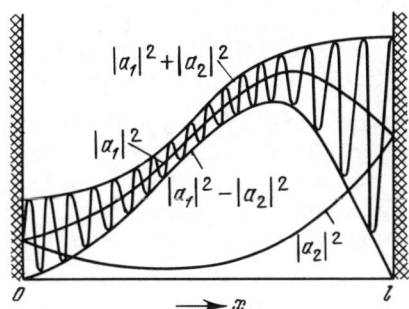

Fig. 21.7. An asymmetric spatially homogeneous regime in a resonator with ideal reflection filled by a nonlinear medium.

The smallness of the nonlinearity obviously does not mean that the interacting waves must be sinusoidal. We have seen that the form of a stationary wave is also dependent on dispersion. If the dispersion and nonlinearity are of the same order of magnitude, then the waves will be considerably nonsinusoidal and relaxational for a vanishingly small dispersion. If the dispersion is strong in comparison to the nonlinearity, then the waves will be quasisinusoidal.

Because of the space-time analogy, which we have already discussed, between the interactions of normal oscillations over time and stationary interactions between waves in space, the classical oscillational effects are frequently simply translated into wave processes. For example, an illustration is given in Fig. 21.6 of a spatial analogy of competition between oscillations in an active nonlinear medium with viscosity (high-frequency or low-frequency viscosity). This process is described by the equations from Chapter 16 in which the time t is replaced by the coordinate x. Starting with the effect of spatial competition, it is possible, for example, to construct an interesting wave device which can select from two or more unknown quasiharmonic signals the one with the greatest (or least) frequency [1].

This effect of wave competition explains the seemingly incredible establishment in a spatially symmetric distributed self-excited generator (for example, with ideal reflection at the boundaries) of a stationary field distribution that is asymmetric along the x-coordinate with the predominance of one of the meeting waves (Fig. 21.7). The equation for the amplitude $a_{1,2}(x,t)$ of these waves given a simple idealization [1] is written in the form

$$\partial a_{1,2}/\partial t \pm v \partial a_{1,2}/\partial x - \mu h \left[1 - \alpha(|a_{1,2}|^2 + 2|a_{2,1}|^2 \right] a_{1,2} = 0 \tag{21.9}$$

with boundary conditions $|a_1(x,t)| = |a_2(x,t)|$ at $x = 0, l$ where l is the length of the resonator. The intensity distribution $|a_{1,2}(x)|^2$ in a stationary regime can easily be re-established in the form of trajectories on the phase plane constructed using (21.9) for $\partial/\partial t \equiv 0$ (Fig. 21.8). In a short resonator, where competition cannot arise, only the usual regime of standing waves is possible. On the phase plane in Fig. 21.8 this corresponds to the equilibrium condition on the straight line $|a_1|^2 = |a_2|^2$. However, in a long resonator waves moving in opposite directions and taking energy from a common source will suppress one another over most of the resonator length and only equalize close to the reflecting walls. As the result, the standing wave regime becomes unstable and one of several spatially nonhomogeneous regimes, which correspond in Fig. 21.8 to trajectories such as 1-3,[1] become established.

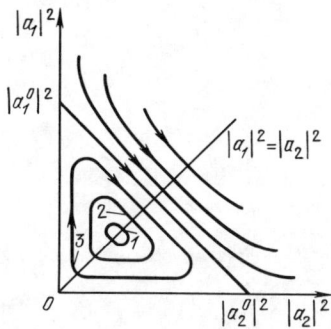

Fig. 21.8. Phase-plane diagram of a system describing stationary waves in a spatially symmetric distributed self-excited generator with ideal reflection at the ends: trajectories 1 - 3 correspond to spatially homogeneous stable regimes.

21.6 Periodic Self-Excited Oscillations in Hydrodynamic Flows

Periodic flows of liquid which develop due to an energy flow or an external source of heat and stabilized by viscosity are frequently encountered in nature. Several of these flows can be described using equations such as (21.1)-(21.3) in the one-dimensional idealization. For example, we have discussed waves on a falling film, and the periodic waves at the interface between two immiscible liquids moving with respect to one another.

Here we shall discuss some simple and quite visual examples of periodic self-excited oscillations in closed two-dimensional flows. These examples are related to the dynamics of a small number of vortices "on the plane" in thin liquid films. The corresponding experiments are interesting from the point of view of modeling global vortex flows in the atmosphere (hurricanes) because our atmosphere may be considered very thin for global vortices.

---

[1] It was noted in [7] that the symmetry of the equations ensured the symmetry of the solution only when the solution was unique. If there is more than one solution, then the equations' symmetry only ensures the presence of a group of symmetry transformations which map the solutions onto each other. Depending on the initial conditions, the system may choose one of the solutions and this will become asymmetric.

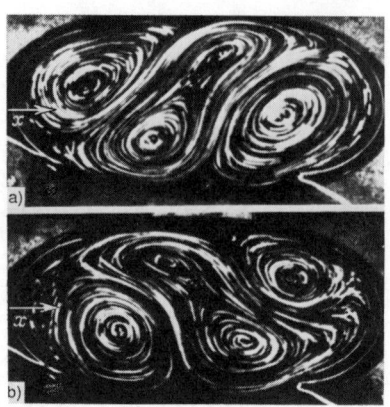

Fig. 21.9. Four-vortex flows excited magnetohydrodynamically in a dish [2]: the periodic self-excited oscillations correspond to alternations over time between flows a) and b).

Four-vortex flows induced by the magnetohydrodynamic method [2] in a petri dish are shown in Fig. 21.9. An electrolyte (aqueous solution of copper sulfate) was poured into a petri dish 0.5-cm deep and 23-cm long through which a direct current was passed in the x-direction. Two permanent magnets were placed above the dish, about the middle of the system and parallel to the x-axis. When current was passed through the liquid, a force acted on the liquid above the magnets causing the liquid to move, in a middle of the dish the liquid flowed from the walls to the x-axis, and beyond the magnetic lines of force the flow returned (the flow was reversed). As the result a flow was established in the form of four similar vortices. These two-dimensional liquid flow patterns look like the phase-plane diagrams of two-dimensional dynamic systems. This is no coincidence.

When the Reynolds number, which in this case is proportional to the current and magnetic field, is increased the stationary four-vortex flow loses its stability and a periodic self-excited oscillational regime is established. This regime is characterized by the pairwise joining of vortices with the same sign (the same rotational direction). When the Reynolds number in the system is increased further, the flow pattern ceases to be symmetric and the nucleus of one of the pairs of interacting vortices decreases and an extended vortex is formed.

An analogous pattern of periodic self-excited oscillation is observed during thermal convection in a liquid occupying a vertical (Hele–Shaw) cell when heated from below (Fig. 21.10) [3]. For convective flows the degree of disequilibrium in a system is characterized by the Rayleigh number $Ra = g\nabla T h^4 \beta/\nu\kappa$ (g is the acceleration of free fall), ∇T is the vertical temperature gradient, h is the height of the layer, β is the thermal expansivity, ν is the viscosity, and κ is the thermal conductivity). In the experiment the following sequence of bifurcations was observed when the Raleigh number Ra was increased: for $Ra > Ra_1$ the

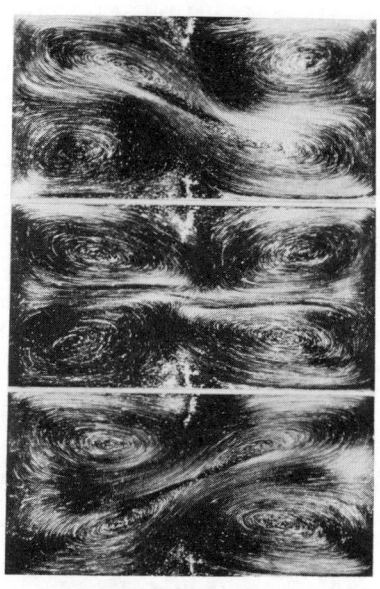

Fig. 21.10. Periodic self-excited oscillations during thermal convection in a liquid in a vertical cell (a Hele-Shaw cell) when heated from below.

hydrodynamic equilibrium lost its stability and was replaced by a stationary one-vortex convection with hot liquid making its way upwards and cold liquid flowing downwards (the direction of the liquid rotation in a one-vortex convection only depended on the initial conditions). When $Ra > Ra_2$, a two or four-vortex convection regime was established. As Ra was increased still further, the stationary cell-like convection was replaced by an self-excited oscillational regime for which periodic pairwise connections between the vortices was characteristic. At large Ra, the connections between the vortices was found to be stochastic over time (Chapter 23).

The main equation of fluid dynamics (the Navier-Stokes equation) cannot be used to describe analytically the establishment of periodic self-excited oscillational flows (even in the two-dimensional approximation). However, the mechanism by which they arise can be understood and the fact demonstrated very rigorously using the theory of bifurcations and some quite general mathematical theorems, primarily the theorem of a central manifold [8]. Here, we cannot delve into the very subtle mathematical theory [5, 8, 12], but we can note that the application of the theorem of the central manifold allows us to reduce the investigation of bifurcations in an infinite-dimensional system to the analysis of a finite-dimensional system. In particular, this theorem allows us for the onset of periodic flows (i.e., the onset of cycles) which we are interested in to operate with a dimensionality of two without any loss of information about the stability ([8], Chaps. 2 and 8). Similarly the theorem applies to the onset of quasiperiodic flows from periodic flows (i.e., the onset of an invariant torus), only the reduced dimensionality is now three (for two incommensurable flow frequencies) and not two.

We shall dwell in some detail on another approach to the investigation in fluid dynamic flows, an approach associated with the approximate description of the flow by a finite-dimensional dynamic system. The most widely used and natural description is called mode description (or Galerkin method) in which a hydrodynamic field $u(x,t)$ (velocity, temperature, etc.) is represented in the form of a linear combination of a finite number of coordinate functions $\phi_n(x)$ (these are usually called the basis functions):

$$u(x, t) = \sum_{n=1}^{N} a_n(t)\phi_n(x) , \qquad (21.10)$$

where $a_n(t)$ is the expansion coefficient for which we obtain a finite-dimensional system of ordinary differential equations. If the initial equations have, say, the form

$$(\partial/\partial t + \hat{M})u(x, t) = f(x, t) , \quad u(x, 0) = u_0(x) , \qquad (21.11)$$

where $\hat{M}$ is some differential operator (in the general case, nonlinear), and the field $u(x,t)$ is smooth, determinate in a bounded region, and satisfies homogeneous boundary conditions at the edge of the region, then the equation for $a_n(t)$ from the condition of orthogonal closure

$$\Delta_n \equiv [(\partial/\partial t + \hat{M})u_n(x, t) - f]$$

of the basis functions $\phi_1, \ldots, \phi_n$. Naturally, if these functions are mutually orthogonal, then the whole procedure is considerably simplified.[2]

We shall demonstrate the derivation of similar finite-dimensional equations using the example of thermal convection in a Hele-Shaw cell [10]. The initial equations in the Bussinesque approximation, in which the compressibility of the liquid is neglected in the continuity equation, have the form

$$\partial \mathbf{v}/\partial t + (\mathbf{v}\nabla)\mathbf{v} = -\nabla p/\rho + \nu\nabla\mathbf{v} - g\beta T$$

$$\partial T/\partial t + (\mathbf{v}\nabla)T = \kappa\Delta T , \qquad (21.12)$$

$$\text{div } \mathbf{v} = 0 .$$

Here g is the acceleration of free fall, β is the coefficient of volumetric expansivities, $T(x,t)$ is the temperature field, and κ is the thermal conductivity. The boundary conditions are:

$$\mathbf{v} = 0 , \quad T = 0 \quad \text{when} \quad z = \pm 1 ;$$

$$\partial v_y/\partial x = 0 , \quad v_x = 0 , \quad \partial T/\partial x = 0 \quad \text{when} \quad x = 0, L ; \qquad (21.13)$$

$$\partial v_x/\partial y = 0 , \quad v_y = 0 , \quad T = 0 \quad \text{when} \quad y = 0, H .$$

Because $L, H \gg 1$, we can assume that the liquid velocity across the layer is approximately zero, $v_z \approx 0$ (the approximation of planar trajectories). It is natural, therefore, as was the case for two-dimensional flows, to introduce a flow function $\Psi(x,y,z,t)$ associated with the velocity components v_x and v_y by the relations $v_x = -\partial\Psi/\partial y$, $v_y = -\partial\Psi/\partial x$. Then the Bussinesque equations (21.12) may be formulated in terms of the flow function Ψ and vorticity $\omega = -(\partial^2/\partial x^2 + \partial^2/\partial y^2)\Psi$, i.e.,

---

[2] Although the Galerkin method seems similar to the asymptotic method for partial differential equations, there is a crucial difference. In the asymptotic method, the approximate solution in the form of a series with a finite number of terms becomes exact as the small parameter tends to zero. This sort of convergence does not exist due to the absence of a small parameter and the accuracy of the method may only be increased by including new basis functions. The absence of a small parameter makes the Galerkin method very difficult to justify mathematically [9].

$$\frac{1}{\text{Pr}}\left[\frac{\partial \omega}{\partial t} + \left(\frac{\partial \Psi}{\partial x}\frac{\partial \omega}{\partial y} - \frac{\partial \Psi}{\partial y}\frac{\partial \omega}{\partial x}\right)\right] = \Delta\omega + \text{Ra}\,\frac{\partial T}{\partial x},$$

$$\frac{\partial T}{\partial t} + \left(\frac{\partial \Psi}{\partial x}\frac{\partial \omega}{\partial y} - \frac{\partial \Psi}{\partial y}\frac{\partial \omega}{\partial x}\right) = \Delta T + \frac{\partial \Psi}{\partial x}$$

(21.14)

(where $\Delta = \partial^2/\partial x^2 + \partial^2/\partial y^2 + \partial^2/\partial z^2$). Here **Ra** is, as before, the Rayleigh number, $\text{Pr} = \nu/\kappa$ is the Prandtl number, and T is the deviation of the temperature from the equilibrium distribution $T_0 = -y$ (maintained by an external heat source). The boundary conditions to (21.14) have the form

$$\Phi = \qquad T = 0 \qquad \text{when } z = \pm 1 \;;$$
$$\Psi = \partial^2\Psi/\partial x^2 = \partial T/\partial x = 0 \qquad \text{when } x = 0, L \;; \qquad (21.15)$$
$$\Psi = \partial^2\Psi/\partial y^2 = T^2 = 0 \qquad \text{when } y = 0, H \;.$$

The units in (21.14) and (21.15) for distance, time, velocity, temperature, and pressure are chosen to be the film thickness d, d^2/κ, κ/d and $\nabla T \cdot d$, $\nabla p \cdot d$. A finite-dimensional approximation of the velocity and temperature fields such as (21.10) is taken for our boundary conditions in the form

$$\Psi(x, y, z, t) = \sum_{n,m=1}^{N,M} \Psi_{nm}(t)\sin\left(\frac{n\pi}{L}x\right)\sin\left(\frac{m\pi}{H}y\right)\cos\left(\frac{\pi}{2}z\right),$$

(21.16)

$$T(x, y, z, t) = \sum_{n=0,m=1}^{N,M} T_{nm}(t)\cos\left(\frac{n\pi}{L}x\right)\sin\left(\frac{m\pi}{H}y\right)\cos\left(\frac{\pi}{2}z\right).$$

When these expressions are substituted into (21.14) and the functions $\Psi_{nm}(t)$ and $T_{nm}(t)$ are orthogonalized, we obtain a system of equations of a form

$$\dot{a}_k = -\nu_k a_k + \sum_{i,j}\sigma_{kij}a_i a_j, \qquad \text{where } a \sim \Psi, T. \qquad (21.17)$$

The main problem arising when constructing a Galerkin approximation for the hydrodynamic equations is how many modes to consider in the expansion. There is no distinct algorithm for this and the only criterion for the correctness of the finite-dimensional description is a comparison with an exact solution (if one is known) or with experiment. Therefore, it is only ever sensible to construct such a finite-dimensional approximation when it is clear which flow pattern we wish to describe. This method of obtaining a finite-dimensional truncation of the equations of fluid mechanics is not unique and possibly it is not always optimal. Finite-dimensional models may be constructed, for example, to simulate the basic properties of these equations, e.g., quadratic form, symmetry, conservation laws (such a system being called a hydrodynamic one [4]). Thus for the four-vortex convection in a Hele-Shaw cell, it is natural to restrict our description to the first three modes in the velocity and temperature field ((nm) = (11), (12), (21), (31), (22)) and to the two modes T_{01} and T_{02} which are spatially homogeneous in the x-direction and which account for changes in the equilibrium temperature distribution to the convection. In order to demonstrate the mathematical difficulties involved, we present the system of equations for these modes [10] (together with the coefficients for the first seven equations, the remaining positive coefficients being omitted):

$$\dot{\Psi}_{11} = -9\frac{1-\varepsilon}{1+\varepsilon}\Psi_{12}\Psi_{21} - 4\frac{5-3\varepsilon}{1+\varepsilon}\Psi_{31}\Psi_{22}$$

$$- \mathbf{Pr}\left[p(1+\varepsilon) + \frac{1}{4}\right]\Psi_{11} - \frac{\mathrm{Ra}\,\mathbf{Pr}}{\pi^4(1+\varepsilon)}T_{11},$$

$$\dot{T}_{11} = -3(T_{21}\Psi_{12} + T_{12}\Psi_{21}) - 4(T_{31}\Psi_{22} + T_{22}\Psi_{31})$$

$$+ 2T_{01}\Psi_{12} - (1+4T_{02})\Psi_{11} - \left[p(1+\varepsilon) + \frac{1}{4}\right]T_{11},$$

$$\dot{T}_{02} = -\left(4p\varepsilon + \frac{1}{4}\right)T_{02} + 2T_{11}\Psi_{11} + 4T_{21}\Psi_{21} + 6T_{31}\Psi_{31},$$

$$\dot{\Psi}_{22} = -\mathbf{Pr}\left[4p(1+\varepsilon) + \frac{1}{4}\right]\Psi_{22} - \frac{\mathrm{Ra}\,\mathbf{Pr}}{2\pi^4(1+\varepsilon)}T_{22}$$

$$+ \frac{8}{1+\varepsilon}\Psi_{11}\Psi_{31},$$

$$\dot{T}_{22} = -\left[4p(1+\varepsilon) + \frac{1}{4}\right]T_{22} - 2\Psi_{22} + 4(T_{11}\Psi_{31} + T_{31}\Psi_{11})$$
$$+ 4T_{01}\Psi_{21},$$

$$\dot{\Psi}_{31} = -\Pr\left[p(9+\varepsilon) + \frac{1}{4}\right]\Psi_{31} - \frac{3\text{Ra Pr}}{\pi^4(9+\varepsilon)}T_{31}$$

$$- 12\frac{1+\varepsilon}{9+\varepsilon}\Psi_{22}\Psi_{11} + 15\frac{1-\varepsilon}{9+\varepsilon}\Psi_{12}\Psi_{21},$$
(21.18)

$$\dot{T}_{31} = -\left[p(9+\varepsilon) + \frac{1}{4}\right]T_{31} - 3(1+4T_{02})\Psi_{31}$$
$$+ 4(T_{11}\Psi_{22} - T_{22}\Psi_{11}) + 5(T_{21}\Psi_{12} - T_{12}\Psi_{21}),$$

$$\dot{\Psi}_{12} = -\Psi_{12} - T_{12} + \Psi_{11}\Psi_{21} + \Psi_{21}\Psi_{31},$$

$$\dot{\Psi}_{21} = -\Psi_{21} - T_{21} - \Psi_{12}\Psi_{11} - \Psi_{31}\Psi_{12},$$

$$\dot{T}_{12} = -T_{12} - \Psi_{12} + T_{01}\Psi_{11} + T_{21}\Psi_{31}$$
$$+ T_{11}\Psi_{21} + T_{21}\Psi_{11},$$

$$\dot{T}_{21} = -T_{21} - \Psi_{21} + T_{01}\Psi_{22} + T_{11}\Psi_{12} - T_{12}\Psi_{11}$$
$$- T_{31}\Psi_{12} - T_{12}\Psi_{31},$$

$$\dot{T}_{01} = -T_{01} - T_{11}\Psi_{12} - T_{21}\Psi_{22} - T_{12}\Psi_{11} - T_{22}\Psi_{21}$$

($p = 1/L^2$, $\varepsilon = L^2/H^2$). For convenience, we have introduced a new unit of time, a new flow function and a new temperature which differ from the old ones by the factors $1/\pi^2$, $(3/2)LH\pi$, and $(3/2)H$, respectively.

We emphasize again that a consideration of a finite number of basis functions to account for the stabilizing effect of viscosity (depriving the smaller-scale perturbations of their "independence" in that they must follow the larger perturbations) naturally restricts the range of Rayleigh numbers within which the system (21.18) can be used.

A general analysis of (21.18) requires computer methods. However, several conclusions may be made from a direct analysis of the structure of these equations. For example, it can be seen that Ψ_{11}, T_{11}, T_{02} (these perturbations describe the one-vortex flow) do not generate other perturbations, i.e., the solution of the system[3]

$$\dot{\Psi}_{11} = -c\,\Pr\,\Psi_{11} - (\text{Ra}\,\Pr/\pi^2 a)T_{11},$$

$$\dot{T}_{11} = -cT_{11} - (4T_{02} + 1)\Psi_{11}, \qquad (21.19)$$

$$\dot{T}_{02} = -bT_{02} + 2T_{11}\Psi_{11}$$

($a = 1 + \varepsilon$, $b = 4p\varepsilon + 1/4$, $c = pa + 1/4$) is a solution of the full system (21.18). This solution naturally only has sense if it is stable with respect to increases in the remaining perturbations. The appropriate analysis is quite simple for a stationary solution of (21.19):

$$\Psi_{11} = \left[\frac{b}{8ac\pi^4}(\text{Ra} - \text{Ra}_1)\right]^{1/2},$$

$$T_{11} = -\frac{\pi^2}{\text{Ra}}\left[\frac{abc}{8}(\text{Ra} - \text{Ra}_1)\right]^{1/2}, \qquad (21.20)$$

$$T_{02} = -\frac{\text{Ra} - \text{Ra}_1}{4\text{Ra}}.$$

We suggest the reader satisfies himself that, for example, when the Prandtl number is $\Pr = 7$ and the Rayleigh number is $\text{Ra}_{cr} \approx 1.4\text{Ra}_1$ begin to grow against the background of (21.20) the disturbances Ψ_{22}, T_{22}, T_{31}. A numerical analysis of (21.18) shows [10] that stable periodic self-excited oscillations arise in this situation for $\text{Ra}_2 > \text{Ra}_{cr}$. These self-excited oscillations correspond to the periodic interconnection of the vortices that we have covered.

---

[3] This system is the same as that of E. Lorentz, which we shall discuss in detail in the next chapter.

CHAPTER 22

STOCHASTIC DYNAMICS IN SIMPLE SYSTEMS

22.1 How Randomness Appears in a Dynamic System

Naturally all the phenomena and effects considered in the previous chapters were regular in that the oscillations or waves in the systems or media occurred without fluctuations and did not require any statistical method to describe them. Our experience and intuition indicates that in a dynamic system describable by regular equations, nothing irregular or random or stochastic should occur. So where does the randomness come from if we give an unambiguous algorithm that uniquely defines for concrete initial conditions the system for however long in the future?[1] Obviously, if the system is very complex and has a large number of degrees of freedom (for example, a gas in a vessel), we know that a deterministic description becomes senseless (though in principle possible), if only because it is impossible to specify the initial coordinates and velocities of all of the, say, 10^{19}, molecules in one cubic centimeter of gas. Moreover, there is no computer that can calculate the trajectories of such a number of particles and account for their collisions. In a simple system, when the number of degrees of freedom is small (e.g., $n \leq 10$) this problem does not arise. Whence we specify the $2n$ numbers describing the initial state of the system, it would seem that we can calculate (even if only by computer) its state at any point in the future. We ask therefore what is meant by the stochastic behavior of a simple system and how does randomness, and consequently unpredictability, become manifest in spite of the existence and uniqueness of a solution which guarantees the unambiguous deterministic behavior for given initial conditions.

Before addressing these topics, we must formulate the concept of the random behavior of a deterministic system.

The randomness of a motion is usually associated with

----

[1] We are only discussing systems for which a unique solution has been proved to exist.

two things: a very "sensitive" dependence on the initial conditions, which in fact means the conditions have an unpredictable effect, and the existence of variables averaged over time. We shall explain this.

Let us suppose that we have a generator of random oscillations, whose parameters we cannot change. Whenever we turn on the generator we get different oscillograms. However, if we repeat the experiment a large number of times, then some statistical pattern will emerge. This pattern must be independent of the probability distribution of the generator's initial states. This initial distribution is not universal and changes from generator to generator, depending both on the actual generator circuits and on the way it was turned on.

The existence of average quantities that are independent of the initial conditions is very important for defining stochasticity. That is consider some function of the instantaneous state x of our deterministic system $F(x)$. Given initial conditions $x(t_0) = x_0$, this function changes over time as $F(t, t_0, x_0)$. Suppose the function changes over time in an irregular way for most x_0 and even small changes in x_0 lead to significant changes in the form of function $F(t, t_0, x_0)$. We are interested in the case when there is an average quantity

$$\langle F(x_0) \rangle = \lim_{T \to \infty} \frac{1}{T} \int_0^T dt \cdot F(t, t_0, x_0) \qquad (22.1)$$

such that $\langle F(x_0) \rangle$ is independent of x_0 for most initial conditions in a given area of the phase space.

We shall consider how the unpredictability of an individual motion can become manifest in a deterministic system and at the same time how this leads to a statistical description.

Let us assume that there is a bounded region in the phase space of the system from which phase trajectories do not leave. We assume that all transition processes finish and all the trajectories in the region are Lyapunov unstable. This assumption is unreasonable for a system with one degree of freedom because if the trajectories do not leave any region on the phase plane then because they cannot intersect they must either be closed or tend to a simple attractor (a limiting cycle for equilibrium), and then all the trajectories inside the region will be stable. However, our assumption is already realistic for systems with at least one and a half degrees of freedom. Later we shall study actual stable regions in which unstable trajectories do not exist.

Now we shall only list trajectories which may exist inside such the region: unstable equilibria, unstable cycles, and unclosed trajectory which may wander for ever inside our bounded region but not leave it. Since the phase volume is bounded, any unclosed trajectory will after a sufficiently long time approach itself. However, the trajectory is unstable and it does not in any way follow from this closeness that the next stage of the motion will be similar to the previous one. By contrast, a small perturbation will grow and the subsequent locus of a tracing point cannot be predicted.

It also follows from these arguments that another manifestation of instability is the impossibility of reproducing an unstable dynamic system by giving the initial conditions to a high, but finite, accuracy. This idea was expressed more distinctly by N.S. Krylov and then Max Born. For instance, Born's definition of determinism is that since every state is always measured with at least some finite inaccuracy ε it is not a number that is determined but some probability distribution. The task is to predict the distribution at a time t from the known initial distribution. If a given solution is stable and the initial perturbations do not increase, then a later state can be predicted by a theory that is called deterministic. Born emphasized that this definition of determinism differed from the traditional one by the change in a sequence of limiting transitions as $\varepsilon \to 0$ and $t \to \infty$. Usually, the region of initial scatter is constricted to a point at first and then the behavior as $t \to \infty$ is considered (and obviously, complete predictability is obtained). This approach, however, is not one of physics, and it must be replaced. Initially, for a given ε the behavior of the trajectories and the region of finite scattering (i.e., the cross-sectional area of a tube of trajectories) is determined for any t and then the way a finite scattering behaves as $t \to \infty$ and how the initial scattering tends to a point is considered. If the finite scatter of the trajectories grows as $t \to \infty$, then the system's behavior is unpredictable.

Let us look at two examples. We first turn to the phase-plane diagrams that we know for certain dynamic second-order systems (Fig. 22.1). The system in Fig. 22.1a has a single asymptotic stable equilibrium (focus). Clearly, the system's motion is absolutely predictable in that any region of initial deviations ε contracts to a point as $t \to \infty$. The situation in Fig. 22.1b is similar, in that as $t \to \infty$ the motion is completely defined; it is periodic with known amplitude and period and it corresponds to a stable limiting cycle on the phase plane. When there is scatter in the initial deviations ε only the phase of the final motion (the point at which the trajectory leaves the limiting cycle) remains undetermined. In Fig. 22.1c, the motion remains predictable as $t \to \infty$ if the initial deviations lie in the region ε_1. However, if they lie in the region ε_2 then various motions may exist even though this is not complete unpredictability.

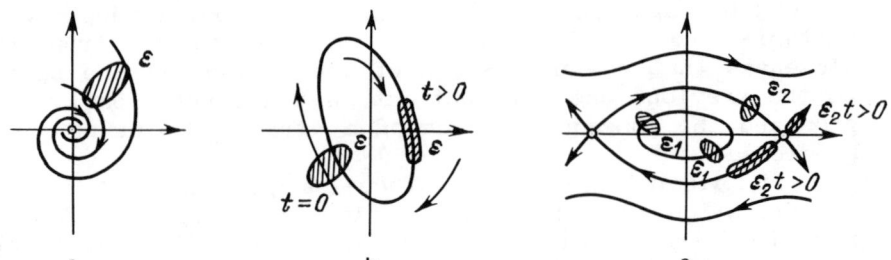

Fig. 22.1. Evolution of an elementary phase volume on the plane when there is a) a stable equilibrium; b) a limiting cycle; c) a separatrix going from a saddle to a saddle.

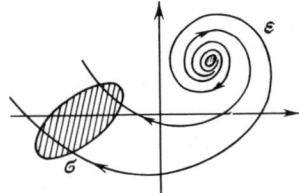

Fig. 22.2. Evolution of an elementary phase volume on a plane when there is an unstable equilibrium.

The situation changes if the trajectories on the phase plane cease to be Lyapunov stable. For example, in the case of an unstable focus (Fig. 22.2) a small scatter in the initial deviations means that the state of the system cannot be determined over a long period of time t (it may lie at any point in the region σ).

Thus, instability is necessary for unpredictability. However, it is not sufficient for stochasticity. There must also be a blurring of the trajectories and to get this it is necessary that they remain in a finite region in phase space, i.e., the phase trajectories must return. We have met examples of returning trajectories in phase space, namely a point moving along a closed trajectory close to a separatrix and leaving the vicinity of a saddle only to return to it. However, there was no randomness there. In order for there to be randomness, the tracing point must be able to move on opposite sides of a separatrix, first along a closed trajectory, and then move away from it. This is not possible on the plane because phase trajectories may not intersect. However, in a three-dimensional phase space (a system with one and a half degrees of freedom) such situations are possible.

Thus, in order for stochastic motions to appear in a dynamic system it is necessary that a) all (or nearly all) neighboring trajectories inside some region diverge, and b) they all remain inside some bounded phase volume inside the phase space of the system. We emphasize that it is common for all or nearly all the trajectories in a bounded region of a phase space to be unstable and this serves as a mathematical criterion of stochasticity.

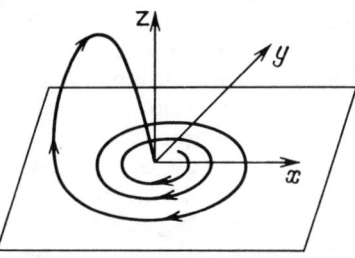

Fig. 22.3. Simple example of a returning unstable trajectory: a trajectory curving around a spiral and whose tail jumps to the origin to start turning again.

In a three-dimensional phase space such behavior by trajectories is easy to imagine, in that first they diverge on a two-dimensional surface and then return in the third dimension. The trajectory may look, for example, like a curving planar spiral whose tail returns to the origin to start curving again (Fig. 22.3). Such a trajectory would fill a bounded volume without ever closing and always behaving in a complicated and confusing way. Bearing in mind the complexity of individual established trajectories and the completely different behaviors of trajectories having infinitesimally close initial conditions, we see how the statistical nature of a dynamic system arises from two situations: first, almost every unclosed trajectory inside the bounded volume is in a certain sense random, and second, the concept of an ensemble, with which we are familiar from the theory of probability, appears in a natural manner. The ensemble is a set of trajectories inside the unstable volume. This ensemble is usually defined by giving a probability distribution density in the phase space. Physically, specifying the probabilities corresponds to a consideration of the evolution of an ensemble of identical systems with different initial conditions. We emphasize that the transition to an ensemble does not mean that some random factor has been added to our dynamic system; it is simply a method of determining the number of trajectories with some particular property.

In order to investigate an actual example of a statistical property in a dynamic system we must make some hypothesis that will allow averaging. In applied probability theory the true probability is determined from the statistically observed frequencies of some events. The existence of a limit to these frequencies is a consequence of the law of large numbers. For dynamic systems we can consider properties of the trajectories averaged over time instead of a series of statistical observations. Such a property might be the fraction of time spent following a section of trajectory of length T in some cell of phase space. When, for any cell and most trajectories (excluding, perhaps, the set of trajectories with zero metric) there is a limit as $T \to \infty$ for the time fraction spent by infinitely long sections of trajectories in the cell and when this limit is independent of the trajectories, the ensemble of the trajectory sections is called ergodic. The ergodic property can be used to substitute averaging over time by averaging over an ensemble.

In conservative systems, in which energy is conserved, the existence of time averages follows from the ergodic theory of dynamic systems. However, the independence of the average from the trajectory still remains an hypothesis in the general case, and one which dates from Boltzmann.

Ergodicity is clearly not randomness, and more than that, even very simple quasiharmonic motions

$$u(t) = a_1 \sin \omega_1 t + a_2 \sin \omega_2 t,$$

where ω_1 and ω_2 are incommensurable (i.e., $n_1 \omega_1 \neq n_2 \omega_2$, where n_1 and n_2 are numbers) are ergodic. Such a movement in phase space corresponds to a nowhere closed winding about a torus. Averaging the trajectories over the ensemble is here equivalent to averaging over time, however the trajectories do not scatter.

A frequent measure of the stochasticity of a system's motions is the rate of decrease in the autocorrelation function

$$K(t) = \lim_{T \to \infty} T^{-1} \int_0^T f(x(t+\tau))f(x(\tau))d\tau . \qquad (22.2)$$

Here, as before, ergodicity is assumed. The existence in $K(t)$ of a periodic or quasiperiodic component means that there are periodic or quasiperiodic components in the motion (e.g., unclosed trajectories around a torus). The development of stochasticity means that the functions $f(x(t+\tau))$ and $f(x(\tau))$ very rapidly become independent, i.e., $K(t)$ quickly tends to zero. The implementation spectrum in this case is continuous.

Note that the correlation function $K(t)$ characterizes a relation between the variable values upon point in time with the values at another, and it is always a real even function with a maximum at $t = 0$. This function may be either positive or negative. A function $K(t)$ with a sharp peak and a rapid fall to zero characterizes a wide and random process with zero average value (if the average $\langle u(t) \rangle$ is nonzero than $K(\infty) = \langle u(t) \rangle^2$). For white noise, i.e., the random process whose energy is equally distributed among all the frequencies, $K(t)$ has the form of a δ-function.

If we are considering a stationary random process, then the Fourier transform of the autocorrelation function is the process's spectral density and is equal to the average of the squares of the realization values, passed through a frequency filter with a filter band $\Delta\omega$:

$$S_u(\omega) = \lim_{\Delta\omega \to 0} (\Delta\omega)^{-1} \left[\lim_{T \to \infty} T^{-1} \int_0^T u^2(t, \omega, \Delta\omega) dt \right],$$

or

$$S_u(\omega) = 2 \int_{-\infty}^{\infty} K(t) e^{i\omega t} dt = 4 \int_0^{\infty} K(t) \cos \omega t \, dt . \qquad (22.3)$$

The spectrum $S_u(\omega)$ is always a real non-negative function.

If the fall in $K(t)$ (to the average) is exponential, then it is said that the system has mixing. Mixing is an undoubted attribute of stochasticity in a dynamic system [1]. Mixing in phase space can be imagined quite graphically in the following way. Take an ensemble of trajectories with initial conditions inside a small phase volume, a "droplet of phase liquid". Let this "droplet" differ in color from the remaining liquid inside the volume of phase space. If, for example, there is a stable limiting cycle in the area, then after some period of time our droplet will be smeared around the limiting cycle (Fig. 22.4) and it will only color a narrow band in the circle of the cycle. However, if all the

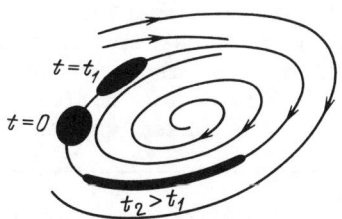

Fig. 22.4. Evolution of a "droplet of phase liquid" in a circle of a limiting cycle.

trajectories inside the bounded region are unstable, then the droplet will be continuously smeared and it will take on a more and more complex shape until at the limit as $t \to \infty$ it evenly colors the whole region, i.e., it is mixed with the uncolored liquid. Thus, the initial probability distribution in a system with unstable trajectories tends to some established, invariant probability distribution, which is what defines the statistical properties of the stochastic motions in a deterministic system.

Thus, we shall say that a dynamic system is stochastic if 1) there is a limiting probability distribution in the phase space of the system to which any initial nonequilibrium distribution tends (here we shall for simplicity) assume that the distribution is unique); 2) the system's behavior is ergodic, namely the value of an arbitrary function defined in the phase space and averaged over time is equal to the average of some limiting (invariant) distribution; and 3) the system's motion is characterized by a continuous spectrum, i.e., a falling autocorrelation function [1].

Verifying these conditions for every concrete system is extremely difficult mathematically. Therefore, we usually restrict ourselves to the verification of very weak conditions. In particular, we shall use the stochasticity criterion, which mainly means determining a quantity h which characterizes the scattering of neighboring trajectories in the linear approximation, and if this quantity is positive then the motion is stochastic.[2] Mathematically, a stochastic motion in a dynamic system is a stochastic set of trajectories in its phase space. These sets have different properties for Hamiltonian systems and dissipative systems.

According to Liouville's theorem, the phase liquid of a Hamiltonian system is incompressible, i.e., the initial flow of a trajectory conserves its volume in phase space. Poincare's theorem is then valid. This states that nearly all the trajectories in a bounded phase volume will pass infinitesimally close to the initial points an infinite number of times because the incompressibility of the phase liquid does not allow them to do anything else. The boundary of a stochastic set may be complex in structure and the stochastic set itself may "break up" into an arbitrary number of regions in which motion is regular, the "islands" of stability.[3] The situation is different in dissipative systems. By definition, the phase volume in these systems is

----

[2] This h is called the entropy metric or Kolmogorov–Sinai entropy.

[3] An example of a stochastic set in a Hamiltonian system is the homoclinic structure arising in the region of homoclinic trajectories (Chapter 15). We shall meet other examples at the end of this chapter and in Chapter 23, for more see [3, 4, 36-38].

STOCHASTIC DYNAMICS IN SIMPLE SYSTEMS

compressible on average, i.e., on average we have div $\mathbf{u} < 0$ throughout the phase space (**u** is the vector field in the phase space). Although the compressibility of a phase volume is a local property of the phase flow (it may be verified at any time and any point), it often brings with it in practically encountered systems with friction or viscosity a global property, namely the existence in the phase space an attractor, i.e., a closed set to which all the surrounding trajectories tend as $t \to \infty$ while remaining in it. Stable equilibria and stable limiting cycles are examples of regular attractors with which we are already acquainted. Since the phase volume in a dissipative system is compressible, the attractor has a zero phase volume. Any trajectory that does not belong to the attractor is a transient one.

Thus, a stochastic set in a dissipative system is a closed attracting set of trajectories inside which all the trajectories belonging to it are unstable. This source of set is called a strange attractor (see [34,35]). The dimensionality of a strange attractor is always less than the dimensionality of the phase space.

We shall now look at why most physical dissipative system with strange attractors do not, strictly speaking, satisfy the definition of a stochastic system we have above. In fact, a strange attractor may contain both a set of unstable trajectories and a set of stable periodic trajectories, although the region in which they are attracted is so small that they do not affect the behavior of the system in either numerical or physical experiments. This is why we shall call any dissipative system with such an attractor stochastic.

22.2 The Stochastic Dynamics of One-Dimensional Mappings

We saw in Chapter 15 that the investigation of the behavior of a dynamic system described by differential equations (section 15.3) is considerably simplified if we go from a system with continuous time to a system with discrete time. This is done by introducing a mapping of the secant surface cutting the phase flow into itself. We thus go from differential equations to difference equations. The method of point mappings is especially useful when analyzing the stochastic behavior of dynamic systems. First, as we saw in Chapter 15, we can effectively reduce the dimensionality of the phase space and, moreover, we can exclude from consideration regular components which do not yield stochasticity but which complicate the description. For example, these components are motions along trajectories belonging to the stochastic set. We should add that a special method, symbolic dynamics [5,6], has been developed in mathematics for analyzing stochastic behavior using mappings.

The basic idea behind symbolic dynamics is the coding of the trajectory by a sequence of symbols from some particular set, i.e., not only does the time in which the system's state is determined become discrete but so do the states themselves.

We shall restrict ourselves to a discussion of one-dimensional mappings. We do this for two reasons: first, they can be investigated in some detail without requiring numerical simulations on a computer, and second, the study of a two-dimensional mapping in one of whose directions the secant surface Σ is severely compressed by the mapping and in the other it is extended (a so-called hyperbolic mapping) (Fig. 22.5) reduces to the investigation of a one-dimensional, or rather almost one-dimensional, mapping. In a system with such a mapping, if we wait long enough, nearly all the points collect close to one or a few lines, and their behavior can later be described by analyzing the one-dimensional mappings of these lines into themselves.

Let as consider a nonbijective extension mapping of a segment into itself. Using purely qualitative considerations and a careful analysis of the phase-plane diagrams, we discovered at the beginning of this chapter that in order for a complex blurred behavior to exist in a bounded region of the phase space, it is necessary that on one hand all or nearly all the trajectories are unstable, and on the other hand that the tracing point cannot escape the region, i.e., the trajectories must return [10]. The simplest way for these conditions to be satisfied is to require that the system be described by a nonbijective extension mapping of a segment into itself, such as the one shown in Fig. 22.6a. The instability of any trajectory is associated here with the fact that everywhere $|dx_k/dx_{k-1}| > 1$, i.e., the mapping is an extension.

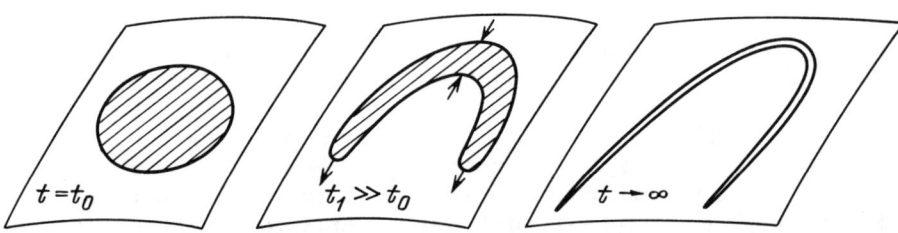

Fig. 22.5. Compression of the initial phase volume in one direction and extension in the other.

We shall show that the motion of a dynamic system described by an extension mapping of a segment into itself may be represented as a random sequence. To make the notation simple, we shall not consider the mappings in Fig. 22.6a, and instead we will deal with an analogous mapping in Fig. 22.7a.

We use the methods of symbolic dynamics. To do this, we

divide up the phase space into a finite number of regions $\Delta_0, \Delta_1, \Delta_2, \ldots, \Delta_n$, and assume that a physical device can indicate the region in which a trace point is located at a given moment. Then, each initial point will correspond to a sequence of regions through which the trajectory passes in a sequence of time intervals. If the motion is periodic, then different Δ_i will alternate periodically as well. If the motion is stochastic, then the sequence Δ_i must be random. We may only choose two regions for the case of the mapping in Fig. 22.7, viz., Δ_0 for the interval $0 \le x_k \le 1/2$ and Δ_1 for the interval $1/2 < x_k \le 1$.

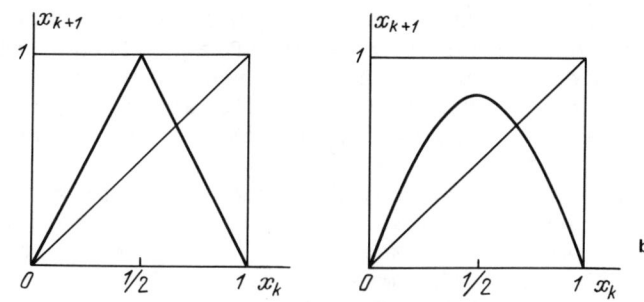

Fig. 22.6. Nonbijective mapping of a segment into itself: a) extension piecewise linear; b) smooth.

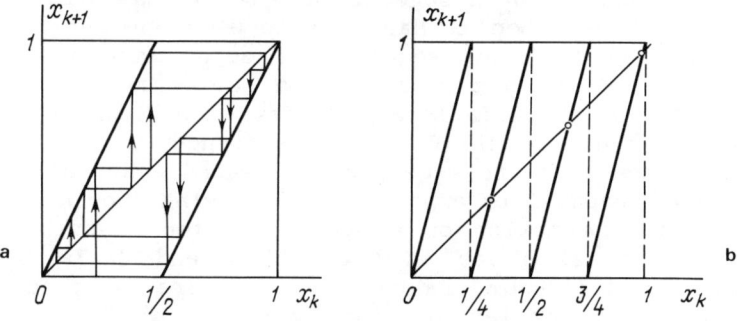

Fig. 22.7. Discontinuous extension mapping of a segment into itself: a) initial mapping; b) double mapping.

Now we note that if the numerical coordinate of the point $0 \le x_k \le 1$ is written in binary rather than decimal form, then our mapping may be written analytically as

$$x_k = \{2x_{k-1}\}, \tag{22.4}$$

where the {} brackets indicate the fractional part of the number. Sometimes another notation is used, viz., $x_k = x_{k-1}$ (mod 1). For example, this mapping acts on the number 0.1001011... simply by shifting it (a Bernoulli shift) and transforms it into the number 0.001011... . That the number is infinite in only one direction and the shift is consequently only one-sided is due to the irreversibility of the transformation.[4]

If the x-coordinate is a rational number, then at some symbol (e.g., the n-th) the sequence of zeros and ones will begin to repeat itself: it will be an n-tuple periodic point of the mapping. It is not difficult to verify that the set of periodic points possessed by our mapping is compact and infinite and that the points of the set are all unstable. This reveals a property typical of all strange attractors, namely, inside a boundary region from which the trajectories may not exist, there is a countable set of unstable cycles which cause the tracing point to move from one cycle to another.

It is easy to see that an extension mapping of a segment into itself has a countable set of unstable periodic points by constructing the sequential iterations of the set (Fig. 22.7b). When the mapping is applied twice there are four stationary points, and when it is applied three times there are 2^3 stationary points etc. There is in fact a mathematical theorem on this topic from which, in particular, it follows that if a continuous (not necessarily smooth) extension mapping of a segment into itself has a cycle period equal to three, then it has a cycle with any period [8]. It is known [9] that the sequence of zeros and ones given by (22.4) will only be periodic for a set of rational numbers, and that for all irrational numbers, i.e., the majority of points in the segment (0,1), the sequence will be random in the same sense as the sequence of "heads" and "tails" in the classical probability experiment of tossing a coin is random.

Thus, the motion of a dynamic system describable by a mapping such as those in Figs. 22.6a and 22.7a do indeed reduce to a random sequence, i.e., are stochastic. The stochastic properties of the mappings in Fig. 22.6a (or 22.7a) are found quite simply. It follows directly from the formulae for the mappings $x_{k+1} = F(x_k)$ that after a single mapping,, the initial probability density given on the segment $\rho_j(x)$ is transformed into the density

---

[4] An analogy is the two-dimensional "baker's transformation", so called because of the process of rolling dough: a square sheet is rolled along one coordinate and then turned over and rolled on the other side etc. [7].

STOCHASTIC DYNAMICS IN SIMPLE SYSTEMS

$$\rho_{j+1}(F(x)) = \sum_j P_j(x)|dF(x)/dx|^{-1}, \qquad (22.5)$$

where the summation is over all arms of the function $F(x)$. This relation means that the initial distribution becomes dF/dx-fold less dense (the extension mapping) but points from several sections of the initial segment fall into the same intervals dx of the segment after the transformation (the transformation is not bijective). A mapping such as those in Figs. 22.6a and 22.7a have an invariant probability distribution $P(x)$, which clearly may be found from the conditions $\rho_{j+1} = \rho_j = P$, i.e., $P(x)$ must satisfy the following

$$P(F(x)) = \sum p(x)|dF(x)/dx|^{-1}. \qquad (22.6)$$

For a piecewise mapping of the form $x_{k+1} = \{2x_k\}$, as can be confirmed by a direct substitution, we have $P(x) = $ const. Assuming (from the norming condition of the whole density to unity) that $P = 1$, and using (22.2) and (22.3) we find, for the mapping (22.4), the mean

$$\langle x \rangle = \int_0^1 x\, dx = 1/2,$$

variance

$$D = \langle(x - \langle x \rangle)^2\rangle = 1/12,$$

and correlation function [3]

$$K(j) = D^{-1}\langle[(x_i - \langle x \rangle)(x_{i+j} - \langle x \rangle)]\rangle$$

$$= 12 \int_0^1 (x - 1/2)(\{2^j\} - 1/2)dx = \exp[-(\ln 2)j].$$

It can be seen that in our case the correlation over time falls exponentially. The argument of the exponential, i.e., the Lyapunov index characterizing the rate at which the correlation falls (the rate at which the trajectories scatter is simultaneously found), is the Kolmogorov–Sinai entropy. In this case the entropy is $h = \ln 2$.

One might ask whether stochasticity is possible in systems which do not reduce to discontinuous mappings, such as the one in Fig. 22.6a, and instead reduce to smooth mappings, such as the one in Fig. 22.6b. The answer is yes, but not always.

Let us turn to the mapping, more accurately the family of mappings, $x_{k+1} = F(x_k)$ which depends on a parameter b i.e.,

$$x_{k+1} = bx_k(1 - x_k) .\qquad(22.7)$$

When the parameter value is $b = 4$, the maximum is $x = 1/2$ and is the inverse image of an unstable stationary point $x = 0$ (the point $x = 0$ is the next one for $x = 1/2$). If we change the variable to $y = \phi(x) = (2/\pi)\arcsin\sqrt{x}$ [6], then the mapping (Fig. 22.6b) for $b = 4$ turns into a piecewise linear mapping (Fig. 22.6a):

$$F(y) = \begin{cases} 2y, & 0 \le y \le 1/2, \\ 2(1 - y), & 1/2 \le y \le 1, \end{cases}$$

for which we know there is an invariant probability distribution. It follows that when $b = 4$ and for the mapping (Fig. 22.6b), there is also an invariant probability distribution. The density of this distribution is $(\pi\sqrt{x(1-x)})^{-1}$.

22.3 Noise Generator. Qualitative Description and Experiment

The stochasticity of concrete dynamic systems using the theory of oscillation is investigated to discover the structure of the stochastic set, to understand the mechanism by which chaos arises, to find criteria for its existence, and finally approximately to describe (on the basis of separating some small parameter) the behavior of the system in the stochastic region.[5] This program can only be implemented for comparatively simple systems with three-dimensional phase spaces, which admit of a description using two-dimensional, and approximately even one-dimensional, Poincare mappings. Let us consider, by way of example, a simple radio generator of stochastic oscillations.

We already understand what periodic self-excited oscillations are (Chapters 14 and 16). Stochastic self-excited oscillations are irregular, random motions of nonconservative dynamic systems that arise due to the action of nonrandom sources of energy. Mathematically, stochastic self-excited oscillations in phase space are strange

---

[5] Here and below we shall use the terms stochasticity and chaos as synonyms.

attractors, about which we have spoken at the beginning of this chapter. We add here that the term "strange" was invented by the mathematicians Ruel and Takens due to the very complicated Cantor [11] structure of the attractors. Now the term is associated simply with the complex irregular behavior of trajectories in an attractor.

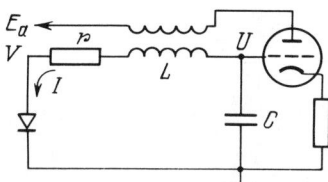

Fig. 22.8. Diagram of a simple noise generator. The circuit parameters are: $C = 1.5$ μF; $L = 5.7$ MHz; the dimensionless circuit parameters are: $g = 2.4$; $\varepsilon \approx 4.8 \times 10^{-5}$; the minimum energy loss is determined by the resistance $r_0 \approx 8.2$ Ohm.

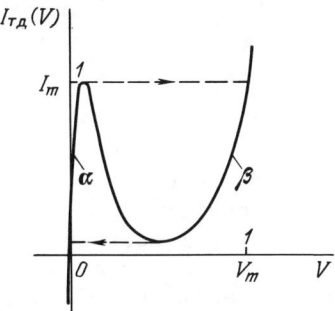

Fig. 22.9. The voltage-current diagram of the tunneling diode used in the noise generator.

The diagram of the noise generator we shall be discussing is given in Fig. 22.8. It only differs from the Van der Pol generator circuit, with which we are familiar (Fig. 14.1b), by the tunneling diode in series with the coil. The circuit is described by [12]

$$LC\, dI/dt = (MS - rC)I + C(U - V),$$
$$C\, dU/dt = -I, \quad C_1 dV/dt = I - I_{td}(V).$$
(22.8)

Here C_1 is the capacitance of the diode, S is the lamp characteristic, and M is the reciprocal inductance. We shall assume that the valve's current-voltage diagram is linear. This is valid because the oscillation regime we are interested in is bounded by the nonlinear characteristics of the tunneling diode $I_{td}(V)$ (Fig. 22.9) to a level at which the nonlinearity of the valve is no longer appreciable.

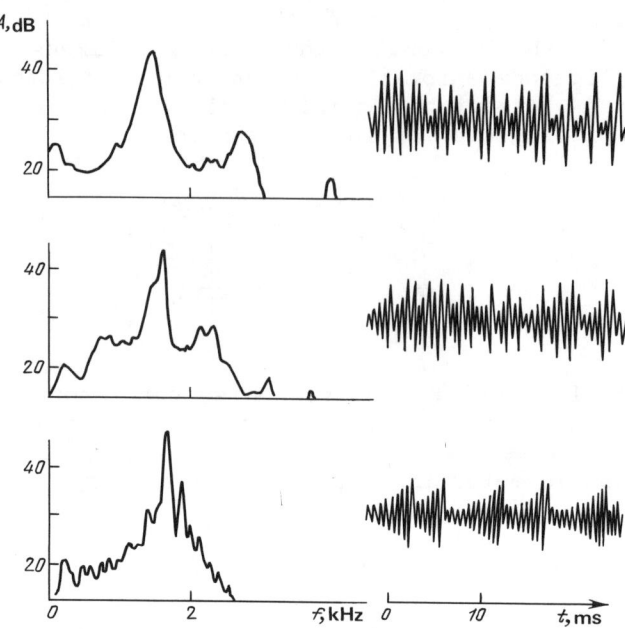

Fig. 22.10. Spectra and oscillograms of the output signal from an self-excited generator of noise for various values of $R < 11$ Ohm (the upper diagram corresponds to the least R).

The qualitative operation of the generator can be described as follows. As long as the current I and voltage U are small, the tunneling diode does not significantly influence the oscillations in the circuit, and they grow. A current flows through the tunneling diode and the voltage in it is determined by the section α of the curve $I_{td}(V)$. When, however, the current I reaches a value I_m, the tunneling diode is almost instantaneously turned on (the rapidity of this switching is due to the smallness of the capacitance C_1) and there is a step change in the voltage to V_m. Now, the current through the tunneling diode is decreased and the reverse switching takes place, from the β section to the α section. As the result of these two switchings, the tunneling diode almost completely absorbs the energy in the circuit and the oscillations once again begin to grow.

Thus, the generated signal $U(t)$ is a sequence of wave trains of increasing oscillations, and at the end of each train there is a pulse with voltage $V(t)$. It is not clear from this description, however, whether a periodic or stochastic regime becomes established. We can discover this by investigating (22.8) and we shall do this later. For the moment we shall look at the experimental evidence [12].

The diagram in Fig. 22.8 was implemented on half a 6N1P triode and on four 3I306G tunneling diodes connected in parallel. The increment in the oscillations in the circuit, i.e., the value of h, can easily be changed by altering the resistance r. The minimum possible energy loss in the circuit was determined by the resistances of the components in the circuit $r_0 \approx 8.2$ Ohm. When $R = r - r_0 \approx 14.5$ Ohm, purely periodic oscillations were induced in the circuit, the oscillations being restricted by the nonlinearity of the valve at such a low level that the diode was not switching ($I < I_m$). When $R \approx 13.5$ Ohm the amplitude of the oscillations reached the threshold value and the signal $U(t)$ became a long packet of oscillations, with rare interruptions by diode switching. Only when $R < 11$ Ohm did the nonlinearity become insignificant and a signal in the form of wave trains, inside each of which the oscillations grew exponentially, was generated. The transition from one wave train to another was accompanied by a voltage pulse at the tunneling diodes $V(t)$. At no value of the resistance below 11 Ohm was it a possible to get a periodic regime and the generator produced a random signal with a continuous spectrum. It can be seen from the spectra and oscillograms in Fig. 22.10 that as R was decreased, the increment in the oscillations h increased and the average length of the wave train was decreased. The peaks in this spectrum were smoothed at the repeat train frequencies. Most of the energy was contained at the main maximum, which corresponds to the circuit's frequency.

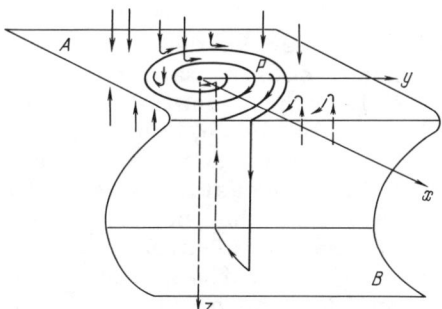

Fig. 22.11. Phase space of the system described by (22.9).

When investigating (22.8) we go to dimensionless variables $x = I/I_m$, $z = V/V_m$, $y = U\sqrt{C}/I_m\sqrt{L}$, and $\tau = t\sqrt{LC}$. As the result we obtain

$$\ddot{x} = 2hx + y - gz, \quad \dot{y} = -x, \quad \mu\dot{z} = \dot{x} - f(z), \qquad (22.9)$$

for $\mu \ll 1$. Here $h = 0.5(MS - rC)\sqrt{LC}$ is the increment in the oscillations in the circuit in the absence of a diode, and $g = V_m\sqrt{C}/I_m\sqrt{L}$ is a parameter determining the influence of the tunneling diode on the process in the circuit; $\mu = gC_1/C$ is a small parameter proportional to the tunneling diode's capacitance, and $f(z) = I_{td}(V_m z)/I_m$ is the normed characteristic of the diode (Fig. 22.9).

The system (22.9) has a small parameter μ multiplied by a derivative. Therefore all the motions in the phase space (Fig. 22.11) may be divided into fast-diode switching (the straight lines x = const and y = const) and slow movements for which the voltage across the diode follows the current corresponding to the trajectory line on the surface A ($x = 0$) and B ($x = f(x)$ and $f'(z) > 0$) corresponding to the sections α and β of the diode diagram).

The system has one unstable (at $2h > g/f'(0)$) equilibrium $x = y = z = 0$ that is a saddle. The trajectory's line on the A plane are untwisted about an unstable focus and at the end the trajectory moves to the surface. Here the tracing point breaks away and follows the line of the fast motions on the B surface. After travelling along B, the tracing point breaks back onto the A surface and reaches the vicinity of the equilibrium. A new wave train of increasing oscillations begins. This description corresponds to reality as shown in Fig. 22.10.

22.4 Statistical Description of a Simple Noise Generator

Our noise generator can be described, as we shall show, for $\mu = 0$, by a nonbijective mapping of a segment into itself. However, it is significantly more complex than the mapping in Fig. 22.7. Therefore, the invariant probability distribution cannot be found analytically by solving (22.9). In order to prove the stochasticity and to determine the statistical parameters of the noise generator for given parameters, we use the method of symbolic dynamics [5].

Thus, we construct a point mapping equivalent to (22.9) for $\mu \to 0$. Let us consider the point mapping of the halfplane $x = 0$, $y > 0$ into itself (Fig. 22.11). As $\mu \to 0$, this halfplane only intersects trajectories laying on the surface of the slow motions, and therefore the mapping $x = z = 0$ of the halfplane $y > 0$ into itself: $y_{j+1} F(y_j)$ is one-dimensional. For arbitrary nonlinearity of the "switching" component (e.g., the tunneling diode) this mapping cannot be analytically described. Therefore, we use the piecewise linear approximation

$$f(z) = \begin{cases} \alpha^{-1}z\,, & z < \alpha \\ (1 - \alpha - z)/(1 - 2\alpha)\,, & \alpha < z < (1 - \alpha) \\ (z - 1 + \alpha)/\alpha\,, & (1 - \alpha) < z \end{cases} \qquad (22.10)$$

In this approximation A and B are halfplanes, the equations of the slow motions on which have the form (compare with (22.9))

a) $\dot{x} = 2\nu x + y$, $\quad \dot{y} = -x \quad$ on plane A,

b) $\dot{x} = 2\nu x + y - b$, $\quad \dot{y} = -x \quad$ on plane B. $\qquad (22.11)$

Here $\nu = h - \alpha g/2$, $b = g/(1 - \alpha)$. These equations are linear and so it is easy to obtain an explicit form for the mapping linking the sections of the trajectories laying on the planes A and B.

The mapping will consist of two parts: the function $F_1(y_j)$ describing that part of the mapping which does not yield trajectories laying on the B halfplane (Fig. 22.12a), and the function $F_2(y_j)$ describing the part giving trajectories laying on both planes (Fig. 22.12b). We immediately obtain from (22.11a)

$$y_{j+1} = F_1(y_j) = \exp(2\pi\nu)y_j \equiv ky_j\,. \qquad (22.12)$$

The function $F_2(y_j)$ is not so easily derived from (22.11b) and so we approximate it by a formula that qualitatively describes the behavior of the trajectories in the regime of stochastic oscillations, i.e.,

$$y_{j+1} = F_2(y_j) = y_0 - (y_j - y_0)^{1/2}\,. \qquad (22.13)$$

Thus, when $y_j < y_0$ we use the section of the mapping in (22.12) and when $y_j > y_0$ we use the section in (22.13). The exponent $1/2$ in (22.13) reflects the fact that the trajectories approach the line of departure $x = 1$ almost along a tangent. The constant y_0 describes the shift in the trajectory when moving on the B plane. Uniting (22.12) and (22.13) yields the mapping $y_{j+1} = F(y_j)$, which is shown in Fig. 22.13. This mapping has an attracting region, i.e., an attractor: $y_0 - (ky_0 - y_0)^{1/2} < y < ky_0$. If $0 < k - 1 < (4y_0)^{-1}$, then the mapping inside the attractor is protracting, i.e., $|dy_{j+1}/dy_j| > 1$.

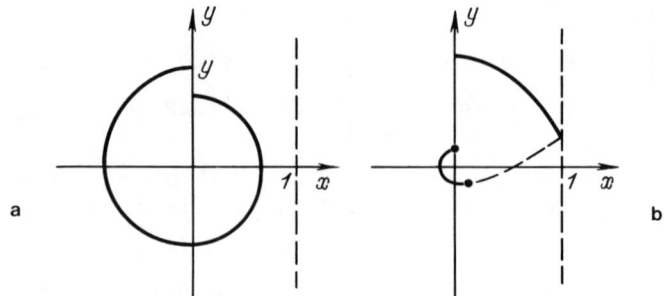

Fig. 22.12. Construction of a Poincare mapping for the system of equations (22.11): a) the trajectories lie on one plane of the slow motions; b) the trajectories break away onto the second plane of the slow motions and return back again.

Thus, there is a stochastic attractor in a parameter region in which (22.9) is described by the mappings (22.12) and (22.13) as $\mu \to 0$ in its phase space, and inside the attractor there is an invariant probability distribution and the motion has a mixing property.

To prove the stochasticity it is necessary to confirm that all the motions inside the attractor are unstable. This is certainly fulfilled if the mapping is protracting, i.e., $|dy_{j+1}/dy_j| > 1$. However, this condition is a little too rigid: the motions do not have to be unstable at each iteration, it is sufficient that they are unstable on average.

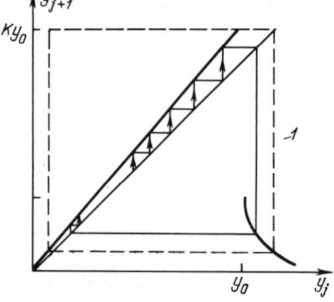

Fig. 22.13. Poincare mapping for a system described by (22.9) when $\mu = 0$: 1) boundary of the attractor.

Let us now turn to calculating the statistical properties of the output signal [13]. This signal, as can be seen from the oscillograms in Fig. 22.10, consists of a sequence of groups of pulses with a random number of maxima in each group. In terms of the Poincare mapping (22.12) and (22.13), the number of maxima in a group is the number of iterations of the mapping with $y < y_0$. We can approximate mapping by a piecewise mapping, as in Fig. 22.14, and divide the whole attractor into segments Δ_i. We are now not interested in the exact coordinates of the point, more in the number of segments in which this point falls. Each trajectory will then correspond to a specific sequence of sections.

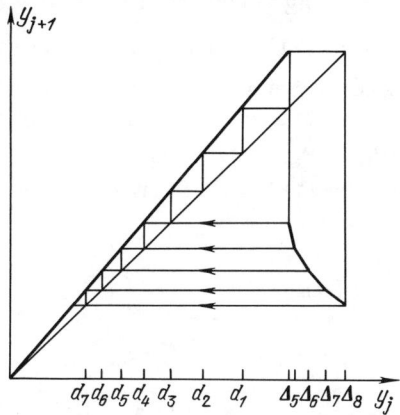

Fig. 22.14. Piecewise linear approximation of the mapping in Fig. 22.13: the Lamerey diagram.

In order to find the signal statistics we must find the invariant probability distribution, i.e., determine the probability of going from one segment to another. In our case (Fig. 22.14), this is not difficult to determine: if only one particular transition is possible from the segment, then the probability of the transfer is unity. This is true of transitions beginning on all the sections except d_1. Several routes are possible from d_1 and the manner in which the probabilities are distributed among them is so far unknown to us. We may conclude from physical considerations that the probability of a transition $d_1 \to \Delta_i$ must be proportional to the length of the section Δ_i. Now, using an expression for the probability of the transition (a diagram is given in Fig. 22.15), i.e.,

$$\mu(d_i) = \mu(d_{i+1}) \quad (i = 1, 2, 3) ,$$

$$\mu(d_i) = \mu(d_{i+1}) + \mu(\Delta_{i+1}) \quad (i = 4, 5, 6) , \quad (22.14)$$

$$\mu(d_7) = \mu(\Delta_8), \quad \mu(\Delta_5 + \Delta_6 + \Delta_7 + \Delta_8) = \mu(d_1)$$

we can determine the probability of a number of steps in the Lamerey diagram (Fig. 22.14), i.e., determine the probability of the number of pulses in a group (the oscillogram in Fig. 22.10). It can be seen from the diagram that in our case the number of pulses may change from 5 to 8. If when a train ends y falls in the interval Δ_5, then in the next train there will be 5 pulses long (the point goes upon the sections d_4, d_3, d_2, d_1 and then falls into some section Δ_i and the train is finished), etc. Therefore, the probability of there being n pulses in a train is the conditional probability $P(n) = \mu(\Delta_n)/\mu(\Delta_5 + \Delta_6 + \Delta_7 + \Delta_8)$. Since this probability is proportional to the length of interval Δ_n, we may approximate it to

$$P(n) \sim (n - 4)^{1/2} k^{-n}. \qquad (22.15)$$

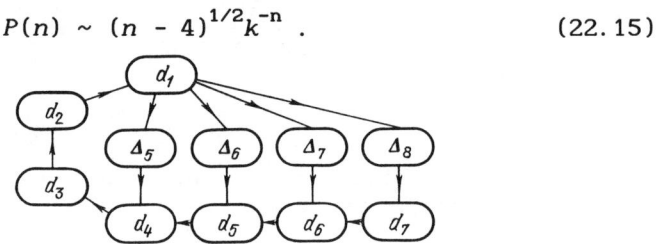

Fig. 22.15. Graph for the mapping shown in Fig. 22.14.

This result was obtained for the situation in Fig. 22.14. If the minimum number of pulses in the train was n_0 (and not 5 as in our case), then instead of (22.15) we would have

$$P(n, n_0) \sim (n - n_0 + 1)^{1/2} k^{-n}.$$

The distribution of the number of pulses in a train constructed using this formula describes the statistics of the pulses in a real generator very well.

A large number of self-excited stochastic generators have recently been suggested and investigated (see [35], ch. 9). The generator with inertial nonlinearity has, for example, been studied in some detail both theoretically and experimentally (it was first suggested in [39]). In this generator self-excited oscillations are induced due to an inertialess positive feedback, which leads to negative resistance, and to their limitation by the nonlinear inertial interactions between the dynamics variables (see [35] and its bibliography).

22.5 Ways in which Strange Attractors Arise

In this section we shall be considering the most typical ways a strange attractor can appear in a system with a three-dimensional phase space.

Many of the transitions that interest us can be described by one-dimensional mappings. We shall thus start our discussion with these mappings, remembering that several multi-dimensional systems can be reduced to one-dimensional mappings close to the boundaries of the occurrence of stochasticity (Chapter 23).

Successive Period Doubling. We return to the mapping (Fig. 22.6b), i.e., $x_{k+1} = bx_k(1 - x_k)$, where the parameter b lies in the interval $0 \le b \le 4$.

Many three-dimensional systems reduce to this mapping, for example, the system whose attractor has the form of a widening band that forms a pleat and then returns to itself (Fig. 22.16). A mapping such as in Fig. 22.6b is obtained for the coordinate ϕ in the secant.

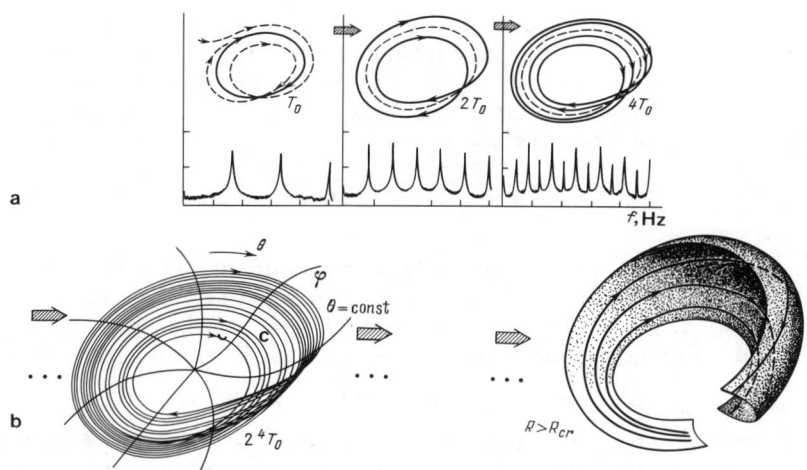

Fig. 22.16. Appearance of a strange attractor in a three-dimensional system via sequence of bifurcations doubling in period (the initial motion has period T_0): a) sequence of doubling in phase space (above) and on a spectrogram (below); b) strange attractor in the form of a doubly-folded and "band" closing in on itself. It arises after the loss of stability of motions with period 2^∞. The band has a Cantor structure in the cross-section [33].

For any b there is a stationary point $x_{k+1} \equiv x_k = x^* = 0$ in this mapping, and for $b > 1$ there is another stationary point: $x^* = 1 - 1/b$. This point is stable up to values of $b = 3$. When $b > 3$, the nontrivial stationary point becomes unstable: the multiplicator dx_{k+1}/dx_k at this point goes through -1 and a stable periodic motion with period 2 arises. This corresponds to the appearance of two real routes in the equation $x_{k+2} = x_k$. The one-fold stationary point does not disappear, but it does become unstable. The two-fold cycle is stable for $3 < b < 3.45$. When $b \approx 3.45$, the two-fold cycle loses stability and a stable four-fold cycle arises. Increasing b causes this cycle to lose stability and a stable cycle with period 2^3 arises followed by one with period $2^4, \ldots, 2^n, 2^{n+1}, \ldots$. Finally, when $b_\infty \approx 3.57$ no further stable periodic motions remain and there is a transition to stochasticity. In a three-dimensional phase space this corresponds to the appearance of a strange attractor (Fig. 22.16). Note that when $b > 3.57$, this mapping may have stable periodic points, for example, for $b = 3.83$ there is a stable three-fold cycle [14].

A remarkable feature of the transition to chaos via an infinite chain of doubling bifurcations is its universality [15]. It turns out that the interval of the parameter b inside which a cycle of period 2^n exists follows a geometric progression with increasing n

$$(b_n - b_{n-1})/(b_{n+1} - b_n) = \delta , \qquad (22.16)$$

where $\delta = 4.66920\ldots$ is the universal Feygenbaum constant. It immediately follows that by determining experimentally the boundaries of the first few doublings it is possible to determine from the formula $(b_\infty - b_n) \sim \delta^{-n}$ the value of the parameter b_∞, at which point cycles with infinite period $T = 2^\infty$ appear and immediately after which stochastic behavior starts. It turns out that the appearance of stochastic behavior for $b \geq b_\infty$ is also universal [16, 17].

The transition to stochasticity via an infinite chain of doubling bifurcations for periodic motions is quite typical for dissipative systems [18, 19]. This is because many dissipative systems, including those with high orders (multi-dimensional phase spaces), can be accurately described close to the transition boundary by a smooth nonbijective one-dimensional mapping (Fig. 22.6b). We shall look at this phenomenon in the next section. Here we give two examples illustrating the route taken by dissipative systems to stochastic behavior that we have been considering.

These examples, which describe resonance interactions between oscillators, are of independent interest for the theory of nonlinear waves.

A resonance interaction between the waves is characteristic of the nonlinear properties of a variety of media. We know (Chapter 20) that the nonlinear phenomena arising under such interactions (harmonic and subharmonic generation, self-modulation and self-focusing of waves, and the various types of parametric processes) are observed in dispersing media even when there is very small nonlinearity if the synchronicity conditions $\sum n_i \omega_i = 0$ and $\sum n_i \mathbf{k}(\omega_i) = 0$ are fulfilled, where ω_i is the frequency, and $\mathbf{k}(\omega_i)$ is the wave vector of an interacting wave. The amplitudes of such waves are slowly changing functions of the space coordinates and of time. A nonlinear interaction between quasiharmonic waves, as we have seen, is very important in the physics of plasma, in fluid mechanics, in nonlinear optics, and in the physics of the condensed state. If the number of elementary excitations in a medium is very large, then usually the wave field's behavior becomes irregular.

In the absence of energy sources or sinks, the spectrum of such waves is such that the energy is equally distributed over all the degrees of freedom (the Rayleigh--Jeans distribution) (Chapter 20). A self-consistent description of real wave turbulence requires an account of dissipation and the pumping of energy from a source (external field when heating a plasma, wind for waves on water, etc.). Such a description leads to a consideration of the dynamics of an ensemble of interacting oscillators. i.e., modes, some of which take energy from a source and others give energy to the temperature-maintaining mechanism. Let us now consider the simplest such models without first requiring that the phases be chaotic (cf. Section 20.4).

In a medium with nonlinearity quadratic with respect to the field, the interaction between three waves is the elementary one (the synchronicity condition is $\omega_1 + \omega_2 - \omega_3 + \delta = 0$, $\mathbf{k}_1 + \mathbf{k}_2 - \mathbf{k}_3 = 0$) (17.30), i.e.,

$$\dot{a}_1 = \sigma_1(a_j) + a_2^* a_3 \exp(i\delta t), \quad \dot{a}_2 = \sigma_2(a_j) + a_1^* a_3 \exp(i\delta t),$$

$$\dot{a}_3 = \sigma_3(a_j) - a_1 a_2 \exp(-i\delta t). \tag{22.17}$$

Here the a_j are the complex wave amplitudes, which are assumed to be spatially homogeneous (the norming is chosen such that the coefficients of the interactions are unity), the σ_i are linear terms describing the pumping of energy and dissipation, and δ is the detuning.

The nature of the energy exchange between the unstable wave ω_3 and the damping pair ω_1 and ω_2, i.e., when, $\sigma_3 =$

$\gamma_3 a_3$, $\sigma_1 = -\nu_1 a_1$ and $\sigma_2 = -\nu_2 a_2$ depend strongly on the ratios γ_3, ν_1 and ν_2. A numerical analysis [20] has shown that the chaotic exchange of energy between such modes arises when the parameter region is wide. Chaos arises due to the appearance of a chain of sequential bifurcations of period doubling. A graphical investigation of a structure including a strange attractor is difficult because the differential equation system resulting from (22.17) when $\nu_1 \neq \nu_2$ is fourth order. A more promising analysis in this respect is to look at the degenerate case of $\nu_1 = \nu_2$. Since the amplitudes of identically damping low-frequency waves as $t \to \infty$ are equalized (it is not difficult to show this using (22.17)), then (22.17) may be represented in the form

$$\dot{X} = Z - 2Y^2 + \delta'Y + X,$$

$$\dot{Y} = 2XY - \delta'X + Y, \qquad (22.18)$$

$$\dot{Z} = -2Z(X + \nu).$$

Here $X = (|a_3|/\gamma_3)\cos \psi$, $Y = (|a_3|/\gamma_3)\cos \psi$, $Z = |a_{1,2}|^2/\gamma_3$, $\psi = \arg a_3 - 2 \arg a_{1,2} - \delta$, $\nu = \nu_{1,2}/\gamma_3$, and $\delta' = \delta/\gamma_3$. When there is exact synchronicity ($\delta = 0$) and $\nu > 1/2$, all the trajectories in the phase space of (22.18) as $t \to \infty$ tend to the plane $Z = 0$ or $Y = 0$. This is because the function $P = ZY$ satisfies equation $dP/dt = (1 - 2\nu)P$, i.e., we have $P \to 0$ as $t \to \infty$. There are no stable equilibria or limiting cycles on the $Z = 0$ and $Y = 0$ planes and all the trajectories on them leave to infinity. The stabilization of an unstable mode due to transfer of energy to an equivalent low-frequency mode must, in this case, be impossible. However, stabilization is possible when there is a nonzero, although very small, detuning. The energy flow then turns out to be either constant over a time (a stable equilibrium in the phase space), or periodic (a limiting cycle), or a random fluctuation (a stochastic attractor), depending on the parameters.

Thus, when $\delta = 2$ and $\nu \leq 3$, the absorption into the low-frequency modes is still not sufficient to stabilize the instability. When $3 \leq \nu \leq 8.5$, there is stabilization and a simple periodic regime with energy exchange becomes established, then when $\nu > 8.5$ a sequence of bifurcation starts with the period of the periodic motion doubling (at some $\nu \geq 11.9$ a four-fold cycle occurs and at some $\nu \geq 13.5$ an eight-fold cycle occurs, etc.). A chaotic regime ensues at large values of the damping [19].

Analogous bifurcations with doubling of the period yield stochastic behavior and have been observed in a system describing a four-wave interaction $2\omega_0 = \omega_1 + \omega_2$, $2\mathbf{k}_0 = \mathbf{k}_1$. The linear unstable mode ω_0 is stabilized by a transfer of energy to the damping satellites ω_1 and ω_2. If $|\omega_1 - \omega_2| \ll \omega_0$, then this regime corresponds to stochastic modulated oscillations with a carrier frequency ω [21].

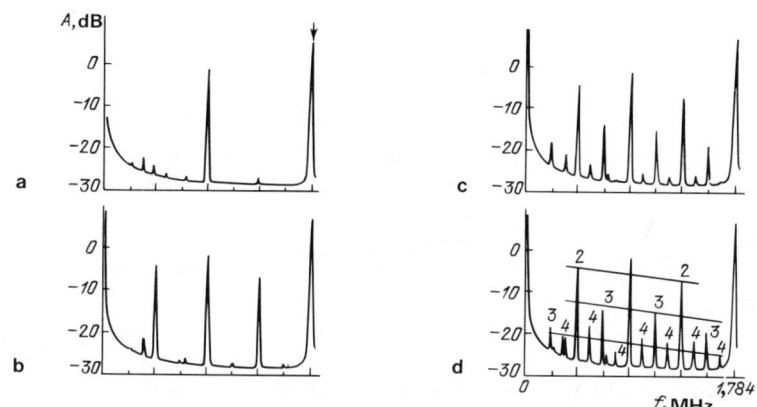

Fig. 22.17. Spectra illustrating bifurcation with doubling of the period at the transition to stochastic behavior in a nonlinear oscillator excited by a periodic force. A is the amplitude and f is the frequency plotted for various amplitudes of the external force, with the amplitude increasing from a) to d).

To conclude this topic we present the results of a physical experiment with a simple dissipative system (an RLC-circuit), in which the regime of stochastic self-excited oscillation arises due to a sequence of doublings [22]. The oscillations were investigated in a series, nonlinear, RLC-circuit in which a periodic signal with a frequency equal to the fundamental frequency of the circuit in the linear approximation (f_0 = 1.784 MHz) was induced. The nonlinear component that was used was a semiconductor diode whose capacitance depended on the voltage thus $C(U) = C_0(1 - U/U_0)^{-0.44}$. As the amplitude of the external force is increased, the subharmonics $f_0/2$, $f_0/4$, $f_0/8$ and $f_0/16$, which correspond to the onset of stable periodic motions with periods $2T_0$, $4T_0$, $8T_0$ and $16T_0$ (Fig. 22.17) appear one by one in the oscillation spectrum. When the amplitude of the external force exceeded a critical point (the transition point), the discrete peaks widened and a plateau was formed. The spectrum of oscillations observed in the parameter range corresponding to developed stochastic motion is shown in Fig. 22.18.

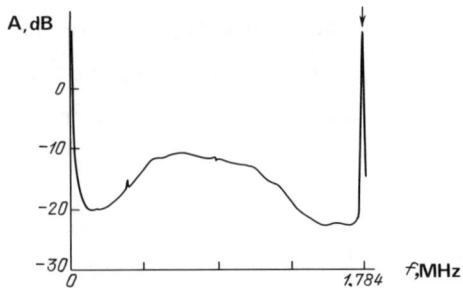

Fig. 22.18. Spectrum of oscillations of a nonautonomous nonlinear oscillator in the stochastic oscillation regime.

Rigid Regime in which Stochastic Self-Excited Oscillations Arise.
One of the mechanisms for the appearance of a strange attractor with a continuous change in the parameters is illustrated for a concrete example, the Lorentz system. Lorentz noticed a "deterministic aperiodic flow" [23] in a simple dissipative system with a three-dimensional phase space. The system, which comes from fluid mechanics, has many different applications [7], as will become apparent, and its dynamics are investigated in detail using qualitative and numerical methods.

The Lorentz system is obtained, for example, from the Bussinesque equations, which described thermal convection in a horizontal layer heated from below, if the analysis is restricted to two-dimensional motions and the current function ψ and changing temperature T are represented (cf. (21.4)) as

$$\psi(x, z, t) = \psi_{11}(t)\sin(\pi\, xa/l)\sin(\pi z/l) ,$$

$$T(x, z, t) = T_{11}(t)\cos(\pi\, xa/l)\sin(\pi z/l) - T_{02}(t)\sin(2\pi z/l) .$$

(22.19)

This representation takes into account three coupled spatial modes, of which two ψ_{11} and T_{11} for $Ra > Ra_1$ grow due to convective instability and the third T_{02} dampens. The parameter $a = 1/\sqrt{2}$ is a characteristic scale of the modes which lose their stability earliest, when $Ra \geq Ra_1$. A solution to (22.19) describes the convection in the form of waves or rollers that do not change along the third coordinate.

The limitation of the instability in this case takes place due to energy transfer from the growing modes to the mode T_{02} which corresponds to a change in the basic temperature profile such that as much energy transfers to the mode ψ_{11} and T_{02} on average as is spent due to viscosity and thermal conductivity. The following system of Lorentz equations is obtained for the amplitudes of these modes $X \sim \psi_{11}$, $Y \sim T_{11}$, $Z \sim T_{02}$:

$$\dot{X} = -\text{Pr}\, X + \text{Pr}\, Y,$$
$$\dot{Y} = -Y + rX - XZ. \qquad (22.20)$$
$$\dot{Z} = -bZ + XY.$$

Here **Pr** is the Prandtl number, $r = \text{Ra}/\text{Ra}_1$ is the Rayleigh number reduced by the critical Rayleigh number, and $b = 4/(1 + a^2) = 8/3$ (Chap. 21). The first column contains the terms corresponding to linear mode damping, the second column contains terms corresponding to all the parametric excitations (terms proportional to Y and X enter with the same signs as those in the equations for $\dot{X}$ and $\dot{Y}$, respectively), and the third column includes terms for the nonlinear pumping of energy into the damping mode Z. Thus a seemingly simple system demonstrates aperiodic behavior (one of its trajectories, belonging to the attractor, is shown in Fig. 22.19 [24]).

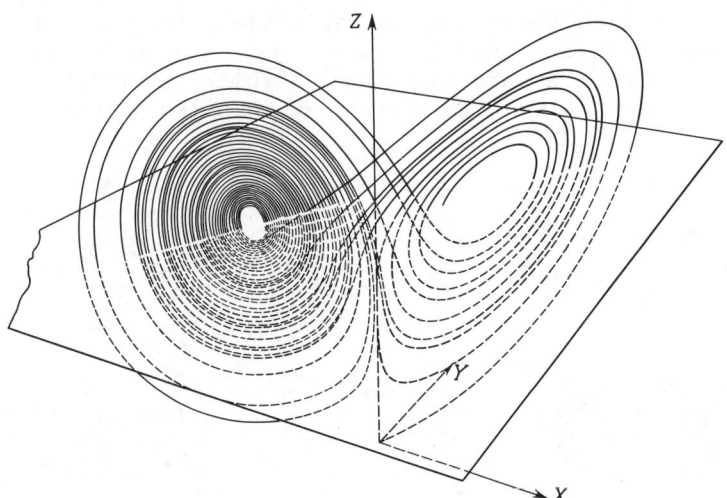

Fig. 22.19. Trajectories leading to the Lorentz attractor (leaving the origin). The horizontal plane corresponds to $z = 27$, $r = 28$.

First, we shall consider the simplest features of (22.20). 1) The system is unstable at infinity and there is a region in phase space to which all the trajectories tend. Assuming $u = X^2 + Y^2 + (Z - r - \text{Pr})^2$, we find from (22.20) that $\dot{u} \leq - C_1 u + C_2$ ($C_{1,2} > 0$), i.e., all the trajectories enter a sphere of radius $u \leq 2C_2/C_1$. 2) The phase volume of (22.20) is uniformly compressed

$$\partial \dot{X}/\partial X + \partial \dot{Y}/\partial Y + \partial \dot{Z}/\partial Z = - (1 + \text{Pr} + b) , \qquad (22.21)$$

i.e., an attracting set has zero volume. 3) The system is symmetric with respect to the substitutions $X \to -X$, $Y \to -Y$, $Z \to -Z$.

We shall look at how the behavior of the system depends on the parameter r (a Rayleigh number). When $r < 1$ a stable node at the origin $O(0,0,0)$ is the only equilibrium. When $r > 1$, the origin becomes a saddle and two stable equilibria are born from it $C^{\pm} = (\pm \sqrt{b(r - 1)}, \pm \sqrt{b(r - 1)}, r - 1)$. These correspond to stationary convection in the form of waves with positive directions of liquid rotation. These nontrivial equilibria exist for any $r > 1$, but are only stable (for $r^* = \infty$) when

$$r < r^* = \text{Pr}(\text{Pr} + b + 3)/(\text{Pr} - b - 1).$$

When $r = r^*$ unstable cycles existing in the vicinity of the equilibrium conditions C^+ and C^- float into them and transfer to them their own instabilities. When $r > r^*$ these equilibria become saddle-focus type states and the one-dimensional separatrix is stable. However, in a two-dimensional situation they are uncurled spirals. Thus,

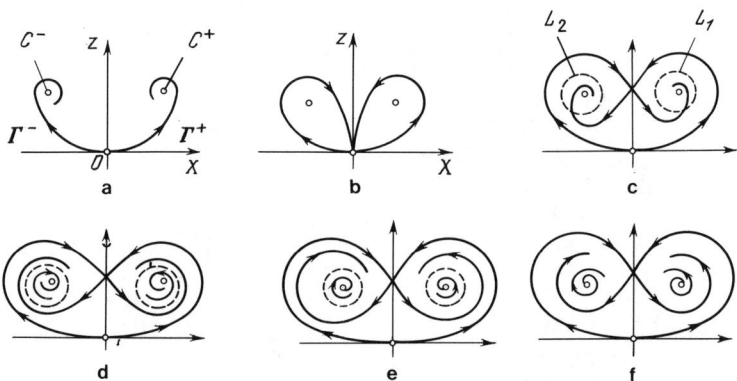

Fig. 22.20. Illustration of sequential bifurcations in the Lorentz system as the parameter r is increased: a) $1 < r < r_1$; b) $r = r_1$; c) $r_1 < r < r_2$; d) $r = r_2$; e) $r_2 < r < r^*$; f) $r^* \leq r$.

when $r > r^*$ all the equilibria inside the region in phase space of (22.20) are unstable. To discover what the trajectories tend to requires considerable nonlocal consideration and requires a numerical investigation [24, 25].

The change in the partition structure of phase space of (22.20) on the trajectories is best explained using Fig. 22.20, where the relative positions of the main components — the separatrix of the saddle $(0,0,0)$, the equilibrium position, and the limiting cycles — are all shown. These results were obtained for **Pr** = 10 and $b = 8/3$ and a variable r.

1. When $1 < r < r_1$, where $r_1 = 13.92$, the system has an equilibrium at the origin and two more equilibrium positions C^+ and C^-. The equilibrium at 0 is a saddle and has a two-dimensional stable manifold W and two unstable one-dimensional separatrices, Γ^+ and Γ^-, which tend to the equilibria C^+ and C^-.

2. When $r = r_1$, each of the separatrices becomes doubly asymptotic to the saddle 0 (Fig. 22.20b). When r becomes r_1, an unstable (saddle) periodic motion arises out of the closed separatrix loops, viz., the limiting cycles L_1 and L_2. In addition to these unstable cycles, a limiting set with a very complex organization is born; however it is not attracting (an attractor) and when $r_1 < r < r_2$, where $r \approx 24.06$, all the trajectories, as before, tend to $C^\pm$. The situation in Fig. 22.20c differs from the previous case in that now the separatrices Γ^+ and Γ^- do not tend to "their" equilibria, i.e., C^+ and C^-, respectively. When $r = r_2$ the separatrices Γ^+ and Γ^- are coiled around the saddle trajectories L_1 and L_2 (Fig. 22.20d).

3. When $r_2 < r < r^*$, where $r^* = 24.74$, the system has both stable equilibria $C^\pm$ and another attracting set with very complex trajectory behaviors, i.e., the Lorentz attractor (Fig. 22.20e).

4. When $r \to r^*$, as we have seen, the saddle cycles L_1 and L_2 are contracted to the equilibria C^+ and C^-, which for $r = r^*$ lose stability and when $r \geq r^*$, the Lorentz attractor is the only attracting set for system (22.20).

Thus, if r tends to r^* from small values, then stochasticity arises immediately by a step change in the Lorentz system, i.e., a rigid stochastic self-excited oscillation arises.

Transition through Alternation. In applications (Chapter 23) we often meet with the transfer to stochasticity which appears on the oscillogram as a gradual (with changing parameter) disappearance of the periodic oscillations due to a discontinuity in the stochastic flashes, that is alternation (Fig. 22.21a). This transition may also be described using a nonbijective mapping of a segment into itself. Let there be some mapping (Fig. 22.21b) whose characteristic feature is the presence of both attracting sections 1 and 2, and section 3. The intersection between this section of the mapping and the nonspectral one corresponds to two stationary states, one stable and the other unstable.

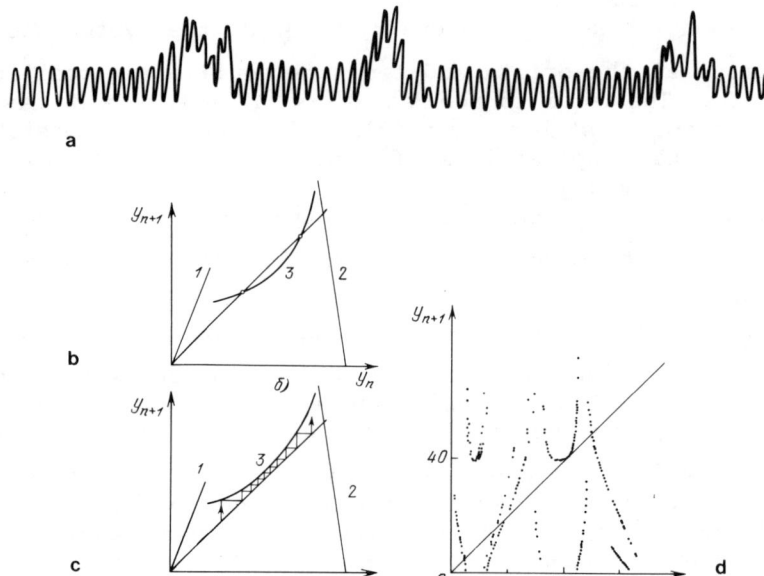

Fig. 22.21. Transfer to stochasticity through alternation: a) oscillogram of stochastic oscillations that arise directly after the transition to stochasticity; b) a model one-dimensional mapping corresponding to a pre-turbulent regime ($r < r_{cr}$); c) the mapping for r r_{cr} ; d) the mapping corresponding to the Lorentz model for $r = 166.2$.

Given that the mapping is protracting in most of the sections, the transitions in the system are very complicated. However, as $t \to \infty$, all the trajectories tend to a single attractor, viz., a stable stationary point which corresponds to a a stable periodic motion. Now, let section 3 rise above the nondescreteness as the parameter is changed. The stable and unstable stationary points merge and disappear, the stable periodic motion also disappearing (Fig. 22.21c). If a mapping which is deformed in this way turns out on average to be protracting, then new (or rather greater-fold) stable periodical points will not arise and the system will move stochastically.

Directly after the merger and disappearance of the stationary points (i.e., strictly periodic motion) a long transitional process is characteristic for the system and corresponds to trajectories passing through the region close to the periodic motions that have just disappeared ("the laminar" phase). After passing through this region, the system moves randomly (the "turbulent" phase) until it again enters this region, etc.

The mapping in Fig. 22.21d corresponds to the Lorentz system we have already discussed at large Rayleigh numbers, $r = 166.2$, $\sigma = 10$, $b = 8/3$. It follows from the form of this mapping that a transition to stochasticity through alternation is also possible in a Lorentz regime [26].[1]

Thus, the transition to stochasticity through alternation in this example is associated with the merger and consequent disappearance of a stable and an unstable periodic trajectory. This transition also occurs in multi-dimensional systems. The corresponding bifurcation causing the appearance of a complex behavior described in [27].

The Appearance of Stochasticity due to the Breakup of Quasiperiodic Motions. In self-excited oscillational systems with several degrees of freedom beats are observed outside the mutual synchronicity band. In the spectrum, these self-excited oscillations contain several incommensurable frequency (usually no more than two or three), and in the phase space they correspond to attracting unclosed windings about a torus, (respectively two or three-fold). When the system's parameters fall in the synchronicity region, a limiting cycle appears on the torus. Losses of stability by these limiting cycles due to one of the methods considered above may also cause the appearance of a strange attractor. We add that a strange attractor may arise as shown in [28] and due directly to breakup of a three-dimensional torus (see also [29]).

Here it is not possible to look into the corresponding mathematics. We leave the description of the physical processes corresponding to the disintegration of toruses to section 23.2, in which we consider the mechanisms by which turbulence arises in fluid mechanic flows.

---

[1] For motions from large numbers $Ra \geq 250$ the appearance of stochasticity due to sequential bifurcations of period doubling is also observed in a Lorentz system.

22.6 Dimensionality of Stochastic Sets

We mentioned at the beginning of this chapter that the dimensionality of the stochastic sets of a Hamiltonian system is the same as the dimensionality of the phase space of the system. However, the dimensionality of stochastic attractors may be considerably smaller than the dimensionality of the phase space of the dissipative system. This explains why both a very simple system, such as a nonlinear oscillator with friction excited by a periodic force, and the very complex system, such as a hydrodynamic flow in a cell (section 23.2), demonstrate the same transition properties.

We have already said that all the trajectories inside a stochastic set must be unstable. They cannot be unstable at once in all directions because this would lead to an unbounded growth in volume, i.e., the attractor would cease to be an attractor. This means that the unstable trajectories inside a bounded phase volume can only be saddle because they are unstable in one direction and stable in another (these directions changing, perhaps, along the trajectory). The rate at which trajectories scatter in each direction is characterized by the positive Lyapunov indices averaged over the trajectories λ_j ($j = 1, 2, \ldots, s$ where s is the number of unstable direction), and the rate at which the trajectories approach each other is characterized by negative indices, λ_j ($s < j \leq n$, where n is the dimension of the phase space). Let us recall (Chapter 15) that the λ_j are equal to the $\ln[l(\tau)/l(0)]$ averaged over the trajectories, where $l(0)$ and $l(\tau)$ are respectively the distances from the disturbed trajectory to the initial trajectory at times 0 and τ (Fig. 22.22).

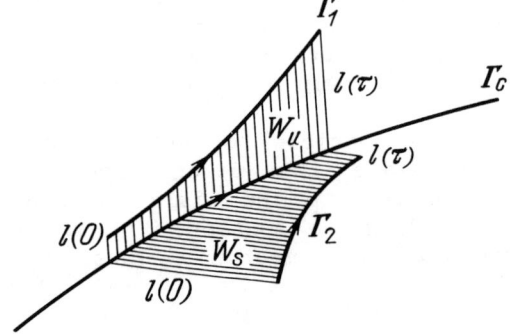

Fig. 22.22. A definition of Lyapunov indices (Γ_s is a saddle trajectory, $\Gamma_{1,2}$ is a disturbed trajectory, and W_s, W_u are the stable and unstable manifolds).

Because the system is dissipative we have

$$\sum_{j=1}^{s}\lambda_j + \sum_{j=s+1}^{n}\lambda_j = \text{div } u < 0.$$

We arrange all the indices in decreasing order: $\lambda_1 > \lambda_2 > \ldots > \lambda_n$, then the characteristic of a stochastic set, called its dimensionality, is determined thus [30]: $D = m + d$, where $\lambda_1 + \ldots + \lambda_{m-1} > 0$, $\lambda_1 + \ldots + \lambda_m < 0$, and d is determined from the relation $\lambda_1 + \lambda_2 + \ldots + \lambda_{2m-1} + d\lambda_m = 0$ (clearly $0 \le d \le 1$). The quantity d is called the attractor's dimensionality fraction (d is sometimes called a fractal dimension [31]).

It can be seen that the dimension of a strange attractor is dependent also on the number of unstable directions and on the summed rate at which the trajectories diverge from them.

It is important to find the link between the dimension of a stochastic set and the parameter characterizing the degree to which the system is nonequilibrial (e.g., the Reynolds number in fluid mechanics). However, so far we only have initial, very upper estimates of this.

If $D \ge 2$, then the phase trajectories forming an attractor lie in a thin layer close to some surface.[2] Motion on an attractor can, therefore, be described approximately (omitting the thickness of the attractor) by a one-dimensional Poincare mapping linking the coordinate of one intersection between a trajectory belonging to the attractor and the secant plane with the coordinate of the next intersection $x_{k+1} = F(x_k)$. The attractor in the Lorentz system is one with $D - 2 = d < 1$. This is why all the known bifurcations in the system are so well described by one-dimensional mappings.

Thus, any dissipative system the dimensionality of whose stochastic matrix is greater or equal to two must demonstrate transitions to stochasticity which are described by one-dimensional mappings independent of the dimensionality of the phase space.

The quantity D characterizes the proximity of a strange attractor in a weakly dissipative system to the stochastic set corresponding to a Hamiltonian system. These proximities, including proximity by statistical parameters, occurs when $D < n$ (n is the dimension of the phase space).

We present another example. Consider the stochastic

---

[2] In reality, the trajectories lie on an infinite number of surfaces because the attractor's structure is Cantor.

self-excited oscillations in a parametrically excited nonlinear oscillator [32]:

$$\ddot{x} + h\dot{x} + (1 - b \cos \Omega t)x + x^3 = 0 . \qquad (22.22)$$

Here h characterizes the dissipation and b is the size of the external field.

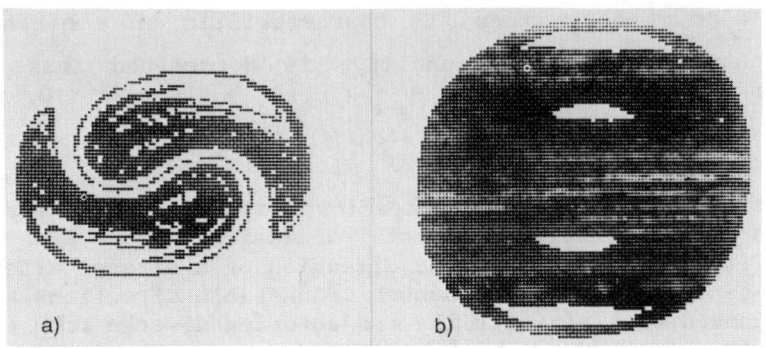

Fig. 22.23. Phase plane diagrams of the stochastic set of (22.22) on the secant plane: a) a strange attractor for $h = 0.12$, $b = 25$; b) the stochastic set corresponding to the Hamiltonian system ($h = 0$).

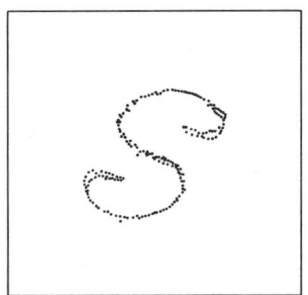

Fig. 22.24. The strange attractor of (22.22) approximately described by a one-dimensional mapping ($D = 2.22$, $h = 0.9$, $b = 17$, $\Omega = 2.04$).

A numerical investigation of this system can easily be done by constructing a Poincare point mapping of the secant plane $t = \text{const}$ into itself with period $T_0 = 2\pi/\Omega$ (Chapter 15). Recall that N points correspond to stable periodic motions with period NT_0 on the secant plane $(x, \dot{x})$. The stochastic set in $x, \dot{x}, T$-phase space of (22.22) corresponds to a complex point set on the secant plane. When $h > 0$ this set is an attractor. The phase-plane diagrams on the secant of one such attractor (for $h = 0.12$, $b = 25$) and the stochastic set of the Hamiltonian system ($h = 0$, $b = 25$)

are shown in Fig. 22.23. The dimensionality of the attractor arising from the stochastic set of the Hamiltonian system is ($D = 2.78 \approx 3$), which explains why the phase-plane diagrams in Fig. 22.23a are so close.

In another limiting case $(D - 2) \to 0$, the strange attractor of (22.22) is approximately described by a one-dimensional mapping [32] (Fig. 22.24).

CHAPTER 23

THE ONSET OF TURBULENCE

23.1 General Remarks

In the last chapter we discussed the onset of stochasticity only for simple systems, that is systems with few degrees of freedom. It seems obvious that the existence of stochastic motions unrelated to the effect of fluctuations or noise must be even more widespread in distributed systems. Indeed, stochastic motions of media or fields is very widespread in nature. Possibly the most important example of such motion is the random wandering flow of liquid that arises when the flow rates are large and there are no random external forces or fields (hydrodynamic turbulence).

At the same time, the presence of an infinite number of (or even simply very many) degrees of freedom in the system makes an understanding of the mechanism and nature of stochasticity in such a case very difficult, if only because there are many different nonlinear regimes in such systems which might arise for very close initial conditions. The effect in these situations of even weak noise causes very complicated and confusing system motions, the statistics of which only weakly depend on the statistics of the activating noise. These flows are observed experimentally, in particular in fluid mechanics. We shall not discuss this type of flow in this chapter and shall instead concentrate on random motions in deterministic distributed systems, particularly on the mechanisms by which hydrodynamic turbulence arises, the mathematical image of which being the strange attractor.

We shall specify straight away that turbulence is a stochastic self-excited oscillation in a distributed system, i.e., a random motion of a nonlinear dissipative medium or field induced by a nonrandom energy source.

The study of turbulence arouse in the middle of the last century when a number of contradictions were noted between theoretical fluid mechanics (with the Navier--Stokes equation) and applied problems on the flow of fluids. For example, it is known experimentally that the resistance to motion of a liquid in a pipe grows as the square of the average (over cross-section) velocity for large flow rates

(Chezy's law). It followed from the theory, however, that the resistance should grow in proportion to the velocity (Poiseuille's law). The first attempt to resolve these contradictions was made by Reynolds, who in 1883 published the results of experiments with dyed streams in flows and he introduced the number $Re = VD/\nu$ (D is the diameter, V is a velocity, and ν is the kinematic viscosity). He was the first to associate Poiseuille's law with laminar flow and Chezy's law with turbulent flow. He established that laminar flow is stable only up to $Re < 2000$ and that at large values of Re turbulence occurs. Thus, for water flowing in a pipe 1 cm in diameter at room temperature, laminar flow usually ends at the average flow rate of ~ 30 cm/s.

Even though turbulence was first defined almost a century ago, its study, which includes an investigation into the nature of random flows of a nonlinear medium and the establishment of methods for describing it consistently, has remained one of the most attractive and intriguing topics in classical physics.

The main topic associated with turbulence, namely the unordered chaotic motion of a continuous medium in all of its physical and other manifestations, has always been and remains its nature, i.e., the reasons for and mechanisms of the appearance of chaos.

From time to time various models have been enthusiastically championed and they were purported to explain the mechanism by which turbulence arises in nonlinear media. However, their inadequacies were discovered relatively quickly. The longest surviving model was the Landau–Hopf model which represented the appearance of turbulence as a long chain of sequential instabilities, with new degrees of freedom being excited at each step such that eventually the motion becomes very complex and confusing.

This idea, viz., that in order for an self-excited oscillation system to transfer to a turbulent state a very large number, if not infinite, degrees of freedom has to be excited, is very widespread. It is clearly associated with the concept, which we have discussed, of the stochasticity of dynamic systems, which was formulated in statistical mechanics. Although the motion of each individual particle in a gas is in principle known and predictable, the motion of a system consisting of a large number of particles (even noninteracting ones) is so complicated that a dynamic description is senseless; a statistical description is therefore required. The self-excited oscillation nature of motion in a field or medium is significant, according to this idea, only at the stage where stationary fluctuations are established, that is, an equilibrium between the removal of energy from sources (e.g., the average flow) and dissipation determines the intensity of the "self-excited oscillational modes". When the regime of such "gas self-excited generators" has become established it would seem that it should not differ from an ideal gas. Similar ideas lay at the heart of the models for the onset of turbulence suggested in 1944 by Landau [1] and independently in a somewhat different form in 1948 by Hopf [2].

According to the Landau–Hopf model, turbulence arises as the Reynolds number is increased due to a chain of sequential bifurcations, which lead to the establishment of quasiperiodic motions $u(t) = F(\omega_1 t, \ldots, \omega_N t)$, where the function F has period 2π in each argument, and the ω_i are incommensurable frequencies. The first bifurcations in this chain are very simple: initially the stable equilibrium becomes unstable and simultaneously a stable limiting cycle forms in its vicinity (thus ω_1 appears), and then the resultant periodic motion loses its stability. In the vicinity of the stable cycle that has just disappeared a two-dimensional manifold, i.e., a torus, the frequency of the windings about which is incommensurable with the fundamental frequency (and so ω_2), appears. After this the two-periodic motion becomes unstable and a three-dimensional torus arises (thus ω_3 appears), and so on. For large N this quasiperiodic process would indeed appear to be random, in particular, its autocorrelation function would decay rapidly (as $1/\sqrt{N}$), and the time to the next maximum (the Poincare return period) is $T \sim \exp(\alpha N)$, where $\alpha \sim 1$ [3].

Although natural from the point of view of accepted ideas, modeling turbulence in the form of a "gas" of self-excited oscillational modes with incommensurable frequencies nevertheless turns out to be only partially valid. The problem is that allowing for even weak interaction between the "particles" in such a "gas" can lead to the instability of the multifrequency, quasiperiodic motions we are interested in. The result of the breakdown of this motion, which can be represented in phase space by an unbroken winding about a torus, both periodic motions, namely a limiting cycle, and true stochasticity, namely a strange attractor, may occur. That small changes in the parameters of an self-excited oscillational system can cause a quasiperiodic motion to become a periodic one was known quite a long time ago and we have come across it as the phenomenon of synchronization (section 16.3). That a strange attractor can arise when the quasiperiodic motions breakdown, i.e., that a motion with a discrete spectrum can be replaced as the result of the usual bifurcations by a motion with a continuous spectrum – stochastic motion – was shown recently by Ruelle and Takens [4].

Hydrodynamic turbulence described by the Navier–Stokes equation has much in common with the motion of a dynamic system described by ordinary differential equations, which we discussed in the previous chapter. The link is due to the viscosity, which deprives highly numbered modes of independence.[1] Hopf even hypothesized that the whole set of trajectories of the Navier–Stokes equation (whose phase space is infinite-dimensional) are attracted to a finite-dimensional set. It would immediately follow that as $t \to \infty$ the motion of a liquid must be described by finite-dimensional equations. This hypothesis has not yet been proved, but it seems quite natural if one takes into account that the viscosity hinders the existence of small-scale perturbations. We add that the basic bifurcations observed for the Navier–Stokes equation have a finite-dimensional character [5]. This, for example, is true of the transition from a stationary stable flow to a periodic one (the inception of a limiting cycle equilibrium), and the establishment of a two-period flow (the inception of a two-dimensional torus) etc. Therefore there is every reason to believe that the following bifurcations, and the transition to unordered flow will also be finite-dimensional for many fluid mechanic problems.

23.2 The Occurrence of Stochastic Self-Excited Oscillations in Experimental Fluid Mechanics

We have seen that even in simple systems there is nearly always a problem of distinguishing between truly inherent stochasticity that is caused by the system's dynamics and the stochasticity due to external noise. The problem becomes especially acute in complex (number of degrees of freedom ≥ 10) and distributed systems. In the end, a final decision can only be made by comparing theory (in which the real stochasticity is observed) with experimental data. Since inherent stochasticity arises due to completely determined bifurcations complicating the spectra, and because the individual realizations in developed stochasticity and close at $t = 0$ exponentially diverge as t increases, it is at this point in time that we must pay most attention in the experiments.

We shall now describe several experiments demonstrating the qualitatively various ways in which hydrodynamic turbulence arises and which correspond to the various ways strange attractors arise, as discussed in the previous chapter.

<u>Sequence of Doublings</u>. The power spectrum of a heat flow in a layer of liquid helium heated from below is shown in

---

[1] Remember that such modes correspond to small-scale fluctuations.

Fig. 23.1 [6]. When $Ra > Ra_1$, a rolling convection arises and when $Ra > Ra_2$ a simple periodic regime with modulated heat flows becomes established. As the heating is increased, there is a sequence of regimes and the subharmonics which are the multiples $1/2f$, $1/4f$ etc. (f being the frequency of the periodic motion) appear in the spectrum (Figs. 23.1a and b). Then, when $Ra > Ra_\infty$, the discrete spectrum becomes continuous but a large number of peaks with frequencies $mf/2n$ remain. This continuous spectrum also evolves in discrete steps as Ra gradually increases. The inverse bifurcation of doubling is observed and at each bifurcation the number of peaks decreases, the remaining ones widen (the peaks in the spectrum disappear faster with respect to Ra the higher the number of the corresponding subharmonic). The term "inverse bifurcation of doubling" can be explained as follows: after a critical point Ra_∞ an attractor appears in the phase space of the system being investigated and as $Ra_{-2n} > Ra_\infty$ the attractor is positioned as if on $2n$ coils of a continuously widening "tape", and thus the exponential divergence of the trajectories belonging to the attractor is realized. At some point in time the tape gets folded in two (Fig. 22.16) and closes in on itself (thus the returnability is completed).

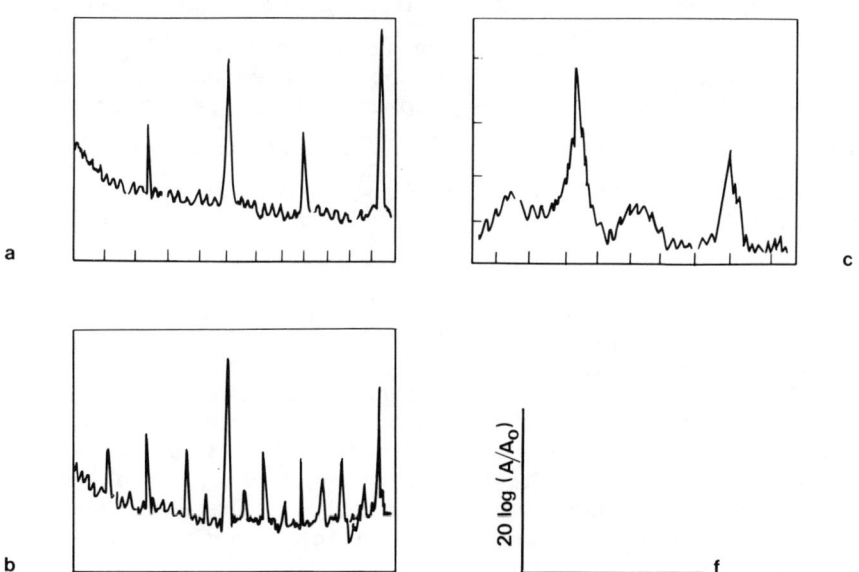

Fig. 23.1. Power spectrum of a heat flow with convection in a layer of liquid helium: a,b) doubling periods as the Rayleigh number at the moment of inception of turbulent convection increases; c) noise spectrum (beyond the transition point).

When $Ra_{-2(n-1)} > Ra_{-2n}$, the number of turns on the tape halves and when $Ra_{-2(n-2)} > Ra_{-2(n-1)}$ it halves again etc., i.e., it is as if a "blurred cycle" with double the period transfers its stability to "cycles" that are blurred twice as much but have half or less the period. These inverse bifurcations also have universality [7]: $(Ra_{-2n} - Ra_\infty) \sim (4.669...)^{-n}$.

The fluid mechanics, if only in a very narrow parameter region, reduces to one-dimensional mapping in the form of a parabola. When the parameters are changed, they frequently become more complex or get higher dimensions (cf. section 22.6). Therefore it is not surprising that in real flows parallel to the chain of doubling bifurcations with one period other motions with incommensurable periods appear and disappear. Such a possibility is illustrated in Fig. 23.2 [8], in which the spectrum of the rate of convective flow at a point is shown.[2] Figures 23.2 a-d) show how turbulence convection arises due to a sequence of periodic doublings of a motion with period f_2.[3] A regime of considerable aperiodic convection is shown in Fig. 23.2e (Ra/Ra_{cr} = 36.9). However, Fig. 23.2f is now the most interesting for us as it contains the spectrum of a flow at the same Rayleigh number that is in Fig. 23.2c (Ra/Ra_{cr} = 27.0), which arose under different initial conditions, namely going from large Rayleigh numbers downwards. It can be seen that another frequency f_* that is incommensurable with f_2 has appeared and its interactions with f_2 and its harmonics and subharmonics have considerably complicated the spectrum of the flow. A two-dimensional torus would exist in the phase space and its bifurcations would

---

[2] Various dynamic regimes can be easily analyzed using the realization spectra (oscillograms) during numerical or laboratory investigations of a concrete system. This convenience arises because the changes in the nature of the flow, which are difficult to see on an oscillogram, e.g., the inception of a new spectral component, the transition from quasiperiodic to stochastic regimes, etc., are easily discovered in a spectrum and easily compared. The most frequently used are energy realization spectra, i.e., the square of the Fourier image of $u(t)$.

[3] This process is elegantly described by Feygenbaum's theory [7], which predicted that a ratio between the spectral intensities of the harmonics and subharmonics would be of the order of 8.2 dB. This is the ratio that was observed experimentally (Fig. 23.2d).

describe the changes in the character of the flow in this case.

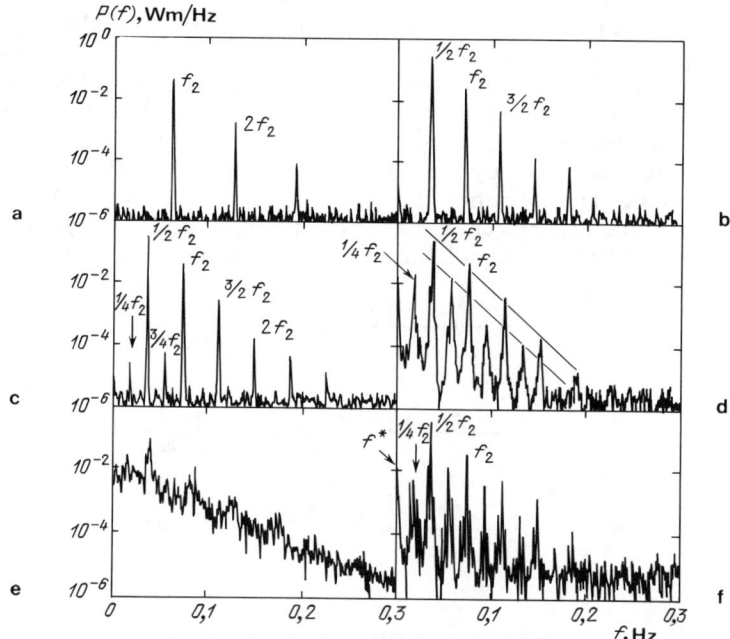

Fig. 23.2. Period doubling and hysteresis observed at the transition to turbulence during thermal convection in a cell; Spectra c and f were obtained for the same Rayleigh number but for different initial conditions. The values were Ra/Ra_{cr}: a) 21.0; b) 26.0; c) 27.0; d) 28.0; e) 36.9; f) 27.0.

Transition to Stochasticity from a Beat Regime. Breakdown of a Two-Dimensional Torus. The nature of the transition to turbulence, as we have noted, considerably depends on the geometry of the flow. In particular, for convective flow in a cell the dimensions are crucial. Thus, in experiments [8] with water (at temperatures from 10°C to 90°C and Prandtl numbers from 9 to 2) several very different routes to chaotic convection were observed when the geometry of a flat horizontal cell was changed. These routes are shown schematically in Fig. 23.3 [8]. As the Rayleigh number was increased, various types of transition other than a doubling sequence were observed, thus: stationary state – periodic convection – quasiperiodic convection (with two or three incommensurable frequencies) – chaotic convection.

A two-frequency quasiperiodic regime usually transforms into the chaotic one via a mode synchronization regime with incommensurable frequencies and then the disappearance or loss of the stability of the resultant periodic motion. Here, two routes are known: 1) the appearance on a two-dimensional torus of a limiting cycle due to synchronization and the cycle then undergoes a sequence of period doubling bifurcations, this route being investigated experimentally in [10] and theoretically noted in [11]; and 2) the appearance on a two-dimensional torus, due to synchronization, of a stable and saddle cycle that merge and disappear. The properties of the stochastic set are determined either by the homoclinic structure that belongs to the saddle cycle or the complex multifolded structure of the torus itself [12].

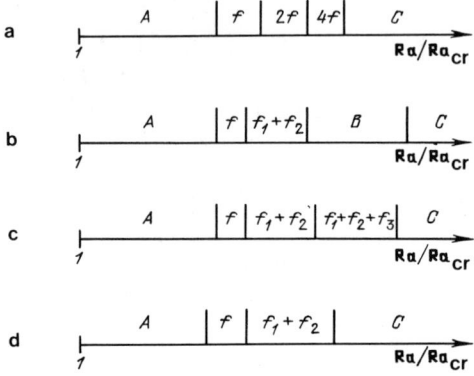

Fig. 23.3. Various routes to the inception of turbulent convection in a cell. The Prandtl number and the cell dimensions will change. A) static regime; B) regime of mode synchronization; C) chaotic regime. a) Onset of turbulence as the result of a sequence of doubling bifurcations ($Pr = 2.5$); b) inception of a two-dimensional torus - appearance of a stable cycle on the torus - disappearance of the cycle and onset of turbulence ($Pr = 5$); c) breakdown of three-dimensional torus ($Pr = 5$); d) breakdown of two-dimensional torus ($Pr = 5$).

Onset of Turbulence with the Breakdown of a Three-Frequency Quasiperiodic Regime. The breakdown of three-dimensional torus is one of the possible routes to turbulence in closed flows, namely in cells and channels. Besides occurring in thermal convection [9,17], which we have already covered, this transition has also been observed [12] in Couette flows between the cylinders with a rotating inner cylinder.[4] When the rotation speed of the cylinder is

---

[4] This flow is very like convection with the buoyancy force being replaced by the centrifugal force. When the parameter $T = \Omega^2 r^2 l^2 / \nu^2$ (T is the Taylor number, Ω and r are the

increased (the Taylor number, or the Reynolds number, which is proportional to it, is increased) perturbations develop on the Taylor vortices in the form of bending azimuthal waves with $\exp(imz - in\theta)$ (Fig. 23.4), where m is the number of Taylor vortices (the number of modes along the vertical), and n is the number of wavelength on the toroidal vortex. It can be seen from the power spectra of the speed shown in Fig. 23.5 that when the rotational velocity is increased a periodic regime of azimuthal oscillations replaces the two-periodic (beat) regime and then a third frequency arises. The frequency spectrum widens. The bifurcation leading to the decay of the three-dimensional torus into an attracting stochastic set, i.e., the strange attractor corresponding to this transition, has now been found by mathematicians [14].

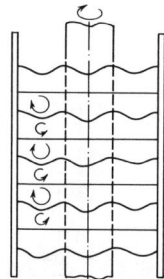

Fig. 23.4. Modulation waves on the Taylor vortices in a Couette flow between cylinders with the internal cylinder rotating.

Transition through Alternation. Similarly in the experiments on thermal convection in a cell, but at large Prandtl numbers (the Prandtl number for transformer oil, for example, can be several hundred), a completely different route to an irregular flow has been observed [15]. When the Rayleigh number is increased, a periodic convection regime is replaced by one with sharp random peaks that occur at intermittent lengths along the regular section and then (as **Ra** increases) these flashes become more and more frequent and the flow turns into an irregular one (the oscillograms in Fig. 22.21a).[5] Mathematically, this transition is a

---

rotation speed and radius of inner cylinder, and l is the gap between the cylinders) exceeds a critical value T_1, the flow becomes unstable and a structure arises in the form of threads of Taylor vortices inside the cylinder of a torus [13].

[5] A similar phenomenon of intermittent "turbulent" and "laminar" phases in a convective flow has also been observed at moderate Prandtl numbers [9].

bifurcation with the merger of a stable and unstable limiting cycle accompanied by the appearance of a strange attractor (Chapter 22).

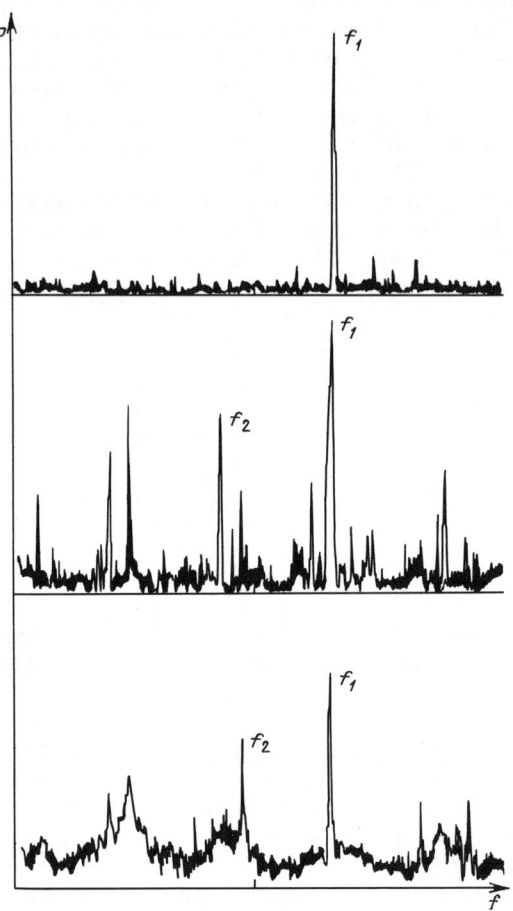

Fig. 23.5. Transition to turbulence in a cylindrical Couette flow (by the breakup of a quasiperiodic regime of azimuthal oscillation).

The routes to the occurrence of turbulence that have been illustrated do not exhaust all the possibilities, even for flows in channels and cells, namely internal flows. It is easy to imagine that any complication of the geometry, for example, a thick convective layer rather than a thin convective layer or a spherical flow rather than a cylindrical flow, must lead to the appearance of new oscillating modes for the flow, which generally speaking will not always synchronize. The spectrum of the pre-turbulent regime may then contain not just three, but four or more incommensurable frequencies [16]. Thus, Landau's model for

THE ONSET OF TURBULENCE

the onset of turbulence due to a sequence of new incommensurable frequencies appearing in the spectrum as **Re**, **Ra**, or **T** are increased is justifiable at the first stage, although the turbulence arises anyway due to the breakdown of the quasiperiodic motions (n-dimensional toruses) and the appearance of attractors, which characterize the exponential instability of nearly all the trajectories in the phase space, rather than due to the complication of the motion, as suggested by Landau.

23.3 Stochastic Modulation

Up to now we have been discussing the onset in a distributed system of stochasticity characterized by a continuous spectrum and including low frequencies, including those as $\omega \to 0$. A situation often encountered in experiments is the appearance of stochastic fluctuations against the background of harmonic oscillations, i.e., stochastic modulation. Since this phenomenon has a variety of applications, we shall dwell on it in some detail.

Stochastic modulation may arise in an autonomous wave system due to the development of inherent instability. An example of such a system is the backwards-wave tube. A transition to an oscillation regime with stochastic modulation has been observed in such an electron generator [17]. The block diagram of the generator is given in Fig. 23.6.

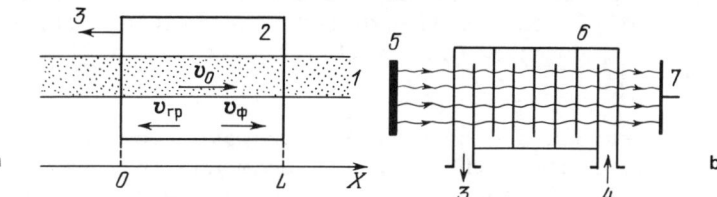

Fig. 23.6. a) The model being analyzed and b) a schematic diagram of the backwards-wave generator: 1) electron beam; 2) medium; 3) output device; 4) input device; 5) electron gun; 6) delay system; 7) collector.

The electron beam moves past a delay system along which waves propagate with longitudinal electric fields. The parameters of the system are such that the phase velocity of the waves coincide at some frequency Ω with the velocity of the beam $v_{ph}(\Omega) \approx v_0$, and the group velocity is directed in the opposite direction. The output signal is taken from the end of the delay system into which the input signal was fed. Thus, when the wave perturbations of frequency $\omega \approx \Omega$

interacts with the electron beam, a distributed feedback is implemented and an absolute instability arises leading to a stationary degeneration regime (Chapter 7). The nature of this regime is determined by a single parameter similar to the Reynolds number for a hydrodynamic flow: $\mathcal{L} = \beta l (IK/4U)^{1/3}$, where β is the wave number of the wave synchronous with the beam, l is the length of the interaction, I is the constant component of the beam current, U is the accelerating potential, and K is a system parameter with dimensions of resistance. The sequence of bifurcations observed in the system on the route to stochastic modulation (the parameter $\mathcal{L}$ is increased) is given in Fig. 23.7. When $\mathcal{L} \geq \mathcal{L}_{cr}$, a stochastic regime characterized by a continuous spectrum arises.

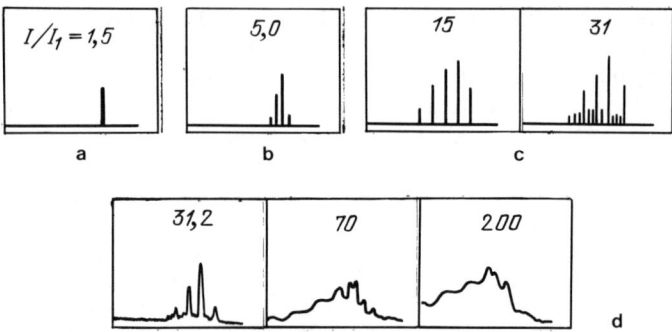

Fig. 23.7. Spectra of the output signal from a backwards-wave tube for different self-excited oscillating regimes: a) single frequency; b,c) multifrequency; d) stochastic oscillations.

In experiments with a backwards-wave tube the parameters of the delay system, the electron beam and the feed etc. were changed and it was observed that the nature of the transition to stochastic modulation did not change qualitatively and was determined in the various experiments by the single parameter $\mathcal{L}$. This similarity indicates that fluctuations (e.g., the noise of the electron beam) was not crucial for the appearance of the stochastic regime in a backwards-wave tube. The stochastic self-excited oscillations could be broken up using a synchronizing external signal [26]. This synchronization had the greatest effect if the periodic external force was at the frequencies corresponding to the left-hand satellites in the spectrum of the pre-turbulent regime. The backwards-process was observed, namely, when the periodic signal affected the backwards-wave tube in the periodic regime the discrete spectrum corresponding to the periodic modulations became a continuous spectrum at a large detuning between the frequency of the input signal and the frequency of the satellites. All these changes took place at the same beam current (i.e., for the same fluctuations in the electron beam), which also indicated that the stochastic regime was dynamic in origin.

A series of calculations and experiments into the processes leading to the appearance of oscillations in a Backwards-wave tube [27] established a clear link between the onset of chaos and the appearance of unstable motions of the system with respect to the perturbations in the initial conditions. As was shown in Chapter 22, a quantitative expression of this is a significant positive Kolmogorov entropy (the attractor must have at least one positive Lyapunov index). Consistent theoretical and experimental estimates of the Kolmogorov entropy are given in [27] and it is shown that the level of instability on the attractor grows with increasing $\mathcal{L}$.

A numerical experiment on the output signal from a backwards-wave tube showed that the resultant stochasticity corresponded to a strange attractor with a finite fractional dimensionality α, and for instance for $\mathcal{L} = 1/2\pi$, the value $\alpha = 5.5$ was obtained.

In a system containing an electron beam and a backwards-wave tube, stochastic modulation can be described in detail by averaged equations obtained from the equations for the field and beam [18]. The interaction between the beam and field of the backwards wave can only be effective if some of the spatial harmonics had velocities close to the electron velocity. Thus, if the field of the harmonics could be written as

$$E(x, t) = \text{Re}\{\mathcal{E}(x, t)\exp[i\Omega(t - x/v)]\} ,$$

where the frequency Ω is determined from the synchronicity condition $v_{ph}(\Omega) = v_0$ (v_0 is the beam velocity), then the following equation could be obtained for the normed slowly changing amplitude $F \sim \mathcal{E}$:

$$\frac{\partial F}{\partial t} - \frac{\partial F}{\partial x} = -\frac{\mathcal{L}}{\pi} \int_0^{2\pi} e^{-i\omega(\alpha)} d\alpha , \quad \frac{\partial^2 \theta}{\partial x^2} = -\mathcal{L}^2 \text{Re}(Fe^{i\omega(\alpha)}) \quad (23.1)$$

with boundary and initial conditions

$$\theta(\alpha)\big|_{x=0} = \alpha , \quad \partial\theta/\partial x\big|_{x=0} = 0 , \quad F\big|_{x=1} = 0 , \quad F\big|_{t=0} = F_0(x) . \quad (23.2)$$

These equations indeed contain only one parameter, viz., $\mathcal{L} = (\Omega l/v_0)(IK/4U)^{1/3}$.

A numerical analysis of (23.1) and (23.2) for $2.0 \leq \mathcal{L} \leq 2.9$ established an unmodulated regime and for $\mathcal{L} \geq 2.9$ a periodic modulation arose, after which when $\mathcal{L} > \mathcal{L}^*$ the process became aperiodic. Spectral processing of the results indicated that the power spectrum in the regime for stochastic modulation was in agreement with the experimental results and the autocorrelation function of the process fell relatively quickly.

Even though in this example the model was simplified (by averaging over the high-frequency oscillations) with respect to the initial equations, the main feature of the equations, that is the infinite number of degrees of freedom, was retained. True, in a numerical account the system was replaced by a finite-dimensional one although it had a large number of modes.[6]

When investigating the complex dynamics arising due to the development of secondary instabilities against the background, for example, of a periodic oscillation, constructing the mode models directly from the initial equations becomes extraordinarily complex. The model itself must frequently be constructed by computer. The development of qualitative ideas, or the construction of a theory, on a physical level becomes much more difficult. In similar situations, it is useful to have a purely phenomenological model based on elementary physical concepts and on experiment. We shall discuss one such model now [19]. It was constructed to describe the onset of chaotic modulations in Taylor vortices in a cylindrical Couette flow.[7]

In the experiment the following sequence of power spectra of the flow were observed as **Re** was increased. When **Re** ≤ 1200 azimuthal waves were excited against the background of Taylor vortices (the boundaries of the vortices bent). Increasing the angular velocity of the inner cylinder caused a series of gradually more complicated spectra and at **Re** ≈ 1270 the peaks widened into a power spectrum

---

[6] The number of degrees of freedom which should be taken into account to construct the model depends on the question being posed. If it is necessary to establish whether chaos can arise in principle with self-excited modulation, then only three modes are necessary. If, however, a detailed description of all the observed supercritical transitions and the evolution of the spectrum of the stochastic process is attempted, then the dimensions of the model must be considerably increased.

[7] The Couette flow was investigated in [19] with a much larger gap between cylinders than in [12], in which no modulation was observed for the parameters of the Taylor vortices along the vertical.

corresponding to a chaotic flow. As **Re** grows these peaks continued to widen and finally the spectrum became almost continuous.

In a phenomenological description of an experiment using a model we can use experiments directly to determine $a_n(t)$, the amplitude of the boundary bending between the vortices in pairs with numbers n [19], i.e.,

$$\frac{da_n}{dt} = \gamma a_n + (i\eta - \rho)|a_n|^2 a_n + \frac{\alpha + i\beta}{4}\left(a_{n+1} + a_{n-1} - 2a_n\right) \quad (23.3)$$

($n = 1, 2, \ldots, N$). These equations were compiled by taking the first two terms of each equation from the right-hand side of Landau's well-known equation [1]. If $\gamma > 0$, then this equation describes the growth and stabilization of the bending oscillations in the vortices due to the self-influencing and interactions with the averaged flow. The model (23.3) also takes into account interactions between the vortices. Because the experiments showed that these interactions were small, it is natural to limit the terms to first order ones in the amplitude. The coefficients of the model must, in principle, be determined directly from experiment. We add that (23.3) is the differential-difference analog of a nonlinear Schrodinger equation for a nonequilibrial medium [20].

When the small parameters are supercritical, the 30 equations in system (23.3) can be cut down to three and the model in [20] can be used. This model also shows chaotic modulation (Chapter 22). Further increases in the supercriticality lead to five equations, and so on. A strange attractor exists in all these models of turbulence, although as the Reynolds number is increased, the dimensionality of the model in which they are observed also increased.

In the investigation of distributed systems there is the question as to whether the similitude and universality principles are valid for their behavior close to the threshold of chaos and in the scenarios of the transition to chaos, that is for behavior which has been established for simple systems (see Chapter 22). The observations of these scenarios in backwards-wave tubes has been covered above. Careful experiments with the self-excited stochastic oscillation generator suggested by V.Ya. Kislov et al. [28] have shown (see [29] in which the generator was a backwards-wave tube, resonance filter, and acoustic delay line connected into a closed loop) that depending on the depth of feedback and detuning of the filter this distributed system demonstrates all the scenarios for the onset of chaos known for simple systems, viz., successive period doubling bifurcations, the breakdown of quasiperiodic motions, the doubling torus bifurcation, and alternation.

23.4 Ideal Flow and Turbulence

The examples so far presented show that the transition from laminar flow to turbulent flow takes place when the Reynolds number (respectively the Taylor or Rayleigh numbers for rotating or convective flows) are increased, or, what is equivalent, the viscosity is decreased. At the same time, a practically inviscid flow (**Re** → ∞) may be laminar but turn into a turbulent one when any of the parameters are changed or there is an external perturbation, even if it is regular.

In the strict sense, turbulence, i.e., a stochastic self-excited oscillation, cannot exist in an ideal liquid because an attracting set (attractor) cannot exist in the phase space of the flow due to the absence of dissipation. However, investigations of stochastic ideal flows are of undoubted interest because several of their properties, for instance, the reaction to an external disturbance, model real flows at large Reynolds numbers.

We emphasize that the flow with **Re** → ∞ is very easy to transform to a turbulent regime, it being sufficient for a small flow perturbation to arise either as the result of an interaction with some other flow or external field.

Let us consider the interaction of a two-dimensional shear flow with an acoustic wave. In the absence of viscosity the liquid motion is described by

$$\frac{\partial \rho}{\partial t} + \mathrm{div}(\rho \mathbf{v}) = 0 \; , \quad \frac{\partial \mathbf{v}}{\partial t} + (\mathbf{v}\nabla)\mathbf{v} + \frac{1}{\rho}\,\mathrm{grad}\,p = 0 \; , \qquad (23.4)$$

where ρ is the liquid density, p is the pressure, and $\mathbf{v}$ is the velocity. The first of these is the continuity equation, and the second is the momentum balance for the element of liquid (Euler's equation). When searching for a solution, it is convenient to represent the total velocity $\mathbf{v}$ as the sum of two velocities, i.e., $\mathbf{v} = \mathbf{v}_1 + \mathbf{v}_2$, where $\mathbf{v}_1 = \mathrm{curl}\,\mathbf{A}$ and $\mathbf{v}_2 = \mathrm{grad}\,\phi$.

Physically this representation is easy to understand by considering two limiting cases: $\mathbf{v}_2 = 0$ and $\mathbf{v}_1 = 0$. In the first, (23.4) describes the flow of an incompressible liquid, and in the second, it describes the acoustic field. If $\mathbf{v}_1$ and $\mathbf{v}_2$ are simultaneously nonzero but small, than there will only be weak interaction between the acoustic and hydrodynamic velocity fields, and the interaction can be accounted for by the method of sequential approximations.

We shall restrict ourselves to one of the simplest models of a hydrodynamic flow, namely the periodic chain of point vortices. A similar chain was used to model the periodic distribution of vorticity arising in shear layers due to the development of instability [21]. This chain is unstable, and the double-period perturbations have a large increment. These perturbations caused two chains to be formed moving relative to each other. Using a well-known result from the theory of point vortices [22], it is possible to obtain equations describing this motion, i.e.,

$$\frac{dx}{d\tau} = -H \frac{\sinh y}{\cosh y - \cos x}, \quad \frac{dy}{d\tau} = +H \frac{\sin x}{\cosh y - \cos x}. \tag{23.5}$$

Now we use dimensionless variables x, $y = (2\pi/l)x'$, y', which are proportional to the components of the vector uniting the chosen pair of vortices from the two chains, $\tau = t 2\Gamma\pi/Hl^2$, Γ is the intensity of one vortex, l the period of the undisturbed chain, and H is so far an arbitrary parameter.

The nonlinear system of (23.5) is Hamiltonian with integral $\cosh y - \cos x = \text{const}$, and consequently only simple motions are possible in it. By equating the constant H to this integral, motion of one route into another (for $H = \text{const}$) can be described by the equation of a compound pendulum

$$\ddot{z} + \sin z = 0, \tag{23.6}$$

where $z = x + iy$. In order to derive this equation we must multiply the second equation in (23.5) by i, differentiate both equations with respect to τ, and add. The solution to (23.6) can be written in the form

$$z(\tau) = \begin{cases} 2 \, \text{am}(H\tau/2 + \theta_0; 2/H), & H > 2, \\ 2 \, \text{arcsin}[h \, \text{sn}(\tau + \theta_0)2; H/2 & H < 2, \\ 2 \, \text{arcsin} \, \tanh(\tau + \theta_0), & H = 2, \end{cases} \tag{23.7}$$

where am and sn are Jacobi functions, the phase $\theta_0 = \theta'_0 + i\theta''_0$ is defined from the initial conditions, and $H > 2$, $H < 2$ and $H = 2$ correspond to motions outside the separatrix, inside the separatrix, and on the separatrix, respectively. Thus, this shear flow in a gas is described by a completely integrable system of equations and demonstrates one simple type of dynamics, namely the vortex streets which either curl around each other (embrace each other), or slide in opposite directions (for $H > 2$) (the phase-plane diagram is shown in Fig. 23.8).

We now allow for weak interactions between these

oscillations of a shear flow and an acoustic wave propagating normally to the shear flow. Let us consider the simple situation when the sound wave can be considered harmonic and given as

$$\mathbf{u} = \mathbf{y}_0 u_0 \sin(2ky - 2\omega t) \;.$$

Then by assuming that the spatial period of the chain is smaller than the wavelength of the sound wave ($kl \ll 1$) and that the amplitude of the sound wave is small ($M = u_0/c \ll 1$, where $c = \omega/k$ is the speed of sound, and M is the Mach number), the motion of one vortex chain in the field of another will be determined by the relative velocity of the chain in a field of the acoustic wave [23], i.e.,

$$\frac{dx}{d\tau} = -H \frac{\sinh y}{\cosh y - \cos x} \;, \tag{23.8}$$

$$\frac{dy}{d\tau} = H \frac{\sin x}{\cosh y - \cos x} + 2M \frac{\omega H}{\Omega} y \cos\left[\frac{2\omega H}{\Omega} \tau\right] \;, \tag{23.9}$$

where $\Omega = \Gamma/l^2$. Thus, we have come to a study of a nonlinear oscillator (23.6) on which a periodic field is acting. We have already discussed this situation (Chapters 15 and 22) and regions with a complex trajectory behavior can exist in the phase spaces (x, y, t) of such systems. The complexity of the motions in these regions is usually associated with homoclinic structures (Chapter 15). The existence of a homoclinic structure in our case may be determined using the Mel'nikov criterion, i.e., from the condition of the sign variable function

$$\Delta(t_0) =$$

$$= \int_{-\infty}^{\infty} \{u[x_0(t - t_0), y_0(t - t_0), t]V[x_0(t - t_0), y_0(t - t_0)]$$

$$- v[x_0(t - t_0), y_0(t - t_0), t]U[x_0(t - t_0), y_0(t - t_0)]\}dt,$$

where u, V, U, v are the right hand sides of (23.8), written in the form

$$dx/dt = U(x, y) + \mu u(x, y, t) = \partial\psi/\partial y + \mu u(x, y, t) \;,$$

$$dy/dt = V(x, y) + \mu v(x, y, t) = -\partial\psi/\partial x + \mu v(x, y, t) \;,$$

and $x_0(t - t_0)$, $y_0(t - t_0)$ is the solution to (23.8) when $\mu = 0$, which corresponds to the separatrix ($H = 2$). After

THE ONSET OF TURBULENCE

substituting (23.7) into $\Delta(t_0)$ and integrating, we find (for moderate $\sigma\omega/\Omega$)

$$\Delta(\tau_0) \approx M_0 \frac{5\sigma\omega}{\Omega} \text{ch}^{-1}\left(\frac{\sigma\omega}{\Omega}\right) \cos\left(\frac{4\omega}{\Omega}\tau_0\right) \equiv \delta_0 \cos\left(\frac{4\omega}{\Omega}\tau_0\right). \quad (23.9)$$

Thus, in our case the function $\Delta(\tau_0)$ is sign-variable, and in the phase space of (23.8) and (23.9) there is a region with stochastic behavior. The width of the stochastic layer (on the x,y secant) is determined by the value of δ, which is at a maximum at $\sigma\omega/\Omega \approx 1.2$ (the dimensions of the stochastic region close to the saddle are proportional to $\sqrt{\delta}$ (Fig 23.8)).

In concluding this section, we note that in a system of point vortices, which model two-dimensional flows in an ideal liquid, stochasticity arises even in the absence of external fields. The system becomes stochastic, for example, even from four vortices if their positions are asymmetric [24].

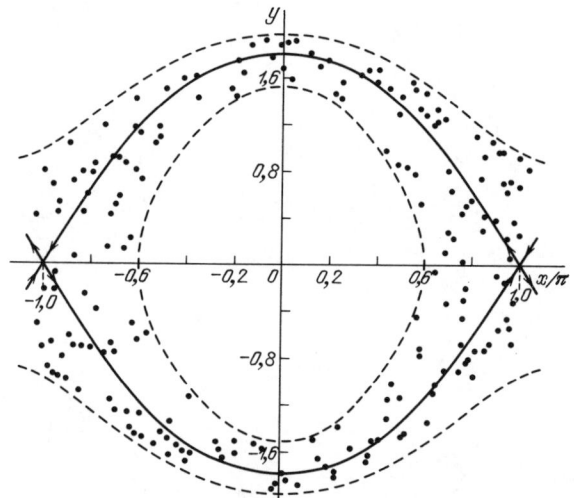

Fig. 23.8. Random wandering of the trajectories of systems (23.8) and (23.9) about the separatrix; the points on the (x,y) secant plane were obtained in terms of the period of the external field.

CHAPTER 24

SELF-ORGANIZATION

24.1 Main Phenomena, Models, and Mathematical Forms

The most widespread phenomena associated with self-organization (the appearance of spatial order from disorder, the formation of complex spatial structures in a homogeneous medium etc.) were first considered in the 1950s and 1960s in work on chemical kinetics and biology. For instance, waves in heart muscle were qualitatively described [1], models of morphogenesis [2] and autocatalytic Belousov–Zhabotinskii reactions [3] were formulated. At about the same time a theory for structure in hydrodynamic flows was constructed (Bernard cells under thermal convection and Taylor vortices between rotating cylinders [4]).

It was fairly quickly established that the occurrence of complex formations in nonlinear media or spatial ensembles with different natures required similar mathematical models and solutions [5,6,9]. This made it possible to transfer experience and knowledge (not for the first time in the theory of oscillations and waves) from, say, the investigation of combustion reactions to the analysis of population propagation in ecological situations or to the propagation of perturbations in heart muscle. As a result, new concepts and forms were worked out, e.g., dissipative structures, travelling pulses, and reverberators, and basic universal models describing the occurrence and existence of these structures began to crystallize [7,8,15,19-21,29]. In fact, a new field in the "nonlinear sciences" had arisen and was variously called nonequilibrium thermal dynamics [5,2], synergetics [6,28], the theory of self-organization [9,27], and self-excited wave theory [7,30].

A keen interest in the phenomenon of self-organization had been stimulated in physicists by biological problems. Self-organization is even observed in ensembles of relatively simple biological objects such as amoeba cells [10]. These cells produce the hormone cAMP once every five minutes although when there is a sufficient quantity of food, the cells do not respond to the hormone and live independently. Under very severe conditions one cell begins to accelerate its output of the hormone cAMP and synchronizes its output with its closest neighbors which in turn synchronize their output of the hormone with neighbors etc. Once excited by the hormone a cell begins to move in the direction of the exciter. Thus, two opposite motions arise i.e., diverging waves of stimulation or synchronization, and a converging motion of cells. This process finishes in the formation of aggregates or spores which can survive in extreme conditions.

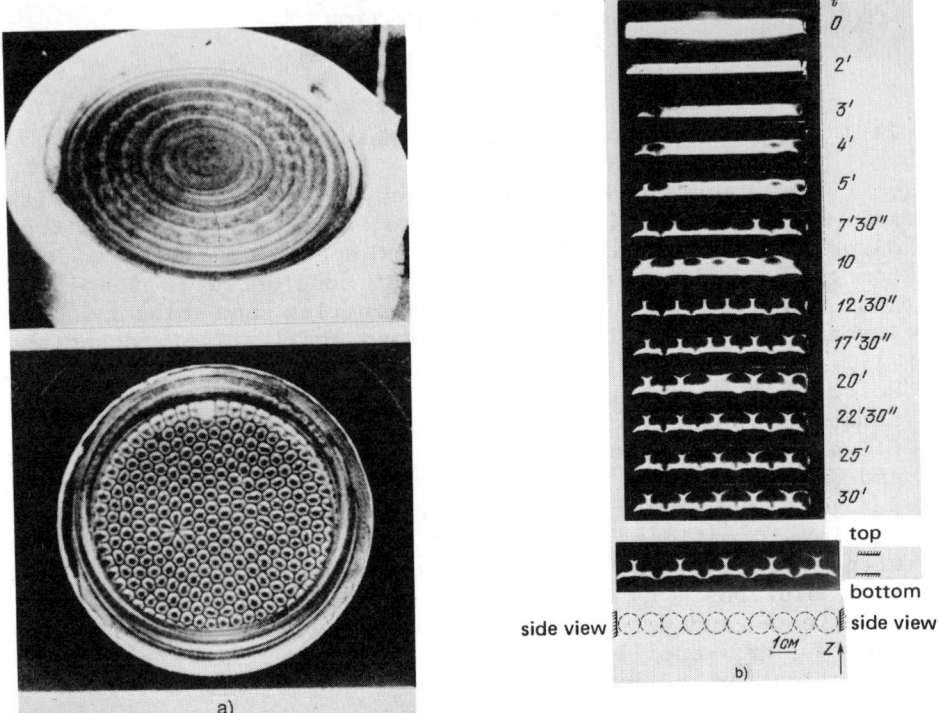

Fig. 24.1. Cell convection: a) structure of Bernard cells; b) the occurrence and establishment of a rolling structure under Bernard convection in a right-angled cell (side view).

The traditional physical example of self-organization is the occurrence in a layer of liquid heated from below of structures with hexagonal prismatic cells (Bernard cells, Fig. 24.1a). In order for this structure to form there must be disequilibrium in the nonlinear medium and its dissipation, as a result of the development of convective instability perturbations in the velocity and temperature fields grow in some interval of the spatial scales and then due to competition between the scales (this is only possible in the presence of dissipation) a grid can survive at some particular scale (Fig. 24.1b). The hexagons form due to synchronization between the grid phase and the spatial

SELF-ORGANIZATION

orientation (section 24.4). This synchronization is possible in liquids when its viscosity (surface tension or diffusion coefficient) depends on the temperature. A formal description of synchronization between different spatial modes is given in Section 24.4. Neither the scale of the grid nor the structure of the cell are in practice dependent on the conditions at the sides of the layer if the dimensions on the horizontal are large.

We shall define self-organization to be the establishment in a dissipative nonequilibrium medium of spatial structures (in general evolving over time) whose parameters are determined by the properties of the media and are only weakly dependent on either the spatial properties of the source of the disequilibrium (energy, mass, etc), or the initial condition of the medium, or the conditions at the boundaries. Thus, the loss of memory concerning the initial conditions and the direct relation between the structural parameters and the medium's properties are crucial for self-organization.

It can be seen from the examples that self-organization is a result of the development of spatially nonhomogeneous instabilities with a subsequent stabilization due to a balance between dissipative losses and incoming energy from the source of the disequilibrium. The occurrence of self-organization is similar to the establishment of self-excited oscillation. However, the result of the development of instability leading to self-organization may be purely "static" in that the spatial formations which arise do not change over time, that is they are dissipative structures (we add that they may also be stochastic [12]). Another difference is that the conditions for self-organization at the periphery of the nonequilibrium dissipative medium are not as significant as they are for self-excited oscillations.

Self-organization phenomena, even within the bounds of our definition, are very diverse. For example, we may include dissipative structures, single fronts (combustion waves [11], population waves [16,7]), pulses (in nerve fibers [13,14] and autocatalytic reactions [9]), and travelling centers and reverberators (heart muscle [17], cooperation between amoeba [10], depression waves in brain tissue and eye retinae [18]). This variety means that self-organization phenomena do not have a single mathematical form, such as a strange attractor for stochastic self-excited oscillations, or limiting cycles for periodic oscillations, instead there may be a limiting cycle for periodic dissipative structures, a strange attractor for stochastic structures, or separatrices going from one equilibrium to another for propagating fronts.

Nevertheless many phenomena can be described by the theory of self-organization using a single model which is mathematically expressed by the following nonlinear kinetic diffusion-like equations:

$$\partial u/\partial t = f(\mathbf{u}) + D\Delta \mathbf{u} . \qquad (24.1)$$

where **u** is the set of physical (chemical etc.) variables which determine its nonlinear kinetics in the absence of diffusion, and D is the matrix of the diffusion coefficients (in general D is also dependent on **u**, i.e., nonlinear diffusions).

We begin our discussion of self-organization phenomena with the topic of solitary fronts. To take a particular case, we shall discuss the establishment of the stationary propagation of a flame. In this case, we have an oxidation reaction during which heat is evolved. The combustion takes place in a comparatively thin region in which the chemical reaction proceeds, i.e., a region separating the cold fuel from the combustion products and moving with respect to the hot products at a constant velocity that is independent of the initial conditions. The combustion wave front corresponds to the particular solution of a system of ordinary differential equations for a stationary wave. In phase space, this solution traces a separatrix uniting two equilibria (Fig. 24.2), one of which corresponds to the variables in front of the front (the reaction hasn't started) and the other corresponds to variables behind the front (the reaction has finished).

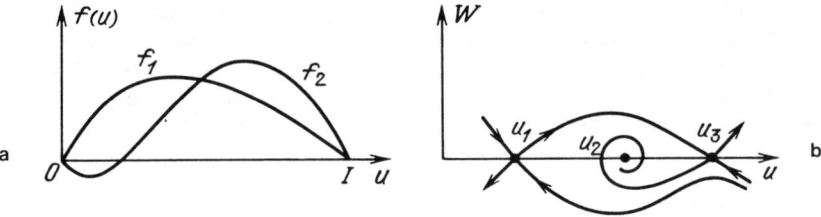

Fig. 24.2. a) Rate of change in u in a "point" system versus u for a wavefront propagating without limit (f_1) and with a limit (f_2). If $f(u)$ has five (or more) zeros, then several stable waves with various amplitudes may be excited in (24.2). b) The trajectories on the (W,u) phase plane for $f(u) = -\text{const} (u - u_1)(u - u_2)(u - u_3)$.

For an analytical description the simplest case is that of a one-dimensional combustion, e.g., the propagation of a flame along a Bickford fuse. We shall consider that the process can be described by one variable u, in which case (24.1) can be replaced by a kinetic equation, viz.,

$$\partial u/\partial t = f(u) + D\partial^2 u/\partial x^2 . \qquad (24.2)$$

In this equation $u(x,t)$ may be a temperature, the population of animals, the concentration of hot fuel, etc. The rate of change in u in a small system without diffusion ($\partial^2 u/\partial x^2 = 0$), which is called a point system, is determined by the function $f(u)$. For this class of nonequilibrium medium, $f(u)$ has the form of the curves in Fig. 24.2a.

We introduce a "travelling" variable $\xi = x + vt$. Then we find the following from (24.2) for a stationary wave:

$$v\, du/d\xi = f(u) + D d^2 u/d\xi^2 \quad \text{or}$$

$$DW\, dW/du - vW + f(u) = 0, \qquad (24.3)$$

when $W = du/d\xi$. If $f(u)$ and the boundary conditions for W are given, then we can find v, the wave propagation velocity from (24.3).

This problem was first formulated and solved in [22] for the following biological problem. Suppose some large territory is occupied by a species with a concentration W close to unity. Along the borders of the territory there is a region of intermediate concentration and along the borders of this region W is close to zero. As a result of "positive" selection, the territory occupied by the species will enlarge, i.e., its border will gradually move in the direction not occupied by the species. We wish to know the motion of the border of the region occupied by the species. Mathematically, the situation is described by (24.3) with the conditions $f(0) = f(1) = 0$, $f(u) > 0$ for $0 < u < 1$ and $df/du = f'(0) = \alpha > 0$, $f'(u) < \alpha$ for $0 < u \leq 1$ being satisfied (the curve $f_1(u)$ in Fig. 24.2a). It is necessary to find a link between v, D and $f'(0)$ at which the solution to (24.3) will be such that $0 \leq u(\xi) \leq 1$, $u(\xi) \to 1$ as $\xi \to 0$ and $u(\xi) \to 0$ as $\xi \to -\infty$. We find from (24.3)

$$dW/du = (DW)^{-1}[vW - f(u)]. \qquad (24.4)$$

Only the integral curves of (24.4) which pass between the straight lines $u = 0$ and $u = 1$ on the (W,u)-plane approaching the points $u = 0$, $W = 0$ and $u = 1$, $W = 0$ [22] are of interest. These are critical points for (24.4) to which the integral curves must approach without intersecting the straight lines $u = 0$ and $u = 1$, i.e., without curving. However, this means that in order for the characteristic plane to exist, the characteristic equation for each of these points must have real roots. If, as in [22], we assume that close to a singular point we have $f(u) = \alpha u$ and $u \sim \exp(p\xi)$, then the characteristic equation can be written as follows for the points $u = 0$ and $W = 0$:

$$p^2 - (v/D)p + (\alpha/D) = 0. \qquad (24.5)$$

Equation (24.5) has real roots when

$$v^2 \geq 4Df'(0). \qquad (24.6)$$

(we leave the reader to find and investigate the characteristic equation for the point $u = 1$, $\omega = 1$). It follows that a stationary wave must have a velocity in the interval $v_M \leq v < \infty$, where the minimum velocity is determined from (24.6), that is $v_M = 2(Df'(0))^{1/2}$. The instability of the initial homogeneous state leads to the appearance of a wave velocity v larger than v_M and only a wave moving at a velocity v_M will be asymptotically stable.

There is no general method for solving the boundary-value problem for an arbitrary $f(u)$ for (24.3), although if $f(u)$ is an antisymmetric polynomial, then $W = u^k(2 - u)^l$. For example, when $f(u) = 2\xi[u^2(1 - u) - \gamma(1 - u)u]$, γ and ξ are constants (the function starts from zero and as u increases it becomes negative, then decaying to zero, and becoming positive to attain a maximum, and finally dying away to zero at $u = 1$ (curve f_2 in Fig. 24.2a). We find a single velocity for a stationary wave [23] by substituting $W = u(1 - u)$ into (24.3), i.e.,

$$v = (1 - 2\gamma)(\xi D/2)^{1/2} \qquad (24.7)$$

When $u > \gamma$ this solution describes the propagation of nerve pulses or "flare ups", etc. The phase trajectory corresponding to the solution is a separatrix going from a saddle to a saddle (Fig. 24.2b).

24.2 Travelling Pulsations

A propagating single front, a switching wave, transfers a medium from one state to another. After the pulse's passage, the medium returns to its initial condition. For a one-dimensional medium, such pulses (like solitons) have closed loop separatrices in the phase space of the differential equation describing the stationary wave (Fig. 24.2b).

We shall present a simple example. In an active transmission line with tunneling diodes, nonstationary processes are described in the one-wave approximation by

$$\partial u/\partial t = \alpha u^2 - v_1 u - v_0 \partial u/\partial x + v_2 \partial^2 u/\partial x^2. \qquad (24.8)$$

Here it is assumed that the operating point of the tunneling diode is at the maximum of the voltage-current diagram and the diagram can be approximated by a parabola. Searching for a solution in the form of stationary waves that only depends on the travelling coordinates $\xi = x - vt$, we obtain an equation for a nonlinear oscillator with friction

$$v_2 d^2u/d\xi^2 + (v - v_0)du/d\xi + \alpha u(u - v_1/\alpha) = 0 . \qquad (24.9)$$

It immediately follows that the travelling pulse may only propagate at the velocity of the linear perturbation $v = v_0$ (this is a consequence of the medium not having dispersion). The solution corresponding to the boundary conditions $du/dx = 0$ ($x \to \pm \infty$) has the form of a soliton, i.e.,

$$u(x, t) = (3v_1/\alpha)\cosh^{-2}\left[\sqrt{v_1/2v_2}(x - v_0 t)\right] .$$

This is called a dissipative soliton.[1]

Dissipative solitons, have been observed in two-dimensional nonequilibrial media, e.g., on a falling film of a viscous liquid. We can obtain an approximate equation [24] for the deviation of the film surface from its undisturbed position. This describes changes in the thickness of a film h flowing down a surface at an angle α to the x-axis, i.e.,

$$\frac{1}{V_0}\frac{\partial h}{\partial t} + \frac{h}{h_0}\frac{\partial h}{\partial x} + \frac{8}{15} Re\, h_0 \frac{\partial^2 h}{\partial x^2} - \frac{2}{3} h_0 \cot \alpha \cdot \Delta h + \frac{2}{3} h_0^2 W \Delta^2 h = 0.$$

$$(24.10)$$

where h_0 is the undisturbed film thickness, $V_0 = h_0^2 g \sin(\alpha/2v)$ is the undisturbed flow rate, v is the viscosity, $Re = V_0 h_0/v$ is the Reynolds number, and $W = \sigma/\rho h_0^2 g \sin \alpha$ where σ is the coefficient of surface tension. This equation is valid for $h/h_0 \ll 1$ and $Re \cdot (h_0/L) \ll 1$ (L is the characteristic dimension of the

---

[1] Note that in this model a dissipative soliton is unstable with respect to perturbations with a nonzero average. Its existence in real transmission lines with tunneling diodes can be ensured by having a constant interval between the diodes, which prevents the occurrence of such perturbations (see [26]).

perturbation). It can be seen that a perturbation develops for **Re** > 5 cot α/4. It has not been possible to trace through analytically the nonlinear development of a perturbation. Numerical solutions show that this model contains several stationary solutions in the form of one-dimensional solitons. However, these are unstable and decay into horse-shoe shaped solitary waves (Fig. 24.3). Such solitons have been observed experimentally on falling films of viscous liquid. A numerical solution of the equation with boundary conditions $u = 0$ as $x \to \pm \infty$ is shown in Fig. 24.3. The solution is a horse-shoe soliton, with an oscillating leading edge and a decaying trailing edge.

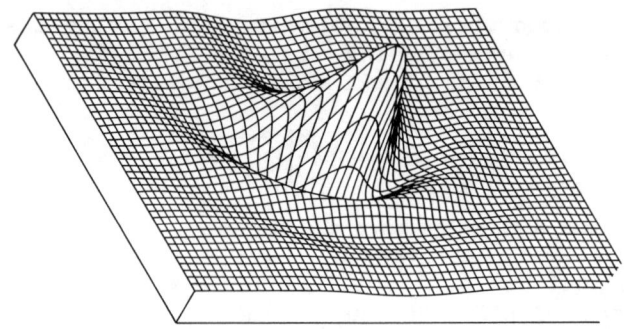

Fig. 24.3. Soliton on a falling film of liquid [25].

Travelling wave formations are typical of many active media with restoration, i.e., media the conductance properties of which are restored a finite time after the passage of a perturbation, the so-called refractory period. A very general model of such a medium is a two-component one:

$$\partial u/\partial t = F(u, v) + D_1 \Delta u \, , \quad \partial v/\partial t = \phi(u, v) + D_2 \Delta u \, . \quad (24.11)$$

These equations describe the excitation of heart muscle (when u is the potential difference at the cell membrane, and v is the transmembrane conductance), or the evolution of disturbances in nerve tissue, consisting of the exciting and breaking cells (i.e., neurons) [14].

An analytical investigation of travelling pulses using this model (24.11) has only been done for special cases, e.g., when the characteristic times for the changes in u and v are significantly different. Then the pulse can be divided into fast and slow changes and the method of discontinuous oscillations is used for the analysis (Chapter 14). We shall illustrate this for a one-dimensional, two-component medium whose equations explicitly contain a small parameter next to a derivative:

$$\mu \partial u/\partial t = f(u, v) + \partial^2 u/\partial x^2 , \quad \partial v/\partial t = g(u, v) , \quad \mu \ll 1.$$
(24.12)

In the region of fast changes in the variables, the quantity v may be considered constant, i.e., the pulse region in which u rapidly changes can be described by

$$\partial u/\partial \tau = f(u)\big|_{v=\text{const}} + \partial^2 u/\partial x^2 ,$$

which we encountered in our analysis of the fronts of switching waves ($\mu\tau = t$). Thus, when there is a small parameter, the travelling pulse may be considered to be two switching waves coming one after the other. The switch in u is governed by the slowly changing v, in one direction v switches small u in the forward direction and in the other region, it switches it in the other direction. In order to determine the boundaries of these regions, it is necessary to discover the forms of the functions $f(u,v)$ and $g(u,v)$ [7,15]. In the interval between the leading and trailing edges of the pulse, there is a small change in v which is "followed" by the rapidly change in u.

24.3 Spiral and Cylindrical Waves. Travelling Centers

Only comparatively simple self-organization phenomena can be analytically investigated in any detail. These phenomena are usually described by self-simulating solutions[2] of (24.1). A detailed analysis of most phenomena requires a computer, and often a numerical experiment does not yield simple solutions for which analytical solutions can be found, and instead they are complex combinations of the simple solutions.

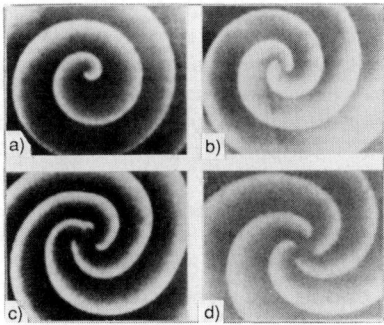

Fig. 24.4. Spiral vortices in a two-dimensional active medium: a-d) vortices with topological charges of one, two, three, and four, respectively.

---

[2] Self-simulating solutions of a partial differential equation depend on only one variable, which is a combination of all the variables.

Fig. 24.5. Fluctuating spiral vortices with a topological charge $N = 3$.

We shall briefly consider structures in the form of spiral wave, which have been experimentally observed in heart tissue during arhythmia [17], in chemical disturbed media [3], and in various other biological media.

The known solution of (24.1) for spiral waves in a two-dimensional medium has the form $u = F(N\theta - \omega t, r)$, where θ and r are polar coordinates and N determines the number of elementary waves rotating together (it is also called the topological charge). A spiral wave with $N > 1$ on a plane looks like a multiarmed spiral (Fig. 24.4). Spiral vortices of this form, which correspond to a rigid rotation of a spiral around a stationary point, have not been observed in experiment. However, it is possible that they are unstable. Recently, spiral waves with topological charges of two, three, and four were experimentally observed [25], but they were not steady states.

The experiment involved a medium remaining an autocatalytic Belousov-Zhabotinskii reaction. It can be seen form Fig. 24.5 that the waves at the center of the vortex are fluctuating, first converging, then diverging. However, the structure as a whole is unstable.

That spiral waves can exist in a homogeneous waiting medium, i.e., a medium consisting of active components with a finite restoration time (refraction), can be qualitatively shown using quite a simple argument. Suppose there is a local disturbance in a medium, namely a hole inside which elements cannot be disturbed. Suppose that waves are excited around the disturbance. The propagation of such a wave may be imagined as that of pulses in concentric circles. It is clear that because of the limiting rate of a pulse's propagation, their motion along different circles must be asynchronous, with perturbation close to the local inhomogeneity moving faster than those at the periphery. Thus a wave rotating around a spontaneous inhomogeneity may only be a spiral wave.

A waiting medium (e.g., a two-dimensional reactor containing an autocatalytic Belousov-Zhabotinskii reaction) demonstrates another self-organization phenomenon, that is the spontaneous appearance of travelling centers. A travelling center is a pulsating source of concentric

SELF-ORGANIZATION

diverging waves. The existence of such a source (like that of a source of spiral waves, a reverberator) is difficult to derive from (24.1) analytically. Although there are papers now which do this [7], it is quite simple to explain the situation qualitatively.

Let us consider a section of a homogeneous medium consisting of two coupled excitable elements. Each of these undergoes one oscillation cycle in response to an external pulsation and consequently they leave their rest state for an excited state. Then they enter the refractory state, after which they return to the rest state. The dynamics of such coupled elements is clearly dependent on the phase of the oscillations and the ratio between the excitation and refractory times. If the excitation time exceeds the refractory period and the state of the elements are correspondingly shifted in phase, then an self-excited oscillation regime will become established in this two-element system. The elements will alternately excite each other, and in a homogeneous medium a source of concentric waves will arise.

Several such sources, travelling centers, may arise in a medium. The concentric waves will extinguish each other (Fig. 24.6).

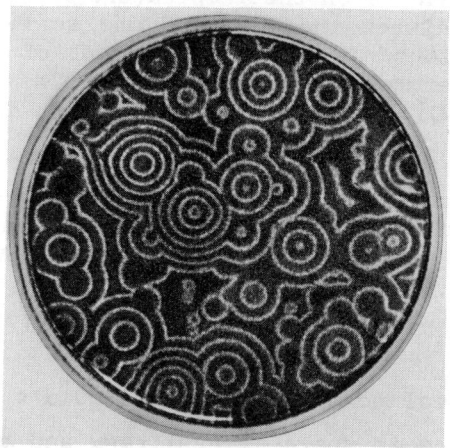

Fig. 24.6. Interactions between concentric waves when a medium contains several travelling centers.

24.4 Concerning Self-Organization Mechanisms

Usually, a wide spectrum of spatial perturbations grows against a linear background. However, when there is resonance instability, i.e., only perturbations with a particular spatial scale may grow, the scale of the resultant structure is often not determined by the perturbations themselves, but rather their interactions. Thus, although important, the instability properties are not as decisive as the selection mechanism and structure formation at the nonlinear stage. However, there are very few specific results, and so we shall restrict ourselves to the formation of various spatial scales and their synchronization. We do this for the example of Bernard convection.

We shall consider convective instabilities just above the threshold $\mathbf{Ra}/\mathbf{Ra}_{st} \geq 1$ (Chapter 21) in a liquid with a viscosity that is quadratic in temperature. The velocity field can then be represented as a large number of sinusoidal modes ignoring their spatial harmonics, this being possible when there is small supercritical nonlinearity. If there is no additional nonlinearity due to the viscosity being related to the temperature, a simple convective structure in the form of waves will become established in a planar layer heated from below (Chapter 21). The spatial scale of these waves is determined, as we have seen, by the competition between modes with close spatial scales. When the boundary conditions can be ignored, the orientation of these wave on the plane is arbitrary and only determined by the initial conditions.

The extra quadratic nonlinearity arising out of the viscosity's dependence on the temperature $\nu = \nu(T)$ leads to a resonance link between the modes of one scale and different spatial orientations. The simplest variant of such a link is one between three modes with the same modulus but whose wave vectors are at $60°$ to each other, i.e., $\pm \mathbf{k}_1 \pm \mathbf{k}_2 = \pm \mathbf{k}_3$ where $|\pm \mathbf{k}_i| = k$ ($i = 1, 2, 3$). The nonlinear interaction leads to a steady state convection with different amplitudes between these modes and the synchronized phases (as in the case just discussed). As a result, the velocity field takes the form

$$v_z(x, y) \sim \cos(kx/2)\cos(1/4)(kx + \sqrt{3}ky)\cos(1/4)(kx - \sqrt{3}ky)$$

(v_z is the vertical component of the liquid's velocity). The cell's orientation in space is arbitrary and only depends on the initial conditions. In order to find a solution describing this structure we must represent v_z in the form

$$v_z = f(z)\left[\alpha_1(t)\cos kx + \alpha_2(t)\cos(kx/2 - \sqrt{3}ky/2) + \alpha_3(t)\cos(kx/2 + \sqrt{3}ky/2)\right].$$

We can get the following for the real amplitudes

SELF-ORGANIZATION

$$\dot{A}_1 = hA_1\left[1 + \sigma\dot{A}_2 - \beta A_1^2 + \delta\left(A_2^2 + A_3^2\right)\right]$$

$$\dot{A}_2 = hA_2\left[1 + \sigma\dot{A}_3 - \beta A_2^2 + \delta\left(A_3^2 + A_1^2\right)\right] \quad (24.14)$$

$$\dot{A}_3 = hA_3\left[1 + \sigma\dot{A}_1 - \beta A_3^2 + \delta\left(A_1^2 + A_2^2\right)\right].$$

Thus, a linear instability turns into an explosive one caused by an interaction between the parametric coupled modes on the dissipative nonlinearity ($\sigma \sim \partial v/\partial T$). The interaction is limited due to the cubic nonlinearity in the viscosity--temperature function. System (24.14) has a stable stationary solution $A_1 = \pm A_2 = \pm A_3$, which corresponds to hexagonal prismatic cells (Fig. 24.1a).

It can, therefore, be seen from this example that synchronicity between modes determines the form of the spatial structures that arise due to instabilities. Competition ensures the stability of the structures with respect to the nonresonance perturbations.

Besides a search for new types of structure and the investigation of the mechanisms leading to their formation, a fascinating new field has recently opened up in self-organization theory, namely, the directed organization of structures by external fields. In order to illustrate a nontrivial problem, we present a relatively simple case. Consider the effect of a static spatially periodic field on a dissipative structure in a one-dimensional medium. The initial equation is that of diffusion

$$\partial u/\partial t = - u\partial u/\partial x + \gamma u + \nu \partial^2 u/\partial x^2 + U_0 \sin qx .$$

When $U_0 = 0$ a dissipative structure will exist in the medium and will be described by the oscillator equation

$$d^2 u/dx^2 - d(u^2/2)/dx + \gamma u = 0.$$

In the presence of periodic inhomogeneities $\sim U_0$, it is natural to expect the imposition of a periodic structure with the given period. However, even when the nonlinearity is weak (small U_0), the structure turns out to be stochastic [25].

The behavior of dissipative structures or travelling pulses in external fields is a special case of the behavior of coherent formations in each other's fields, i.e., a type of interaction. This also concerns problems involving nerve pulses, combustion fronts, and cylindrical and spiral waves. The analysis of interactions between structures of various types and nature is clearly of interest. The field has seen some success and we note, in particular, the experiment of

Agladze and Krinskii [25], in which interactions were observed between spiral vortices with structures resembling Bernard cells in a system with a two-dimensional Belousov-Zhabotinskii reaction. As a result of the interaction, the chemical reaction moves into a stochastic regime leading to "chemical" turbulence.

The formation of self-excited structures (localized formations independent of the initial and boundary conditions) is considered in [29] in nonequilibrial dissipative media and the dynamics of a spatial ensemble of such structures is analyzed. For instance, a simple model consisting of a one-dimensional ensemble of nonmutually linked structures, namely a chain of components whose dynamics are described by one-dimensional parabolic mappings, was studied. Remember that such mappings can describe the dynamics of a wide variety of physical systems that undergo period doubling bifurcations when their parameters are changed. Suppose the parameters of the chain are chosen such that there is a regular oscillation with period T in the first component. For some j, a one-frequency oscillation in the j-th element will cease to be stable and there will be period doubling, and then it will lose stability, etc. Tis continues until the oscillation is chaotic. If each component (self-excited generator) is working in a stochastic oscillation regime, then the development of chaos will be seen as one moves along the chain, the intensity of the oscillation increases and the peaks in the spectrum get smaller (the spectrum gets smoother). The spatial transition to chaos in a chain of Van der Pol type self-excited generators is via quasiperiodicity: at first a quasiperiodic regime is observed, which is then replaced by a beat regime with a large number of harmonics. Further down the chain the regime becomes stochastic. The chaos develops, the intensity of the oscillations rises, but at some large j there is almost no change since a spatially homogeneous chaos regime has developed.

It is suggested in [31] that such models can be used to explain the development of chaos in systems other than hydrodynamic ones (chains of interconnected Taylor vortices in which azimuthal modes have been induced and ensembles of spiral vortices in the boundary layer or a rotating body, etc.), namely, electron beams. This has been confirmed experimentally [32] with cylindrical electron beams drifting in a longitudinal constant magnetic field.

REFERENCES

Chapter One

1. A. A. Andronov, "L. I. Mandel'shtam and the theory of nonlinear oscillations" in: Academician Mandel'shtam, Essays to Mark the Centenary of His Birth [in Russian], Nauka, Moscow (1979), p. 127.

2. A. A. Andronov, A. A. Vitt, and S. E. Khaikin, Oscillation Theory [in Russian], Fizmatgiz, Moscow (1959).

3. A. Lotka, Elements of Physical Biology, Baltimore (1925).

4. Yu. M. Romanovskii, N. V. Stepanovna, D. S. Chernavskii, What is Mathematical Biophysics: Kinetic Models in Biophysics [in Russian], Prosveshchenie, Moscow (1971), Ch. 1, Sec. 1; Ch. 4, Sec. 2.

5. V. Volterra, Lecons sur la Theorie Mathematique de la Lutte pour la Vie [Russian translation], Nauka, Moscow (1976).

6. V. N. Shevchik, Fundamentals of Microwave Electronics [in Russian], Sovetskii Radio, Moscow (1959), Ch. 5.

7. V. N. Shevchik and D. I. Trubetskov, Analytical Calculation Methods in Microwave Electronics [in Russian], Sovetskii Radio, Moscow (1970).

8. C. Townes and A. Schawlow, Microwave Spectroscopy [Russian translation], Izd. Ino. Lit., Moscow (1959).

9. W. H. Louisell, Coupled Mode and Parametric Electronics [Russian translation], Izd. Ino. Lit., Moscow (1963).

10. W. H. Louisell, Radiation and Noise in Quantum Electronics [Russian translation], Nauka, Moscow (1972).

11. P.A.M. Dirac, Principles of Quantum Mechanics [Russian translation], Nauka, Moscow (1979).

12. R.P. Feynman, R.B. Leighton, and M. Sands, The Feynman Lectures on Physics [Russian translation], Mir, Moscow (1965), Vol.2, pp.135-136.

Chapter Two

1. L.A. Vainstein, Open Resonators and Open Waveguides [in Russian], Sovetskii Radio, Moscow (1966).

2. F. Bertein, "Sinusoidal methods of oscillations in electromagnetic resonators" in: A. Blaquiere, Nonlinear Systems Analysis [Russian translation], Mir, Moscow (1969), pp.377-379.

3. L.I. Mandel'shtam, Lectures on Oscillation Theory [in Russian], Nauka, Moscow (1972).

4. L.D. Landau and E.M. Lifshits, Mechanics [in Russian], Nauka, Moscow (1965), Sec.3.

5. W.H. Louisell, Coupled Mode and Parametric Electronics [Russian translation], Izd. Ino. Lit., Moscow (1963).

6. W.H. Louisell, "Correspondence between Pierce's coupled mode amplitudes and quantum operators," J. Applied Physics, 33, 2435-2436 (1962).

7. L.A. Vainstein, Electromagnetic Waves [in Russian], Sovetskii Radio, Moscow (1957), Sec.77.

8. J.P. Den Hartog, Mechanical Vibrations (McGraw-Hill, 1956) [Russian translation], Gostekhizdat, Moscow (1942).

Chapter Three

1. D.B. Sivukhin, General Course of Physics, Vol. IV, Optics [in Russian], Nauka, Moscow (1980), Ch.8.

2. L.V. Postnikov and V.P. Korolev (eds), Collection of Problems in Oscillation Theory [in Russian], Nauka, Moscow (1978).

3. A.G. Marshall, Biophysical Chemistry (J. Wiley and Sons) [Russian translation], Mir, Moscow (1981),

Vol. 2.

4. J. Ziman, Principles of the Theory of Solids [Russian translation], Mir, Moscow (1966), Ch. 8.

5. Ya. B. Zel'dovich and L. D. Myshkis, Elements of Applied Mathematics [in Russian], Nauka, Moscow (1972), Ch. 5, Secs. 8, 9.

6. I. L. Fabelinskii, Molecular Scattering of Light, Nauka, Moscow (1965).

7. E. I. Yakubovich, "On the dynamics of processes in media with nonhomogeneous line widening in the working medium," Zh. Exp. Teor. Fiz., **55**, 304-311 (1968).

Chapter Four

1. J. Ziman, Principles of the Theory of Solids [Russian translation], Mir, Moscow (1966).

2. L. Brillouin and M. Parodi, Propagation des Ondes less Milieux Periodiques [Russian translation], Izd. Ino. Lit., Moscow (1959).

3. L. V. Postnikov and V. P. Korolev (eds), Collection of Problems in Oscillation Theory [in Russian], Nauka, Moscow (1978).

4. V. S. Stal'makhov, "Magnetostatic spinor waves in microwave engineering" in: Lectures on Microwave Electronics and Radiophysics, 5-th Summer School-Seminar for Engineers [in Russian], Izd. Saratovskogo Univ., Saratov (1981), Vol. 4, pp. 37-41.

5. V. L. Ginzburg, Theoretical Physics and Astrophysics [in Russian], Nauka, Moscow (1975), Ch. 10.

6. V. M. Agranovich and V. L. Ginzburg, Crystal Optics Given Spatial Dispersion and Exiton Theory [in Russian], Nauka, Moscow (1979).

7. K. E. Lonngren, H. C. S. Hsuan, L. D. Landt, et al., "Properties of plasma waves defined by dispersion relations," IEEE Trans. on Plasma Sci., **Ps-2**, 93-108 (1974).

8. V. I. Kalinin and G. M. Gershtein, Introduction to

Radio Physics [in Russian], Gostekhizdat, Moscow (1957), Sec. 9.1.

9. R. Kompfner, The Invention of the Travelling-Wave Tube, San Francisco Press (1964).

10. V.N. Shevchik and D.I. Trubetskov, Analytical Calculation Methods in Microwave Electronics [in Russian], Sovetskii Radio, Moscow (1970), Ch. 1.

11. L.A. Vainstein and V.A. Solnetsev, Lectures in Microwave Electronics [in Russian], Sovetskii Radio, Moscow (1973), Lecture 3.

12. C. Kittel, Quantum Theory of Solids (J. Wiley and Sons, 1963) [in Russian], Nauka, Moscow (1967).

13. E. Henley and W. Thirring, Elementary Quantum Field Theory [Russian translation], Izd. Ino. Lit., Moscow (1963).

14. M.I. Kaganov, Electrons, Phonons, Magnons [in Russian], Nauka, Moscow (1979).

15. B. Wunderlich and H. Baur, Heat Capacities of Linear High Polymers [Russian translation], Mir, Moscow (1972).

16. E.M. Lifshits and L.P. Pitaevskii, Statistical Physics [in Russian], Nauka, Moscow (1978), Pt. 2.

also L.D. Landau and E.M. Lifshits, Statistical Physics [in Russian], Nauka, Moscow (1964), Pt. 1, Sec. 65.

Chapter Five

1. L.D. Landau and E.M. Lifshits, Fluid Mechanics [in Russian], Nauka, Moscow (1986), Ch. 1.

2. L.I. Sedov, Mechanics of Continuous Media [in Russian], Nauka, Moscow (1970), Vol. 1, Ch. 4, Sec. 1.

3. L.M. Brekhovskikh, V.V. Goncharov, K.A. Naugol'nykh, and S.A. Rybak, "Ocean waves," Izv. Vuzov: Radiofiz., **19**, 842-852 (1976).

4. A.S. Monin and M.N. Koshlyakov, "Synoptic vortices or Rossby waves in the ocean" in: A.V. Gaponov (ed), Nonlinear Waves [in Russian], Nauka, Moscow (1979), pp. 281-287.

REFERENCES

5. M. J. Lighthill, Waves in Fluids [Russian translation], Mir, Moscow (1981).

6. D. B. Sivukhin, General Course of Physics, Vol. II, Thermodynamics and Heat Physics [in Russian], Nauka, Moscow (1980), Sec. 40.

7. M. A. Lavrent'ev and B. V. Shabat, Problems of Hydrodynamics and Their Mathematical Models [in Russian], Nauka, Moscow (1973).

8. S. Flugge, Practical Quantum Mechanics [Russian translation], Mir, Moscow (1974), Vols. 1-2, pp. 292-300.

9. I. M. Khalatnikov, Theory of Superfluidity [in Russian], Nauka, Moscow (1971).

10. S. J. Putterman, Superfluid Hydromechanics (Amsterdam, 1974) [Russian translation], Mir, Moscow (1978).

11. E. M. Lifshits and L. P. Pitaevskii, Statistical Physics [in Russian], Nauka, Moscow (1978), Pt. 2, Ch. 3.

12. L. A. Artsimovich, "Selected works: atomic physics and plasma physics" in: Lectures on Plasma Physics [in Russian], Nauka, Moscow (1978), pp. 164-245.

13. B. B. Kadomtsev, Group Phenomena in Plasmas [in Russian], Nauka, Moscow (1976), Ch. 2.

14. N. A. Krall and A. W. Trivelpiece, Principles of Plasmaa Physics (Springer Verlag, 1971) [Russian translation], Mir, Moscow (1975), Ch. 4.

15. K. E. Lonngren, H. C. S. Hsuan, L. D. Landt, et al., "Properties of plasma waves defined by dispersion realtions," IEEE Trans. on Plasma Sci., **Ps-2**, 93-108 (1974).

16. L. A. Artsimovich and P. Z. Sagdeev, Plasma Physics for Physicists [in Russian], Atomizdat, Moscow (1979), Sec. 10.

17. W. H. Louisell, Coupled Mode and Parametric Electronics [Russian translation], Izd. Ino. Lit., Moscow (1963), Chs. 2-3.

18. M. C. Steele and B. Vural, Wave Interactions in

Solid State Plasmas [in Russian], Atomizdat, Moscow (1973).

19. Yu. K. Pozhela, Plasma and Current Instabilities in Semiconductors [in Russian], Nauka, Moscow (1977).

20. V. V. Vladimirov, A. F. Volkov, and E. Z. Melikhov, Plasma Semiconductors [in Russian], Atomizdat, Moscow (1979).

21. O. M. Phillips, The Dynamics of the Upper Ocean [Russian translation], Mir, Moscow (1969).

22. P. H. LeBlond and L. Mysak, Waves in Oceans (Elsevier, 1978) [Russian translation], Mir, Moscow (1981).

23. L. M. Brekhovskikh, Waves in Stratified Media [in Russian], Nauka, Moscow (1973).

24. Z. L. Miropol'skii, Dynamics of Internal Gravity Waves in the Ocean [in Russian], Gidrometeoizdat, Leningrad (1981).

25. E. L. Andronikashvili, Notes on Liquid Helium [in Russian], Ganatelba, Tibilisi (1980).

26. L. M. Brekhovskikh and L. M. Chyncharov, Introduction to the Mechanics of Continuous Media [in Russian], Nauka, Moscow (1982), Pt. 2.

Chapter Six

1. I. G. Aramanovich, G. L. Lunts, and L. E. El'sgol'ts, Functions of a Complex Variable. Operator Calculus. Stability Theory [in Russian], Nauka, Moscow (1965), Pt. 43, Ch. 9; Pt. 2, Chs. 7-8.

2. A. V. Netushil (ed), Theory of Automatic Control [in Russian], Vysshaya Shkola, Moscow (1972), Pt. 2.

3. L. V. Postnikov and V. P. Korolev (eds), Collection of Problems in Oscillation Theory [in Russian], Nauka, Moscow (1978).

4. M. L. Krasnov and G. I. Makarenko, Operator Calculus. Stability of Motion [in Russian], Nauka, Moscow (1964), Ch. 2.

5. V. G. Doetsch, Anleitung zum Praktischen Gebranch

REFERENCES

der Laplace-Transformation (Handbook on the Practical Use of the Laplace Transform) [Russian translation], Fizmatgiz, Moscow (1958), Sec. 11.

6. "The Gann Effect and Elements of its Phenomenological Theory" in: V.N. Shevchik and M.A. Grigor'ev (eds), High Frequency Electronic Devices [in Russian], Izd. Saratovskogo Univ., Saratov (1980), Ch. 12.

also M.E. Levenshtein, Yu.K. Pozhela, and M.S. Shur, The Gann Effect [in Russian], Sovetskii Radio, Moscow (1975).

7. Yu.M. Romanovskii, N.V. Stepanovna, D.S. Chernavskii, Mathematical Modelling in Biophysics [in Russian], Nauka, Moscow (1975), Ch. 8.

8. Y. Rocard, L'instabilite en Mechanique [Russian translation], Izd. Ino. Lit., Moscow (1959).

Chapter Seven

1. K. Rohlfs, Lectures on Density Wave Theory (Berlin, 1977) [Russian translation], Mir, Moscow (1980).

2. M.V. Vol'kenstein, General Physics [in Russian], Nauka, Moscow (1978), Ch. 8, Sec. 8.4.

3. A.M. Zhabotinskii, Concentration Oscillations [in Russian], Nauka, Moscow (1974), pp. 164-167.

4. H. Haken, Synergetics [Russian translation], Mir, Moscow (1980), pp. 342-358.

5. G. Nicolis and I. Prigogine, Self-Organisation in Nonequilibrium Systems [Russian translation], Mir, Moscow (1979).

6. V.N. Shevchik, G.N. Shvedov, and A.V. Soboleva, Wave and Oscillation Effects in Electron Beams at High Frequencies [in Russian], Izd. Saratovskogo Univ., Saratov (1962), Ch. 1.

7. V.N. Shevchik and D.I. Trubetskov, Analyticall Calculation Methods in Microwave Electronics [in Russian], Sovetskii Radio, Moscow (1970).

8. J.R. Pierce, Travelling-Wave Tube [Russian translation], Sovetskii Radio, Moscow (1952).

9. J.H. Piddington, "Growing electromagnetic waves. Growing electric waves of space charge and a twin-beam tube" in: Topics on Radio Location [Russian translation], Izd. Ino. Lit., Moscow (1956), No.6(36), pp.53-66. (from Phys. Rev., **101**, No.1, 9, Jan.1, 1956).

10. V.N. Shevchik and D.I. Trubetskov, Electron Tube with Reverse Wave [in Russian], Izd. Saratovskogo Univ., Saratov (1975).

11. Yu.G. Al'tshuler and A.S. Taratenko, Low-Power Tube with Reverse Wave [in Russian], Sovetskii Radio, Moscow (1963), Table 1.1, p.35.

12. V.Ya. Arsenin, Methods of Mathematical Physics and Special Functions [in Russian], Nauka, Moscow (1974).

13. R. Courant, Partial Differential Equations; R. Courant, D. Hilbert, Methods of Mathematical Physics [Vol.2. [Russian translation], Mir, Moscow (1964), Ch.5, Sec.1.

14. V.M. Lopukhin, Exciting Electromagnetic Oscillations and Waves by Electron Beams [in Russian], Gostekhizdat, Moscow (1953), Ch.8.

15. P.A. Sturrock, "Kinematics of growing waves" in: Microwave Oscillations in Plasmas [Russian translation], Izd. Ino. Lit., Moscow (1961), pp.72-112.

16. E.M. Lifshits and L.P. Pitaevskii, Physical Kinetics [in Russian], Nauka, Moscow (1979), Vol.10, Ch.6.

17. R.J. Briggs, Electron Stream Interaction with Plasmas, MIT Press, Cambridge, Mass. (1964).

18. R.J. Briggs, "Twin-Beam Instability" in Advances in Plasma Physics [Russian translation], Mir, Moscow (1974), pp.132-171.

19. L.D. Landau and E.M. Lifshits, Mechanics of Continuous Media [in Russian], Gostekhizdat, Moscow (1954), Sec.30.

20. A.M. Fedorchenko and N.Ya. Kotsarenko, Absolute and Convective Instabilities in Plasmas and Solids [in Russian], Nauka, Moscow (1981).

21. A.I. Akhiezer and R.V. Polovin, "Criteria of Wave Growth," UFN, **104**, 185 (1971).

22. Yu.K. Pozhela, Plasma and Current Instabilities in Semiconductors [in Russian], Nauka, Moscow (1977), Ch.3.

23. A.S. Monin and A.M. Yaglom, Statistical Fluid Mechanics [in Russian], Nauka, Moscow (1967).

24. L.A. Vainshtein, "Pulse propagation," UFN, **118**, 338- 367 (1971).

25. J.R. Pierce, "An interesting wave amplifier," IRE Trans., **ED-7**, 73-74 (1960).

26. R.L. Kyhl and H.R. Webster, "Breakup of hollow cylindrical electron beams," IRE Trans., **ED-3**, 172-183 (1956).

Chapter Eight

1. L.I. Mandel'shtam, "Lectures on some topics of oscillation theory" in: Lectures on Optics, Relativity Theory, and Quantum Mechanics [in Russian], Nauka, Moscow (1972), pp.401-437.

2. L.A. Vainstein, Electromagnetic Waves [in Russian], Sovetskii Radio, Moscow (1957), Ch.8.

3. M.J. Lighthill, Waves in Fluids [Russian translation], Mir, Moscow (1981), pp. 293-319, 375-409.

4. L.A. Ostrovskii and Nonlinear and Nonstationary Waves (Fourth All-Union School-Seminar on Wave Diffraction and Propagation) [in Russian], Izd. Ryazansk. Radiotekhn. Inst., Ryazan (1975), Lect.1.

5. L.A. Vainshtein, "Pulse propagation," UFN **118**, 339-367 (1971).

6. B.B. Kadomtsev, Group Phenomena in Plasmas [in Russian], Nauka, Moscow (1976), Ch.2, Sec.5.

7. L.M. Brekhovskikh, V.V. Goncharov, K.A. Naugol'nykh, and S.A. Rybak, "Ocean waves," Izv. Vuzov: Radiofiz., **19**, 843-852 (1976).

8. G.B. Whitham, Linear and Nonlinear Waves [Russian translation], Mir, Moscow (1977).

9. M.L. Levin, "How light overcomes dark: W.R. Hamilton and the concept of group velocity", UFN, **125**, 565-567 (1978).

10. L.N. Loshakov and Yu. N. Pchel'nikov, "On the relation between phase and group velocities in transmission lines for electromagnetic energy," Radiotekhnika, **36**, No.6, 71-72 (1981).

11. V.G. Polevoi and S.M. Rytov, "On four-dimensional group velocities," UFN, **125**, 549-565 (1978).

Chapter Nine

1. J.R. Pierce, Almost All About Waves (MIT Press, 1974) [Russian translation], Mir, Moscow (1976).

2. V.V. Beletskii, Essays on the Motion of celestial Bodies [in Russian], Nauka, Moscow (1977).

3. V.G. Whitham, Linear and Nonlinear Waves [Russian translation], Mir, Moscow (1977).

4. B.B. Kadomtsev, Group Phenomena in Plasmas [in Russian], Nauka, Moscow (1976), Ch.2.

5. L.D. Landau and E.M. Lifshits, Electrodynamics of Continuous Media [in Russian], Fizmatgiz, Moscow (1959), Ch.9, Sec.61.

6. M.B. Vinogradov, O.V. Rudenko, and A.P. Sukhorukov, Wave Theory [in Russian], Nauka, Moscow (1979), Ch.2.

7. J. Weiland and H. Wilhelmsson, Coherent Nonlinear Interactions of Waves in Plasmas [Russian translation], Energoizdat, Moscow (1981), Ch.4.

8. J. Asken and O. Nil'son, "Electrostatic relations for waves in systems with temporal and spatial dispersion," TIIEP, **57**, No.8, 83-84 (1968).

9. V.G. Polevoi and S.M. Rytov, "On four-dimensional group velocities," UFN, **125**, 549-565 (1978).

REFERENCES

Chapter Ten

1. L.D. Landau and E.M. Lifshits, Electrodynamics of Continuous Media (2nd ed) [in Russian], Fizmatgiz, Moscow (1982).

2. M.I. Rabinovich and A.L. Fabrikant, "Nonlinear Waves in Nonequilibrial Media," Izv. Vuzov: Radiofiz., **19**, 721-766 (1976).

3. P.A. Sturrock, "In what sense do slow waves carry negative energy?," J. Appl. Phys., **31**, 2052-2056 (1976).

4. J.R. Pierce, Almost All About Waves [Russian translation], Mir, Moscow (1976), Chs. 3-4.

5. M.V. Nezlin, "Waves with negative energy and anomalous Doppler effects," UFN, **120**, 481-495 (1976).

6. B.B. Kadomtsev, Group Phenomena in Plasmas [in Russian], Nauka, Moscow (1976), pp. 88-90.

7. J. Weiland and H. Wilhelmsson, Coherent Nonlinear Interaction of Waves in Plasmas [Russian translation], Energoizdat, Moscow (1981), Ch. 8.

8. V.N. Shevchik, "Fundamentals of microwave electronics," Sovetskii Radio, Moscow (1959), pp. 87-92.

9. W.H. Louisell and J.R. Pierce, "Power flow in electron beam devices," Proc. IRE, **43**, 435-427 (1955).

10. C.K. Birdsall, G.R. Brewer, and A.V. Haeff, "The resistive wall amplifier," Proc. IRE, **41**, 865-874 (1953).

11. V.N. Tsytobich, Nonlinear Effects in Plasmas [in Russian], Nauka, Moscow (1967).

12. W.H. Louisell, Coupled Mode and Parametric Electronics [Russian translation], Izd. Ino. Lit., Moscow (1963).

13. V.N. Shevchik and D.I. Trubetskov, Analytical Calculation Methods in Microwave Electronics [in Russian], Sovetskii Radio, Moscow (1970).

14. C.C. Cutler, "Mechanical travelling wave oscillator," Bell Lab. Record, 134-138 (1954).

15. V.L. Ginzburg, "On the radiation of an electron travelling near a dielectric," Dokl. Akad. Nauk SSSR, **6**, 145 (1947).

16. V.L. Ginzburg, "On the radiation of a microwave and its absorption in air," Izv. Akad. Nauk SSSR, Fiz, **11**, 165 (1947).

17. V.L. Ginzburg, "Various topics on radiation theory for superlight travel in a medium," UVN, **69**, 537 (1959).

18. R.J. Briggs, "Twin-Beam Instability" in: A. Simon, W.B. Thompson (eds,) Advances in Plasma Physics (Wiley Interscience 1969) [Russian translation], Mir, Moscow (1974), pp. 132-171.

19. T.B. Benjamin, "The threefold classification of unstable disturbances in flexible surfaces bounding inviscid flows," J. Fluid Mech., **16**, No.3, 436-450 (1963).

20. L.A. Ostrovskii, Yu.A. Stepanyants, and L.Sh. Tsimring, "Interactions between inner waves and flows and turbulence in the ocean" in: Nonlinear Waves. Self- Organization [in Russian], Nauka, Moscow (1963)..

21. L.J. Chu, "The kinetic power theorem" in: IRE Electron Devices Conference, Univ. of New Hampshire, (June 1951).

22. M.V. Nezlin, Dynamics of Beams in PLasmas [in Russian], Energoizdat, Moscow (1982).

23. G. Bekefi, Radiation Processes in Plasmas [Russian translation], Mir, Moscow (1971).

24. N.M Frank, "Einstein and optics," UFN, **129**, 694-703 (1979).

Chapter Eleven

1. L. Brillouin and M. Parodi, Propagation des ondes les milieux periodiques [Russian translation], Izd. Ino. Lit., Moscow (1959), Ch. 9.

REFERENCES

2. L. I. Mandel'shtam, Lectures on Oscillation Theory [in Russian], Nauka, Moscow (1972), Lects. 18-19.

3. N. W. Ashcroft and N. D. Mermin, Solid State Physics (Holt Reinhart, and Winston) [Russian translation], Mir, Moscow (1979), Vol. 1.

4. L. D. Landau and E. M. Lifshits, Mechanics [in Russian], Nauka, Moscow (1965), Sec. 27.

5. W. Louisell, Coupled Mode and Parametric Electronics [Russian translation], Izd. Ino. Lit., Moscow (1963), Chs. 4-9.

6. P. Penfield jr., Frequency-Power Formulas, MIT Press, Cambridge, Mass. (1960).

7. M. B. Vinogradova, O. V. Rudenko, and A. P. Sukhorukov, Wave Theory [in Russian], Nauka, Moscow (1979), Ch. 4.

8. Sh. Elashi, "Waves in active and passive periodic structures: Review," TIIER, **64**, No. 14, 22-59 (1976).

9. R. A. Silin and V. P Sazonov, Damping Systems [in Russian], Sovetskii Radio, Moscow (1966).

10. A. Hessel, "General characteristics of travelling wave antennas" in: Antenna Theory Pt. 2, McGraw Hill, New York (1969).

11. A. Oliner (ed), Surface Acoustic Waves [in Russian], Mir, Moscow (1981).

12. L. N. Magdich and V. Ya. Molchanov, Acoustic Devices and their Applications [in Russian], Sovetskii Radio, Moscow (1978).

13. P. L. Kapitsa, "Dynamic stability of a pendulum with a weight oscillating about a point," Zh. Exp. Teor. Fiz., **21**, 588-607 (1951).

14. P. L. Kapitsa, "Pendulum with Vibrating weight," UFN, **44**, 7-20 (1951).

15. M. A. Muller, "Motion of charged particles in high frequency electromagnetic fields," Izv. Vuzov:Radiofiz., **1**, 110-123 (1958).

16. M. A. Muller, "On a principle for generating high frequency oscillations," Izv. Vuzov:Radiofiz., **1**,

166- 167 (1958).

17. A.V. Gaponov-Grekhov (ed), Relativistic High Frequency Electronics [in Russian], Izd. IPF, Akad. Nauk SSSR, Gorky (1979).

18. V.L. Bratman, N.S. Ginzburg, and M.I. Petelin, "Nonlinear theory of forced wave scattering on relativistic electron beams," Zh.Exp.Teor.Fiz., **76**, 930-943 (1979).

19. V.L. Bratman, N.S. Ginzburg, and M.I. Petelin, "Energetc possibilities for a relativistic Compton laser," Zh.Exp.Teor.Fiz., **78**, 207-211 (1979).

20. N.N. Bogolyubov and Yu.A. Mitropol'skii, Asymptotic Methods in the Theory of Nonlinear Oscillations [in Russian], Nauka, Moscow (1963).

21. A.H. Nayfeh, Introduction to Perturbation Techniques [Russian translation], Mir, Moscow (1976).

Chapter Twelve

1. N.N. Moiseev, Asymptotic Methods in Nonlinear Mechanics [in Russian], Nauka, Moscow (1969), ch.4.

2. J. Heading, An Introduction to Phase Integral Methods (London 1962) (VKB method) [Russian translation], Mir, Moscow (1965).

3. L.D. Landau and E.M. Lifshits, Mechanics [in Russian], Nauka, Moscow (1965), Sec.49.

4. M. Kruskal, Asymptotic Theory of Hamiltonian and other systems with all solutions nearly periodic (Princeton 1961) [Russian translation], Izd. Ino. Lit., Moscow (1962).

5. Din Yu Hsich, "Variational method and nonlinear oscillations and waves," J. Maths. Phys., **16**, No.8, 1630-1636 (1975)..

6. J.B. Whitham, "Nonlinear dispersive waves," Proc. Roy. Soc, Series A, **283**, 238-261 (1965).

7. M.L. Krasnov, G.I. Makarenko, and A.I. Kiselev, Variational Calculus. Selected Chapters on Advanced Mathematics for Engineers and Students at

REFERENCES

Polytechnics [in Russian], Nauka, Moscow (1975).

8. Yu. A. Kravtsov and Yu. I. Orlov, Geometric Optics of Nonhomogeneous Media [in Russian], Nauka, Moscow (1980).

9. O. Buneman, R. H. Levy, and L. M. Linson, "Stability of crossed-field electron beams," J. Appl. Phys., **37**, No. 8, 3203 (1966).

also M. V. Gavrilov and D. I. Trubetskov, "Wave Phenomenona and low density electron beams in crossed fields with a fall in the single beam state" in: Intercollege Scientific Collection on Questions of Microwave Electronics, Izd. Saratovskogo Univ., Saratov (1977), No. 10, pp. 156-181.

10. V. M. Lopukhin, V. G. Magalinskiim V. P. Martynov, and A. S. Poshal', Noise and Parametric Phenomena in High-Frequency Electronic Devices [in Russian], Nauka, Moscow (1966), pp. 75-80.

11. G. I Haddad and R. M. Bevensee, "Start oscillations of tapered backward-wave oscillator," IRS Trans. on EA., **10**, No. 6, 389-393 (1963).

12. G. M. Zaslavskii, V. P. Meitlis, and N. N. Filonenko, Wave Interactions in Nonhomogeneous Media [in Russian], Nauka, Novosibirsk (1982).

13. L. M. Brekhovskikh, Waves in Stratified Media [in Russian], Nauka, Moscow (1973).

14. V. M. Babich and V. S. Bundyrev, Asymptotic Methods in Problems Concerning the Diffraction of Short Waves [in Russian], Nauka, Moscow (1972).

15. V. A. Borovikov and B. E. Kinber, Geometric Theory of Diffraction, Svyas', Moscow (1978).

16. L. A. Vainshtein and D. E. Vakman, Separation of Frequency in the Theory of Oscillations and Waves [in Russian], Nauka, Moscow (1983).

17. M. B. Vinogradova, O. V. Rudenko, and A. P. Sukhorukov, Wave Theory [in Russian], Nauka, Moscow (1979), Ch. 7.

18. V. L. Ginzburg, Propagation of Electromagnetic Waves in Plazmas [in Russian], Nauka, Moscow (1967).

19. V.V. Zheleznyakov, V.V. Kocharovskii, and V.V. Kocharovskii, "Linear interactions between electromagnetic waves in nonhomogeneous, weakly anisotropic media," UFN, **141**, No.2, 257-310 (1983).

20. K.G. Budden, Radio Waves in the Ionosphere, Cambridge University Press, Cambridge (1961).

21. V.V. Zheleznyakov, Radio Radiation from the Sun and Planets [in Russian], Nauka, Moscow (1964).

22. B.Z. Katsenelembaum, High-Frequency Electrodynamics [in Russian], Nauka, Moscow (1966).

23. A.W. Snuder and J.D. Love, Optical Waveguide Theory [in Russian], Radio i Svyaz', Moscow (1987).

Chapter Thirteen

1. A.V. Gaponov-Grekhov and M.I. Rabinovich, "L.I. Mandel'shtam and the modern theory of nonlinear oscillations," UFN, **128**, 579-624 (1979).

2. L.I. Mandel'shtam, Lectures on Oscillation Theory [in Russian], Nauka, Moscow (1972).

3. A. Blaquiere, Nonlinear Systems Analysis [Russian translation], Mir, Moscow (1969).

4. V.L. Bratman, N.S. Ginzburg, and M.I. Petelin, "Theory of free electron lasers and masers" in: Lectures on Microwave Electronics and Radiophysics, 5-th Summer School-Seminar for Engineers [in Russian], Izd. Saratovskogo Univ., Saratov (1981), Vol.4, pp. 69-172.

5. A.V. Gaponov, M.I. Petelin, and V.K. Yulpatov, "Induced radiation of excited classical oscillators and its application to microwave electronics," Izv. Vuzov: Radiotekh., **10**, 1414-1453 (1967).

6. B.B. Kadomtsev, Group Phenomena in Plasmas [in Russian], Nauka, Moscow (1976).

7. L.A. Vainstein, "Electron waves in breaking systems. On the nonlinear equations of the travelling wave tube,"" Radio i Elekt., **2**, 688-695 (1957).

REFERENCES

8. L. A. Vainstein, "Nonlinear theory of travelling wave tubes," Radio i Elekt., **2**, 883-894, 1027-1047 (1957).

9. V. T. Ovcharov (ed), Travelling Wave Tube, Collection of Translated Articles [Russian translation], Gosenergoizdat, Moscow (1959).

10. G. M. Zaslavskii and B. V. Chirikov, "Stochastic Instability of nonlinear oscillations," UFN, **105**, 7-13 (1971).

11. L. D. Landau and E. M. Lifshits, Mechanics [in Russian], Nauka, Moscow (1965).

12. G. M. Zaslavskii, Statistical Irreversibility in Nonlinear Systems [in Russian], Nauka, Moscow (1970).

13. B. V. Chirikov, Nonlinear Resonance [in Russian], Izd. Novosibirskogo Univ., Novosibirsk (1977).

14. V. P. Reutov, "Plasma hydrodynamics analogy and the nonlinear stage of vortex wave instability," Izv. Akad. Nauk SSSR, Fiz. Atm. i Okeana, **16**, 1266-1275 (1980).

15. A. A. Andronov and A. L. Fabrikant, "Landau damping, vortex waves, and surf" in: Nonlinear Waves [in Russian], Nauka, Moscow (1979), p.68.

16. R. Betchov and W. O. Criminale, Stability of Parallel Flows [Russian translation], Mir, Moscow (1971).

17. V. D. Shapiro and V. I. Shevchenko, "Wave-particle interactions in nonequilibrial media," Izv. Vuzov: Radiofiz., **19**, 767 (1976).

18. R. V. Khokhlov, "On the propagation of waves in nonlinear dispersing lines," Radio i Elekt., **6**, 1116 (1961).

19. R. V. Khokhlov, "Towards the theory of shock waves in nonlinear lines," Radio i Elekt., **6**, 917 (1961).

20. G. M. Zaslavskii, Stochasticity of Dynamic Systems [in Russian], Nauka, Moscow (1984), chs. 1-4.

Chapter Fourteen

1. L. I. Mandel'shtam, Collected Works [in Russian], Izd. Akad. Nauk SSSR, Moscow (1950).

2. A. A. Andronov, Collected Works [in Russian], Izd. Akad. Nauk SSSR, Moscow (1955).

3. A. H. Nayfeh, Introduction to Perturbation Techniques [Russian translation], Mir, Moscow (1976).

4. A. A. Andronov, A. A. Vitt, and S. E. Khaikin, Oscillation Theory [in Russian], Fizmatgiz, Moscow (1959).

5. N. N. Bogolyubov and Yu. A. Mitropol'skii, Asymptotic Methods in the Theory of Nonlinear Oscillations [in Russian], Nauka, Moscow (1963).

6. E. N. Lorentz, "Deterministic nonperiodic flow" in: Strange Attractors [Russian translation], Mir, Moscow (1981), 88-116.

Chapter Fifteen

1. A. A. Andronov, A. A. Vitt, and S. E. Khaikin, Oscillation Theory [in Russian], Fizmatgiz, Moscow (1959).

2. A. A. Andronov, E. A. Leontovich, I. I. Gordon, A. G. Maier, Qualitative Theory of Second-Order Dynamic Systems [in Russian], Nauka, Moscow (1966).

3. H. Poincare, On the Curves Determinable by Differential Equations [Russian translation], Gostekhizdat, Moscow (1947).

4. A. V. Gaponov-Grekhov and M. I. Rabinovich, "L. I. Mandel'shtam and the modern theory of nonlinear oscillations," UFN, **128**, 579-624 (1979).

5. N. V. Butenin, Yu. I. Neimark, N. A. Fufaev, Introduction to Nonlinear Oscillations [in Russian], Nauka, Moscow (1976), Ch. 7, Sec. 4.

6. Yu. I. Neimark, Method of Point Mappings in Nonlinear Oscillation Theory [in Russian], Nauka, Moscow (1972).

REFERENCES

7. L.P. Shil'nikov, "On the birth of periodic motion from trajectories bisymmetric about a saddle type equilibrium," Matem. Sbor., **77**, No. 119, 461-472 (1968).

8. V.M. Alekseev and M.V. Yakobson, "Symbolic dynamics and hyperbolic dynamic systems" in: R. Bowen (ed), Methods of Symbolic Dynamics [Russian translation], Mir, Moscow (1979).

9. V.K. Mel'nikov, "On the stability of a center during temporally periodic disturbances," Tr. Mosk. Matem. Ob., **12**, 3-52 (1963)..

10. V.I. Arnol'd, Additional Chapters to the Theory of Ordinary Differential Equations [Russian translation], Nauka, Moscow (1978).

11. A.D. Morozov, "On a complete qualitative investigation of Duffing's equation," ZhVMMF, **12**, 1134-1152 (1973).

12. J.E. Marsden and M. MacCracken, The Hopf Bifurcation and its Applications (1976) [Russian translation], Mir, Moscow (1980).

13. R. Balescu, Equilibrium and Nonequilibrium Statistical Mechanics (Wiley Interscience) [Russian translation], Mir, Moscow (1978).

14. V.M. Alekseev, Symbolic Dynamics (XI Summer Mathematics School) [in Russian], Inst. Matem. Akad. Nauk Ukrain. SSR, Kiev (1976), p.212.

15. H. Poincare, Selected Works [Russian translation], Nauka, Moscow (1972), Vol.2, Ch.33.

16. N.V. Butenin, Yu.I. Heimark, N.A. Furaev, Introduction to the Theory of Nonlinear Oscillations [in Russian], Nauka, Moscow (1987), Ch.3, Sect.2.

17. V.C. Aframovich, "Short essay on the qualitative theory of dynamic systems" in: Lectures on Microwave Electronics and Radio Physics (Sixth Summer School-Seminar for Engineers [in Russian], Izd. Saratovskogo Univ., Saratov (1983), Vol.2, pp. 75-89.

Chapter Sixteen

1. I.I Blekhman, Synchronization in Nature and Technology [in Russian], Nauka, Moscow (1981); also Synchronization of Dynamic Systems [in Russian], Nauka, Moscow (1971).

2. C. Huygens, Three Memoirs on Mechanics [Russian translation], Izd. Akad. Nauk SSSR, Moscow (1951).

3. A.A. Andronov and A.A. Vitt, "Towards a theory of Van- der Pol damping" in: Collection of Works by A.A. Andronov [in Russian], Izd. Akad. Nauk SSSR, Moscow (1956), pp. 51-64.

4. C. Hayashi, Nonlinear Oscillations in Physical systems [Russian translation], Mir, Moscow (1968), Pt. 4.

5. A.A. Andronov and A.A. Vitt, "Towards a mathematical theory of self-excited oscillation systems with two degrees of freedom," ZhTF, **4**, 122 (1934).

6. Van der Pol, Nonlinear Theory of Electrostatic Oscillation [Russian translation], Svyaz'izdat, Moscow (1935).

7. M. Williamson, The Analysis of Biological Populations (1972) [Russian translation], Mir, Moscow (1976).

8. J.M. Smith, Models in Ecology [Russian translation], Mir, Moscow (1976).

9. M.I. Rabonovich, "Stochastic self-excited oscillations and turbulence," UFN, **125**, 138-140 (1978).

10. Y. Aizawa, "Synergetic approach to the phenomenon of mode-locking on nonlinear systems," Prog. Theor. Phys., **56**, 703 (1976).

11. P.S. Landa, Self-Excited Oscillations in Systems with a Finite Number of Degrees of Freedom [in Russian], Nauka, Moscow (1980).

12. V.A. Zharkov (ed), Tides in the Solar System [Russian translation], Mir, Moscow (1979).

13. S.V. Kiyashko and M.I. Rabinovich, "On wave spectrum transformations in active nonlinear media," Izv. Vuzov: Radiofiz, **15**, 1807-1814 (1972).

14. A.M. Molchanov, "On the resonance structure of the Solar system" in: Modern Problems of Celestial Mechanics and Astrodynamics [in Russian], Nauka, Moscow (1973).

15. V.V. Beletskii, Essays on the Motion of celestial Bodies [in Russian], Nauka, Moscow (1977).

Chapter Seventeen

1. R.V. Khokhlov, "On the propagation of waves in nonlinear dispersing lines," Radio i Elekt., **6**, 1116 (1961).

2. L.D. Landau and E.M. Lifshits, Mechanics [in Russian], Nauka, Moscow (1965).

3. V.V. Beletskii, Essays on the Motion of celestial Bodies [in Russian], Nauka, Moscow (1977).

4. F. Kaczmarek, Wstep do fizyki laserow (Warsaw 1979) [Russian translation], Mir, Moscow (1981).

5. A. Yariv, Quantum Electronics and Nonlinear Optics [in Russian], Sovetskii Radio, Moscow (1973).

6. O.M. Phillips, "Wave interaction" in: Nonlinear Waves [Russian translation], Mir, Moscow (1977), Ch.7.

7. J. Weiland and H. Wilhelmsson, Coherent Nonlinear Interactions bewteen Waves and Plasma [Russian translation], Energoizdat, Moscow (1981).

8. S.A. Akhmanov and R.V. Khokhlov, Problems of Nonlinear Optics [in Russian], Izd. VINITI, Moscow (1964).

9. N. Blombergen, Nonlinear Optics. A Lecture Note. [Russian translation], Mir, Moscow (1966).

10. M.I. Rabinovich and V.P. Reutov, "Interaction between parametrically coupled waves in nonequilibrial media," Izv. Vuzov: Radiofiz., **16**, 825-826 (1973).

11. M. I. Rabinovich and V. P. Reutov, "Explosive instability and the generation of solitons in active media," ZhTF, **42**, 2458-2465 (1972).

12. M. I. Rabinovch, "On the asymptotic method in the theory, of the oscillations of distributed systems," Dokl. Akad. Nauk SSSR, **191**, 1253-1255 (1971).

Chapter Eighteen

1. A. V. Gaponov, L. A. Ostrovskii, and M. I. Rabinovich, "One-dimensional waves in linear systems with dispersion," Izv. Vuzov: Radiofiz, **13**, 164-213 (1970)..

2. B. B. Kadomtsev and V. I. Karpman, "Nonlinear waves," UFN, **103**, 193-232 (1971).

3. V. I. Karpman, Nonlinear Waves in Dispersing Media [in Russian], Nauka, Moscow (1976).

4. B. B. Kadomtsev, Group Phenomena in Plasmas [in Russian], Nauka, Moscow (1976).

5. A. C. Scott, Active and Nonlinear Wave Propagation in Electronics (Wiley Interscience, 1970) [Russian translation], Sovetskii Radio, Moscow (1977).

6. G. B. Whitham, Linear and Nonlinear Waves [Russian translation], Mir, Moscow (1977).

7. L. A. Vainstein and V. A. Solnetsev, Lectures in Microwave Electronics [in Russian], Sovetskii Radio, Moscow (1973).

8. V. N. Shevchik, "Fundamentals of microwave electronics," Sovetskii Radio, Moscow (1959).

9. L. D. Landau and E. M. Lifshits, Mechanics of Continuous Media [in Russian], Gostekhizdat, Moscow (1953).

10. A. V. Gaponov and G. I. Freidman, "On electromagnetic shock waves in ferrites," Zh. Exp. Teor. Fiz., **36**, 957 (1959).

11. I. G. Kataev, Electromagnetic Shock Waves [in Russian], Sovetskii Radio, Moscow (1963).

REFERENCES

12.   R.V. Polovin, "Nonlinear magnetohydrodynamic waves," Differ. Urav., **1**, 499-522 (1965).

13.   B.L. Rozhdestvenskii and N.N. Yanenko, Systems of Quasilinear Equations [in Russian], Nauka, Moscow (1968).

14.   R.V. Khokhlov, "Towards a theory of radio shock waves in nonlinear lines," Radio i Elekt., **6**, 1116 (1961).

15.   R.V. Khokhlov, "On the propagation of waves in nonlinear dispersing lines," Radio i Elekt., **6**, 1116 (1961).

16.   O.V. Rudenko and S.I. Soluyan, Theoretical Fundamentals of Nonlinear Acoustics [in Russian], Nauka, Moscow (1975).

Chapter Nineteen

1.   P. Courant and K. Friedricks, Supersonic flow and Shock Waves [Russian translation], Izd. Ino. Lit., Moscow (1950).

2.   Ya.B. Zel'dovich and Yu.P. Raizer, Physics of Shock Waves and High Temperature Hydrodynamic Phenomena [in Russian], Nauka, Moscow (1966).

3.   L.I. Sedov, Similarity and Dimensional Analysis in Mechanics [in Russian], Nauka, Moscow (1977).

4.   B.B. Kadomtsev, Group Phenomena in Plasmas [in Russian], Nauka, Moscow (1976).

5.   B.B. Kadomtsev and V.I. Karpman, "Nonlinear waves," UFN, **103**, 193-232 (1971).

6.   A.V. Gaponov, L.A. Ostrovshkii, and G.I. Freidman, "Electromagnetic shock waves," Izv. Vuzov: Radiofiz, **10**, 1371 (1967).

7.   Yu.K. Bogastyrev, Impulse Devices with Nonlinear Distributed Parameters [in Russian], Sovetskii Radio, Moscow (1974).

8.   A.M. Belyantsev, A.V. Gaponov, E.Ya. Daume, and G.I. Freidman, "Experimental investigation of the propagation of electromagnetic waves with finite amplitudes in wave guides filled with ferrite,"

Zh. Exp. Teor. Fiz., **47**, 1699 (1964).

9. V. I. Karpman, Nonlinear Waves in Dispersing Media [in Russian], Nauka, Moscow (1976).

10. A. C. Scott, F. Y. E. Chu, and D. W. Mclaughlin, "Soliton: A new concept in applied science" in: A. C. Scott, Active and Nonlinear Wave Propagation in Electronics (Wiley Interscience 1970) [Russian translation], Sovetskii Radio, Moscow (1977).

11. V. E. Zakharov, S. V. Manakov, S. P. Novikov, and L. P. Pitaevskii, Soliton Theory. The Method of Inverse Problems [in Russian], Nauka, Moscow (1980).

12. K. E. Lonngren and A. C. Scott (eds), Solitons in Action (Acad. Press 1978) [Russian translation], Mir, Moscow (1981).

13. K. Rebbi, "Solitons," UFN, **130**, 329-356 (1980).

14. L. D. Landau and E. M. Lifshits, Quantum Mechanics [in Russian], Nauka, Moscow (1974).

15. N. J. Zabusky and M. D. Kruskal, "Interaction of solitons in a collisionless plasma and recurrence of initial states," Phys. Rev. Lett., **15**, 240-243 (1965)..

16. K. A Gorshkov, L. A. Ostrovskii, and V. V. Papko, "Interaction and coupled states between solitions as classical particles," Zh. Exp. Teor. Fiz., **71**, 585 (1976).

17. V. I. Petviashvili, "Inhomogeneous solitons" in: A. V. Gaponov (ed), Nonlinear Waves [in Russian], Nauka, Moscow (1979), p. 5.

18. V. I. Petviashvili, "Jupiter's red spot and drift solitons in plasmas," Let. to Zh. Exp. Teor. Fiz., **32**, 632 (1980).

19. S. V. Antipov, M. V. Nezlin, E. N. Snezhkin, and A. S. Trubetskov, "Rossby solitons in the laboratory," Zh. Exp. Teor. Fiz., **82**, 145 (1982).

20. V. M. Kamenkovich and A. S. Monin, Fluid Mechanics of the Ocean [in Russian], Nauka, Moscow (1978), Ch. 8.

21. C. S. Gardner, J. M. Green, and R. M. Miara, "Method for solving the Korteveg-de Vries equation," Phys.

Rev. Lett., **19**, 1095-1097 (1967).

22. V.E. Zakharov, "Instability in nonlinear soliton oscillations," Let. to Zh.Exp.Teor.Fiz., **22**, 364 (1975).

23. M.V. Nezlin, E.N. Snezhkin, and A.S. Trubetskov, "Kelvin-Helmholtz instability and Jupiter's big red spot," Let to Zh.Exp.Teor.Fiz., **36**, 190-193 (1982).

24. E.A. Zabolotskaya and R.V. Khokhlov, "Converging and diverging beams in nonlinear media," Akust. Zh., **16**, 49 (1970).

25. A.T. Filipov, The Multifaced Soliton [in Russian], Nauka, Moscow (1986).

Chapter Twenty

1. G.A. Askar'yan, "Effect of the gradient of the field from an intense electromagnetic beam on electrons and atoms," Zh.Exp.Teor.Fiz., **42**, 1567 (1962).

2. M.J. Lighthill, "Contributions to the theory of waves in nonlinear dispersive systems," J.Inst.Math.Appl., **1**, 269 (1965).

3. B.Ya. Zel'dovich, V.I. Popovichev, V.V. Ragul'skii, and F.S Faizullov, "On the link between the wave fronts of reflected and exciting light in forced Mandel'shtam--Brillion scattering," Zh.Exp.Teor.Fiz., **62**, 872 (1972).

4. A.V. Gaponov (ed), Nonlinear Waves: Propagation and Interaction [in Russian], Nauka, Moscow (1981).

5. M.B. Vinogradova, O.V. Rudenko, and A.P. Sukhorukov, Wave Theory [in Russian], Nauka, Moscow (1979).

6. G.B. Whitham, Linear and Nonlinear Waves [Russian translation], Mir, Moscow (1977).

7. K.A Gorshkov, L.A. Ostrovskii, and E.H. Pelinovskii, "Questions on the asymptotic theory of nonlinear oscillations," TIIER, **62**, No.11, 113-120 (1974).

8. L.A. Ostrovskii and L.V. Soustov, "Self-organization of electromagnetic waves in nonlinear transmission lines," Izv. Vuzov:Radiofiz, **15**, 242-248 (1972).

9. A.V. Gaponov, L.A. Ostrovshkii, and M.I. Rabinovich, "One-dimensional waves in linear systems with dispersion," Izv. Vuzov: Radiofiz, **13**, 164-213 (1970).

10. L.A. Ostrovskii and M.I. Rabinovich, Nonlinear and Nonstationary Waves (Fourth All-Union School-Seminar on Wave Diffraction and Propagation) [in Russian], Izd. Ryazansk. Radiotekhn. Inst., Ryazan (1975).

11. H.C. Yuen and B.M. Lake, "Theory of nonlinear waves and nonstationary waves over deep water" in: K.E. Lonngren and A.C. Scott (eds), Solitons in Action (Acad. Press 1978) [Russian translation], Mir, Moscow (1981).

12. P. De Gennes, Superconductivity of Metals and Alloys [Russian translation], Mir, Moscow (1968).

13. B.B. Kadomtsev and V.I. Karpman, "Nonlinear waves," UFN, **103**, 193-232 (1971).

14. V.E. Zakharov, S.V. Manakov, S.P. Novikov, and L.P. Pitaevskii, Soliton Theory. The Method of Inverse Problems [in Russian], Nauka, Moscow (1980).

15. L.A. Ostrovskii, "Propagation of wave packets and space- time self-focussing in a nonlinear medium," Zh.Exp.Teor.Fiz., **51**, 1189 (1966).

16. A.C. Scott, F.Y.E. Chu, and D.W. Mclaughlin, "Soliton: AA new concept in applied science" in: A.C. Scott, Active and Nonlinear Wave Propagation in Electronics (Wiley Interscience, 1970) [Russian translation], Sovetskii Radio, Moscow (1977), p.276.

17. B.M. Lake, H.C. Yuen, H. Rungaldier, and W.E. Ferguson, "Nonlinear deep-water waves: Theory and experiment. Part II: Evolution of a continuous wave train," J. Fluid Mech., **83**, 49 (1977).

18. H.C. Yuen and W.E. Ferguson, "Relationship between Benjamin-Feir instability and recurrence in the nonlinear Schrodinger equation," Phys. Fluids, **21**, 1275 (1978).

REFERENCES

19. R. Hirota and K. Suzuki, "Studies on lattice solitons by using lumped networks," J. Phys. Soc. Japan, **28**, 1366 (1970).

20. E. Fermi, Scientific Works [Russian translation], Nauka, Moscow (1972), Vol. 2.

21. V. E. Zakharov and S. V. Manakov, "On the complete integrability of nonlinear equations," TMF, **19**, 333 (1974).

22. M. I. Rabinovich and A. L. Fabrikant, "Stochastic wave self-simulation in nonequilibrial media," Zh. Exp. Teor. Fiz., **77**, 617-629 (1979).

23. S. V. Kiyashko, V. V. Papko, and M. I. Rabinovich, "Model experiments on the interaction between Langmuir and ion-sound waves," Fizika Plazmy, **1**, 1013 (1975).

24. V. E. Zakharov and A. M. Rubenchik, "On nonlinear interactions between low and high frequency waves," PMTF, No. 5, 84 (1972).

25. V. I. Talanov, "On the self-focussing of wave beams in nonlinear media," Let. to Zh. Exp. Teor. Fiz., **2**, 218-222 (1965).

26. S. A. Akhmanov, A. P. Sukhorukov, and R. V. Khokhlov, "Self-focussing and diffraction of light in a nonlinear medium," UFN, **93**, 19 (1967).

27. S. P. Vlasov, V. I. Talanov, and V. A. Petrishchev, "Averaged description of wave packets in linear and nonlinear media (moments method)," Izv. Vuzov: Radiofiz, **14**, 1353 (1971).

28. V. S. L'vov, Lectures on the Physics of Nonlinear Media [in Russian], Izd. Novosibirsk. Univ., Novosibirsk (1977).

29. P. Cuiti, G. Jernetti, and M. S. Sagoo, "Optical visualization of nonlinear acoustic propagation in cavitating liquids," Ultrasonics, **18**, 111 (1980).

30. M. I. Rabinovich, V. P. Reutov, and V. A. Tsvetkov, "On the merging of pulses and beams during explosive instability," Zh. Exp. Teor. Fiz., **67**, 525 (1974).

31. Yu. N. Karamzin, A. P. Sukhorukov, and T. S Fillipchuk, "On a new class of coupled solitons in a

dispersing medium with quadratic nonlinearity," Vestn. Mosk. Univ., Fiz. Astron., **19**, No. 4, 91 (1978).

32. B. Ya. Zel'dovich, O. Yu. Nosach, V. I. Popovichev, V. V Ragul'skii, and F. S Faizullov, "Inversion of wave front of light during its forced scattering," Vestn. Mosk. Univ., Fiz. Astron., **19**, No. 4, 137 (1978).

33. V. I. Bespalov, A. A. Betlin, A. I. Pyatlov, et al., "Nonlinear interactions between light waves with complex space-time structures in cubic media" in: A. V. Gaponov (ed), Nonlinear Waves: Propagation and Interaction [in Russian], Nauka, Moscow (1981).

34. V. E. Zakharov and S. V. Manakov, "On resonance interactions between wave packets in nonlinear media," Let. to Zh. Exp. Teor. Fiz., **18**, 413 (1973).

35. B. B. Kadomtsev, Group Phenomena in Plasmas [in Russian], Nauka, Moscow (1976).

36. B. B. Kadomtsev and V. M Kontorovich, "Theory of turbulence in hydrodynamics and in plasmas," Izv. Vuzov: Radiofiz., **17**, 509 (1974).

37. K. Hasselman, "Description of nonlinear interactions by the methods of theoretical physics (with application to the formation of waves by wind)" in: M. J. Lighthill (ed.) Discussion on Nonlinear Theory of Wave Propagation in Dispersive systems (Proc. Roy. Soc. Ser. A. Math. & Phys.. Sci., No. 1456, V. 299, 1967) [Russian translation], Mir, Moscow (1970).

38. A. A. Galeev and R. Z Sagdeev, "Nonlinear theory of plasmas" in: M. A Leontovich (ed) Questions on Plasma Theory [in Russian], Atomizdat, Moscow (1973), No. 3.

39. B. S. Abramovich and V. V. Tamoikin, "Diffusion approximation in the theory of nonlinear wave interactions in chaotically nonhomogeneous media" in: A. V. Gaponov (ed), Nonlinear Waves: Propagation and Interaction [in Russian], Nauka, Moscow (1981).

40. L. D. Landau and E. M. Lifshits, Statistical Physics [in Russian], Nauka, Moscow (1964), Pt. 1.

41. E. M. Lifshits and L. P. Pitaevskii, Statistical Physics [in Russian], Nauka, Moscow (1979).

REFERENCES

42. G.M. Zaslavskii and B.V. Chirikov, "Stochastic Instability of nonlinear oscillations," UFN, **105**, 7-13 (1971).

43. V.I. Karpman, Nonlinear Waves in Dispersing Media [in Russian], Nauka, Moscow (1976).

44. V.E. Zakharov and N.N. Filonenko, "Energy spectra for stochastic oscillations on a liquid surface," Dokl. Akad. Nauk SSSR, **170**, 1292-1303 (1966).

45. H.C. Yuen and B.M. Lake, Nonlinear Dynamics of Deep- Water Gravity Waves, Academic Press (1982).

46. B.Ya. Zel'dovich, H.F. Polipetskii, and V.V. Shkunov, Inversion of a Wave Front [in Russian], Nauka, Moscow (1985).

Chapter Twenty One

1. M.N. Rabinovich, "Self-excited oscillations of distributed systems," Izv. Vuzov: Radiofiz., **17**, 477 (19744).

2. M.Z. Gak, "Laboratory investigation of self-excited oscillation in a system of four vortices," Izv. Akad. Nauk SSSR, Fiz. Atm. i Okean, **17**, 201 (1981).

3. G.F. Putin and E.A. Tkacheva, "Experimental investigation of supercritical convective motion in a Hele-Shaw cell," Izv. Akad. Nauk SSSR, Mekh. Zhid. i Gaza., 3 (1979).

4. E.B. Gledzer, F.B. Dolzhanskii, and A.M. Obukhov, Hydrodynamic-Like Systems and Their Applications [in Russian], Nauka, Moscow (1981).

5. P.D. Joseph, Stability of Fluid Motions (Berlin 1976) [Russian translation], Mir, Moscow (1981).

6. V.I. Petviashvili, "Nonhomogeneous solitons" in Nonlinear Waves [in Russian], Nauka, Moscow (1979).

7. Ya.B. Zel'dovich and B.A. Malomed, "Complex wave regimes in distributed dynamics systems," Izv. Vuzov, Radiofiz., **25**, 591 (1982).

8. J.E. Marsden and M. MacCracken, The Hopf Bifurcation and its Applications (1976) [Russian translation], Mir, Moscow (1981).

9. L.V. Kantorovich and V.I. Krylov, Approximation Methods in Higher Mathematics [in Russian], Fizmatgiz, Moscow (1962), Ch.4, Sec.2.

10. D.V. Lyubimov, G.F. Putin, and V.I. Chernotynskii, "On convective flows in Hele-Shaw cells," Dokl. Akad. Nauk SSSR, **235**, 554 (1977).

11. G. Roskers, "Three dimensional long waves on a liquid film," Phys. Fluids, **13**, 1440 (1970).

12. V.I. Yudovich, "On the occurrence of self-excited oscillations in a liquid," PMM, **35**, 638 (1971).

13. P.S. Landa, Self-Excited Oscillations in Distributed Systems [in Russian], Nauka, Moscow (1983).

Chapter Twenty Two

1. Ya.G. Sinai, "Stochasticity of Dynamic Systems" in: A.V. Gaponov (ed), Nonlinear Waves [in Russian], Nauka, Moscow (1979).

2. V.I. Arnol'd, Mathematical Methods of Classical Mechanics [in Russian], Nauka, Moscow (1974).

3. G.M. Zaslavskii, Statistical Irreversibility in Nonlinear Systems [in Russian], Nauka, Moscow (1970).

4. G.M. Zaslavskii and B.V. Chirikov, "Stochastic Instability of nonlinear oscillations," UFN, **105**, 7-13 (1971).

5. V.M. Alekseev, Symbolic Dynamics (XI Summer Mathematics School) [in Russian], Inst. Matem. Akad. Nauk Ukrain. SSR, Kiev (1976), p.212..

6. R. Bowen (ed), Methods of Symbolic Dynamics [Russian translation], Mir, Moscow (1979).

7. M.I. Rabonovich, "Stochastic self-excited oscillations and turbulence," UFN, **125**, 138-140 (1978).

8. A.N. Shapkovskii, "Coexistence of cycles continuously mapped onto themselves," Ukr. Matem. Zh., **13**, No.3, 86- 94 (1961).

9. P. Billingsley, Ergodic Theory and Information [Russian translation], Mir, Moscow (1969).

10. V.M. Alekseev and M.V. Yakobson, "Symbolic dynamics and hyperbolic dynamic systems" in: R. Bowen (ed), Equilibrium States and the Ergodic Theory of Anosov Diffeomorphisms, (Springer Verlag 1975) [Russian translation], Mir, Moscow (1979).

11. M. Henon, "A two-dimensional mapping with a strange attractor" in: Strange Attractors [Russian translation], Mir, Moscow (1981), 152-163.

12. S.V. Kiyashko, A.S. Pikovskii, and M.I. Rabinovich, "Radio spectrum autogenerator with stochastic behavior," Radio i Elekt., **25**, 336-343 (1980).

13. A.S. Pikovsky and M.I. Rabinovich, "Stochastic oscillations in dissipative systems," Physica, **2D**, 8-24 (1981).

14. J.A. Yorke and E.D. Yorke, "Chaotic behavior and fluid dynamics" in: H.L. Swinney and J.P Gollub (eds), Hydrodynamic Instabillities and the Transitions to Transion Turbulence (Topics in Applied Physics), Springer (1981), Vol.45, pp.77-96.

15. M.J. Feigenbaum, "Universal behavior in nonlinear systems," Los Alamos Sci., **1**, 4-27 (1980).

16. B.A. Huberman and J. Rudnic, "Scaling behavior of chaotic flows," Phys. Rev. Let., **45**, 154-156 (1980).

17. A.S. Pikovskii, "On the stochastic properties of simple models of stochastic self-excited oscillation," Izv. Vuzov, Radiofiz., **23**, 883-884 (1980).

18. M.I. Rabinovich (ed), Nonlinear Waves: Stochasticity and Turbulence [in Russian], Nauka, Gorky (1981).

19. E. Ott, "Strange attractors and chaotic motions of dynamical systems," Rev. Mod. Phys., **83**, 655-671 (1981).

20. S.Ya. Vyshkind, "On the occurrence of stochasticity given nondegenerate interaction between wave in a medium with amplification," Izv. Vuzov, Radiofiz., **21**, 850-856 (1978).

21. M.I. Rabinovich and A.L. Fabrikant, "Stochastic wave self-simulation in nonequilibrial media," Zh.Exp.Teor.Fiz., **77**, 617-629 (1979).

22. P.S. Lindsay, "Period doubling and chaotic behavior in a driven anharmonic oscillator," Phys. Rev. Let., **47**, 1349-1352 (1981).

23. E.N. Lorentz, "Deterministic nonperiodic flow" in: Strange Attractors [Russian translation], Mir, Moscow (1981), 88-116.

24. J.A. Yorke and E.D. Yorke, "Metastable chaos: the transition to stable chaotic behavior in the Lorentz model" in: Strange Attractors [Russian translation], Mir, Moscow (1981), 193-212.

25. V.S. Afraimovich, V.V Bykov, and L.P. Shil'nikov, "On the occurrence and structure of a Lorentz attractor," Dokl. Akad. Nauk SSSR, **234**, 336-339 (1977).

26. P. Manneville and Y. Pomeau, "Different ways to turbulence in dissipative dynamical systems," Physica, **1D**, 219 (1980).

27. V.S. Afraimovich and L.P. Shil'nikov, "On some global bifurcations associated with the disappearance of stationary saddle-node points," Dokl. Akad. Nauk SSSR, **219**, 1281-1285 (1974).

28. D. Ruelle and F. Takens, "On the nature of turbulence" in: Strange Attractors [Russian translation], Mir, Moscow (1981), 116-151.

29. J.P. Eckmann, "Roads to turbulence in dissipative dynamical systems," Rev. Mod. Phys., **53**, 643-654 (1981).

30. F. Ledrappier, "Some relations between dimension and Lyapunov exponents," Commun. Math. Phys., **81**, 229 (1981)..

31. B.B. Mandelbrot, Fractals, Form, Chance, and Dimension, W.H. Freeman and Co., San Francisco (1977)

32. F.M. Izrailev, M.I. Rabinovich, and A.D. Ugodnikov, "Approximate description of three-dimensional dissipative systems with stochastic behavior," Phys Let., **68A**, 321- 325 (1981).

REFERENCES

33. J. Cratchfield, D. Farmer, N. Packard, et al., "Power spectral analysis of a dynamical system," Phys. Let., **76A**, 1 (1980).

34. L. D. Landau and E. M. Lifshits, Fluid Mechanics [in Russian], Nauka, Moscow (1986), Sect. 31.

35. Yu. I. Neimark and P. S. Landa, Stochastic and Chaotic Oscillations [in Russian], Nauka, Moscow (1987).

36. G. M. Zaslavskii, Stochasticity of Dynamic Systems [in Russian], Nauka, Moscow (1984).

37. A. V. Gaponov-Grekhov and M. I. Rabinovich, Physics of the XX Century: Development and Prospects [in Russian], Nauka, Moscow (1984), pp. 219-280.

38. A. J. Lichtenberg and M. A. Lieberman, Regular and Stochastic Motion [Russian translation], Mir, Moscow (1984).

39. K. F. Teodorik, Self-Excited Oscillatory Systems [in Russian], Gostekhizdat, Moscow (1952).

Chapter Twenty Three

1. L. D. Landau, "On the problem of turbulence," Dokl. Akad. Nauk SSSR, **44**, 339-342 (1944).

2. E. Hopf, "A mathematical example displaying the features of turbulence," Pure Appl. Math., **1**, 303-322 (1948).

3. V. I. Arnol'd, Mathematical Methods of Classical Mechanics [in Russian], Nauka, Moscow (1974).

4. D. Ruelle and F. Takens, "On the nature of turbulence," Comm. Math. Phys., **20**, 167 (1971).

5. V. I. Yudovich, "On the occurrence of self-excited oscillations in a liquid," PMM, **35**, 638-655 (1971).

6. A. Libchaber, "Rayleigh-Bernard experiment in liquid helium" in: T. Riste (ed) NATO Study of Nonlinear Phenomena and Phase Transitions, Plenum Press, New York (1981).

7. M. J. Feigenbaum, "Universal behavior in nonlinear systems," Los Alamos Sci., **1**, 4-27 (1980).

8. J.P Gollub, S.V. Benson, and J.A. Steinman, "A subharmonic route to turbulent convection," Annals New York Acad. of Science, **357**, 22 (1981).

9. J.P Gollub and S.V. Benson, "Many routes to turbulent convection," J. Fluid Mech., **100**, 449-470 (1974).

10. V.S. Afraimovich and L.P. Shil'nikov, "On some global bifurcations associated with the disappearance of stationary saddle-node points," Dokl. Akad. Nauk SSSR, **219**, 1281-1285 (1974).

11. V.S. Afraimovich, "On the collapse of torus" in: Proc. 9-th Int. Conf. on Nonlinear Oscillations [in Russian], Izd. IM Akad. Nauk Ukrainsk. SSR, Kiev (1983), pp.118- 120.

12. P.R. Fenstermaher, H.L. Swinney, and J.P Gollub, "Dynamical instabilities and transition to chaotic Taylor vortex flow," J. Fluid Mech., **94**, 103-128 (1979).

13. M.I. Rabonovich, "Stochastic self-excited oscillations and turbulence," UFN, **125**, 138-140 (1978).

14. S. Newhouse, D. Ruelle, and F. Takens, "Occurrence of strange axiom attractors near quasiperiodic flows in T^m, $m \geq 3$," Comm. Math. Phys., **64**, 35 (1978).

15. P. Berge, M. Dubois, P. Mannville, and Y. Pomeau, "Intermittency in Rayleigh-Bernard convection," J. Phys. Let., **41**, L341 (1980).

16. Yu.N. Belyev and I.M. Yaborskaya, "Transition to stochastic regimes in the flow between two rotating spheres" in: M.I. Rabinovich (ed), Nonlinear Waves: Stocasticity and Turbulence [in Russian], Nauka, Gorky (1981).

17. Yu.N. Bezruchko, S.P. Kuznetskov, and D.I. Trubetskov, "Experimental observation of stochastic self-excited oscillation in a dynamic system of electron beams - the inverse electromagnetic wave," Let. to Zh.Exp.Teor.Fiz., **29**, 180-184 (1979)..

18. H.S. Ginzburg, S.P. Kuznetskov, and T.N. Fedoseeva, "Theory of transition processes in relativistic travelling wave tubes," Izv. Vuzov, Radiofiz., **21**, 1037 (1978).

REFERENCES

19. V.S. L'vov and A.A Predtechenskii, "Transition to turbulence in a Coutte flow between cylinders" in: M.I. Rabinovich (ed), Nonlinear Waves: Stocasticity and Turbulence [in Russian], Nauka, Gorky (1981).

20. M.I. Rabinovich and A.L. Fabrikant, "Stochastic wave self-simulation in nonequilibrial media," Zh. Exp. Teor. Fiz., **77**, 617-629 (1979).

21. R. Betchov, "Transition" in: Turbulence. Principles and Application [Russian translation], Mir, Moscow (1980), p.164.

22. N.E. Kochin, N.A. Kibel', and N.V. Roze, Theoretical Fluid Mechanics [in Russian], Gostekhizdat, Moscow (1948), Vol.1.

23. M.I. Rabinovich and M.M. Sushchik, "Coherent structures in turbulent flows" in: A.V. Gaponov and M.I. Rabinovich (eds), Nonlinear Waves, Self-Organization [in Russian], Nauka, Moscow (1983), pp.58-81.

24. S.P. Novikov, "Dynamics and statics of vortices," Zh. Exp. Teor. Fiz., **68**, 1868 (1975).

25. V.A. Katz, "Mechanism for the appearance of chaos in a distributed backwards-wave generator" in Various Questions in Modern Physics [in Russian], Izd. Saratov. Gos. Univ., Saratov (1984), pp.28-33.

26. B.P. Bezruchko, L.V. Bulgakova, S.P. Kuznetsov, and D.I. Trubetskov, "Experimental and theoretical investigation of stochastic self-excited oscillations in a backwards- wave tube" in: Lectures in Microwave Electronics and Radio Physics [in Russian], Izd. Saratov. Gos. Univ., Saratov (1984), Book 5, pp.28-33.

27. B.P. Bezruchko, L.V. Bulgakova, S.P. Kuznetsov, and D.I. Trubetskov, "Stochastic self-excited oscillations and instability in a backwards-wave tube," Radiotekhnika i Electronika, **28**, 1136-1139 (1983).

28. V.L. Kislov, N.N. Zalonin, and E.A. Myasin, "Investigation of stochastic self-excited oscillatory processes in a self-excited generator with delay," Radiotekhnika i Electronika, **24**, No.6, 1118-1130 (1979).

29.   V.A. Katz and D.I. Trubetskov, "Stochastization of Nonstationary structures in distributed oscillations with delay" in: V.I. Krinsky (Ed) Self-Organization Autowaves and Structures of Equilibrium, Springer Verlag (1984), pp. 35-38.

Chapter Twenty Four

1.   N. Wiener and A. Rosenblueth, The Mathematical Formulation of the Problem of Conduction in a Network of Connected Excitable Elements, Specifically in Cardiac Muscle, Arch. Inst. Cardiologia de Mexico V. XVI No. 3-4 (1946) [Russian translation], Izd. Ino. Lit., Moscow (1961).

2.   V. Ebeling, Strukturbildung bei Irreversiblen Prozessen (Teubner 1976) [Russian translation], Mir, Moscow (1979).

3.   A.M. Zhabotinskii, Concentration Oscillations [in Russian], Nauka, Moscow (1974).

4.   P.D. Joseph, Stability of Fluid Motions (Berlin 1976) [Russian translation], Mir, Moscow (1981).

5.   P. Glansdorff and I. Prigogine, Thermodynamic Theory of Structure, Stability, and Fluctuations [Russian translation], Mir, Moscow (1973).

6.   H. Haken, Synergetics [Russian translation], Mir, Moscow (1980).

7.   Autowave Processes in Systems with Diffusion [in Russian], Izd. IPF Akad. Nauk SSSR, Gorky (1981).

8.   Ya.B. Zel'dovich and B.A. Malomed, "Complex wave regimes in distributed dynamics systems," Izv. Vuzov, Radiofiz., **25**, 591-648 (1982).

9.   G. Nicolis and I. Prigogine, Self-Organisation in Nonequilibrium Systems [Russian translation], Mir, Moscow (1979).

10.   G. Gerish, "Cell aggregation and differentiation inn Dictyostelium discoideum," Curr. Top. Deo. Biol., **3**, 157 (1968).

11.   Ya.B. Zel'dovich and D.A. Frank-Kamenetskii, "Theory of even flame propagation," Dokl. Akad. Nauk SSSR, **19**, 693 (1938).

REFERENCES

12. A. N. Zaikin and A. M. Zhabotinsky, "Concentration wave propagation in two-dimensional liquid-phase self- oscillating systems," Nature, **225**, 535 (1970).

13. A. C. Scott, Active and Nonlinear Wave Propagation in Electronics (Wiley Interscience, 1970) [Russian translation], Sovetskii Radio, Moscow (1977).

14. V. S. Markin, V. F. Pastushenko, and Yu. A. Chizmadzhev, Theory of Excitable Media [in Russian], Nauka, Moscow (1981).

15. V. A. Vasil'ev, Yu. M. Romanovskii, and V. G. Yakhno, "Autowave processes in distributed kinetic systems," UFN, **128**, 625-666 (1979).

16. Yu. M. Svirezhev and D. O. Logofet, The Stability of Biological Communities [in Russian], Nauka, Moscow (1978).

17. V. A Ivanitskii, V. I. Krinskii, E. E. Sel'kov, Mathematical Biophysics of Cells [in Russian], Nauka, Moscow (1978).

18. V. I. Koroleva and G. D. Kuznetsova, "Properties of the propagation of depression when bipotassium foci are created in rat cortex" in: Electrical Activity of the Brain [in Russian], Nauka, Moscow (1971), p. 130.

19. Yu. M. Romanovskii, N. V. Stepanovna, D. S. Chernavskii, Mathematical Modelling in Biophysics [in Russian], Nauka, Moscow (1975).

20. M. V. Vol'kenstein, General Physics [in Russian], Nauka, Moscow (1978), Ch. 8.

21. M. V. Vol'kenstein, Biophysics [in Russian], Nauka, Moscow (1981), Ch. 16.

22. A. N. Kolmogorov, I. G. Petrovskii, N. S. Piskunov, "Investigation of diffusion equations coupled with an increase in mass, and its application to a biological problem" in: Cybernetics Problems [in Russian], Izd. Akad. Nauk SSSR Moscow (1975), No. 12, pp. 3-20.

23. D. A. Frank-Kamenetskii, Diffusion and Heat Transfer in Chemical Kinetics [in Russian], Nauka, Moscow (1967).

24. V.I. Petviashvili, "Inhomogeneous solitons" in: A.V. Gaponov (ed), Nonlinear Waves [in Russian], Nauka, Moscow (1979), p.5-19.

25. K.I. Agladze and V.I. Krinsky, "Multi-armed vortices in an active chemical medium," Nature, **296**, 424 (1982).

26. S.V. Kiyashko, M.I. Rabinovich, and V.P. Reutov, "Explosive instability and the generation of solitons in an active medium," Zh.Exp.Teor.Fiz., **42**, 2458-2465 (1972).

27. Nonlinear Waves. Self-Organization [in Russian], Nauka, Moscow (1983).

28. H. Haken, Advanced Synergetics (Berlin, 1983) [Russian translation], Mir, Moscow (1983).

29. Nonlinear Waves, Structures, and Bifurcations, [in Russian], Nauka, Moscow (1987).

31. V.A. Vasiliev, Yu.M. Romanovskii, and V.G. Yakhno, Self- Excited Wave Processes [in Russian], Nauka, Moscow (1987).

30. D.V. Sokolov and D.I. Trubetskov, "Nonlinear waves, dynamic chaos, and some problems of microwave electronics" in: Collection of Scientific Articles on the Problems of Physical Electronics [in Russian], Leningrad. Fiz. Tekh. Inst., Leningrad (1986), pp. 141-177.

32. V.R. Ampilogoba, A.V. Zdorovokii, D.I. Trubetskov, and K.V. Khudzik, "On verifying one hypothesis on the onset of chaos from structures in electron beams" in: Lectures on Microwave Electronics and Radio Physics (Seventh Summer School-Seminar for Engineers [in Russian], Izd.

Index

Acoustic curve 55
Alternation 495

Bussinesque 86
    geometric optics 177,
      247-249, 255, 269
        Wentsel-Kramers-
      Brillouin 241, 268
    Attractor 301
  strange 308, 473, 479
Belusov-Zhabotinskii reaction
    4, 523, 532
Bifurcation 313
    birth of invariant
      torus 321
    doubling 321, 507
Bruscellator 146

Cell 524, 536
    Hele-Shaw 549
Chain of
    identical particles 56
    linked particles 50
    point vortices 532
Criterion
    Briggs 166
    Raus-Gurvits 124, 127

Debye radius 110
Dispersion 49, 60
    anomalous 41
    normal 41
    spatial 63
    temporal 63
Dissipative structure 525
    crude d.s. 313

Equation
    Burgers 375, 395
    Bussinesque 492
    characteristic 11, 73
    disperion 71
    Ginsburg-Landau 423
    Hills 215
    Kadomtsev-Petviashvili
      411
    Khokhlov-Zabolotskii
      411

    Klein-Gordon 61, 68,
      185, 375
    Korteveg-de Vries 273,
      375, 404, 406, 411,
      425
    Landau 517
    Mathieu 217
    non-linear parabolic
      422
    Rayleigh 302, 305
    Rickatti 240
    Schrodinger 407, 422
    Van der Pol 302

Homoclinic structure 323

Index
    Lyapunov characteristic
      index 216
    Poincaré 313, 316
Independent system 2
Instability 14, 121, 142
    absolute 142, 153
    convective 142, 153,
      534
    explosive 372, 535
    Helmholtz 157, 165
    Jean's 144
    stochastic 297
    Turing 145
Internal wave 90
    discontinuous 447
    gravity 91, 384
    kinematic 375
    modulated 417
    Riemann 386
    Rossby 90, 413
    shock 390, 395
    simple 381
    stationary 390, 447
    with positive and
      negative energy 160,
      196
Ion sound 118

Kapitsa's pendulum 232
Kolmogorov-Sinai entropy 477
Klystron 15, 376

Lamere diagram 318, 485
Landau damping 118, 283
Langmuir oscillation 114
Limiting cycle 299
Linkage 30
Lorentz system 492, 497

Method
    asymptotic 223, 240
    averaging with motion in a rapidly 232
    coupled wave 203
    D-partition 308
    direct variational 242
    Galerkin 458
    inverse scattering 406
    Langmuir multiplier 19
    perturbation in theory of oscillations and waves 218, 291, 382
    Van der Pol 261, 287, 290, 320, 332, 344
    Wentsel-Kramers-Brillouin 264
    Whitham's 188, 242
Miller's force 237
Mode
    baroclinic 102
    barotropic 102
Model
    Landau-Hopf 504
    Lorentz 496
    Lotkas 3
    predator-prey 4
    two-liquid hydrodynamic 112
    Volterra 4, 347
Modulation 417
    stochastic 513
Multiplicator 321

Number
    Mach 520
    Prandtl 493, 510
    Rayleigh 493
    Reynolds 504, 511
    Taylors 511

Optic curve 55
Oscillation
    linked form 31
    normal form 31
Phase
    plane 6
    space 125, 126, 309, 4688
Poincaré mapping 317, 321, 484

Relaxational self-excited oscillation 305
Resistive amplification 201
Resonance 13
    interaction between waves 354, 364
    interaction between oscillators 267, 353
    nonlinear 286
    of wave systems 71
    overlap 290
Restoration (refraction) time 532

Separatrix 7, 277, 325, 495
Soliton 278, 395, 403, 425
    as particle 409
    dissipative 517
    Rossby 413
Stability 121, 141
    asymptotic 123
    Lyapunov 122
    of nonautonomous systems 131
    orbital 123
    structural 309
Synchronization
    of waves 76, 203, 374
    condition 76, 364, 496
Synergetics 523

Theorem
    Liouville's 472
    Tchu 196, 204
Travelling center 533
Tube
    backward wave 108, 151
    travelling wave 76, 148, 150, 284
    twin beam 157

Turbulence 442, 503
    hydrodynamic 503
    transition to 509, 510
    wave 442
    weak 442

Velocity
    energy propagation 178
    group 66, 171, 178
    phase 62, 66, 171
Venn diagram 344

Wave packet 172, 186
Waveguide 197